RIEMANN SURFACES

PRINCETON MATHEMATICAL SERIES

Editors: MARSTON MORSE AND A. W. TUCKER

1. The Classical Groups, Their Invariants and Representatives. By HERMANN WEYL.
2. Topological Groups. By L. PONTRJAGIN. Translated by EMMA LEHMER.
3. An Introduction to Differential Geometry with Use of the Tensor Calculus. By LUTHER PFAHLER EISENHART.
4. Dimension Theory. By WITOLD HUREWICZ and HENRY WALLMAN.
5. The Analytic Foundations of Celestial Mechanics. By AUREL WINTNER.
6. The Laplace Transform. By DAVID VERNON WIDDER.
7. Integration. By EDWARD JAMES MCSHANE.
8. Theory of Lie Groups: I. By CLAUDE CHEVALLEY.
9. Mathematical Methods of Statistics. By HARALD CRAMÉR.
10. Several Complex Variables. By SALOMON BOCHNER and WILLIAM TED MARTIN.
11. Introduction to Topology. By SOLOMON LEFSCHETZ.
12. Algebraic Geometry and Topology. Edited by R. H. FOX, D. C. SPENCER, and A. W. TUCKER.
13. Algebraic Curves. By ROBERT J. WALKER.
14. The Topology of Fibre Bundles. By NORMAN STEENROD.
15. Foundations of Algebraic Topology. By SAMUEL EILENBERG and NORMAN STEENROD.
16. Functionals of Finite Riemann Surfaces. By MENAHEM SCHIFFER and DONALD C. SPENCER.
17. Introduction to Mathematical Logic, Vol. I. By ALONZO CHURCH.
18. Algebraic Geometry. By SOLOMON LEFSCHETZ.
19. Homological Algebra. By HENRI CARTAN and SAMUEL EILENBERG.
20. The Convolution Transform. By I. I. HIRSCHMAN and D. V. WIDDER.
21. Geometric Integration Theory. By HASSLER WHITNEY.
22. Qualitative Theory of Differential Equations. By V. V. NEMICKII and V. V. STEPANOV. Translated by SOLOMON LEFSCHETZ.
23. Topological Analysis. By GORDON T. WHYBURN.
24. Analytic Functions. By L. AHLFORS, H. BEHNKE and H. GRAUERT, L. BERS, et al.
25. Continuous Geometry. By JOHN VON NEUMANN. Edited by ISRAEL HALPERIN.
26. Riemann Surfaces. By LARS V. AHLFORS and LEO SARIO.

RIEMANN SURFACES

BY

LARS V. AHLFORS
PROFESSOR OF MATHEMATICS
HARVARD UNIVERSITY

AND

LEO SARIO
PROFESSOR OF MATHEMATICS
UNIVERSITY OF CALIFORNIA AT LOS ANGELES

PRINCETON, NEW JERSEY
PRINCETON UNIVERSITY PRESS
1960

L.C. Card 59-11074

Printed in the United States of America

Preface

In his efforts to build up a solid foundation for the theory of analytic functions of one complex variable Riemann came to recognize that for full generality it is necessary to abandon the complex plane as the carrier of functions. The reason is that analyticity is a local property, and to impose a global carrier with special properties leads to unnecessary restrictions. When only the local properties are retained, the carrier becomes an abstractly defined Riemann surface.

The theory of Riemann surfaces has a geometric and an analytic part. The former deals with the axiomatic definition of a Riemann surface, methods of construction, topological equivalence, and conformal mappings of one Riemann surface on another. The analytic part is concerned with the existence and properties of functions that have a special character connected with the conformal structure, for instance, subharmonic, harmonic, and analytic functions.

The importance of the topological foundations can hardly be overemphasized. Accordingly, the book opens with a thorough treatment of the topology of surfaces. In the classical manner we have kept this treatment strictly independent of the existence of a differentiable structure. This is not merely desirable from the point of view of an orderly presentation, but it is mandatory if one wants to include the important result that Riemann surfaces are not subject to any topological restrictions other than orientability and countability.

We have made an effort to make our book self-contained and as concise as possible. For instance, we use only such methods in topology that can be developed briefly *ab ovo*. Similarly, we presuppose only rudimentary knowledge of algebra, and in the analytic part no advanced knowledge of Hilbert space theory is required. We hope that the accomplished mathematician will take no offense at our attempts to make the presentation more palatable for the novice.

We are conscious that many notable omissions are difficult to defend, except by reference to space and time. We regret, above all, that we were not able to include the already classical results of H. Behnke, K. Stein, and their students—results which generalize the theorems of Runge, Weierstrass, and Mittag-Leffler. Our best excuse is that these questions have received very adequate treatment in a recent book of H. Behnke and F. Sommer [1].

References are very scarce throughout the book. We have, of course, acknowledged ideas and results whose first appearance is easy to locate. In many more cases the same idea has occurred to several persons, more or less simultaneously, and it seemed meaningless to pinpoint the day it has reached the printed page.

Concerning the mechanics of the joint authorship, both authors have contributed their full share to the planning and organization of the book, to constant rewriting, and unending search for improvements. The first two chapters, on the foundations, and Chapter IV, the theory of differentials, are an outgrowth of lectures by the senior author during the last decade. Chapter III on principal functions is based on research by the junior author, and he shares, in equal parts, the responsibility for the material in Chapter IV, the classification theory.

Our deepest gratitude goes to Mrs. Kaija Sario who has typed and mimeographed countless versions of the manuscript. We are also obliged to many friends who have read the manuscript at various stages and given their valued advice, among them M. Heins, E. Schlesinger, R. Accola, G. Weill, B. Rodin, and K. Oikawa who also assisted with the bibliography and index.

We are indebted to the Office of Ordnance Research for substantial financial support through many years. The junior author also gratefully acknowledges his Guggenheim Fellowship during the academic year 1957–1958 and earlier support by the Institute for Advanced Study.

LARS V. AHLFORS
Harvard University
LEO SARIO
University of California, Los Angeles

Contents

RIEMANN SURFACES

CHAPTER I

Surface Topology

A Riemann surface is, in the first place, a surface, and its properties depend to a very great extent on the topological character of the surface. For this reason the topological theory of surfaces belongs in this book.

There is a great temptation to bypass the finer details of topology in favor of a more rapid and sometimes more elegant treatment in which the topological properties are derived from analytical considerations. The main demerit of this approach is that it does not yield complete results. For instance, it cannot be proved by analytical means that every surface which satisfies the axiom of countability can be made into a Riemann surface. In other instances the analytical method becomes so involved that it no longer possesses the merit of elegance.

The classical way, which we shall follow, begins with an independent derivation of the topological properties of surfaces. For complete results this derivation must be based on the method of triangulation. On the other hand, it is much easier to obtain superficial knowledge without use of triangulations, for instance, by the method of singular homology. It so happens that this superficial knowledge is adequate for most applications to the theory of Riemann surfaces, and our presentation is influenced by this fact.

Since we strive for completeness, a considerable part of the first chapter has been allotted to the combinatorial approach. We have tried, however, to isolate this part from the considerations that do not make use of a triangulation. This is done by formulating the combinatorial theory as a theory of triangulated surfaces, or polyhedrons. At the very end of the chapter it is then shown, by essential use of the Jordan curve theorem, that every surface which satisfies the second axiom of countability permits a triangulation.

In §1 we give a brief survey of the topological concepts that will be used throughout the book. This elementary section has been included for the sake of completeness and because beginning analysts are not always well prepared on this point. The section can of course be omitted by readers who are already familiar with the concept and properties of topological spaces.

The definition of manifolds and surfaces is given in §2. The fundamental group is introduced, and the notion of bordered surface (surface with boundary) is defined.

In §3 we discuss covering surfaces. This topic is of course of special importance for the theory of Riemann surfaces.

§4 deals with the purely combinatorial aspects of homology theory, while §5 centers around the notions of singular homology and simplicial approximation. In the next section, §6, the topological nature of the ideal boundary of an open surface is analyzed in considerable detail. §7 is again combinatorial and leads to the classification of finite polyhedrons. The chapter closes with the construction of a triangulation.

§1. TOPOLOGICAL SPACES

A set of objects is often called a space, and the objects are called points. If there are no relations between the points, pure set theory exhausts all possibilities. As soon as one wants to go beyond set theory to limits and continuity it becomes necessary to introduce a topology, and the space becomes a topological space.

1. Definition

1A. A set S of points p becomes a *topological space* if a family $\mathscr{T}$ of subsets O, to be called *open sets*, is distinguished. The following conditions shall be fulfilled:

(A1) *The union of any collection of open sets is open.*

(A2) *The intersection of any finite collection of open sets is open.*

The requirements are interpreted to hold also for the empty collection. According to usual conventions the union of an empty collection of sets is the empty set 0, and the intersection of an empty collection is the whole space S. Hence (A1) and (A2) imply that 0 and S are open.

In most cases a third requirement is added:

(A3) *Given any two points* $p_1 \neq p_2$ *there exist open sets* O_1, O_2 *such that* $p_1 \in O_1$, $p_2 \in O_2$ *and* $O_1 \cap O_2 = 0$.

A topological space which satisfies (A3) is called a *Hausdorff space.*

1B. For the actual construction of topological spaces the definition that we have given is not very practical, for it requires us to name all open sets. It is therefore convenient to introduce the notion of a *basis for the open sets* (briefly: a *basis*). Such a basis is a system $\mathscr{B}$ of subsets of S which satisfies condition

(B) *The intersection of any finite collection of sets in* $\mathscr{B}$ *is a union of sets in* $\mathscr{B}$.

Again, this shall hold also for the empty collection; the whole space S

is thus a union of sets in $\mathscr{B}$. The basis $\mathscr{B}$ generates a topology $\mathscr{T}$ whose open sets are the unions of sets in $\mathscr{B}$. It is a Hausdorff topology if and only if any two distinct points belong to disjoint sets in $\mathscr{B}$.

1C. A *neighborhood* of a set $A \subset S$ is a set $V \subset S$ which contains an open set O with $A \subset O \subset V$. Mostly, we consider only neighborhoods of points, and we use the notation $V(p)$ to indicate that V is a neighborhood of p.

An open set is a neighborhood of any subset, and a set is open if it contains a neighborhood of every point in the set.

1D. A space with more than one point can be topologized in different manners. A topology $\mathscr{T}_1$ is said to be *weaker* than the topology $\mathscr{T}_2$ if $\mathscr{T}_1 \subset \mathscr{T}_2$. The weakest topology is the one in which S and 0 are the only open sets. The strongest topology is the *discrete* topology in which every subset is open.

1E. The simplest nontrivial topological space is the n-dimensional Euclidean space R^n. Its points $x=(x_1, \cdots, x_n)$ are given by n real coordinates, and its topology is defined by a basis consisting of all solid spheres $\{x: |x-y|^2 = \sum_1^n (x_i-y_i)^2 < \rho^2\}$. Condition (B) is a consequence of the triangle inequality, and R^n is a Hausdorff space.

For the points of the *plane* R^2 we shall frequently use the complex notation $z=x+iy$. The sphere S^2, also referred to as the *extended plane*, is obtained by adding a point ∞ to R^2, and the sets $\{z: |z| > \rho\} \cup \{\infty\}$ to the basis.

2. Subsets

2A. The complement of an open set is said to be *closed*. It follows from (A1) and (A2) that the intersection of an arbitrary collection and the union of a finite collection of closed sets are closed. The empty set and the whole space are simultaneously open and closed.

The *closure* of a set $P \subset S$ is defined as the least closed set which contains P, and it is denoted by $\bar{P}$ or $\operatorname{Cl} P$. It exists, for it can be obtained as the intersection of all closed sets which contain P. A point p belongs to $\bar{P}$ if and only if every $V(p)$ intersects P.

2B. Frequent use is made of the trivial relation

$$\overline{A \cup B} = \bar{A} \cup \bar{B}.$$

This is to be contrasted with the formula

$$\overline{A \cap B} \subset \bar{A} \cap \bar{B}$$

which is weaker inasmuch as it gives only an inclusion.

2C. The *interior* of P is the largest open set contained in P, and the exterior is the complement of the closure, or the interior of the complement.

The *boundary* of P is formed by all points which belong neither to the interior nor to the exterior. In other words, p belongs to the boundary of P if and only if every $V(p)$ intersects P as well as the complement of P.

The interior and boundary of a set P are denoted by Int P and Bd P respectively.

2D. From a given topological space others can be constructed by means of certain standard procedures. One of these is the process of *relativization.*

On any set $S' \subset S$ a topology can be introduced in which the open sets are intersections $S' \cap O$ of S' with the open sets of S. In fact, (A1)–(A2) are obviously satisfied; (A3) holds if it holds on S. We call this topology on S' the *relative topology* induced by the topology on S.

The open and closed sets on S' are often referred to as relatively open or closed sets. We shall always understand that the topology on a subset is its relative topology.

2E. The *sum* of two topological spaces S_1, S_2 is their union $S_1 \cup S_2$ on which the open sets are those whose intersections with S_1 and S_2 are both open. This definition has an obvious generalization to the case of an arbitrary collection of topological spaces.

The *topological product* $S_1 \times S_2$ is defined as follows: its points are ordered pairs (p_1, p_2), $p_1 \in S_1$, $p_2 \in S_2$, and a basis for the open sets is formed by all sets of the form $O_1 \times O_2$ where O_1 is open in S_1 and O_2 is open in S_2.

2F. Another method for constructing new topological spaces is the process of *identification.* We assume that a topological space S is represented as the union of disjoint subsets P, and we consider a space whose points are the sets P. It amounts to the same thing if we identify all points p which belong to the same P.

A topology in the space of points P can be introduced by the agreement that a set is open if and only if the union of the corresponding sets P is an open set in S. Again, (A1)–(A2) are trivially fulfilled. It must be observed, however, that the new space is not necessarily a Hausdorff space even if S is one.

The space obtained by identification can be referred to as the *quotient-space* of S with respect to the equivalence relation whose equivalence classes are the sets P. The construction of a quotient-space is sometimes preceded by forming the sum of several spaces. This means that points can be identified which were not initially in the same space.

3. Connectedness

3A. Certain characteristic properties which may or may not be present in a topological space are very important not only in the general theory, but in particular for the study of surfaces. Foremost is the property which characterizes a connected topological space.

Definition. *A space is connected if it cannot be represented as the union of two disjoint open sets neither of which is void.*

This is the most useful form of the definition for a whole category of proofs. If we wish to prove that a certain property holds for all points in a connected space, we try to show that the set of points which have the property and the set of points which have not are both open. If this is so, one of the sets must be empty, and we conclude that the property holds for all points or for no points. Examples of this type of proof will be abundant.

3B. The definition applies also to subsets in their relative topology, and we can hence speak of connected and disconnected subsets. An open connected set is called a *region*, and the closure of a region is referred to as a *closed region*. A closed connected set with more than one point is a *continuum*. The empty set and all sets with only one point are trivially connected.

Theorem. *The union of any collection of connected sets with a common point is itself connected.*

Proof. Denote the given sets by P_α where α runs through an arbitrary set of indices, and suppose that $P = \cup P_\alpha = O_1 \cup O_2$ where O_1 and O_2 are disjoint relatively open subsets of P. If O_1 is not empty it meets at least one P_α. This P_α has a decomposition $P_\alpha = (P_\alpha \cap O_1) \cup (P_\alpha \cap O_2)$ into disjoint subsets. They are relatively open with respect to P_α, for O_i, $i = 1, 2$, has the form $O_i = P \cap \Omega_i$ with absolutely open Ω_i, and thus $P_\alpha \cap O_i = P_\alpha \cap P \cap \Omega_i = P_\alpha \cap \Omega_i$. From the connectedness of P_α we conclude that $P_\alpha \cap O_1 = P_\alpha$, and so $P_\alpha \subset O_1$. Hence O_1 contains the common point of all P_α. This reasoning applies equally well to O_2, and we find that O_1 and O_2 cannot both be nonempty and at the same time disjoint. Hence $P = \cup P_\alpha$ is connected.

3C. Theorem. *If P is a connected set and $P \subset Q \subset \bar{P}$, then Q is also connected.*

Proof. Suppose that $Q = O_1 \cup O_2$ where O_1, O_2 are relatively open and disjoint. Then P has the corresponding decomposition $P = (P \cap O_1) \cup (P \cap O_2)$, and by the connectedness of P we must have, for instance, $P \subset O_1$. But O_1 is also relatively closed in Q. Therefore, the relative closure of P with respect to Q, which is $\bar{P} \cap Q = Q$, is still contained in O_1. From $Q \subset O_1$ it follows that O_2 is empty, and hence Q is connected.

3D. As a consequence of 3B every point in a topological space belongs to a maximal connected subset, namely the union of all connected sets which contain the given point. This maximal connected set is called the *component* of the space determined by the point. By 3C every component is closed. Any two components are either disjoint or identical. Hence there exists a unique decomposition of any topological space into components.

A space is connected if and only if it consists of a single component. The opposite extreme is a *totally disconnected* space in which every component reduces to a point.

3E. Definition. *A space is locally connected if it has a basis consisting of connected sets.*

This property neither implies nor is implied by connectedness in the large. Every open subset of a locally connected space is itself locally connected, for it has a basis consisting of part of the basis for the whole space.

Theorem. *The components of a locally connected space are simultaneously open and closed.*

Proof. We have already pointed out that the components are closed, independently of the local connectedness. Suppose that p belongs to the component C, and let $V(p)$ be a connected neighborhood. Then $C \cup V(p)$ is connected, by 3B, and by the definition of components we obtain $C \cup V(p) \subset C$ or $V(p) \subset C$. This shows that C is open.

The theorem applies in particular to the components of an open subset of a locally connected space. It shows that every open set in a locally connected space is a union of disjoint regions.

3F. *Example.* The line R^1 is a connected space. This is one of the fundamental properties of the real number system. It follows that every R^n is connected, for R^n can be represented as the union of all lines through the origin. R^n is also locally connected.

4. Compactness

4A. We proceed to the definition of *compact* spaces. It has been found most convenient to base the definition on the consideration of *open coverings*. An open covering is a family of open sets whose union is the whole space, and a covering is finite if the family contains only a finite number of sets.

Definition. *A space is compact if and only if every open covering contains a finite subcovering.*

In other words, every family of open sets which cover the space must contain a finite number of sets $O_1, \cdots, O_N$ whose union is the whole space. For instance, a compact space can thus be covered by a finite number of sets from an arbitrary basis.

A compact subset is of course one which is compact in the relative topology. In the case of a subset it is convenient to replace the relatively open sets of a covering by corresponding open sets of the whole space. An open covering of a subset is then a family of open sets whose union contains the given subset, and if the subset is compact we can again select a finite subcovering.

4B. Sometimes it is more convenient to express compactness in terms of closed sets. We shall say that a family of closed sets has the *finite intersection property* if any finite number of sets in the family have a nonvoid intersection. Evidently, this is equivalent to saying that the family of complements contains no finite covering. The following theorem is thus merely a rephrasing of the definition.

Theorem. *A space is compact if and only if every family of closed sets with the finite intersection property has a nonvoid intersection.*

4C. *In a Hausdorff space every compact subset is closed.* Let C be a compact subset of a Hausdorff space, and let p be a point in the complement of C. Every point $q \in C$ belongs to an open set O_q which does not meet a certain neighborhood $V(p)$, depending on q. All sets O_q with this property form an open covering of C, and we can select a finite subcovering. The intersection of the corresponding neighborhoods $V_1(p), \cdots, V_N(p)$ is a neighborhood of p which does not meet C. Hence the complement of C is open, and C is closed.

The proof implies more than has been stated. Indeed, the union of the sets O_q in the finite subcovering is an open set which contains C and does not meet a certain neighborhood of p. In other words, a compact set and a point outside the compact set have disjoint open neighborhoods.

This result can be generalized as follows:

Theorem. *In a Hausdorff space any two disjoint compact sets have disjoint open neighborhoods.*

The proof is an obvious repetition of the previous argument and will be left to the reader.

4D. In the opposite direction we can prove that *every closed subset of a compact space is compact*. In fact, to any open covering of the subset we need only add the complement of the given set to obtain an open covering of the whole space. A finite subcovering of the whole space will contain a finite subcovering of the subset, extracted from the original covering. It follows that the subset is compact. In a compact Hausdorff space compact and closed subsets are thus identical.

It is convenient to use the term *relatively compact* for a subset whose closure is compact. It follows from what we have said that a subset of a

Hausdorff space is relatively compact as soon as it is contained in a compact set.

4E. It is frequently important to conclude that the intersection of a family of connected sets is connected (cf. Ch.IV, 7F). In this respect we prove:

Theorem. *Let Φ be a nonempty family of compact connected sets in a Hausdorff space. Suppose that, whenever $C_1, C_2 \in \Phi$, there exists a $C_3 \in \Phi$ such that $C_3 \subset C_1 \cap C_2$. Then the intersection of all sets in Φ is nonvoid, compact, and connected.*

The hypothesis implies, by induction, the finite intersection property. Therefore the intersection is nonvoid by Theorem 4B. The compactness follows from the remarks in 4D.

Suppose that the intersection E were disconnected. Then we could write $E = E_1 \cup E_2$ where E_1, E_2 are compact, disjoint, and nonvoid. By Theorem 4C it is possible to enclose E_1, E_2 in disjoint open sets O_1, O_2. Let B be the boundary of O_1. Then each $C \in \Phi$ intersects B, by the connectedness. The sets $B \cap C$ are compact, and it follows from the hypothesis that they have the finite intersection property. Hence E intersects B. But this is impossible, since the points of E_1 lie in the interior and the points of E_2 in the exterior of O_1. Hence E is connected.

4F. A space is *locally compact* if every point has a compact neighborhood. For a locally compact Hausdorff space we can show that *every point has arbitrarily small compact neighborhoods.* Indeed, let V be an arbitrary and U a compact neighborhood of p. Consider $V_0 = U \cap V$. The boundary of V_0 is contained in U, and hence compact. By Theorem 4C the point p and Bd V_0 have disjoint neighborhoods. Hence we can find a neighborhood V_1 of p whose closure does not meet Bd V_0. This implies $\bar{V}_1 \cap \bar{V}_0 \subset V_0 = U \cap V$. We conclude that $\bar{V}_1 \cap \bar{V}_0$ is a compact neighborhood contained in V.

A locally compact Hausdorff space S can always be made compact in the following manner: We add a point, denoted by the symbol ∞, and to the family of open sets we add all sets consisting of ∞ together with the complement of a compact subset of S. The postulates for a topological space are easily seen to be satisfied.

The resulting space S_∞ is compact, for in an open covering of S_∞ there is one set which contains ∞. Its complement is compact and can be covered by a finite number of the sets, whence S_∞ has itself a finite subcovering. The local compactness of S implies that S_∞ is a Hausdorff space.

S_∞ is called the *Alexandroff compactification* of S. In the theory of surfaces the point ∞ will frequently be referred to as the *ideal boundary* of the surface.

4G. In the theory of surfaces (37C) we shall need a theorem on totally disconnected spaces (or subsets) whose proof depends decisively on compactness. We recall that a space is said to be totally disconnected if each component reduces to a point (3D).

Theorem. *Let a and b be distinct points in a compact totally disconnected space S. Then S permits a decomposition $S = A \cup B$ into closed subsets with $a \in A$, $b \in B$ and $A \cap B = 0$.*

We say that a decomposition with the stated properties *separates* a and b. More generally, it is also said to separate any subset of A from any subset of B. Note that A and B are simultaneously open and closed.

It is convenient to start by proving a lemma:

Lemma. *If a can be separated from each point of a closed set $C \subset S$, then it can be separated from the whole set C.*

Let $S = A(c) \cup B(c)$ be a decomposition which separates a and $c \in C$. All sets $B(c)$ form an open covering of C. Select a finite subcovering by $B(c_1), \cdots, B(c_N)$, and set $A = A(c_1) \cap \cdots \cap A(c_N)$, $B = B(c_1) \cup \cdots \cup B(c_N)$. Then $S = A \cup B$ separates a and C.

Let M be the set of points that can *not* be separated from a. The theorem will be proved if we show that M reduces to a. It is evident that M is closed. Indeed, $S - M$ is the union of all sets B that occur in decompositions $S = A \cup B$ with $a \in A$. Since each B is open, the same is true of the union, and hence M is closed.

We prove that M is connected. If not, let $M = M_1 \cup M_2$ be a nontrivial decomposition into closed subsets with $a \in M_1$. By Theorem 4B it is possible to enclose M_1, M_2 in disjoint open sets U_1, U_2. Then Bd U_1 meets neither M_1 nor M_2. Hence each point on Bd U_1 can be separated from a, and by the lemma there exists a decomposition $S = A_1 \cup B_1$ with $a \in A_1$, Bd $U_1 \in B_1$. The set $A = U_1 \cap A_1$ is open, for it is the intersection of two open sets. But $U_1 \cap A_1 = \bar{U}_1 \cap A_1$, because the boundary of U_1 does not meet A_1. Hence A is also closed, and $S = A \cup (S - A)$ is a decomposition. Moreover, $a \in A$ and $M_2 \subset S - A$, since M_2 does not meet U_1. Thus a is separated from M_2, which is impossible. Hence M is connected. Since S is totally disconnected it follows that M must reduce to a point, and the proof is complete.

Remark. The theorem is known to be false without the assumption of compactness. The proof that we have given is taken from W. Hurewicz—H. Wallman [1].

5. Countability

5A. Important simplifications occur if the space S has a basis which consists of a countable number of sets. Such spaces will be called *countable*. It is obvious that any subset of a countable space is countable.

We prove first:

Theorem. *Every open covering of a countable space contains a countable subcovering.*

Let $\mathscr{B}$ be a countable basis and $\mathscr{C}$ an open covering. All sets $\Omega \in \mathscr{B}$ which are contained in some $O \in \mathscr{C}$ can be arranged in a sequence $\{\Omega_m\}$, and to each Ω_m we can find an $O_m \in \mathscr{C}$ so that $\Omega_m \subset O_m$. Every $O \in \mathscr{C}$ is a union of sets Ω_m. Therefore, every point is in an Ω_m, and consequently in the corresponding O_m. It follows that the O_m constitute a countable subcovering.

Corollary. *A locally connected countable space has countably many components.*

The covering by the components is open (Theorem 3E), and it has no proper subcoverings.

5B. A compact countable space can be characterized as follows:

Theorem. *A countable space is compact if and only if every infinite sequence $\{p_n\}$ has a limit point.*

By definition, q is a limit point of $\{p_n\}$ if every $V(q)$ contains p_n for infinitely many n.

From an open covering we select a countable subcovering by sets O_m. If S is not covered by a finite number of these O_m, then we can find, for each m, a point p_m outside of the union $O_1 \cup \cdots \cup O_m$. The sequence $\{p_m\}$ has no limit point. In fact, every $q \in S$ has a neighborhood O_n, and O_n contains p_m at most for $m \leqq n$. We have proved that the condition is sufficient.

Suppose now that S is compact, and let $\{p_n\}$ be a given sequence. If q is not a limit point there exists an open neighborhood $V(q)$ which contains at most a finite number of points p_n. If there are no limit points the sets $V(q)$ would constitute an open covering, and we could select a finite subcovering. It would follow that there is only a finite number of distinct n. The contradiction shows that $\{p_n\}$ has a limit point.

5C. In the presence of a countable basis the notion of limit point has the following interpretation:

Theorem. *In a countable space q is a limit point of $\{p_n\}$ if and only if it is the limit of a subsequence $\{p_{n_k}\}$.*

We say that $\{p_n\}$ has the limit q if to every $V(q)$ there exists an n_0 such that $p_n \in V(q)$ for $n \geqq n_0$. In a Hausdorff space a sequence can have at most one limit.

If q is the limit of a subsequence it is obviously a limit point. To prove the opposite, let Ω_k run through the sets in a countable basis which contain the limit point q, and set $\Omega'_k = \Omega_1 \cap \ldots \cap \Omega_k$. For every k we can find, recursively, an $n_k > n_{k-1}$ such that $p_{n_k} \in \Omega'_k$. The sequence $\{p_{n_k}\}$ has the limit q. In fact, every $V(q)$ contains an Ω_{k_0}, and $p_{n_k} \in \Omega_{k_0}$ for $k \geqq k_0$.

5D. Countability has a local and a global aspect. A space is said to be *locally countable* if every point has a countable neighborhood.

It is evident that a compact locally countable space is countable. More generally, the same is true of a locally countable space which can be represented as a countable union of compact sets.

5E. The space R^n is countable, for the solid spheres with rational center-coordinates and rational radii form a countable basis.

A subset of R^n is compact if and only if it is closed and bounded. This is known as the lemma of Heine-Borel, or, in terms of limit points, as the Bolzano-Weierstrass theorem.

6. Mappings

6A. The reason for introducing topologies is that they enable us to define and study *continuous mappings.*

We make no distinction between mappings and functions. A function f from a space S to another space S' assigns a value $f(p) \in S'$ to every $p \in S$. With another notation we speak of the mapping $p \to f(p)$, and this notation is always used when there is an explicit expression for $f(p)$.

According to our definition every function is single-valued. Later on we will also speak of multiple-valued functions to conform with classical usage, and for clarity it will then be necessary to use the pleonastic term "single-valued function". At present, however, we stick firmly to the convention that there is only one value $f(p)$ corresponding to each p.

The space S is the *domain* of f. We can of course consider functions whose domain is a subset $P \subset S$. In particular, if f has domain S it has a restriction to every subset P. In most cases the restriction can again be denoted by f without fear of confusion.

The image $f(P)$ of $P \subset S$ is the set of all values $f(p)$ for $p \in P$. The whole image $f(S)$ is also called the *range* of f or the *image* of f. The *inverse image* $f^{-1}(P')$ of $P' \subset S'$ is the set of all $p \in S$ with $f(p) \in P'$. Observe that $f(f^{-1}(P')) = P'$ while $f^{-1}(f(P)) \supset P$.

6B. A function f from S to S' is said to map S *into* S', and if $f(S) = S'$ we say that f maps S *onto* S'. A mapping is *one to one* if $f(p_1) = f(p_2)$ implies $p_1 = p_2$. In that case the *inverse function* f^{-1} is defined on the range of f.

A composite function $f \circ g$ is defined on the domain of g provided that the range of g is contained in the domain of f. Its value at p is $f(g(p))$.

6C. The introduction of neighborhoods on S and S' makes it easy to generalize the classical definition of a continous function. We say that f is continuous at $p_0 \in S$ if to every neighborhood V' of $p_0' = f(p_0)$ there exists a $V(p_0)$ whose image is contained in V'. A continuous mapping is one which is continuous at each point in the domain.

There are various equivalent ways in which continuous mappings can be characterized, some very applicable and with a striking formulation. For instance, a mapping is continuous if and only if the inverse image of every open set is open, or if and only if the inverse image of every closed set is closed. The proofs of these statements are very simple and will not be reproduced.

6D. If a one to one mapping and its inverse are both continuous the mapping is said to be *topological* or a *homeomorphism*. Two spaces S and S' are said to be homeomorphic or *topologically equivalent* if there exists a topological mapping of S onto S'. This is clearly an equivalence relation.

A topological mapping f of S onto S' maps open sets onto open sets and closed sets onto closed sets, for f is the inverse of a continuous function.

Suppose now that f is a one to one mapping of a subset $E \subset S$ onto $E' \subset S'$. Then f is said to be a homeomorphism of E onto E' if f and f^{-1} are continuous *in the relative topologies of E and E' respectively*. In this situation, if E is closed (or open) in the topology of S it is not necessarily true that E' is closed (or open) in the topology of S'.

6E. Theorem. *The image of a connected set under a continuous mapping is connected.*

We may assume that f maps a connected space S onto S', for after relativization the general situation reduces to this case. If $S' = O_1' \cup O_2'$, where O_1' and O_2' are open and disjoint, then S has a decomposition $S = f^{-1}(O_1') \cup f^{-1}(O_2')$ with the same properties. One of the sets $f^{-1}(O_1')$, $f^{-1}(O_2')$ must be void, and since the mapping is onto S' either O_1' or O_2' is empty. Hence S' is connected.

6F. Theorem. *The image of a compact set under a continuous mapping is compact.*

The initial remark in the proof of 6E applies again, and we use the same notations. If S' is covered by open sets O_α', then S is covered by the open sets $f^{-1}(O_\alpha')$. The finite subcovering of S determines a finite subcovering of S', and we conclude that S' is compact.

6G. In particular, homeomorphic spaces are simultaneously connected and simultaneously compact. We express this by saying that connectedness and compactness are *topological properties*. The same is of course true of local connectedness and local compactness.

We conclude also that *every continuous one to one mapping of a compact*

Hausdorff space onto another is a homeomorphism. Indeed, we need only show that the inverse mapping is continuous, and this is true if the direct mapping carries closed sets into closed sets. The latter is a consequence of 6F, for in a compact Hausdorff space closed and compact sets are identical (4D).

7. Arcs

7A. An arc is defined by a continuous mapping of the closed interval [0, 1] into a topological space S. We write the parametric equation of the arc in the form $p=f(t)$, $0 \leq t \leq 1$. The points $f(0)$ and $f(1)$ are the *end points* of the arc: $f(0)$ is the *initial point* and $f(1)$ is the *terminal point.*

A continuous mapping of the open interval (0, 1) is an *open arc.* By the interior of an arc we shall mean the open arc obtained by restricting f to (0, 1).

An arc is a *Jordan arc* if f is a homeomorphism; if S is a Hausdorff space it is sufficient to assume that f is one to one. An *open Jordan arc* is defined in the same manner.

A *closed curve* is an arc whose initial and terminal points coincide. It is a *Jordan curve* if the mapping becomes a homeomorphism when the end points of [0, 1] are identified (2F).

7B. An arc or open arc can also be considered as a point set, namely as the range of f. As a point set every arc is connected and compact; an open arc is connected. In most cases it is clear from the context whether we mean the mapping or the point set; consequently we can use the same symbol γ to cover both concepts. When the mapping needs to be specified we write $\gamma: t \to f(t)$.

It may happen that two mappings define the same arc in the obvious geometric sense. We shall later formalize the underlying concept of reparametrization, but for the moment it is preferable to agree that different mappings define different arcs. The specialization of the parametric interval to [0, 1] is likewise introduced as a matter of convenience.

7C. A space is said to be *arcwise connected* if any two points are the end points of an arc. It can be proved that any arc with distinct end points contains a Jordan arc with the same end points. In order to avoid the proof we prefer to require, from the start, that any two distinct points can be joined by a Jordan arc.

A space is *locally arcwise connected* if it has a basis consisting of arcwise connected sets.

Theorem. *A locally arcwise connected space is arcwise connected if and only if it is connected.*

If p_0 is joined to p_1 by a Jordan arc γ_1 and p_1 is joined to $p_2 \neq p_0$ by a

Jordan arc γ_2, then p_0 can be joined to p_2. Indeed, it suffices to follow γ_1 from p_0 to the first intersection with γ_2 and continue along γ_2 to p_2. Here, and in similar circumstances, the formal part of the reasoning is so simple that it will be suppressed.

Choose a fixed point p_0 and denote by E_1 the set consisting of p_0 and all points which can be joined to p_0 by Jordan arcs, by E_2 the set of points which cannot be so joined. For $p_1 \in E_1$, let $V(p_1)$ be an arcwise connected neighborhood. It follows from our initial remark that $V(p_1) \subset E_1$. Hence E_1 is open, and a similar reasoning shows that E_2 is likewise open. If the space is connected, E_2 must be void, and the space is arcwise connected.

Conversely, if the space S is arcwise connected it is the union of a point and all Jordan arcs from that point. Since each arc is connected the connectedness of S follows by Theorem 3B.

7D. R^n is locally arcwise connected, and therefore every region in R^n is arcwise connected. By a slight modification of the argument in 7C one sees that any two distinct points in a region can be joined by a polygonal Jordan arc in the region.

§2. SURFACES

We are now ready to leave the extreme generality which has characterized our discussion of topological spaces. By specialization we arrive through manifolds to the notion of surface which is in the center of our interest.

A powerful and completely elementary tool for the study of surfaces is the concept of homotopy. Its definition is so simple that it can just as easily be introduced for arbitrary arcwise connected spaces. Through its connection with the index of a closed curve it achieves particular significance for the theory of surfaces.

8. Definitions

8A. We formulate the following precise definition of an *n-dimensional manifold*:

Definition. *An n-dimensional manifold is a Hausdorff space on which there exists an open covering by sets homeomorphic with open sets in R^n.*

A manifold is evidently locally arcwise connected, and therefore, by 7C, each component is arcwise connected. It is a matter of convenience whether one does or does not require a manifold to be connected.

8B. A manifold may or may not be countable. Countability is an important property, but many proofs can be carried out without this assumption. We shall find that Riemann surfaces are countable, but this is a theorem and not a hypothesis.

8C. The case of 1-dimensional manifolds is almost trivial. Elementary considerations which we leave to the reader show that a compact connected 1-dimensional manifold is homeomorphic with a circle. A noncompact component is topologically an open interval, provided that it is countable. The uncountable case is less perspicuous, but we shall have no occasion to consider uncountable 1-dimensional manifolds.

8D. Our concern is with the case $n=2$. It is convenient to require connectedness, and we agree to make this slight distinction between a 2-dimensional manifold and a *surface*.

Definition. *A surface is a connected 2-dimensional manifold.*

Compact surfaces are conventionally referred to as *closed* surfaces, while noncompact surfaces are said to be *open*. This terminology is time-honored, and we will use it in spite of its ambiguity.

The plane, the sphere, and any plane region are elementary examples of surfaces. The torus is another well-known example. The sphere and the torus are closed surfaces.

For an example of an uncountable surface the reader is referred to the treatise of R. Nevanlinna [23, p. 51].

8E. An open set Δ on a surface F is called a *Jordan region* if its closure can be mapped topologically onto a closed disk, in such a way that Δ corresponds to the open disk. The boundary of a Jordan region is thus a Jordan curve. Since every point in a plane region is the center of a closed disk contained in that region, we see that every surface has an open covering by Jordan regions. The same construction yields an even stronger result, for each Jordan region used for the covering has the additional property that the homeomorphism which maps a disk onto the Jordan region can be extended to a homeomorphism of a larger concentric disk.

A Jordan region which satisfies this stronger condition will be referred to as a *parametric disk*, and its closure is a *closed parametric disk*. It is true that every Jordan region is a parametric disk, but this fact will not be needed.

9. The fundamental group

9A. One of the main tools for the study of the topological properties of a space, and in particular of a surface, is the investigation of arcs and their deformations.

We recall that an arc γ on a topological space S is determined by its equation $t \to f(t)$, $0 \leq t \leq 1$, where f is continuous and has its values on S (7A). The *product* of two arcs $\gamma_1 : t \to f_1(t)$ and $\gamma_2 : t \to f_2(t)$ can be formed if the terminal point of γ_1 coincides with the initial point of γ_2. It is defined as the arc $\gamma_1\gamma_2$ whose equation $t \to f(t)$ is given by $f(t) = f_1(2t)$ for $0 \leq t \leq \frac{1}{2}$ and $f(t) = f_2(2t-1)$ for $\frac{1}{2} \leq t \leq 1$. The *inverse* arc of $\gamma: t \to f(t)$ is $\gamma^{-1}: t \to f(1-t)$.

9B. An arc $\gamma_1: t \to f_1(t)$ is deformed into an arc $\gamma_2: t \to f_2(t)$ with the same initial and terminal points p_1, p_2 if a continuous mapping f of the closed square $0 \leq t \leq 1$, $0 \leq u \leq 1$ into S is constructed which satisfies the following conditions: (1) $f(t, 0) = f_1(t)$, $f(t, 1) = f_2(t)$ for all t, (2) $f(0, u) = p_1$, $f(1, u) = p_2$ for all u. If a deformation exists γ_1 and γ_2 are said to be *homotop*.

It is easily proved that this relation is reflexive, symmetric, and transitive. We can therefore divide the set of all arcs with common end points into homotopy classes. In this section the sign $\approx$ will indicate homotopy.

9C. Let $t \to \tau(t)$, $0 \leq t \leq 1$, be a continuous, real-valued, nondecreasing function with $\tau(0) = 0$, $\tau(1) = 1$. The arc $t \to f(\tau(t))$ is said to result from $t \to f(t)$ by a *reparametrization*.

A reparametrized arc is homotop to the original arc. In fact, a deformation is defined by $(t, u) \to f((1-u)t + u\tau(t))$. It follows that the homotopy class of an arc is determined by the succession of its points, rather than by a specific equation.

Let us prove, for instance, that the homotopy class of a Jordan arc is determined by the arc as a point set as soon as we designate one of the end points as the initial point. Suppose that f and g determine the same Jordan arc γ in this sense. Then $\tau = f^{-1} \circ g$ is a topological mapping of $[0, 1]$ onto itself, and the assumption implies $\tau(0) = 0$, $\tau(1) = 1$. The complement with respect to $[0, 1]$ of a point $t_0 \neq 0, 1$ has two components and t belongs to the same component as 0 if and only if $t < t_0$. The images of these components under the mapping τ must coincide with the complementary components of $\tau(t_0)$. It follows that $t < t_0$ implies $\tau(t) < \tau(t_0)$. Hence τ is increasing, and $g = f \circ \tau$ is a reparametrization of f.

The practical consequence of the preceding remark is that as far as homotopy is concerned the equation of a Jordan arc need not be specified as long as we indicate the initial point. A similar reasoning shows that the homotopy class of a Jordan curve is determined as soon as we indicate the initial point and a positive sense along the curve. The positive sense can be prescribed by marking the initial and terminal points of a subarc which does not pass through the initial point of the Jordan curve.

9D. Let us now restrict the attention to the family of closed curves from a common origin O on S. In this family the product and the inverse are always defined. These operations can be carried over to the homotopy classes, for evidently $\alpha \approx \alpha'$, $\beta \approx \beta'$ imply $\alpha\beta \approx \alpha'\beta'$ and $\alpha^{-1} \approx \alpha'^{-1}$. In view of these facts no confusion will arise if the same letter will sometimes denote an individual closed curve and sometimes its homotopy class.

The multiplication of homotopy classes is associative, for although $(\alpha\beta)\gamma$ and $\alpha(\beta\gamma)$ are not identical closed curves they are easily seen to be reparametrizations of one another. There is a unit element, denoted as 1, namely the homotopy class of the closed curve which reduces to the point O. We have $\alpha 1 = 1\alpha = \alpha$ and $\alpha\alpha^{-1} = 1$. The latter assertion is proved by constructing the deformation

$$(t, u) \to \begin{cases} f(2t) & 0 \leqq t \leqq (1-u)/2 \\ f(1-u) \quad \text{for} & (1-u)/2 \leqq t \leqq (1+u)/2 \\ f(2-2t) & (1+u)/2 \leqq t \leqq 1. \end{cases}$$

Because of these facts the homotopy classes form a group $\mathscr{F}_O(S)$, the *fundamental group* of S referred to the origin O.

9E. From now on we assume that S is arcwise connected. Then we can show that different origins determine isomorphic fundamental groups. Let O' be a second origin, and draw an arc σ from O' to O. To every closed curve α from O there corresponds a closed curve $\alpha' = \sigma\alpha\sigma^{-1}$ from O'. It is easily seen that the homotopy class of α' depends only on the homotopy class of α. Hence we obtain a mapping $\alpha \to \sigma\alpha\sigma^{-1}$ of the homotopy classes. This mapping is product preserving, for $\sigma\alpha\sigma^{-1} \cdot \sigma\beta\sigma^{-1} \approx \sigma(\alpha\beta)\sigma^{-1}$. Finally, the mapping is one to one, for $\alpha' \approx \sigma\alpha\sigma^{-1}$ implies $\alpha \approx \sigma^{-1}\alpha'\sigma$, and onto, for $\sigma^{-1}\alpha'\sigma$ is mapped on α'. We conclude that σ determines an isomorphic mapping of $\mathscr{F}_O(S)$ onto $\mathscr{F}_{O'}(S)$.

The abstract group $\mathscr{F}(S)$ which is isomorphic to all groups $\mathscr{F}_O(S)$ is called *the* fundamental group of S.

9F. Consider a continuous mapping φ of S into another space S'. To every arc $\gamma: t \to f(t)$ on S there corresponds an arc γ' on S' determined by the mapping $\varphi \circ f$. Homotop arcs are carried into homotop arcs. Therefore, the correspondence can be regarded as a mapping of $\mathscr{F}_O(S)$ into $\mathscr{F}_{O'}(S')$ where $O' = \varphi(O)$. This mapping preserves products, and is thus a homomorphism. If φ is topological and onto S' the corresponding mapping is clearly an isomorphism onto $\mathscr{F}_{O'}(S')$. We have thus:

Theorem. *The fundamental group of an arcwise connected space is a topological invariant.*

9G. It can happen that the fundamental group reduces to its unit element. When this is so, any two arcs with the same end points are homotop, and the space is said to be *simply connected.*

Definition. *A space is simply connected if its fundamental group has only one element.*

For short, a simply connected space is said to have a trivial fundamental group.

10. The index of a curve

10A. The topologically invariant character of the fundamental group makes it desirable to compute it explicitly in as many cases as possible. For surfaces this program will be carried out in 43, 44. For the moment we are content with two special, but very important results.

Theorem. *The space R^n is simply connected.*

If $x=(x_1, \cdots, x_n)$ we write $kx=(kx_1, \cdots, kx_n)$ for any real k. We choose the origin as initial point and consider a closed curve $t \to f(t)$, $f(0)=f(1)=O$, in R^n. The deformation $(t, u) \to (1-u)f(t)$ transforms it into the curve $1: t \to O$. In other words, every closed curve is homotop to 1, and the fundamental group has only one element.

The result holds in particular for the plane, and therefore for an open disk and a half-plane. The proof is obviously valid for any star-shaped subset of the plane. For this reason the fundamental group of a closed disk is also trivial.

10B. The plane minus a point is called the punctured plane. It is homeomorphic with an annulus.

Theorem. *The fundamental group of the punctured plane is the infinite cyclic group.*

Using the complex notation we suppose that $z=0$ is omitted, and we take $z=1$ as origin for the closed curves. The unit circle in the parametrization $t \to e^{2\pi i t}$ is denoted by α.

First of all it is clear that any closed curve from 1 can be deformed by central projection into a curve that lies on $|z|=1$. Next, by virtue of uniform continuity any such curve γ can be written as $\gamma_1\gamma_2 \cdots \gamma_n$ where each γ_k either does not pass through 1 or does not pass through -1. It is easily seen that γ_k can be deformed into one of the circular arcs between its end points. We suppose that the deformation has been carried out and use the same notation γ_k for the circular arcs. For $2 \leqq k \leqq n$, let σ_k denote one of the circular arcs on $|z|=1$ from 1 to the initial point of γ_k. On setting $\sigma_1=\sigma_{n+1}=1$ we have $\gamma \approx (\sigma_1\gamma_1\sigma_2^{-1})(\sigma_2\gamma_2\sigma_3^{-1}) \cdots (\sigma_n\gamma_n\sigma_{n+1}^{-1})$. Each factor $\sigma_k\gamma_k\sigma_{k+1}^{-1}$ is composed of three arcs, and we find by examination of all possible cases that it is homotop to 1, α or α^{-1}. Therefore, γ is homotop to a power α^m.

10C. It remains to prove that α^m is not homotop to 1 for $m \neq 0$. The proof will provide us with an opportunity to introduce the *index* or *winding number* of a closed curve γ in the plane with respect to a point z_0 not on γ.

Let the equation of γ be $z=f(t)$, $0 \leqq t \leqq 1$. If $f(t) \neq z_0$ for all t there exists a $\rho>0$ such that $|f(t)-z_0|>\rho$ throughout the interval. By uniform continuity the interval $[0, 1]$ can be divided into a finite number of subintervals $[t_i, t_{i+1}]$ such that $|f(t)-f(t_i)| < \rho$ for $t \in [t_i, t_{i+1}]$. The complex number

$$w_i = \frac{f(t_{i+1})-z_0}{f(t_i)-z_0}$$

has a positive real part, for $|w_i-1|<1$. Therefore there exists a unique value Θ_i of $\arg w_i$ which satisfies $-\frac{\pi}{2}<\Theta_i<\frac{\pi}{2}$. It is clear that $\sum\Theta_i$ is an integral multiple of 2π, and we set

$$n(\gamma, z_0) = \frac{1}{2\pi}\sum\Theta_i.$$

This is the index of γ with respect to z_0.

10D. It must be shown that $n(\gamma, z_0)$ does not depend on the subdivision. We need only show that nothing is changed if we replace $[t_i, t_{i+1}]$ by $[t_i, \tau]$ and $[\tau, t_{i+1}]$ where $t_i<\tau<t_{i+1}$. We denote the new arguments by Θ_i', Θ_i'' and have

$$\Theta_i = \Theta_i'+\Theta_i''+k\cdot 2\pi$$

where k is an integer. Since all three arguments are $<\frac{\pi}{2}$ in absolute value we obtain $|k|<\frac{3}{4}$, and hence $k=0$. The assertion follows.

10E. *As a function of z_0 the index $n(\gamma, z_0)$ is continuous on the complement of γ, and hence constant in each component of the complement.*

For the proof we use the same notations as above and choose z_0' so close to z_0 that $|z_0-z_0'|<\rho$ and $|f(t)-z_0'|>\rho$ for all t. We can then use the same subdivision to define $n(\gamma, z_0)$ and $n(\gamma, z_0')$. Writing

$$v_i = \frac{f(t_i)-z_0'}{f(t_i)-z_0}$$

we have $|v_i-1|<1$, and it is possible to set $\alpha_i=\arg v_i$ with $-\frac{\pi}{2}<\alpha_i<\frac{\pi}{2}$. The new argument Θ_i' is connected with Θ_i by

$$\Theta_i' = \Theta_i+\alpha_{i+1}-\alpha_i+k\cdot 2\pi,$$

and the resulting estimate $|k|<1$ implies $k=0$. It follows that $\sum\Theta_i'=\sum\Theta_i$ as required.

We note that $n(\gamma, z_0)$ *is zero when z_0 lies in the unbounded component of the complement*. To see this, choose $\rho>2\max|f(t)|$ and assume that $|z_0|>\frac{3}{2}\rho$. Then $|f(t)-z_0|>\rho>|f(t)-f(0)|$ for all $t\in[0, 1]$. This means that $n(\gamma, z_0)$ can be computed without subdividing the interval. With $t_0=0$, $t_1=1$ we get $w_0=1$, $\Theta_0=0$ and hence $n(\gamma, z_0)=0$.

10F. We show next that $n(\gamma, z_0)$ *does not change if* γ *is deformed without passing through* z_0. Let $f(t, u)$ define a deformation, and suppose that $f(t, u) \neq z_0$ for $0 \leqq t \leqq 1$, $0 \leqq u \leqq 1$. We can find a $\rho > 0$ and subintervals $[t_i, t_{i+1}]$, $[u_j, u_{j+1}]$ such that $|f(t, u) - z_0| > \rho$ and $|f(t, u) - f(t_i, u_j)| < \rho$ for $t \in [t_i, t_{i+1}]$, $u \in [u_j, u_{j+1}]$. Let Θ_i, Θ_i' be the arguments that correspond to $u = u_j$, $u = u_{j+1}$ respectively, and observe that we can choose

$$\beta_i = \arg \frac{f(t_i, u_{j+1}) - z_0}{f(t_i, u_j) - z_0}$$

so that $-\frac{\pi}{2} < \beta_i < \frac{\pi}{2}$.

We have the relation

$$\Theta_i' - \Theta_i = \beta_{i+1} - \beta_i + k \cdot 2\pi,$$

and we see as before that $k = 0$. Hence $\sum \Theta_i' = \sum \Theta_i$ and we conclude that the index does not change as we pass from u_j to u_{j+1}. Therefore the curves that correspond to $u = 0$ and $u = 1$ have the same index.

10G. We can now return to the proof of Theorem 10B. It is clear that $n(\alpha^m, 0) = m$ and $n(1, 0) = 0$. Hence $\alpha^m \approx 1$ only if $m = 0$, and the theorem is proved.

The proof remains valid if the punctured plane is replaced by a circular annulus, open or closed.

11. The degree of a mapping

11A. Let φ be a complex-valued continuous function whose domain is an open set D in the complex z-plane. We say that φ is *regular* at $z_0 \in D$ if z_0 has a neighborhood V in which $\varphi(z) \neq \varphi(z_0)$ for $z \neq z_0$. When this condition is fulfilled we are going to define a *degree* of the mapping φ at z_0.

11B. We denote by Ω a punctured disk $0 < |z - z_0| < r$ which is contained in V. According to the hypothesis φ maps Ω into the punctured plane Ω' obtained by omitting $w_0 = \varphi(z_0)$.

We know from 9F that φ determines a homomorphic mapping of the fundamental group $\mathscr{F}(\Omega)$ into $\mathscr{F}(\Omega')$. Because both groups are infinite cyclic it is very easy to determine the nature of the homomorphism.

In fact, the groups are generated by circles α and β respectively which we represent by equations $z - z_0 = (a - z_0)e^{2\pi i t}$ and $w - w_0 = (b - w_0)e^{2\pi i t}$. Here a and $b = \varphi(a)$ are corresponding points with $a \in \Omega$. We denote the homomorphism by Φ. Suppose that $\Phi(\alpha) = \beta^d$. This completely determines the homomorphism, for then $\Phi(\alpha^m) = \beta^{md}$. If $d = 0$ the whole group $\mathscr{F}(\Omega)$ is mapped into the unit element of $\mathscr{F}(\Omega')$. If $d \neq 0$ we obtain an isomorphic mapping of $\mathscr{F}(\Omega)$ onto the subgroup generated by β^d, for $\Phi(\alpha^m) = \beta^{md} = 1$ implies $m = 0$.

The number $d=d(\varphi)$ is the *degree* of the mapping at z_0. It is evidently independent of the choice of $a \in \Omega$, and hence independent of the choice of Ω.

11C. If a second mapping ψ is regular at $w_0=\varphi(z_0)$, then $\psi \circ \varphi$ is regular at z_0. We claim that $d(\psi \circ \varphi)=d(\psi)d(\varphi)$ where the degrees are computed at z_0 and w_0 respectively.

This is practically evident. We need only note that Ω' can be replaced by an arbitrarily small punctured disk about w_0 without changing the previous result. For sufficiently small $|a-z_0|$ the image $\varphi(\alpha)$ lies in the punctured disk and belongs to the homotopy class $\beta^{d(\varphi)}$. On the other hand, ψ maps β onto an element of the homotopy class $\gamma^{d(\psi)}$ with respect to a punctured plane, γ being a positively oriented circle. It follows that $\psi(\varphi(\alpha)) \approx \psi(\beta^{d(\varphi)}) \approx \gamma^{d(\varphi)d(\psi)}$, and the multiplicative property of the degrees is proved.

11D. Suppose now that D is a region, and that φ is a homeomorphic mapping of D onto a set $\varphi(D)$. If $\varphi(D)$ is open we can define the degree of the inverse mapping. Since the identity mapping has degree 1, we obtain $d(\varphi)\, d(\varphi^{-1})=1$, and hence $d(\varphi)=\pm 1$.

Actually, $\varphi(D)$ is always open ("invariance of the region"), but we shall avoid the use of this fairly advanced topological theorem. We prove, without using the fact that $\varphi(D)$ is open, that $d(\varphi)$ is constant in D.

For a fixed $z_0 \in D$, let Ω, Ω' and α, β be chosen as in 11B. By assumption, $\varphi(\alpha) \approx \beta^{d(\varphi)}$ in Ω', and hence $n(\varphi(\alpha),\ \varphi(z_0))=d(\varphi)$. Let z_1 be so close to z_0 that $n(\alpha,\ z_1)=n(\alpha,z_0)=1$ and $n(\varphi(\alpha),\ \varphi(z_1))=n(\varphi(\alpha),\ \varphi(z_0))= d(\varphi)$. To define the degree at z_1 we use a punctured disk Ω_1 centered at z_1, the punctured plane Ω_1' which omits $\varphi(z_1)$, and generators α_1, β_1. Because φ is a homeomorphism there is no restriction on Ω_1, other than that it be contained in D. For this reason, if z_1 is sufficiently close to z_0 we can choose Ω_1 so that it contains α. Then $\alpha \approx \alpha_1$ in Ω_1 and $\varphi(\alpha) \approx \varphi(\alpha_1)$ in Ω_1'. It follows that $n(\varphi(\alpha_1),\ \varphi(z_1))=n(\varphi(\alpha),\ \varphi(z_1))=d(\varphi)$, and hence that $\varphi(\alpha_1) \approx \beta_1^{d(\varphi)}$ in Ω_1'. Consequently, the degree at z_1 is equal to $d(\varphi)$, the degree at z_0. We conclude that $d(\varphi)$ is constant in a neighborhood of each point, and hence in each component of D. If D is connected it is constant throughout D.

By the preceding remark, if we use the invariance of the region it follows that $d(\varphi)$ is constantly 1 or -1. However, to reach this conclusion we need merely know that $\varphi(D)$ has an interior point $\varphi(z_0)$. Indeed, let $\Delta \subset \varphi(D)$ be an open disk containing $\varphi(z_0)$. Then φ has the degree ± 1 in $\varphi^{-1}(\Delta)$, and consequently in all of D.

Theorem. *The degree of a topological mapping of a region is constantly* $+1$ *or* -1.

We have proved this theorem under the condition that the image of the region has an interior point. A topological mapping of degree $+1$ is called *sense-preserving*, and one of degree -1 is *sense-reversing*.

12. Orientability

12A. We shall use the term *planar region* for any region V on a surface F which can be mapped topologically onto an open set in the plane. The homeomorphisms h which effect such mappings can be divided into two classes by assigning h_1 and h_2 to the same class if and only if $h_1 \circ h_2^{-1}$ is sense-preserving. For a given h we denote by $\bar{h}$ the mapping obtained by setting $\bar{h}(p)$ equal to the complex conjugate of $h(p)$. Then h and $\bar{h}$ are in different classes, for $h \circ \bar{h}^{-1}$ is sense-reversing. For any other mapping h_1 either $h \circ h_1^{-1}$ or $\bar{h} \circ h_1^{-1}$ is sense-preserving. This shows that there are exactly two classes.

The choice of one of the two classes as the positively oriented one constitutes an *orientation* of V. It is clear that an orientation of V induces an orientation of every subregion $V' \subset V$, namely by restricting a positively oriented mapping of V to V'.

The orientations of two overlapping planar regions V_1 and V_2 are said to be *compatible* if they induce the same orientation in *all* their common subregions.

12B. A surface F is *orientable* if it is possible to define compatible orientations of its planar regions.

Definition. *A surface F is said to be orientable if it is possible to assign a positive orientation to each planar region in such a way that the orientations of any two overlapping planar regions are compatible.*

Orientability is obviously a topologically invariant property. We can thus divide the classes of homeomorphic surfaces into classes of orientable and nonorientable surfaces. A well-known example of a nonorientable surface is the Möbius band.

An orientable surface has always two different orientations, for if a system of compatible orientations is given we get a new compatible system by reversing all the orientations. It follows in an obvious way by the connectedness that there cannot be more than two orientations.

12C. In order to orient a surface it is sufficient to orient all planar regions which belong to some open covering by planar regions. Suppose that the sets V of the open covering have been oriented in a compatible manner. Let V' be an arbitrary planar region. A point $p \in V'$ has a connected open neighborhood $U \subset V'$ which is contained in a V. The orientation of V induces an orientation of U, and we can choose an orientation of

V' which induces the same orientation of U. Because of the compatibility this orientation will not depend on the choice of U or V. Moreover, the points p which determine the same orientation as a fixed point $p_0 \in V'$ form an open set, and so do the points which determine the opposite orientation. Since V' is connected, its orientation will be uniquely determined. Finally, the orientations are compatible, for the two orientations of a component of $V' \cap V''$ can be determined by a point $p \in V' \cap V''$.

13. Bordered surfaces

13A. We have already mentioned that an open surface has an *ideal boundary*, defined as the point ∞ in the Alexandroff compactification (4F). We denote the ideal boundary of a surface F by β or $\beta(F)$.

The complement of any relatively compact set on F is often called a neighborhood of β, even though, strictly speaking, a neighborhood should include the point ∞ itself.

For many purposes it is preferable not to use the Alexandroff compactification but distinguish different points on β. This can be done, in many different ways, by imbedding F as an open set in a compact Hausdorff space M. It is clear that the boundary of F in M may be considered as a realization of β.

13B. The ideal boundary of a disk, and therefore of the plane, can be realized as a circle. Such standard realizations, when possible, are of course very useful.

We generalize the example of the disk by introducing the notion of a *bordered surface.*

Definition. *A connected Hausdorff space $\bar{F}$ is called a bordered surface if it is not a surface, and if there exists an open covering of $\bar{F}$ by sets which can be mapped homeomorphically onto relatively open subsets of a closed half-plane.*

13C. We may choose the homeomorphisms h so that the range of each h lies in the closed upper half-plane $y \geqq 0$ of the complex z-plane. If $h(p_0)$ is an interior point of the upper half-plane for one homeomorphism, then we contend that the same is true for all homeomorphisms defined at p_0. Suppose that p_0 is in the domain of h and h_1. We consider the homeomorphic mapping $\varphi = h_1 \circ h^{-1}$, restricted to a small open disk about $z_0 = h(p_0)$. The image of this disk is relatively open in the closed half-plane, and contains, therefore, an interior point with respect to the topology of the plane. It follows that we can apply Theorem 11D without appealing to the invariance of the region. A sufficiently small circle α about z_0 must be mapped by φ onto a closed curve whose index with respect to $h_1(p_0)$ is ± 1. But if $h_1(p_0)$ lies on the x-axis it belongs to the unbounded component of the complement of $\varphi(\alpha)$, and by 10E the index is zero. Hence $h_1(p_0)$ must be an interior point of the half-plane.

We conclude that it is possible to distinguish two complementary subsets of $\bar{F}$, the set B or $B(\bar{F})$ of points which are always mapped into the x-axis, and the set F of points which are never mapped into the x-axis. F is open, and B is closed, B is not empty, for then $\bar{F}$ would be a surface. Any neighborhood of a point on $\bar{F}$ contains points from F. Hence $\bar{F}$ is the closure and B is the boundary of F.

We call B the *border* and F the *interior* of $\bar{F}$ (in spite of the ambiguity). The components of B are called *contours*.

13D. Every point $p_0 \in B$ has a neighborhood which can be mapped homeomorphically onto a semiclosed half-disk $|z|<1$, $y \geqq 0$ with p_0 corresponding to the origin and points on B to the diameter $y=0$. From this representation we see that p_0 has a relative neighborhood on B which is homeomorphic with an open interval. In other words, B is a 1-dimensional manifold. Every compact contour is a Jordan curve.

We prove next that the interior F is a surface. We need only show that F is connected. Suppose that F has a decomposition into nonvoid open sets F_1 and F_2. Then $\bar{F}=\bar{F}_1 \cup \bar{F}_2$, and since $\bar{F}$ is connected $\bar{F}_1$ and $\bar{F}_2$ must have a common point $p_0 \in B$. But p_0 has a neighborhood V whose intersection with F is homeomorphic with an open half-disk, and consequently connected. The decomposition $V \cap F=(V \cap F_1) \cup (V \cap F_2)$ contradicts the connectedness of $V \cap F$, and we conclude that F must be connected.

If $\bar{F}$ is compact, $B(\bar{F})$ may be considered as a realization of $\beta(F)$. If $\bar{F}$ is not compact, $B(\bar{F})$ realizes only part of $\beta(F)$, and the remaining part is represented by the Alexandroff point $\beta(\bar{F})$ of $\bar{F}$.

Remark. The connection between border and ideal boundary is so intimate that it would be unwise to insist on separate notations under all circumstances. We agree that B will always stand for a concrete border, while the notations β and $\beta(F)$ will be used more freely with various related meanings.

13E. A bordered surface $\bar{F}$ is said to be orientable if and only if its interior F is orientable. We show that an orientation of F induces a positive direction on the border B.

Consider a neighborhood V of $p_0 \in B$ which is homeomorphic with a semiclosed half-disk, for instance $|z|<1, y \geqq 0$. We say that a homeomorphism h of V onto the semiclosed half-disk is positively oriented if its restriction to $V \cap F$ belongs to the class of homeomorphisms by which F is oriented. It is always possible to find a positively oriented homeomorphism, for $p \to h(p)$ and $p \to -\bar{h}(p)$ are oppositely oriented and have the same range.

We can and will assume that h is defined on a region which contains $\bar{V}$. Suppose that h_1 is a second positively oriented homeomorphism of V with the same properties. Then $\varphi = h \circ h_1^{-1}$ is a homeomorphic mapping of the closed half-disk onto itself. It can be extended to a slightly larger half-disk, and it can also be extended by symmetry to a homeomorphic mapping of the corresponding full disk.

Since φ is sense-preserving in the upper half-plane it is also sense-preserving in the whole disk. For this reason the mapping of $|z|=1$ on itself preserves the direction. Inasmuch as the upper half-circles correspond to each other this is possible only so that $\varphi(1)=1$ and $\varphi(-1)=-1$. We can now designate the inverse image of -1 as the intitial point and the inverse image of 1 as the terminal point of the arc on B which corresponds to the diameter; we have just shown that this designation is independent of the choice of h. In other words, a sufficiently small subarc of B has a well-defined positive direction with respect to the orientation of F.

It is easy to extend the determination of the positive direction to arbitrary subarcs of B. We can start from an arbitrary direction of the arc. It induces a direction of every subarc, and for subarcs that already have a positive direction with respect to the orientation of F, the induced direction is either equal or opposite to that direction. For overlapping subarcs the same alternative must occur. It follows that the union of all subarcs with the same direction and the union of all subarcs with opposite direction are both open. Hence one of these sets is void, and the given arc has a well-defined positive direction which agrees with that of all sufficiently small subarcs.

We conclude, in particular, that each compact contour, regarded as a Jordan curve, has a positive sense with respect to the orientation of F.

13F. The notion of bordered surface is closely connected with the idea of a region with a smooth boundary. In order to introduce a convenient terminology we begin by defining a 1-*dimensional submanifold*. The purpose of this auxiliary notion is to bypass certain delicate topological considerations which would lead too far afield.

Definition. *A subset Γ of a surface F is called a* 1-*dimensional submanifold if every $p \in \Gamma$ has an open neighborhood V which can be mapped homeomorphically onto $|z|<1$ in such a way that the intersection $V \cap \Gamma$ corresponds to the real diameter.*

Consider now a subregion Ω of F. We will say that Ω and $\bar{\Omega}$ are *regularly imbedded* if, in the first place, Ω and its exterior have the same boundary, and if, secondly, this boundary is a 1-dimensional submanifold. Let p be a point on Bd Ω, and choose a neighborhood V as in the definition. Since the half-disks $y>0$ and $y<0$ are connected their inverse images are contained either in Ω or in the exterior of Ω. The fact that p lies on the common boundary of Ω and the exterior implies that one of the images must belong to Ω and the other to the exterior. Hence $V \cap \bar{\Omega}$ is mapped on a semiclosed half-disk, and we conclude that $\bar{\Omega}$ is a bordered surface whose border coincides with the boundary of Ω.

13G. For a regularly imbedded region $\Omega \subset F$, let E be a component of $F - \Omega$ or, more generally, a union of such components. *Then $\Omega \cup E$ is also a regularly imbedded region.* To see that this set is open we need only examine a point p on the boundary of E. Such a point is also on the boundary of Ω, and as such it has a neighborhood V whose intersection with $F - \Omega$ is connected. It follows that $V \cap (F - \Omega) \subset E$, and hence $V \subset \Omega \cup E$, proving that $\Omega \cup E$ is open. It is connected because $\Omega \cup (\bar{\Omega} \cap E)$, a set between Ω and $\bar{\Omega}$, is connected. Finally, the boundary of $\Omega \cup E$ consists of the contours of Ω which do not belong to E, and one sees at once that $\Omega \cup E$ is regularly imbedded.

Suppose now that $\bar{\Omega}$ is compact. Then Ω has only a finite number of contours, and hence only a finite number of complementary components. Let E be the union of those components that are compact. Then $\Omega \cup E$ is a relatively compact regularly imbedded region with the additional property that all complementary components are noncompact. Such a region will be called a *regular region*. Because this terminology is used throughout the book we state a formal definition:

Definition. *A region $\Omega \subset F$ is a regular region if* (1) *Ω is relatively compact,* (2) *Ω and $F - \bar{\Omega}$ have a common boundary which is a 1-dimensional submanifold,* (3) *all components of $F - \Omega$ are noncompact.*

We have shown that every relatively compact and regularly imbedded region can be completed to a regular region.

13H. It is often desirable to imbed a bordered surface in a surface. In this respect we prove:

Theorem. *Every bordered surface $\bar{F}$ can be regularly imbedded in a surface. If $\bar{F}$ is compact, it can be regularly imbedded in a closed surface.*

We construct a standard imbedding which will find important use, especially in the theory of Riemann surfaces.

Let $\bar{F}_1$ be a topological space which is homeomorphic with $\bar{F}$, and consider a topological mapping φ of $\bar{F}$ onto $\bar{F}_1$. In the sum $\bar{F} + \bar{F}_1$ we identify each $p \in B(\bar{F})$ with its image $\varphi(p)$, and topologize as in 2F. The resulting space is denoted by $\hat{F}$. It is trivial to verify that $\hat{F}$ is a Hausdorff space, and also that every point in $F \cup \varphi(F)$ has a neighborhood which is homeomorphic to an open set in the plane.

Suppose now that $p_0 \in B$. Then p_0 has a neighborhood V relatively to $\bar{F}$ which can be mapped on a semiclosed half-disk as in 13D, and

$\hat{V} = V \cup \varphi(V)$ is an open neighborhood on $\hat{F}$. We denote the mapping of V by h. A homeomorphic mapping of $\hat{V}$ onto an open disk is obtained by mapping $p \in V$ into $h(p)$ and $p \in \varphi(V)$ into $\bar{h}(\varphi^{-1}(p))$. We have proved that $\hat{F}$ is a surface, and that $\bar{F}$ is regularly imbedded in $\hat{F}$.

The process that we have described is called *duplication*, and $\hat{F}$ is referred to as the *double* of $\bar{F}$. The mapping φ can be extended to an involutory topological self-mapping of $\hat{F}$ by requiring that $\varphi(\varphi(p))=p$. If $\bar{F}$ is orientable, so is $\hat{F}$, and φ is sense-reversing.

The points p and $\varphi(p)$ are said to be symmetric with respect to the border. We shall frequently denote $\varphi(p)$ by p^* and $\varphi(F)$ by F^*.

The double of a disk is a sphere, and the double of an annulus is a torus.

§3. COVERING SURFACES

The reader is familiar with the crude notion of Riemann surfaces which occurs in classical function theory. For the description of such surfaces a pictorial language is used which in spite of its didactic value does not belong in a mathematical treatment.

The notion in question is purely topological, and in this section we are concerned with its axiomatization under the name of covering surface.

A covering surface is smooth if it has no branch points, and a smooth covering surface will be called regular if it has no boundary in a sense that will have to be made more precise. We point out right here that this terminology does not coincide with common usage.

The regular covering surfaces are connected with the fundamental group of the underlying surface by virtue of the monodromy theorem whose consequences we shall have reason to discuss in detail.

14. Smooth covering surfaces

14A. For our basic definition we choose the following:

Definition. *A connected Hausdorff space $\tilde{F}$ and a mapping f of $\tilde{F}$ into a surface F are said to determine a smooth covering surface of F, if every $\tilde{p}_0 \in \tilde{F}$ has a neighborhood $\tilde{V}$ which is mapped topologically by f onto a neighborhood V of $P_0 = f(\tilde{p}_0)$.*

The definition implies that f is continuous, and the mapping may be described as locally topological. It is always possible to replace $\tilde{V}$ and V by open neighborhoods, and even by Jordan regions. To see this, choose open neighborhoods $V_0(p_0) \subset V$ and $\tilde{V}_0(\tilde{p}_0) \subset \tilde{V}$. Then $\tilde{U} = f^{-1}(V_0) \cap \tilde{V}_0$ is open, and $\tilde{U} \subset \tilde{V}_0$, $U = f(\tilde{U}) \subset V_0$. Because f is topological on $\tilde{V}$, $f(\tilde{U})$ is relatively open in V, and since $f(\tilde{U})$ is contained in the open subset V_0 it is absolutely open. We see that the sets $\tilde{U}$ and U are open and correspond to each other topologically. The final step is to replace U by a

Jordan region $\Delta \subset U$, containing p_0, and $\tilde{U}$ by the topological image $\tilde{\Delta} = f^{-1}(\Delta) \cap \tilde{U}$.

Our reasoning shows that $\tilde{F}$ is always a surface. The covering surface is usually denoted by $(\tilde{F}, f)$, but in appropriate circumstances we abbreviate the notation to $\tilde{F}$.

14B. An important property is that *open sets are mapped on open sets.* With the previous notations, assume that $\tilde{p}_0$ belongs to the open set $\tilde{O}$. Since f is topological on $\tilde{U}$ we know that $f(\tilde{U} \cap \tilde{O})$ is open. The assertion follows from $p_0 \in f(\tilde{U} \cap \tilde{O}) \subset f(\tilde{O})$.

In particular, $f(\tilde{F})$ is a region on F, and it is clear that $(\tilde{F}, f)$ may be considered as a smooth covering surface of $f(\tilde{F})$. Conversely, if we start from a region $G \subset F$, then any component $\tilde{G}$ of $f^{-1}(G)$ yields a smooth covering surface $(\tilde{G}, f)$ of G. For the proof we need only recall that $\tilde{G}$ is open.

14C. For the study of covering surfaces it is essential to introduce the notion of *continuation along an arc.* The classical connection would be clearer if we spoke of the continuation of the multiple-valued inverse function f^{-1}, but since we do not wish to consider multiple-valued functions it is preferable to avoid this language.

Consider first an arc $\tilde{\gamma}$ on $\tilde{F}$ with the equation $t \to \tilde{\varphi}(t)$, $0 \leqq t \leqq 1$. The arc $\gamma = f(\tilde{\gamma})$ on F has, by definition, the equation $t \to f(\tilde{\varphi}(t))$. It is called the projection of $\tilde{\gamma}$. Now we reverse the point of view and regard γ as primarily given through its equation $t \to \varphi(t)$. Any arc $\tilde{\gamma}$ with the projection γ and the initial point $\tilde{p}_0$ is called a continuation along γ from $\tilde{p}_0$; naturally, $\tilde{p}_0$ must lie over p_0. In the literature the process of forming the continuation is also referred to as *lifting* the arc.

The following uniqueness theorem holds:

Theorem. *On a smooth covering surface, any two continuations along the same arc and from the same initial point are identical.*

Let the two continuations be given by $\tilde{\varphi}_1$ and $\tilde{\varphi}_2$. We consider the set E of all t such that $\tilde{\varphi}_1(t) = \tilde{\varphi}_2(t)$. This set is closed, for $\tilde{\varphi}_1, \tilde{\varphi}_2$ are continuous and $\tilde{F}$ is a Hausdorff space. We prove that E is also relatively open. Consider $t_0 \in E$ and determine neighborhoods $\tilde{V}$ of $\tilde{\varphi}_1(t_0) = \tilde{\varphi}_2(t_0)$ and V of $\varphi(t_0)$ which are in one to one correspondence by f. Because of the continuity there exists a neighborhood of t_0 in which $\tilde{\varphi}_1(t)$ and $\tilde{\varphi}_2(t)$ are contained in $\tilde{V}$. Since they have the same projection they must coincide, $\tilde{\varphi}_1(t) = \tilde{\varphi}_2(t)$, and we have proved that E is relatively open. By assumption E contains 0; it follows that E must comprise the whole interval $[0, 1]$.

14D. The existence of a continuation cannot be asserted. Therefore, we define a class of *regular* covering surfaces as follows:

Definition. *A smooth covering surface is said to be regular if there exists a continuation along any arc γ of F and from any point over the initial point of γ.*

14E. In order to analyze this definition we examine the case in which a continuation fails to exist. Let γ and $\tilde{p}_0$ be given. The set of all $t_0 \in [0, 1]$ for which there exists a continuation from p_0 along the subarc corresponding to $[0, t_0]$ is easily seen to be relatively open. In fact, if $\tilde{\varphi}(t)$ is defined in $[0, t_0]$, $t_0 < 1$, we can determine neighborhoods of $\tilde{\varphi}(t_0)$ and $\varphi(t_0)$ which are in one to one correspondence, and using this correspondence it is possible to continue $\tilde{\varphi}(t)$ beyond t_0. If we assume that there is no continuation along all of γ it follows that the set under consideration is a half-open interval $[0, \tau)$ with $0 < \tau \leqq 1$.

The function $\tilde{\varphi}$ is defined on $[0, \tau)$. We claim that $\tilde{\varphi}(t)$ tends to the ideal boundary of $\tilde{F}$ as $t \to \tau$. Suppose that this were not the case. Then there exists a compact set $C \subset \tilde{F}$ and a sequence of values $t_n \to \tau$ such that $\tilde{\varphi}(t_n) \in C$. We can extract a subsequence $\{t_{n_k}\}$ with the property that $\tilde{\varphi}(t_{n_k})$ converges to a point $\tilde{a} \in C$. It follows by continuity that $a = f(\tilde{a}) = \varphi(\tau)$. We choose corresponding neighborhoods $\tilde{V}$, V of $\tilde{a}$, a and denote by f^{-1} the inverse function of f restricted to $\tilde{V}$. There exists $\tau' < \tau$ such that $\varphi(t) \in V$ for $\tau' \leqq t \leqq \tau$, and $\tilde{\varphi}$ can be extended from $[0, \tau']$ to $[0, \tau]$ by setting $\tilde{\varphi} = f^{-1} \circ \varphi$ in $[\tau', \tau]$. This contradicts the definition of τ, and the assertion follows.

We say that $\varphi(\tau)$ is an *asymptotic value* of f. The path described by $\tilde{\varphi}(t)$ for $0 \leqq t < \tau$ is the corresponding *path of determination*. A regular covering surface is thus a smooth covering surface on which there is no path of determination. In particular, since a compact surface has no ideal boundary, *a compact smooth covering surface is always regular.*

14F. The preceding considerations permit us to characterize the property of regularity in a different and sometimes more convenient way. In a slightly more general context the equivalent condition that we formulate below will be referred to as *completeness* (21A).

Theorem. *A smooth covering surface is regular if and only if every $p_0 \in F$ has a neighborhood V with the property that each component of $f^{-1}(V)$ is compact.*

For the present we prove only the sufficiency. Suppose that $\tilde{\gamma}$ is a path of determination which leads to the asymptotic value p_0. If V is any neighborhood of p_0, the path $\tilde{\gamma}$ must, from a certain point on, lie in a component of $f^{-1}(V)$. By assumption, $\tilde{\gamma}$ is not contained in any compact set. Hence the component that contains the end part of $\tilde{\gamma}$ cannot be compact. We conclude that the completeness condition cannot be satisfied, and hence that a complete covering surface is necessarily regular.

The necessity will be proved in 15D.

14G. For the purpose of proving that a covering surface is regular the following corollary to the sufficiency part of the preceding theorem is particularly useful:

Corollary. *Given a smooth covering surface* $(\tilde{F}, f)$ *of* F, *let* G *be a region on* F, *and suppose that* $\tilde{G}$ *is a relatively compact component of* $f^{-1}(G)$. *Then* $(\tilde{G}, f)$ *is a regular covering surface of* G.

The proof is immediate. For $p_0 \in G$, let V be a compact neighborhood of p_0 which is contained in G. To avoid confusion we denote the restriction of f to $\tilde{G}$ by g. Then $g^{-1}(V) = f^{-1}(V) \cap \tilde{G} = f^{-1}(V) \cap \mathrm{Cl}\, \tilde{G}$, for since $\tilde{G}$ is a component of $f^{-1}(G)$ its boundary points cannot belong to $f^{-1}(G)$, and much less to $f^{-1}(V)$. But $f^{-1}(V)$ is closed and $\mathrm{Cl}\, \tilde{G}$ is compact. Hence their intersection $g^{-1}(V)$ is compact on $\tilde{F}$, and a fortiori on $\tilde{G}$. It follows that V has the property required in Theorem 14F, and we conclude that $(\tilde{G}, f)$ is regular.

15. The monodromy theorem

15A. Consider a regular covering surface $(\tilde{F}, f)$ of F. Let a, b be two points on F, and $\tilde{a}$ a point over a. We connect a to b by an arc γ. Then the continuation $\tilde{\gamma}$ along γ from $\tilde{a}$ has a terminal point $\tilde{b}$ over b. It is important to determine to what extent $\tilde{b}$ depends on the choice of γ.

A partial answer to this question is furnished by the *monodromy theorem* which occupies a central position in the theory of covering surfaces.

Theorem. *Suppose that* $(\tilde{F}, f)$ *is a regular covering surface of* F. *If* γ_1 *and* γ_2 *are homotop arcs from* a *to* b *on* F, *then the continuations* $\tilde{\gamma}_1$ *and* $\tilde{\gamma}_2$ *from a common initial point* $\tilde{a}$ *over* a *terminate at the same point* $\tilde{b}$. *Moreover,* $\tilde{\gamma}_1$ *and* $\tilde{\gamma}_2$ *are homotop on* $\tilde{F}$.

15B. Let $(t, u) \to \varphi(t, u)$, defined on $0 \leqq t \leqq 1$, $0 \leqq u \leqq 1$, be a deformation of γ_1 into γ_2. The theorem will be proved if we can construct a continuous function $(t, u) \to \tilde{\varphi}(t, u)$ with $f \circ \tilde{\varphi} = \varphi$ and $\tilde{\varphi}(0, 0) = \tilde{a}$. In fact, $\tilde{\varphi}(0, u)$ and $\tilde{\varphi}(1, u)$ must then reduce to constants $\tilde{a}$, $\tilde{b}$, and $t \to \tilde{\varphi}(t, 0)$, $t \to \tilde{\varphi}(t, 1)$ will be the equations of $\tilde{\gamma}_1, \tilde{\gamma}_2$.

For any fixed u we define $\tilde{\varphi}(t, u)$ as the continuation along the curve $t \to \varphi(t, u)$ from the initial value $\tilde{a}$. We must show that $\tilde{\varphi}(t, u)$ is continuous in both variables.

For a given u_0, let E be the set of all $\tau \in [0, 1]$ with the property that the restriction of $\tilde{\varphi}$ to the rectangle $0 \leqq t \leqq \tau$, $0 \leqq u \leqq 1$ is continuous in both variables on the line segment $0 \leqq t \leqq \tau$, $u = u_0$. We are going to show that 1 belongs to E. Since u_0 is arbitrary this will prove that $\tilde{\varphi}$ is continuous in the closed unit square.

The definition of E is such that E is either a semi-open interval $[0, t_0)$ or a closed interval $[0, t_0]$. In either case we choose open neighborhoods $\tilde{V}$, V

of $\tilde{\varphi}(t_0, u_0)$, $\varphi(t_0, u_0)$ respectively which correspond to each other topologically, and we denote by f^{-1} the restricted inverse of f which is defined in V and has values in $\tilde{V}$. There exists a $\delta > 0$ with these properties: (1) $\varphi(t, u) \in V$ for $|t-t_0| < \delta$, $|u-u_0| < \delta$, (2) $\tilde{\varphi}(t, u_0) \in \tilde{V}$ for $|t-t_0| < \delta$. It is tacitly understood that we consider only points (t, u) in the unit square.

Choose $t_1 = t_0$ if E is a closed interval and $t_0 - \delta < t_1 < t_0$ if E is semiopen. $\tilde{V}$ is open, and $\tilde{\varphi}(t_1, u)$ is continuous in u at $u = u_0$. Hence we can find a positive $\eta \leqq \delta$ such that $\tilde{\varphi}(t_1, u) \in \tilde{V}$ for $|u-u_0| < \eta$. For these values $\tilde{\varphi}(t_1, u) = f^{-1}(\varphi(t_1, u))$. In the part of the rectangle $0 \leqq t \leqq t_0 + \delta$, $|u-u_0| < \eta$ which lies in the unit square, define a function $\tilde{\varphi}_1(t, u)$ which is equal to $\tilde{\varphi}(t, u)$ for $t \leqq t_1$ and equal to $f^{-1}(\varphi(t, u))$ for $t > t_1$. Then $\tilde{\varphi}_1$ is continuous in t for fixed u, and $f \circ \tilde{\varphi}_1 = \varphi$. By the uniqueness of continuation $\tilde{\varphi}_1$ must coincide with $\tilde{\varphi}$, and we find that $\tilde{\varphi}$ is continuous in both variables on the part of the segment $0 \leqq t < t_0 + \delta$, $u = u_0$ which lies within the closed unit square. This contradicts the definition of E unless E is closed and $t_0 = 1$. Hence 1 belongs to E, and the theorem is proved.

15C. As an immediate corollary of the monodromy theorem we obtain:

Theorem. *If $(\tilde{F}, f)$ is a regular covering surface of a simply connected surface F, then f is a homeomorphic mapping of $\tilde{F}$ onto F.*

We choose a point $\tilde{a} \in \tilde{F}$ and consider its projection a. An arc γ can be drawn from a to any $b \in F$. The continuation $\tilde{\gamma}$ from $\tilde{a}$ along γ ends at a point $\tilde{b}$ over b. Hence every point on F is the projection of a point on $\tilde{F}$.

Suppose that $\tilde{a}_1, \tilde{a}_2 \in \tilde{F}$ have the same projection a. We connect $\tilde{a}_1$ and $\tilde{a}_2$ by an arc $\tilde{\gamma}$. Its projection is a closed curve γ, and $\tilde{\gamma}$ is the continuation along γ from $\tilde{a}_1$. The simple connectivity implies that γ is homotop to 1, and by the monodromy theorem $\tilde{a}_1 = \tilde{a}_2$. We have proved that f is one to one. It is continuous in both directions because of its local properties.

15D. It is now easy to complete the proof of Theorem 14F. Suppose that $(\tilde{F}, f)$ is a regular covering surface of F. Let Δ be a Jordan region which contains $p_0 \in F$, and let $V \subset \Delta$ be a compact neighborhood of p_0. Consider a component $\tilde{V}$ of $f^{-1}(V)$. The theorem will be proved if we show that $\tilde{V}$ is compact.

$\tilde{V}$ is contained in a component $\tilde{\Delta}$ of $f^{-1}(\Delta)$, and $(\tilde{\Delta}, f)$ is a regular covering surface of Δ. Since Δ is simply connected it follows by the monodromy theorem that f, restricted to $\tilde{\Delta}$, is a homeomorphism. For greater clarity, let the restriction be denoted by g and its single-valued inverse by g^{-1}. We know that $\tilde{V} \subset g^{-1}(V)$, and $g^{-1}(V)$, the continuous image of a compact set, is compact. But $\tilde{V}$ is closed, for it is a component of the closed set $f^{-1}(V)$. We conclude that $\bar{V}$, a closed subset of a compact set, is necessarily compact.

16. Applications of the monodromy theorem

16A. The monodromy theorem is often applied to questions which are only indirectly connected with the theory of covering surfaces. In the following we consider a fairly general situation which covers most of the usual applications.

We suppose that a simply connected surface F is covered by regions V. In each V a nonempty family Φ_V of complex-valued functions is given. The problem that interests us is to define a function on the whole surface F whose restriction to each V belongs to Φ_V; in other words, we want to piece together a global function from a supply of local functions. We will show that this is possible if the families Φ_V satisfy the following conditions:

(A1) If $\varphi \in \Phi_V$, $\varphi' \in \Phi_{V'}$, and $V \cap V' \neq 0$, then the interior of the set of all points p with $\varphi(p)=\varphi'(p)$ is relatively closed in $V \cap V'$.

(A2) To every $p \in V \cap V'$ and every $\varphi \in \Phi_V$ there exists a $\varphi' \in \Phi_{V'}$ such that $\varphi=\varphi'$ in a neighborhood of p.

Condition (A1) implies that in each component of $V \cap V'$ the set where $\varphi(p)=\varphi'(p)$ is either the whole component, or else a set without interior points. It would have been simpler to assume that either $\varphi=\varphi'$ or $\varphi\neq\varphi'$ throughout each component of $V \cap V'$, but this condition would not be sufficiently general for the application in 16F.

16B. We construct a space $\tilde{F}$ whose points are determined by a point $p \in F$ together with a $\varphi \in \Phi_V$ such that $p \in V$; this point is denoted by (p, φ). Two points are identified if and only if they are determined by the same p and by functions φ which coincide in a neighborhood of p. A set of points (p, φ) with a fixed φ is declared to be an open set if the points p form an open set on F. We choose these particular open sets as a basis for the topology on $\tilde{F}$.

To see that $\tilde{F}$ is a Hausdorff space, let (p_0, φ_0) and (p_1, φ_1) be distinct points. If $p_0 \neq p_1$ disjoint neighborhoods are formed by (p, φ_0) and (p', φ_1) where p, p' run through disjoint neighborhoods of p_0, p_1. If $p_0=p_1$ and φ_0, φ_1 are defined in V_0, V_1 respectively, we consider the component of $V_0 \cap V_1$ which contains p_0. The condition $(p_0, \varphi_0)\neq(p_0, \varphi_1)$ means that φ_0 and φ_1 cannot coincide in any neighborhood of p_0, and hence not on the whole component. Therefore, by (A1), $\varphi_0=\varphi_1$ at most on a set with empty interior, and it follows that $(p, \varphi_0)\neq(p, \varphi_1)$ for all p in the component under consideration. We have shown that the condition for a Hausdorff space is fulfilled.

It is quite obvious that the mapping $f:(p, \varphi)\rightarrow p$ defines each component of $\tilde{F}$ as a smooth covering surface of F. We wish to prove, in addition, that this covering surface is regular.

16C. Consider an arc $\gamma: t \rightarrow \omega(t)$ and divide $[0, 1]$ into subintervals $[t_i, t_{i+1}]$ so that $\omega(t)$ belongs to a V_i for $t_i \leqq t \leqq t_{i+1}$. Choose an initial point $\tilde{p}_0 = (p_0, \varphi_0)$ over $p_0 = \omega(0)$; we may assume that $\varphi_0 \in \Phi_{V_0}$.

According to (A2) there exists a $\varphi_1 \in \Phi_{V_1}$ which is equal to φ_0 in a neighborhood of $\omega(t_1)$; similarly, there exists a $\varphi_2 \in \Phi_{V_2}$ which coincides with φ_1 in a neighborhood of $\omega(t_2)$, and so on. This process leads to a continuation along γ, defined by $\tilde{\omega}(t) = (\omega(t), \varphi_i)$ for $t_i \leqq t \leqq t_{i+1}$. Consequently, each component of $\tilde{F}$ is a regular covering surface.

16D. Since F was supposed to be simply connected, we conclude by the monodromy theorem that the component of $\tilde{F}$ which contains the initial point $\tilde{p}_0$ is in topological correspondence with F. To every $p \in F$ there corresponds a unique point $(p, \varphi) = f^{-1}(p)$, and hence a unique value $\varphi(p)$; we denote this value by $\tilde{\varphi}(p)$. Consider a point p_1 and assume that $f^{-1}(p_1) = (p_1, \varphi_1)$, $\varphi_1 \in \Phi_{V_1}$. The points (p, φ_1) with $p \in V_1$ form a neighborhood of (p_1, φ_1). Because f^{-1} is continuous we have $f^{-1}(p) = (p, \varphi_1)$, and hence $\tilde{\varphi}(p) = \varphi_1(p)$, as soon as p is sufficiently close to p_1. Thus $\tilde{\varphi} = \varphi_1$ on an open subset of V_1. Suppose, on the other hand, that $\tilde{\varphi}(p_2) \neq \varphi_1(p_2)$ at a point $p_2 \in V_1$. Then $\tilde{\varphi}(p) = \varphi_2(p)$, say, in a neighborhood of p_2, and $(p_2, \varphi_2) \neq (p_2, \varphi_1)$. Since $\tilde{F}$ is a Hausdorff space it follows that $(p, \varphi_2) \neq (p, \varphi_1)$ when p is near p_2, and we conclude that $\tilde{\varphi} \neq \varphi_1$ on an open set. In view of the connectedness of V_1 we find through this reasoning that $\tilde{\varphi} = \varphi_1$ throughout V_1. In other words: *it is possible to define a single-valued function $\tilde{\varphi}$ on F which in every V is identical with a $\varphi \in \Phi_V$.* This function is uniquely determined if required to coincide with a particular φ in a neighborhood of a given point.

16E. The significance of this result becomes clearer if we give a familiar application. Let it be the *argument principle*:

Theorem. *Suppose that the complex-valued function ψ is continuous and $\neq 0$ on a simply connected surface F. Then it is possible to pick out a single-valued continuous branch of* arg ψ.

Let q be an arbitrary point on F. We can find a connected neighborhood $V(q)$ such that $|\psi(p) - \psi(q)| < |\psi(q)|$ for $p \in V(q)$. Let $\overline{\arg}\, \psi(q)$ be the value of the argument which lies in the interval $[0, 2\pi)$, and define $\overline{\arg}\, \psi(p)$ in $V(q)$ as the branch which satisfies $|\overline{\arg}\, \psi(p) - \overline{\arg}\, \psi(q)| < \frac{\pi}{2}$; this branch is continuous. We choose the functions $\overline{\arg}\, \psi(p) + n \cdot 2\pi$ with integral n as the functions φ in $\Phi_{V(q)}$. The conditions (A1) and (A2) are almost trivially fulfilled. Hence we can determine a single-valued function, called arg ψ, which in every $V(q)$ coincides with one of the branches $\overline{\arg}\,\psi + n \cdot 2\pi$. In particular, this function is continuous.

16F. Another application of the construction principle in 16A, B leads to the classical notion of multiple-valued analytic functions. Let F be the complex plane. For every region $V \subset F$ we consider the family Φ_V of all functions f which are defined and analytic in V according to the classical definition.

With this definition of Φ_V we see that (A1) is fulfilled while (A2) is not. The construction of the space $\tilde{F}$ depended only on (A1). When the construction is carried out we see that each component of $\tilde{F}$ is a smooth covering surface of the plane, but it need not be regular. The components of $\tilde{F}$ are the *analytic functions in the sense of Weierstrass*, except that the algebraic function elements are omitted.

17. The class of regular covering surfaces

17A. By use of the monodromy theorem we are going to show that the regular covering surfaces of a given surface F have a simple relation to the fundamental group of F.

Given F and a regular covering surface $(\tilde{F}, f)$ we choose an origin O on F and a point $\tilde{O}$ over O. Consider two arcs γ_1, γ_2 on F which begin at O and end at a common point a. The covering surface determines continuations $\tilde{\gamma}_1, \tilde{\gamma}_2$ along γ_1, γ_2 from $\tilde{O}$ which terminate at certain points $\tilde{a}_1, \tilde{a}_2$ over a. The monodromy theorem gives a sufficient condition under which $\tilde{a}_1 = \tilde{a}_2$, but this condition is not necessary. We are thus faced with the problem of characterizing the pairs of arcs γ_1, γ_2 which determine the same point $\tilde{a}$ over a.

17B. We can restrict our attention to closed curves γ from O, for γ_1 and γ_2 will determine the same $\tilde{a}$ if and only if the continuation along $\gamma_1\gamma_2^{-1}$ leads back to $\tilde{O}$. If the continuation along γ is a closed curve from $\tilde{O}$, then the same is true, by the monodromy theorem, for any curve which is homotop to γ. It is also true for γ^{-1}, and if it holds for γ_1 and γ_2 it is true for $\gamma_1\gamma_2$. It follows that the homotopy classes of the curves γ with a closed continuation $\tilde{\gamma}$ from $\tilde{O}$ form a subgroup $\mathscr{D}$ of the fundamental group $\mathscr{F}_O(F)$.

17C. The above construction depends on the choice of $\tilde{O}$. In order to clarify the situation we shall say that the triple $(\tilde{F}, f, \tilde{O})$ covers the couple (F, O) if $(\tilde{F}, f)$ is a regular covering surface of F and, in addition, $f(\tilde{O}) = O$. With this terminology it is clear that $\mathscr{D}$ is uniquely determined by $(\tilde{F}, f, \tilde{O})$.

Suppose now that $\mathscr{D}$ and $\mathscr{D}_1$ are determined by $(\tilde{F}, f, \tilde{O})$ and $(\tilde{F}, f, \tilde{O}_1)$

respectively. We join $\tilde{O}$ to $\tilde{O}_1$ by an arc $\tilde{\sigma}$; its projection is a closed curve σ. A given closed curve γ_1 on F determines a closed curve from $\tilde{O}_1$ if and only if $\sigma\gamma_1\sigma^{-1}$ determines a closed curve from $\tilde{O}$. It follows that $\mathscr{D}_1$ consists of all homotopy classes $\sigma^{-1}\gamma\sigma$ with $\gamma \in \mathscr{D}$. In other words, $\mathscr{D}_1$ is a conjugate subgroup of $\mathscr{D}$.

Conversely, if $\mathscr{D}_1 = \sigma^{-1}\mathscr{D}\sigma$, then $\mathscr{D}_1$ corresponds to the triple $(\tilde{F}, f, \tilde{O}_1)$, where $\tilde{O}_1$ is the terminal point of the continuation along σ from $\tilde{O}$.

17D. We agree to identify $(\tilde{F}_1, f_1, \tilde{O}_1)$ and $(\tilde{F}_2, f_2, \tilde{O}_2)$ if there exists a topological mapping φ of $\tilde{F}_1$ onto $\tilde{F}_2$ which is such that $f_2 \circ \varphi = f_1$ and $\varphi(\tilde{O}_1) = \tilde{O}_2$. The identification is legitimate, for it is obviously defined by means of an equivalence relation. We remark that the mapping φ is uniquely determined. In fact, from $f_2 \circ \varphi = f_2 \circ \psi$ and the definition of a smooth covering surface it follows that the set with $\varphi(p) = \psi(p)$ and the set with $\varphi(p) \neq \psi(p)$ are both open. Hence $\varphi(\tilde{O}_1) = \psi(\tilde{O}_1)$ implies $\varphi = \psi$.

With this identification the following theorem holds:

Theorem. *The construction that we have introduced defines a one to one correspondence between identified triples $(\tilde{F}, f, \tilde{O})$ and the subgroups $\mathscr{D}$ of $\mathscr{F}_O(F)$. Two triples can be represented by means of the same $(\tilde{F}, f)$ if and only if the corresponding subgroups are conjugate.*

The fact that identified triples determine the same subgroup is obvious from the construction. Suppose now, conversely, that $(\tilde{F}, f, \tilde{O})$ and $(\tilde{F}_1, f_1, \tilde{O}_1)$ determine the same $\mathscr{D}$. Take a $\tilde{p} \in \tilde{F}$ and join $\tilde{O}$ to $\tilde{p}$ by $\tilde{\sigma}$ with the projection σ. We determine the continuation of $\tilde{F}_1$ along σ from the initial point $\tilde{O}_1$. Its terminal point $\varphi(\tilde{p})$ does not depend on the choice of $\tilde{\sigma}$. In fact, if $\tilde{\sigma}$ and $\tilde{\sigma}'$ both lead from $\tilde{O}$ to $\tilde{p}$, then $\sigma\sigma'^{-1} \in \mathscr{D}$, and the continuation on $\tilde{F}_1$ along σ and σ' must lead to the same point. It is easily seen that φ is a homeomorphism, and that $f_1 \circ \varphi = f$. Hence $(\tilde{F}, f, \tilde{O})$ and $(\tilde{F}, f_1, \tilde{O}_1)$ are to be identified.

The last assertion was proved in 17C. It remains to show that there exists a triple $(\tilde{F}, f, \tilde{O})$ corresponding to every subgroup. This part of the proof requires an explicit construction.

17E. Let $\mathscr{D}$ be a subgroup of $\mathscr{F}_O(F)$. In order to construct a corresponding surface $\tilde{F}$ we proceed as follows: With every arc γ from O to a point p we associate a point $\tilde{p}$. For two arcs γ_1, γ_2 the corresponding points $\tilde{p}_1, \tilde{p}_2$ will be identified if and only if γ_1, γ_2 lead to the same point p and if, in addition, $\gamma_1\gamma_2^{-1}$ belongs to $\mathscr{D}$. The identification is evidently legitimate. The classes of identified points will again be denoted by $\tilde{p}$; they constitute the points of $\tilde{F}$.

In order to introduce a topology on $\tilde{F}$, consider a Jordan region $V \subset F$, and let γ lead from O to a point $p \in V$. We construct the set $\gamma\tilde{V}$ which consists of all points $\tilde{q}$ corresponding to arcs $\gamma\sigma$ where σ is contained in V.

Since V is simply connected, the correspondence between points $\tilde{q} \in \gamma \tilde{V}$ and $q \in V$ is one to one. We choose all sets $\gamma \tilde{V}$ as a basis for the open sets on $\tilde{F}$.

We must prove that the postulate for a basis is satisfied. Suppose that $\gamma_1 \tilde{V}_1$ and $\gamma_2 \tilde{V}_2$ overlap; then V_1, V_2 overlap, and $V_1 \cap V_2$ is a union of Jordan regions V. Consider $\tilde{q} \in \gamma_1 \tilde{V}_1 \cap \gamma_2 \tilde{V}_2$ with $q \in V \subset V_1 \cap V_2$. Then q can be determined by a $\gamma_1 \sigma_1$ and by a $\gamma_2 \sigma_2$ with $\sigma_1 \subset V_1$, $\sigma_2 \subset V_2$. The sets $\gamma_1 \sigma_1 \tilde{V}$ and $\gamma_2 \sigma_2 \tilde{V}$ are seen to be identical and contained in $\gamma_1 \tilde{V}_1 \cap \gamma_2 \tilde{V}_2$. Consequently, the intersection is a union of sets $\gamma \tilde{V}$, and the postulate (B) in 1B is established.

If $\tilde{p}_1$, $\tilde{p}_2$ correspond to different p_1, p_2 we can find disjoint open neighborhoods $\gamma_1 \tilde{V}_1$, $\gamma_2 \tilde{V}_2$ by choosing V_1, V_2 disjoint. If $p_1 = p_2$ we suppose that $\tilde{p}_1$, $\tilde{p}_2$ are determined by γ_1, γ_2 where $\gamma_1 \gamma_2^{-1}$ is not in $\mathscr{D}$. For any V which contains $p_1 = p_2$ the sets $\gamma_1 \tilde{V}$, $\gamma_2 \tilde{V}$ are disjoint, for $\gamma_1 \sigma (\gamma_2 \sigma)^{-1} = \gamma_1 (\sigma \sigma^{-1}) \gamma_2^{-1}$ is homotop to $\gamma_1 \gamma_2^{-1}$. We have shown that $\tilde{F}$ is a Hausdorff space. It is evidently connected.

17F. We set $f(\tilde{p}) = p$ with the same notation as above. The set $\gamma \tilde{V}$ is mapped topologically onto V. This proves that $(\tilde{F}, f)$ is a smooth covering surface of F. To show that it is regular we need only consider arcs γ from O. If the equation of γ is $t \to \omega(t)$ we can define $\tilde{\omega}(t)$ as the point determined by the subarc corresponding to $[0, t]$. This is a continuation from the initial point $\tilde{O}$ determined by the unit curve 1.

It is now clear that $(\tilde{F}, f, \tilde{O})$ corresponds to the subgroup $\mathscr{D}$. In fact, the continuation along γ from $\tilde{O}$ is closed if and only if $\gamma \in \mathscr{D}$.

17G. As a complement to Theorem 17D we prove:

Theorem. *The fundamental group of $\tilde{F}$ is isomorphic with $\mathscr{D}$.*

The projection of a closed curve from $\tilde{O}$ is in $\mathscr{D}$. Homotop curves have homotop projections, and products are preserved. This shows that the projection maps $\mathscr{F}_{\tilde{O}}(\tilde{F})$ homomorphically into $\mathscr{D}$. The mapping is onto, for there is a curve $\tilde{\gamma}$ over any $\gamma \in \mathscr{D}$. Finally, it is an isomorphism, for $\gamma \approx 1$ implies $\tilde{\gamma} \approx 1$ by the monodromy theorem.

18. The partial ordering

18A. If $(\tilde{F}_2, f, \tilde{O}_2)$ covers $(\tilde{F}_1, \tilde{O}_1)$ and $(\tilde{F}_1, f_1, \tilde{O}_1)$ covers (F, O), then it is clear that $(\tilde{F}_2, f_1 \circ f, \tilde{O}_2)$ covers (F, O). In this situation we say that $(\tilde{F}_2, f_1 \circ f, \tilde{O}_2)$ is *stronger* than $(\tilde{F}_1, f_1, \tilde{O}_1)$ over (F, O). In other words, of two triples $(\tilde{F}_1, f_1, \tilde{O}_1)$ and $(\tilde{F}_2, f_2, \tilde{O}_2)$ which cover (F, O) the latter is stronger if and only if there exists an f such that $(\tilde{F}_2, f, \tilde{O}_2)$ covers $(\tilde{F}_1, \tilde{O}_1)$ and $f_2 = f_1 \circ f$. This relationship is evidently transitive.

Let $\mathscr{D}_1$ and $\mathscr{D}_2$ be the subgroups of $\mathscr{F}_O(F)$ that correspond to $(\tilde{F}_1, f_1, \tilde{O}_1)$ and $(\tilde{F}_2, f_2, \tilde{O}_2)$. We prove:

Theorem. *The triple* $(\tilde{F}_2, f_2, \tilde{O}_2)$ *is stronger than* $(\tilde{F}_1, f_1, \tilde{O}_1)$ *if and only if* $\mathscr{D}_2 \subset \mathscr{D}_1$.

Suppose first that $(\tilde{F}_2, f_2, \tilde{O}_2)$ is stronger than $(\tilde{F}_1, f_1, \tilde{O}_1)$, and let f be the mapping whose existence is thus postulated. To $\gamma \in \mathscr{D}_2$ there corresponds a closed curve $\tilde{\gamma}_2$ on $\tilde{F}_2$ from $\tilde{O}_2$. It is projected by f on a closed curve $\tilde{\gamma}_1$ whose projection by f_1 is identical with γ. Hence $\gamma \in \mathscr{D}_1$, and we have proved that $\mathscr{D}_2 \subset \mathscr{D}_1$.

Suppose now that $\mathscr{D}_2 \subset \mathscr{D}_1$. Given $\tilde{p}_2 \in \tilde{F}_2$ we join $\tilde{O}_2$ to $\tilde{p}_2$ by $\tilde{\sigma}_2$, determine the projection $\sigma = f_2(\tilde{\sigma}_2)$ and construct the continuation $\tilde{\sigma}_1$ on $\tilde{F}_1$ along σ from $\tilde{O}_1$. It follows from the hypothesis that the terminal point $f(\tilde{p}_2)$ of $\tilde{\sigma}_1$ depends only on $\tilde{p}_2$. The mapping f has the required properties.

It follows from this theorem together with Theorem 17D that the partial ordering introduced by the relation of being stronger is compatible with the identification of triples, as introduced in 17D, in the sense that two triples are both stronger than the other if and only if they are identifiable.

18B. Our theorem shows that the ordering of regular covering surfaces according to relative strength is isomorphic with the ordering of the corresponding subgroups by inclusion. For any two subgroups there is a largest subgroup contained in both, namely their intersection, and a smallest subgroup containing both. Because of the isomorphism the same is true for regular covering surfaces. In other words, to any two regular covering surfaces there exists a strongest one which is weaker than both and a weakest one which is stronger than both. A partially ordered system with this property is called a *lattice*.

18C. The lattice of regular covering surfaces has the additional property of containing a weakest and a strongest element. The weakest covering surface of (F, O) is evidently $(\tilde{F}, e, O)$ where e denotes the identity mapping. Its corresponding subgroup $\mathscr{D}$ coincides with $\mathscr{F}_O(F)$. The strongest covering surface corresponds to the subgroup which consists only of the unit element. It is called the *universal covering surface* of F and will be denoted by $\tilde{F}_\infty$.

According to Theorem 17G the fundamental group of $\tilde{F}_\infty$ reduces to the unit element.

Theorem. *The universal covering surface is simply connected.*

19. Cover transformations

19A. A *cover transformation* of a regular covering surface $(\tilde{F}, f)$ is a topological mapping of $\tilde{F}$ onto itself with the property that corresponding points have the same projection. It is clear that the cover transformations form a group.

The group of cover transformations is closely related to the subgroup $\mathscr{D}$. The elements $\sigma \in \mathscr{F}_O(F)$ with $\sigma\mathscr{D}\sigma^{-1}=\mathscr{D}$ form a group $\mathscr{N}$ in which $\mathscr{D}$ is a normal subgroup. It is called the *normalizer of* $\mathscr{D}$. We prove:

Theorem. *The group of cover transformations of $\tilde{F}$ is isomorphic with the quotient group $\mathscr{N}/\mathscr{D}$ where $\mathscr{N}$ is the normalizer of $\mathscr{D}$ in $\mathscr{F}(F)$.*

We consider a fixed choice of origins O, $\tilde{O}$. Let σ be an element of $\mathscr{F}_O(F)$, and consider the terminal point $\tilde{O}_1$ of the continuation $\tilde{\sigma}$ from $\tilde{O}$. The subgroup $\mathscr{D}_1$ associated with $(\tilde{F}, f, \tilde{O}_1)$ is $\sigma^{-1}\mathscr{D}\sigma$. If σ is in the normalizer, $\mathscr{D}_1=\mathscr{D}$, and by Theorem 17D the triples $(\tilde{F}, f, \tilde{O}_1)$ and $(\tilde{F}, f, \tilde{O})$ can be identified. This means that there exists a unique cover transformation T_σ which takes $\tilde{O}$ into $\tilde{O}_1$. It is clear that $T_{\sigma\tau}=T_\sigma T_\tau$. Hence we have constructed a homomorphic mapping of $\mathscr{N}$ into the group of cover transformations.

T_σ is the identity mapping if and only if $\tilde{O}_1=\tilde{O}$, and that is so if $\sigma \in \mathscr{D}$. Hence $\mathscr{D}$ is the kernel of the homomorphism, and the quotient group $\mathscr{N}/\mathscr{D}$ is mapped isomorphically.

In order to show that the mapping is onto, let T be a cover transformation and set $\tilde{O}_1=T(\tilde{O})$. We join $\tilde{O}$ to $\tilde{O}_1$ by an arc $\tilde{\sigma}$ with the projection σ. The subgroup $\mathscr{D}_1$ associated with $(\tilde{F}, f, \tilde{O}_1)$ is $\sigma^{-1}\mathscr{D}\sigma$. By Theorem 17D $\mathscr{D}_1=\mathscr{D}$, and hence $\sigma \in \mathscr{N}$. Moreover, $T=T_\sigma$, and the theorem is proved.

19B. The result is particularly simple if $\mathscr{D}$ is a normal subgroup, for then $\mathscr{N}=\mathscr{F}(F)$. A regular covering surface which corresponds to a normal subgroup is called a *normal* covering surface. (In the prevailing terminology such a covering surface is called regular.) Intuitively speaking, a covering surface is normal if points with the same projection cannot be distinguished from each other by properties of the covering. For instance, if there is one closed curve $\tilde{\sigma}$ over σ, then all curves over σ are closed.

In particular, the universal covering surface $\tilde{F}_\infty$ is normal, and its group of cover transformations is isomorphic with the fundamental group $\mathscr{F}(F)$.

19C. We note that no cover transformation other than the identity has a fixed point. In fact, the fixed point could be chosen for origin, and we have already remarked that a cover transformation which takes $\tilde{O}$ into itself is the identity.

In the case of a normal covering surface there is always a cover transformation which carries a given point $\tilde{p}$ into a prescribed point $\tilde{p}_1$ with the same projection. For if we take $\tilde{O}=\tilde{p}$, $\tilde{O}_1=\tilde{p}_1$ and join $\tilde{O}$ to $\tilde{O}_1$ by an arc with the projection σ, then σ is always in the normalizer, and $T_\sigma(\tilde{O})=\tilde{O}_1$.

19D. The *commutator subgroup* of a group $\mathscr{F}$ is defined as the smallest subgroup which contains all elements that can be written in the form

$aba^{-1}b^{-1}$ with $a, b \in \mathcal{F}$. It is usually denoted by $\mathcal{F}^{(1)}$. It is a normal subgroup, for if $d \in \mathcal{F}^{(1)}$, then $cdc^{-1} = (cdc^{-1}d^{-1})d$ is also in $\mathcal{F}^{(1)}$.

The quotient group $\mathcal{F}/\mathcal{F}^{(1)}$ is Abelian. In fact, ab and ba are in the same coset of $\mathcal{F}^{(1)}$ since $ab(ba)^{-1} = aba^{-1}b^{-1}$ is in $\mathcal{F}^{(1)}$.

Let $\mathcal{D}$ be any normal subgroup of $\mathcal{F}$ for which $\mathcal{F}/\mathcal{D}$ is Abelian. Then every element of the form $aba^{-1}b^{-1}$ is in $\mathcal{D}$, so that $\mathcal{F}^{(1)} \subset \mathcal{D}$. We conclude that $\mathcal{F}^{(1)}$ is the smallest subgroup whose quotient group is Abelian.

19E. We apply the preceding to the fundamental group $\mathcal{F} = \mathcal{F}(F)$ of a surface. The commutator subgroup $\mathcal{F}^{(1)}(F)$ determines a normal covering surface $\tilde{F}_{hom}$ which we call the *homology covering surface*. It has an Abelian group of cover transformations, and it is the strongest normal covering surface with this property.

The quotient group $\text{Hom } F = \mathcal{F}(F)/\mathcal{F}^{(1)}(F)$ is called the *homology group* of F. It is isomorphic with the group of cover transformations of $\tilde{F}_{hom}$. We shall find, in 33D and 34A, a different characterization of the homology group which makes it an indispensable tool for the later theory.

20. Ramified covering surfaces

20A. The properties of the mapping $z \to z^m$, where m is a positive integer, are well known from elementary function theory. It maps $|z| < 1$ onto itself so that every $z \neq 0$ has m inverse images $z^{1/m}$ while $z = 0$ corresponds only to itself.

It is customary to say that this correspondence defines a covering surface of $|z| < 1$ with a *branch point* at the origin. The integer m is the *multiplicity* and $m-1$ is the *order* of the branch point.

20B. We wish to generalize this concept of a covering surface with branch points. As before, $\tilde{F}$ will denote a connected Hausdorff space, and F will be a surface. It will also be necessary to assume explicitly that $\tilde{F}$ is locally compact, for this is not implied by the other conditions. Our aim is to study the situation that arises when a mapping f of $\tilde{F}$ into F has the same local properties as the mapping $z \to z^m$. However, instead of making these properties our starting point we prefer to formulate simpler conditions from which they follow.

Definition. *The continuous mapping f defines $\tilde{F}$ as a covering surface of F if every $\tilde{p}_0 \in \tilde{F}$ has a neighborhood $\tilde{V}$ with the property that $(\tilde{V} - \tilde{p}_0, f)$ is a smooth covering surface of $F - f(p_0)$.*

The condition is trivially fulfilled if $(\tilde{F}, f)$ is a smooth covering surface of F. This fact justifies the terminology that we are using.

Remark. It has been shown by S. Stoïlow [14] that an apparently much weaker condition leads to the same result. Stoïlow assumes merely that f

maps open sets on open sets, and that the inverse image of any point is totally disconnected. Because of the intricacies of the proof we are not following this line of reasoning.

20C. We will now analyze some direct consequences of the definition. First, we determine a compact neighborhood $\tilde{U}$ of $\tilde{p}_0$ which is contained in $\tilde{V}$ (see 4F). The boundary of $\tilde{U}$ is mapped by f on a closed set B which does not contain $p_0 = f(\tilde{p}_0)$. We can therefore find a Jordan region Δ on F which contains p_0 but does not meet B. The component $\tilde{\Delta}$ of $f^{-1}(\Delta)$ which contains $\tilde{p}_0$ cannot intersect the boundary of $\tilde{U}$. Therefore it is contained in $\tilde{U}$, and is consequently relatively compact.

We introduce the notations $\Delta_0 = \Delta - p_0$, $\tilde{\Delta}_0 = \tilde{\Delta} - \tilde{p}_0$ and contend that $(\tilde{\Delta}_0, f)$ is a regular covering surface of Δ_0. The smoothness is obvious. To prove regularity we cannot make direct use of Corollary 14G, for although $\tilde{\Delta}_0$ is relatively compact on $\tilde{V}$ it is not relatively compact on $\tilde{V} - \tilde{p}_0$. However, the same method of proof can be applied. Accordingly, we consider a point $q \in \Delta_0$ and choose a compact neighborhood W of q which is contained in Δ_0. The restriction of f to $\tilde{\Delta}$ is denoted by g. Then $g^{-1}(W) = f^{-1}(W) \cap \tilde{\Delta} = f^{-1}(W) \cap \mathrm{Cl}\,\tilde{\Delta}$, for the boundary of $\tilde{\Delta}$ cannot meet $f^{-1}(W)$. We conclude that $g^{-1}(W)$ is compact on $\tilde{F}$, and therefore also as a subset of $\tilde{\Delta}_0$. Consequently $(\tilde{\Delta}_0, f)$ is regular over Δ_0.

We state the result as a lemma:

Lemma. *Any pair of corresponding points $p_0, \tilde{p}_0$ are contained in relatively compact regions Δ, $\tilde{\Delta}$ such that $(\tilde{\Delta} - \tilde{p}_0, f)$ is a regular covering surface of $\Delta - p_0$; Δ can be chosen as a Jordan region.*

20D. According to Theorem 10B the fundamental group of Δ_0 is an infinite cyclic group whose generator we denote by α. By Theorem 17G the fundamental group $\mathscr{F}(\tilde{\Delta}_0)$ is isomorphic to a subgroup $\mathscr{D}$ of $\mathscr{F}(\Delta_0)$.

The subgroups of $\mathscr{F}(\Delta_0)$ are those generated by a power α^m, $m \geqq 0$. For $m = 0$ the subgroup reduces to the unit element, and if $\mathscr{D}$ were this subgroup $\tilde{\Delta}_0$ would be the universal covering surface of Δ_0. It would have infinitely many cover transformations, and there would exist infinitely many points with the same projection. This clearly contradicts the fact that $\tilde{\Delta}_0$ has a compact closure in $\tilde{V}$.

We find that $\mathscr{D}$ consists of all α^{mn} for some $m \geqq 1$, n running through all integers, and that $\mathscr{F}(\tilde{\Delta}_0)$ is infinite cyclic. Moreover, the cover transformations of $\tilde{\Delta}_0$ form a finite cyclic group of order m.

The number m is called the multiplicity of $\tilde{p}_0$, and if $m > 1$ we say that $\tilde{p}_0$ is a branch point of order $m - 1$. Clearly, the branch points are isolated.

20E. Suppose that $\tilde{p}_0$ has the multiplicity m. We consider the mapping $z \to z^m$ which defines the disk $E: |z| < 1$ as a covering surface of itself with the multiplicity m at the origin. Let ω be a homeomorphism of Δ onto E which takes p_0 into 0 and therefore makes Δ_0 correspond to $E_0: 0 < |z| < 1$. Then $(\tilde{\Delta}_0, \omega \circ f)$ is a regular covering surface of E_0 which determines the same subgroup $\mathscr{D}$ as the mapping $z \to z^m$. According to Theorem 17D there exists a homeomorphic mapping φ of $\tilde{\Delta}_0$ onto E_0 which is such that corresponding points have the same projection. This condition reads

$$\omega(f(\tilde{p})) = \varphi(\tilde{p})^m. \tag{1}$$

In other words, if we set $\omega(p) = \zeta$, $\varphi(\tilde{p}) = z$ the correspondence is given by $\zeta = z^m$.

Since $\omega \circ f$ is continuous at $\tilde{p}_0$ and $\omega(f(\tilde{p}_0)) = 0$ it follows from (1) that $\varphi(\tilde{p})$ tends to 0 as $\tilde{p} \to \tilde{p}_0$. Therefore, φ can be extended to a homeomorphism of $\tilde{\Delta}$ onto E.

The existence of homeomorphisms ω and φ which satisfy (1) is the precise expression for our contention that f has the same local properties as the mapping $z \to z^m$. Since the latter mapping is elementary we have complete insight in the local properties of f.

Incidentally, we have proved that $\tilde{F}$ is a surface, for $\tilde{\Delta}$ is an open neighborhood of $\tilde{p}_0$ which is homeomorphic to a disk.

20F. We are now able to prove:

Theorem. *Every covering surface of an orientable surface is orientable.*

For each Δ we choose the homeomorphism ω so that its orientation agrees with that of the surface F, and with $\tilde{\Delta}$ we associate the homeomorphism φ defined above. The mapping $z \to z^m$ will be denoted by π so that the identity (1) can be written in the form $\omega \circ f = \pi \circ \varphi$. The degree of π is 1, except at $z = 0$.

$\tilde{F}$ has an open covering by sets $\tilde{\Delta}$. We must show that φ_1, φ_2 define compatible orientations of $\tilde{\Delta}_1, \tilde{\Delta}_2$. It is sufficient to consider a point $\tilde{q} \in \tilde{\Delta}_1 \cap \tilde{\Delta}_2$ which is not a branch point. Let $\tilde{U} \subset \tilde{\Delta}_1 \cap \tilde{\Delta}_2$ be a neighborhood of $\tilde{q}$ which is so small that π_1 has an inverse when restricted to $\pi_1(\varphi_1(\tilde{U}))$. From $\omega_1 \circ f = \pi_1 \circ \varphi_1$, $\omega_2 \circ f = \pi_2 \circ \varphi_2$ we obtain $\omega_1^{-1} \circ \pi_1 \circ \varphi_1 = \omega_2^{-1} \circ \pi_2 \circ \varphi_2$ and hence $\varphi_1 \circ \varphi_2^{-1} = \pi_1^{-1} \circ (\omega_1 \circ \omega_2^{-1}) \circ \pi_2$ on $\varphi_2(\tilde{U})$. The mappings π_1^{-1}, $\omega_1 \circ \omega_2^{-1}$ and π_2 all have degree 1 in $\varphi_2(\tilde{U})$. Therefore $\varphi_1 \circ \varphi_2^{-1}$ has degree 1, and the orientations are compatible.

21. Complete covering surfaces

21A. Continuations along an arc can be defined for arbitrary covering surfaces in exactly the same manner as for smooth covering surfaces. However, the uniqueness theorem holds only for continuations which do not pass through any branch points.

If there is not always a continuation we proved in 14E that there exists a path of determination which leads to an asymptotic value. The proof remains valid, with minor changes, in the case of a covering surface with ramifications. The opposite conclusion becomes false: even if the continuation is always possible, there may exist asymptotic values.

The lack of uniqueness makes it undesirable to base the study of arbitrary covering surfaces on continuations. However, we can replace regularity by the concept of *completeness*, which in the case of smooth covering surfaces was proved to be equivalent with regularity. We repeat the definition, which is identical with the necessary and sufficient condition in Theorem 14F.

Definition. *A covering surface is said to be complete if every $p_0 \in F$ has a neighborhood V with the property that each component of $f^{-1}(V)$ is compact.*

With this terminology a regular covering surface is one which is smooth and complete.

21B. Complete covering surfaces have the following important property:

Theorem. *A complete covering surface $(\tilde{F}, f)$ of F covers each point of F the same number of times, provided that the branch points are counted as many times as their multiplicity indicates.*

We prove first that every point is covered at least once. If not, the projection $f(\tilde{F})$ would have a boundary point p_0. We choose, first, a neighborhood V of p_0 with the property that each component of $f^{-1}(V)$ is compact, and then an open connected neighborhood $U \subset V$. Let C be a component of $f^{-1}(U)$, a nonvoid open set. Then C is open, and $\bar{C}$ is compact. The mapping f is continuous and takes open sets into open sets. Therefore, $f(C)$ is open and $f(\bar{C})$ is closed. But C does not meet the other components of $f^{-1}(U)$. Hence $f(C) = f(\bar{C}) \cap U$, so that $f(C)$ is relatively closed in U. Since U is connected we would have $f(C) = U$, in contradiction with the assumption that p_0 is a boundary point of $f(\tilde{F})$.

Consider now a point $q \in F$ which is covered at least n times, that is to say we suppose that there are certain points $\tilde{q}_k$ over q whose multiplicities have a sum $\geqq n$. To $\tilde{q}_k$ and q we determine regions $\tilde{\Delta}_k$, Δ_k according to Lemma 20C. Clearly, the $\tilde{\Delta}_k$ can be chosen disjoint from each other. Then every $p \neq q$ which is contained in the intersection of all Δ_k will be covered by at least n points. Hence the set of points that are covered at least n times is open.

The complementary set is formed by all points which are covered at most $n-1$ times. Let q be a point of this kind, covered by certain $\tilde{q}_k$. We construct corresponding $\tilde{\Delta}_k$ and Δ_k. Let V be an open connected neighborhood of q which is contained in all the Δ_k. Every component of $f^{-1}(V)$ is a complete covering surface of V. Therefore, each component projects on the whole of V, and must contain a point $\tilde{q}_k$. It follows that the component is contained in $\tilde{\Delta}_k$. Hence the inverse images of any point $p \in V$ must lie in the $\tilde{\Delta}_k$, and there are at most $n-1$ such points over p. We have proved that the complementary set is open.

Since F is connected, one of the sets must be void. It follows that all points are covered the same number of times. This number is called the *number of sheets* of the covering surface; naturally, it can be infinite, and in that case our reasoning could be used to show that the cardinal numbers are the same.

21C. We proceed to consider certain subregions of covering surfaces. Let $(\tilde{F}, f)$ be any covering surface of F, and Ω a subregion of F. Then f defines each component of $f^{-1}(\Omega)$ as a covering surface of Ω.

Suppose, in particular, that one of these components, $\tilde{\Omega}$, is relatively compact. We assert that $(\tilde{\Omega}, f)$ is a complete covering surface of Ω. In fact, every $p \in \Omega$ has a closed neighborhood $V \subset \Omega$. The components of $f^{-1}(V)$ are either contained in $\tilde{\Omega}$ or disjoint from $\tilde{\Omega}$. The components in $\tilde{\Omega}$ are closed and contained in the compact closure of $\tilde{\Omega}$. Therefore they are compact, and we have shown that $(\tilde{\Omega}, f)$ is complete.

Moreover, $\tilde{\Omega}$ has only a finite number of sheets, for it can be covered by a finite number of neighborhoods in which only a finite number of points have the same projection. Finally, the boundary of $\tilde{\Omega}$ projects onto the boundary of Ω. In fact, since the closure of $\tilde{\Omega}$ is compact it projects on a closed set. The projection contains Ω and is contained in $\bar{\Omega}$; hence it coincides with $\bar{\Omega}$. Since $\tilde{\Omega}$ projects into Ω and Bd $\tilde{\Omega}$ into Bd Ω, it follows that $f(\text{Bd } \tilde{\Omega}) = \text{Bd } \Omega$.

We call $\tilde{\Omega}$ a *complete region* over Ω. We know by Lemma 20C that any two corresponding points have neighborhoods $\tilde{\Delta}$ and Δ with the property that $\tilde{\Delta}$ is a complete region over Δ.

21D. Any complete region has a countable basis. Indeed, as a relatively compact subset of a surface $\tilde{\Omega}$ can be covered by a finite number of Jordan regions, and each Jordan region has a countable basis. For this reason complete regions are useful in proving:

Theorem. *Every covering surface of a countable surface is itself countable.*

Consider a countable basis of F, consisting of open sets O_k. We may assume that the O_k are connected, for if this is not so we replace each O_k by its countably many components (Corollary 5A). Every $\tilde{p} \in \tilde{F}$ is contained in a complete region $\tilde{\Delta}$ over a Δ. The projection $f(p)$ is contained in an $O_k \subset \Delta$, and hence $\tilde{p}$ lies in a complete region over O_k. In other words, the complete regions over all O_k form an open covering of $\tilde{F}$.

We choose the notation so that there exists a complete region $\tilde{O}_1$ over O_1. There are at most a countable number of complete regions $\tilde{O}_2$ over O_2 which meet $\tilde{O}_1$, for the different intersections are disjoint and open, and $\tilde{O}_1$ has a countable basis. The same reasoning applies to any pair O_i, O_j.

Having fixed $\bar{O}_1$ we consider chains $\{\bar{O}_{i_1}, \cdots, \bar{O}_{i_n}\}$ where $\bar{O}_{i_k}$ is a complete region over O_{i_k}, $\bar{O}_{i_1} = \bar{O}_1$, and $\bar{O}_{i_k} \cap \bar{O}_{i_{k+1}} \neq 0$. By what we have just said the number of such chains is countable. On the other hand, the union of all $\bar{O}_k$ that appear in a chain beginning with $\bar{O}_1$ is open, and so is the union of all $\bar{O}_k$ which do not appear in any such chain. These sets are complementary, and because $\bar{F}$ is connected the second set must be empty. Therefore all $\bar{O}_k$ appear in chains, and their number is countable. Since each $\bar{O}_k$ has a countable basis the same must be true of $\bar{F}$.

§4. SIMPLICIAL HOMOLOGY

Complete insight in the topology of surfaces cannot be gained without the use of combinatorial methods. These methods apply most directly in the presence of a triangulation. For this reason we focus our attention on surfaces which permit a triangulation, and for lack of a better name such surfaces will be called polyhedrons.

In the present section we shall be concerned with that part of combinatorial topology which centers around simplicial homology groups. The word simplex refers to the triangles, sides, and vertices which make up the triangulation.

Certain features of simplicial homology which are particularly relevant for the study of open surfaces have received very scant attention. For the purposes of this book it is essential that we include properties of infinite triangulations which have no counterpart when the number of triangles is finite.

22. Triangulations

22A. The idea of *triangulation* appeals so directly to the imagination that a formalization may seem superfluous. Nevertheless, it is necessary to agree on a precise language, and this is most easily accomplished in the framework of a formal treatment.

We begin by defining an *abstract complex*. An abstract complex is a finite or infinite set K together with a family of finite subsets, called *simplices*. The following properties are postulated:

(A1) Every $\alpha \in K$ belongs to at least one and at most a finite number of simplices.

(A2) Every subset of a simplex is a simplex.

The dimension of a simplex is one less than the number of its elements. An n-dimensional simplex is called, briefly, an n-simplex. A 0-simplex can be identified with the element of which it consists, and is called a *vertex*.

The dimension of a complex K is the maximum dimension of its simplices (if there is no maximum, the complex is of infinite dimension). Ultimately, we shall use only 2-dimensional complexes.

22B. An abstract complex K can be used to construct a corresponding *geometric complex* K_g, which is a topological space. A point in K_g is a real-valued function λ on K with these properties:

(B1) $\lambda(\alpha) \geqq 0$ for all $\alpha \in K$.

(B2) The elements α with $\lambda(\alpha) > 0$ form a simplex.

(B3) $\sum_{\alpha \in K} \lambda(\alpha) = 1$.

If s is a simplex we obtain a subset s_g of K_g by imposing on λ the further condition that $\lambda(\alpha) = 0$ whenever α is not in s. It follows from (B2) that K_g is the union of all s_g.

Suppose, for instance, that s is a 2-simplex, consisting of α_1, α_2, α_3, and write $\lambda_k = \lambda(\alpha_k)$. A point of s_g can be represented as a triple $(\lambda_1, \lambda_2, \lambda_3)$ with $\lambda_k \geqq 0$ and $\lambda_1 + \lambda_2 + \lambda_3 = 1$. We obtain a realization of s_g as a triangle in R^3. The generalization to arbitrary dimension of s is immediate.

On each s_g we introduce the topology of its geometric realization. We can then define a topology on K_g by declaring that a set is closed if and only if its intersection with each s_g is closed. Because of (A1) the space is locally compact.

We have constructed a realization of the abstract complex K with the fundamental property that the subsets which are associated with two simplices s_1, s_2 intersect exactly in the subset associated with $s_1 \cap s_2$. In other words, the inclusion relations are preserved. The individual geometric simplices have an immediately recognizable concrete form, and the device of adding a new coordinate for each vertex has saved us from unnecessary difficulties which would be connected with an imbedding in a space of fixed dimension.

22C. A *triangulation* of a surface F or a bordered surface $\bar{F}$ is defined in terms of an abstract 2-dimensional complex K and a law which assigns a subset $\sigma(s)$ of F (or $\bar{F}$) to each simplex s of K. In order to conform with the intuitive idea of a triangulation, we postulate the following conditions:

(C1) $\sigma(s_1 \cap s_2) = \sigma(s_1) \cap \sigma(s_2)$.

(C2) There exists a homeomorphism of s_g onto $\sigma(s)$ which maps every s'_g, $s' \subset s$, onto $\sigma(s')$.

(C3) The union of all $\sigma(s)$ is F (or $\bar{F}$).

(C4) Every point on F ($\bar{F}$) has a neighborhood which meets only a finite number of $\sigma(s)$.

22D. The nature of a triangulation is made clearer by proving:

Theorem. *For a given triangulation of F by a complex K there exists a homeomorphism of K_g onto F which maps each s_g onto $\sigma(s)$.*

For any vertex α we set $\varphi(\alpha_g)=\sigma(\alpha)$. Next, if s is a 1-simplex we define φ on s_g as one of the homeomorphisms whose existence is asserted in (C2). Finally, having thus defined φ on the boundary of each geometric 2-simplex s_g we can extend it to a homeomorphism of the whole simplex, for the homeomorphism given by (C2) can be adjusted on the boundary by composing it with a suitable homeomorphism of s_g onto itself.

The constructed mapping φ is continuous, and as a consequence of (C1) it is one to one. By (C3) the image is all of F. The inverse mapping is continuous by virtue of (C4). Hence φ is indeed a homeomorphism.

As far as topological properties are concerned we conclude that a triangulated surface F can be replaced by the corresponding geometric complex K_g. The same reasoning applies in the case of a bordered surface $\bar{F}$.

22E. Our next concern is to determine under what conditions a geometric complex K_g is a triangulated surface, the triangulation being defined by the complex K and $\sigma(s)=s_g$.

In order to enumerate the characteristic properties of a triangulation we will use letters α, a and A to denote 0-, 1- and 2-simplices respectively. If we wish to specify the vertices which belong to a 1- or 2-simplex we write, for instance, $a=(\alpha_1\alpha_2)$, $A=(\alpha_1\alpha_2\alpha_3)$. For the moment we pay no attention to the order in which the vertices are named.

Theorem. *A 2-dimensional geometric complex K_g is a triangulated surface with $\sigma(s)=s_g$ if and only if K satisfies the following conditions:*

(E1) *Every a is contained in exactly two A.*

(E2) *The a and A which contain a given α can be denoted cyclically as $a_1,\cdots,a_m$ and $A_1,\cdots,A_m$ in such a manner that $a_i=A_i\cap A_{i-1}$ ($a_1=A_1\cap A_m$) and $m\geqq 3$.*

(E3) *K is connected, in the sense that it cannot be represented as the union of two disjoint complexes.*

The sufficiency is practically evident, and the formal proof will be omitted. The necessity of (E3) is also immediate. We turn to the necessity of (E1).

22F. In the first place, every a must belong to at least one A. To see this, let p be an interior point of a_g (i.e., not an end point). By (C4) there exists a neighborhood $V(p)$ which meets only a finite number of s_g, and because each s_g is closed we can find a smaller neighborhood $U(p)$ which meets only those s_g which actually contain p. Finally, we can find a neighborhood $\Delta\subset U(p)$ which is a Jordan region. If a were not contained in any A we would have $\Delta\subset a_g$. Consider a closed curve γ in $\Delta-p$ with index 1 with respect to p. Because γ is connected it must be contained in one of the components of a_g-p, and for this reason γ can be shrunk to a point. It follows that the index would be O, contrary to the assumption. We have shown that a belongs to at least one A.

Assume that a belongs to $A_1, \cdots, A_n$; we have to show that $n=2$. As above we can determine a Jordan region Δ such that $p \in \Delta \subset (A_1)_g \cup \cdots \cup (A_n)_g$, and we choose again a closed curve γ in $\Delta - p$ with index 1 with respect to p. Suppose first that $n=1$. On the triangle $(A_1)_g$ we determine a closed half-disk C_1 with center p, sufficiently small to be contained in Δ. The curve γ can be deformed into one that lies in $C_1 - p$, and a closed curve in $C_1 - p$ can be deformed to a point. We are lead to the same contradiction as in the previous case.

Consider now the case $n>2$. We construct half-disks $C_i \subset \Delta \cap (A_i)_g$ with a common diameter. The curve γ can again be deformed into one that lies in $C_1 \cup \cdots \cup C_n$. On using the geometric structure of the triangles it can be further deformed until it lies on $\sigma_1 \cup \cdots \cup \sigma_n$ where σ_i is the half-circle on the boundary of C_i. We direct the σ_i so that they have the same initial and terminal points. By a simple reasoning, analogous to the one in 10B, it is found that any closed curve on $\sigma_1 \cup \cdots \cup \sigma_n$ can be deformed into a product of curves $\sigma_i \sigma_j^{-1}$. But when $n>2$ we can show that $\sigma_i \sigma_j^{-1}$ has index 0 with respect to p. In fact, the index does not change when p moves continuously without crossing σ_i or σ_j. We can join p to the opposite vertex of a third triangle $(A_k)_g$, $k \neq i, j$, without touching $\sigma_i \cup \sigma_j$. The joining arc must cross the boundary of Δ, and for a point near the boundary the index is 0. Therefore the index vanishes also for the original position of p. We conclude that γ has index 0. With this contradiction we have completed the proof of (E1).

22G. We prove now that the simplices which contain a given α_0 can be arranged in the asserted manner.

A point is not a surface. Therefore α_0 belongs to at least one $a_1 = (\alpha_0 \alpha_1)$. By (E1) this a_1 is contained in an $A_1 = (\alpha_0 \alpha_1 \alpha_2)$, and $a_2 = (\alpha_0 \alpha_2)$ belongs to an $A_2 = (\alpha_0 \alpha_2 \alpha_3) \neq A_1$ so that $\alpha_3 \neq \alpha_1$. When this process is continued we muct come to a first $A_m = (\alpha_0 \alpha_m \alpha_{m+1})$ such that $\alpha_{m+1} = \alpha_1$.

We can find a Jordan region Δ which contains $(\alpha_0)_g$ and does not meet $(\alpha_1\alpha_2)_g \cup (\alpha_2\alpha_3)_g \cup \cdots \cup (\alpha_m\alpha_1)_g$. The intersection of $(A_1)_g \cup \cdots \cup (A_m)_g$ with $\Delta - (\alpha_0)_g$ is open and relatively closed in $\Delta - (\alpha_0)_g$. Because a punctured disk is connected it follows that $\Delta \subset (A_1)_g \cup \cdots \cup (A_m)_g$. This proves that $a_1, \cdots, a_m$ and $A_1, \cdots, A_m$ exhaust all simplices that contain α_0. Condition (E2) is satisfied.

22H. Theorem 22E can be generalized to the case where K_g represents a triangulated bordered surface $\bar{F}$. The previous reasoning can be applied without change as soon as a_g contains a single interior point of $\bar{F}$ (as opposed to the points on the border), and we find again that a belongs to exactly two A. Moreover, all interior points of a_g will be interior points of $\bar{F}$.

Suppose now that a_g is contained in the border $B(\bar{F})$. We assume that a belongs to $A_1, \cdots, A_n$. As before we consider an interior point p of a_g and determine a neighborhood Δ of p which is contained in $(A_1)_g \cup \cdots \cup (A_n)_g$ and does not meet any geometric 1-simplices other than a_g. This time Δ may be chosen homeomorphic with a semiclosed half-disk, and we know that the points on a_g correspond to points on the diameter. It follows from this representation that $\Delta' = \Delta - (a_g \cap \Delta)$ is connected and nonvoid. On the other hand, Δ' is the union of the disjoint open sets $[(A_i)_g - a_g] \cap \Delta$, none of which is void. This is possible only for $n = 1$, and we have shown that a belongs to a single A.

We have found that each a_g belongs to one or two A_g, and that the border $B(\bar{F})$ is composed of those a_g which belong to only one A_g. Since the border is not empty there is at least one such a_g.

Simplices whose geometric counterparts are contained in $B(\bar{F})$ are called *border simplices*. One shows further that a border vertex is contained in simplices which can be denoted as $a_1, \cdots, a_{m+1}$ and $A_1, \cdots, A_m$ with $a_i = A_i \cap A_{i-1}$ for $i = 2, \cdots, m$. Here a_1 and a_{m+1} are border simplices which are contained only in A_1 and A_m respectively.

Conversely, if these conditions are satisfied it is evident that K_g represents a bordered surface.

22I. If a complex is connected it is easy to see that it contains only a countable number of simplices. It follows that a surface or bordered surface which can be triangulated is necessarily countable. We will prove in 46 that this condition is also sufficient.

23. Homology

23A. Consider an abstract complex K. We have already used the explicit notation $(\alpha_0\alpha_1 \cdots \alpha_n)$ for an n-simplex, where up to now the order of the elements has been irrelevant. We take now a different point of view and define an *ordered n-simplex* as $s^n = (\alpha_0\alpha_1 \cdots \alpha_n)$ where the order is essential. We continue to require that the α_i are distinct, and that they are the elements of a simplex in K.

We fix the dimension n and take all ordered n-simplices of K to be the generators of a free Abelian group $C_n(K)$. This means that each element of $C_n(K)$ can be represented in one and only one way as a formal sum

$$\sum x_i s_i^n$$

where the x_i are integers, and only a finite number of them are different from zero. Moreover, the law of composition is given by

$$\sum x_i s_i^n + \sum y_i s_i^n = \sum (x_i + y_i) s_i^n.$$

Here i runs over an arbitrary set of indices, but it is no serious loss of generality to assume that the number of simplices is countable. As a practical matter we agree that terms with the coefficient 0 can be added or omitted at will. If n exceeds the dimension of K the group $C_n(K)$ reduces to the zero element which we denote by 0. By convention, $C_{-1}(K)$ will also be 0.

23B. It is desirable to introduce certain identifications in the groups $C_n(K)$. If two ordered n-simplices s_1^n and s_2^n contain the same elements in different order, we will identify s_1^n with s_2^n if the orders differ by an even permutation, and with the $(-1)s_2^n = -s_2^n$ if they differ by an odd permutation. Theoretically, we are introducing the quotient group of $C_n(K)$ with respect to the subgroup generated by all sums of the form $s_1^n - s_2^n$ and $s_1^n + s_2^n$ respectively. From a practical point of view, however, it is simpler to think of s_1^n and s_2^n or s_1^n and $-s_2^n$ as different notations with the same meaning. Accordingly, we shall not complicate matters by introducing new notations, but continue to denote the groups by $C_n(K)$. The elements of $C_n(K)$, after the identification, are called n-dimensional *chains*.

23C. We will now define the fundamental operation which consists in forming the *boundary* of a chain. It determines a homomorphic mapping ∂ of each $C_n(K)$ into the preceding $C_{n-1}(K)$. For $n=0$ the homomorphism must trivially transform all 0-chains into 0. For $n>0$ the homomorphism is completely determined if we prescribe ∂s_i^n for all generators of $C_n(K)$. We do this by setting

$$\partial(\alpha_0\alpha_1\cdots\alpha_n) = \sum_{k=0}^{n} (-1)^k(\alpha_0\cdots\alpha_{k-1}\alpha_{k+1}\cdots\alpha_n).$$

The boundary of an arbitrary chain is then

$$\partial(\sum x_i s_i^n) = \sum x_i \partial s_i^n$$

where in general the right hand side needs to be rewritten in its simplest form.

It must be verified that the operator ∂ is compatible with our identifications, i.e. that identified chains have identified boundaries. For $n \leqq 2$ the verification is immediate, and these are the only dimensions that occur in this book.

One finds by explicit computation that $\partial\partial s^n = 0$ for every simplex. It follows that $\partial\partial$ carries the whole group $C_n(K)$ into the zero element of $C_{n-2}(K)$.

23D. A chain is called a *cycle* if its boundary is zero. The n-dimensional cycles form a subgroup $Z_n(K)$ of $C_n(K)$, and $Z_n(K)$ is the kernel of the homomorphism ∂. The image of $C_n(K)$ under the same homomorphism is denoted by $B_{n-1}(K)$; it consists of all $(n-1)$-dimensional *boundaries*. From the fact that $\partial\partial=0$ we see that every boundary is a cycle. Hence $B_n(K)$ is a subgroup of $Z_n(K)$.

The greatest interest is attached to the quotient group $Z_n(K)/B_n(K)$. It is called the *homology group* of dimension n, and we denote it by $H_n(K)$. In most cases we do not use special symbols for the elements of a homology group, but we write $x \sim y$ to indicate that x is homologous to y. The relation is meaningful only if x and y are cycles of the same dimension, and it asserts the existence of a chain z such that $x-y=\partial z$. The dimensions are usually clear from the context, but they can also be indicated by superscripts.

24. Abelian groups

24A. Before continuing we append a brief discussion of Abelian groups. Only very elementary properties will be needed.

A set $\{u_i\}$ of elements of an Abelian group G is called a *system of generators* if every element of G can be represented in the form $\sum x_i u_i$ where the x_i are integers and only a finite number are $\neq 0$. The elements in a set $\{v_i\}$ are said to be *linearly independent* if a relation $\sum x_i v_i = 0$ implies $x_i = 0$ for all i. It is proved in elementary linear algebra that the number of elements in any linearly independent set is at most equal to the number of generators.

A set is a *basis* if it is at once a system of generators and linearly independent. Thus $\{u_i\}$ is a basis if and only if every element of G has a unique representation in the form $u = \sum x_i u_i$. All bases of G have the same number of elements. A group is said to be *free* if it has a basis.

The maximum number of linearly independent elements in any Abelian group G is its *rank* $r(G)$. For a free group the rank equals the number of basis elements.

We denote the additive group of integers by Z. A free group of rank r is isomorphic to Z^r, the direct product of r groups Z. The notation nZ will refer to the subgroup of Z formed by all multiples of n. A finite cyclic group can thus be denoted by Z/nZ.

24B. The following theorem is true for arbitrary free groups, but we prove it only for groups with a countable basis.

Theorem. *Every subgroup of a free group is free.*

Proof. Let F be a subgroup of G, and suppose that G has the basis $\{u_i\}$. Let G_k be the subgroup of G that is generated by $u_1, \cdots, u_k$, and set $F_k = F \cap G_k$. The quotient group F_k/F_{k-1} is isomorphic to the group formed by the coefficients of u_k in the elements of F_k. It is therefore either 0 or an infinite cyclic group. In the latter case we choose an element

$v_k \in F_k$ whose coset with respect to F_{k-1} generates F_k/F_{k-1}; if the group is 0 we set $v_k = 0$. The elements v_k generate F, for every element of F is in an F_k. The nonzero v_k are linearly independent, for a relation $x_1v_1 + \cdots + x_kv_k = 0$ implies $x_k\bar{v}_k = 0$ where $\bar{v}_k$ is the coset of v_k with respect to F_{k-1}. If $v_k \neq 0$ it follows that $x_k = 0$, and hence every relation must be trivial. We have shown that the nonzero v_k form a basis of F.

24C. The *order* of an element $u \in G$ is the least positive integer n such that $nu = 0$. If there is no such integer, the order is infinite. In a free group all elements $\neq 0$ have infinite order, but the converse is not true. For instance, in the additive group of rational numbers all nonzero elements have infinite order, but the group is not free since any two rational numbers are linearly dependent. However, the following is true:

Theorem. *If an Abelian group has a finite system of generators, and if all nonzero elements are of infinite order, then the group is free.*

Proof. Let k be the least number of elements by which the group can be generated. Let $u_1, \cdots, u_k$ be a system of generators. If they are not linearly independent we consider all nontrivial relations $x_1u_1 + \cdots + x_ku_k = 0$ which they satisfy. Let h be the smallest absolute value of all nonzero coefficients x_i which occur in these relations. Moreover, assume that the generators have been chosen so as to make h as small as possible. By renumbering the generators we can assume that there is a relation of the form $hu_1 + x_2u_2 + \cdots + x_ku_k = 0$. For $i = 2, \cdots, k$, set $x_i = n_ih + y_i$ with $0 \leqq y_i < h$. Write $v_1 = u_1 + n_2u_2 + \cdots + n_ku_k$. Then $v_1, u_2, \cdots, u_k$ is a system of generators, and there is a relation $hv_1 + y_2u_2 + \cdots + y_ku_k = 0$. This violates the assumption unless all the y_i are 0. But then $hv_1 = 0$, and hence $v_1 = 0$. It follows that $u_2, \cdots, u_k$ is a system of generators, contrary to the definition of k. Hence $u_1, \cdots, u_k$ must be a basis.

24D. In any Abelian group G the elements of finite order form a subgroup T, the *torsion group* of G. The quotient group G/T is without torsion, for if $nu \in T$ for some n, then $mnu = 0$ for some m, and hence $u \in T$.

If G is generated by u_i, then G/T is generated by $\bar{u}_i$ where $\bar{u}_i$ is the coset of u_i. If G is finitely generated, so is G/T, and we conclude by the preceding theorem that G/T has a basis $\bar{v}_1, \cdots, \bar{v}_k$. Choose $v_i \in \bar{v}_i$. Every coset $\bar{u}$ has a unique representation $\bar{u} = x_1\bar{v}_1 + \cdots + x_k\bar{v}_k$, and hence every element of G has a unique representation $u = x_1v_1 + \cdots + x_kv_k + t$ with $t \in T$. We have proved:

Theorem. *Any finitely generated Abelian group G is isomorphic to a direct product $Z^r \times T$ where $r = r(G)$ and T is the torsion group of G.*

24E. A useful theorem concerning ranks is the following:

Theorem. *If F is a subgroup of the Abelian group G, then $r(G) = r(F) + r(G/F)$.*

Let $\{u_i\}$ and $\{\bar{v}_j\}$ be linearly independent in F and G/F respectively. The $\bar{v}_j$ are cosets of F, and we choose elements $v_j \in \bar{v}_j$. Then the system $\{u_i, v_j\}$ is linearly independent in G. In fact, $\sum x_i u_i + \sum y_j v_j = 0$ implies $\sum y_j \bar{v}_j = 0$. Hence $y_j = 0$ for all j, and the remaining relation $\sum x_i u_i = 0$ gives $x_i = 0$ for all i. We conclude that $r(G) \geqq r(F) + r(G/F)$.

For the opposite conclusion we may assume that $r(F)$ and $r(G/F)$ are finite. We choose maximal linearly independent systems $\{u_i\}$ and $\{\bar{v}_j\}$. If $v_j \in \bar{v}_j$ it is easily seen that any $z_k \in G$ satisfies a relation $m_k z_k = \sum x_i u_i + \sum y_j v_j$ with $m_k \neq 0$. Since the free group generated by the u_i and v_j has rank $r(F) + r(G/F)$ it follows that any larger number of elements $m_k z_k$ will be linearly dependent. The same is then true of the corresponding elements z_k. We conclude that $r(G) \leqq r(F) + r(G/F)$, and the theorem is proved.

25. Polyhedrons

25A. We shall call a 2-dimensional abstract complex K, or the corresponding geometric complex K_g, *a polyhedron* if it represents a triangulated surface or bordered surface, i.e., if the conditions of Theorem 22E or the modified conditions of 22H are satisfied. The name is also attached to K_g considered as a topological space, and in this sense any surface or bordered surface which permits a triangulation is a polyhedron. For the moment, however, we prefer to think of a polyhedron as an abstract complex. If the complex is finite we speak of a finite polyhedron. We shall also speak of closed, bordered, and open polyhedrons.

Our aim is to determine the homology groups of a polyhedron. We remark at once that all groups $C_n(K)$, $Z_n(K)$, $B_n(K)$ and $H_n(K)$ with $n > 2$ reduce to 0, so that we need consider only the dimensions $0 \leqq n \leqq 2$.

The rank of $H_n(K)$ is called the *Betti number* of dimension n; it is denoted by p_n or $p_n(K)$.

25B. Every 0-dimensional chain is a cycle. Hence $Z_0 = C_0$ and $H_0 = C_0/B_0$.

In a 0-dimensional boundary $\sum x_i \partial s_i^1$, the algebraic sum of the coefficients is 0. This shows that no nonzero multiple $x_0 \alpha_0$ of a single vertex can be homologous to 0. On the other hand, any two vertices are homologous to each other. For in view of the connectedness of K there exists a finite sequence of ordered 1-simplices $(\alpha_0 \alpha_1)$, $(\alpha_1 \alpha_2)$, $\cdots$, $(\alpha_{n-1} \alpha_n)$ which ends with a prescribed α_n. The sum of these simplices has the boundary $\alpha_n - \alpha_0$, and we deduce that $\alpha_n \sim \alpha_0$.

It follows that $H_0(K)$ is an infinite cyclic group. Its rank is thus $p_0 = 1$.

25C. There are no 2-dimensional boundaries other than 0. Hence $B_2 = 0$ and $H_2 = Z_2$.

In order that a chain $\sum x_i s_i^2$ be a cycle it is evidently necessary that any two 2-simplices with a common 1-simplex have coefficients which are either equal or opposite, depending on the orientations. As a consequence of condition (E1) in Theorem 22E or its analogue for bordered polyhedrons, together with the connectedness of K, it is readily seen that one can pass from any 2-simplex to another through a finite sequence of adjacent 2-simplices. This proves that all coefficients must have the same absolute value $|x_i|$. If K is infinite it follows that the common value must be zero, and 0 is the only 2-cycle. The same is true if K has a border, for the coefficient of a 2-simplex which contains a border 1-simplex must be zero. In all these cases H_2 reduces to 0, and $p_2=0$.

There remains the case in which K represents a closed surface. If there exists a nonzero cycle $\sum x_i s_i^2$, we set $x_i=\epsilon_i|x_i|$ with $\epsilon_i=\pm 1$. Since all the $|x_i|$ are equal we conclude that $\sum \epsilon_i s_i^2$ is a cycle. If $\sum y_i s_i^2$ is any other cycle we can make the coefficient of s_1^2, say, zero by subtracting $\epsilon_1 y_1(\sum \epsilon_i s_i^2)$. This must make all other coefficients zero as well, and we find that $\sum y_i s_i^2$ is a multiple of $\sum \epsilon_i s_i^2$. *Hence* H_2 *is either* 0 *or an infinite cyclic group, so that* $p_2=0$ *or* $p_2=1$.

25D. When K_g is a closed surface we want to show that $p_2=1$ *if and only if* K_g *is orientable*. Far from being obvious, this requires a careful proof.

If $p_2=1$ we can choose the orientations so that $\sum s_i^2$ is a cycle. Let $(\alpha_0\alpha_1\alpha_2)$ be a 2-simplex in this orientation. We recall that the corresponding A_g is a triangle. It can therefore be mapped by an affine transformation f onto a triangle in the plane, for instance so that α_0, α_1, α_2 correspond to 0, 1, i. This mapping determines an orientation of A_g. It is conceivable that the orientation would depend on the particular order of the vertices. However, the affine transformation which effects a cyclic permutation of 0, 1, i can be written explicitly as $(x, y)\to(1-x-y, x)$ and is readily seen to be sense-preserving. Therefore, we obtain a definite orientation of the interior of each 2-simplex.

The *star* $S(\alpha_0)$ of a vertex is the union of all geometric simplices which contain α_0. The interiors of all stars form an open covering of K_g, and we will show that this covering permits a compatible orientation.

It is easy to show that Int $S(\alpha_0)$ is homeomorphic with a disk. Hence every star is orientable, and the orientation of an $A_g\subset S(\alpha_0)$ determines an orientation of the star. It must be shown that different A_g determine the same orientation, and for that purpose it is sufficient to consider two adjacent triangles A_g, A_g' in $S(\alpha_0)$. If the common side is $(\alpha_0\alpha_1)$ the orientations must be of the form $A=(\alpha_0\alpha_1\alpha_2)$, $A'=(\alpha_0\alpha_3\alpha_1)$, and we denote the corresponding affine mappings by f, f'. A topological mapping h of $A_g\cup A_g'$ into the plane can be constructed by setting $h=f$ on A_g and

$h = \sigma \circ f'$ on A'_g where σ is given as $z \to -iz$. The orientations defined by h and f' agree in A'_g, for σ is sense-preserving. It follows that the orientations of $S(\alpha_0)$ defined by f and f' are the same.

We have now obtained orientations of all stars. Moreover, two open stars are either disjoint or have a connected intersection which contains the interior of an A_g. We know that the orientations agree on A_g. Hence K_g is an orientable surface.

Conversely, suppose that K_g is orientable. Then we can choose the affine mapping f of A_g so that it agrees with the orientation of K_g. This determines an order $(\alpha_0\alpha_1\alpha_2)$ of the vertices. If an adjacent A'_g were ordered by $(\alpha_0\alpha_1\alpha_3)$ we could map $A_g \cup A'_g$ by $h = f$ on A_g and $h = \sigma' \circ f'$ on A'_g where σ' is the mapping $z \to \bar{z}$. But σ' is sense-reversing, so that f' would not agree with h and hence not with the orientation of K_g. It follows that A'_g is ordered by $(\alpha_0\alpha_3\alpha_1)$, and the common side $(\alpha_0\alpha_1)$ cancels from the boundary. Thus $\sum s_i^2$ is a cycle, and $p_2 = 1$.

The same reasoning applies to the case of a finite bordered complex. It is found that K_g is orientable if and only if the 2-simplices can be oriented so that the boundary of $\sum s_i^2$ consists of only border simplices.

26. The 1-dimensional homology group

26A. We have seen that the groups $H_0(K)$ and $H_2(K)$ of a polyhedron are very simple and can be determined at a glance. We turn now to the group $H_1(K)$. To begin with we determine its torsion group, denoted by $T_1(K)$.

Theorem. *The torsion group $T_1(K)$ of a polyhedron reduces to* 0, *except when K is nonorientable and closed, in which case it is a cyclic group of order* 2.

Let y be a 1-dimensional cycle, and suppose that $my \sim 0$, i.e., $my = \sum x_i \partial s_i^2$, for some integer $m \neq 0$. If s_i^2 and s_j^2 are adjacent one sees that x_i must differ from x_j or $-x_j$ by a multiple of m, and because of the connectedness the same must be true for arbitrary i and j. If K is infinite each x_i must be divisible by m, and it follows that $y \sim 0$. If K is bordered the same conclusion can be drawn, for the coefficient of an s_i^2 which contains a border 1-simplex must be a multiple of m. Finally, if K is closed and orientable we can assume that $\sum s_i^2$ is a cycle. We have then $my = \sum (x_i - x_1)\partial s_i^2$, and the same argument shows that $y \sim 0$.

If K is closed and nonorientable we choose arbitrary orientations of the s_i^2 and note that the coefficients in $\sum \partial s_i^2$ are 0 or ± 2. Write $2z = \sum \partial s_i^2$. Then $2z \sim 0$, but z cannot be homologous to 0, for from $z = \sum x_i \partial s_i^2$ would follow that $\sum (2x_i - 1)\partial s_i^2 = 0$, contrary to the fact that there are no 2-dimensional cycles other than 0. Consequently, z is of order 2.

Consider again the homology $my = \sum x_i \partial s_i^2$. We have $x_i \equiv x_j \pmod{m}$ or $x_i \equiv -x_j \pmod{m}$ for any i, j, and hence we can write $my = x_1 \sum \epsilon_i \partial s_i^2 +$

$m\sum t_i \partial s_i^2$ with $\epsilon_i = \pm 1$. At least one coefficient of $\sum \epsilon_i \partial s_i^2$ is ± 2. Therefore $2x_1$ is divisible by m, and we find that $2y \sim 0$. If $2y = \sum u_i \partial s_i^2$ the u_i are either all even or all odd. If they are even we get $y \sim 0$, and if they are odd we find $y \sim z$. Hence z is the only element of finite order.

26B. We assume now that K is finite. Then $C_1(K)$ has a finite number of basis elements s_i^1. It follows by Theorem 24B that the subgroup $Z_1(K)$ is free, and by Theorem 24E that it has finite rank. Hence $H_1(K) = Z_1(K)/B_1(K)$ is finitely generated, and we conclude by Theorem 24D that $H_1(K)$ is isomorphic to $\mathbf{Z}^{p_1} \times T_1(K)$, where p_1 is the Betti number of dimension 1. Together with the preceding result we have thus:

Theorem. *The first homology group $H_1(K)$ of a finite polyhedron is a free group of rank p_1, except when K is closed and nonorientable. In the latter case $H_1(K)$ is the direct product of a free group of rank p_1 and a cyclic group of order* 2.

The discussion of the infinite case is deferred to 30.

26C. For the finite case we give an explicit formula for p_1. For this purpose we let n_0, n_1, n_2 denote the number of 0-, 1- and 2-simplices in K.

In the first place, $B_1(K)$ is generated by all ∂s_i^2. They are linearly independent, except in the case of a closed orientable K, in which case they satisfy the relation $\sum \partial s_i^2 = 0$. Hence $B_1(K)$ is a free group of rank n_2 or $n_2 - 1$.

Secondly, $C_1(K)$ is mapped by ∂ onto $B_0(K)$, and the kernel of the homomorphism is $Z_1(K)$. Hence C_1/Z_1 is isomorphic to B_0, and by Theorem 24E we obtain $r(C_1) = r(Z_1) + r(B_0)$. But C_1 is free with rank n_1, and because of the connectedness B_0 is free with rank $n_0 - 1$. We find that $r(Z_1) = n_1 - n_0 + 1$.

From $Z_1/B_1 = H_1$ we get $p_1 = r(H_1) = r(Z_1) - r(B_1) = -n_0 + n_1 - n_2 + 1$ or $-n_0 + n_1 - n_2 + 2$. The number $\rho = -n_0 + n_1 - n_2$ is called the *Euler characteristic of* K.

Theorem. *The* 1*-dimensional Betti number of a finite polyhedron is either $\rho + 1$ or $\rho + 2$ where ρ denotes the Euler characteristic. The second value is assumed in the case of a closed orientable surface.*

Note that the characteristic can therefore never be less than -2.

27. Relative homologies

27A. A complex L is called a *subcomplex* of K if every simplex of L is also a simplex of K. In particular, L is then contained in K. It is possible, however, that a more than 0-dimensional simplex of K is contained in L, as a set of vertices, without being a simplex of L.

The group $C_n(L)$ is a subgroup of $C_n(K)$, and its elements are said to lie in L. The quotient group $C_n(K)/C_n(L)$ is more conveniently denoted as

$C_n(K/L)$. Its elements are *relative chains* modulo L, obtained by identifying any two chains on K whose difference lies in L.

If x lies in L, its boundary ∂x also lies in L. This makes it possible to define ∂ on $C_n(K/L)$ by letting the coset of x be mapped on the coset of ∂x. The kernel $Z_n(K/L)$ of this homomorphism is the group of *relative cycles*, and its image $B_{n-1}(K/L)$ consists of *relative boundaries*. The *relative homology group* is $H_n(K/L) = Z_n(K/L)/B_n(K/L)$.

Thus, a chain on K represents a relative cycle if its boundary lies in L, and x is homologous to 0 mod L if there exists an y on K such that the difference $x - \partial y$ lies in L.

27B. Our purpose in introducing the relative homology groups is to study, in greater detail, the homologies on a bordered polyhedron. We assume that K is a finite bordered polyhedron with q contours, and we take L to be its border B, regarded as a 1-dimensional subcomplex. We are going to determine the groups $H_n(K/B)$ for $n = 0, 1, 2$.

The cases $n=0$ and $n=2$ are quickly disposed of. Every vertex of K can be joined by a polygonal line to a vertex on the border. Hence every 0-chain is homologous to a multiple of a border vertex, and we see at once that $H_0(K/B)$ reduces to 0.

The elements of the group $H_2(K/B) = Z_2(K/B)$ are the 2-chains whose boundary lies on the border. The reasoning in 25C and the remark at the end of 25D show that $H_2(K/B)$ is 0 if K is nonorientable, and an infinite cyclic group if K is orientable.

27C. In order to determine the group $H_1(K/B)$ we need to investigate the absolute homology group $H_1(K)$ more closely. The elements of $Z_1(B)$, i.e., the cycles on B, are referred to as *border cycles*. In $H_1(K)$ we distinguish an important subgroup, denoted by $H_1B(K)$, which consists of all homology classes that contain a border cycle. The cycles whose homology classes lie in $H_1B(K)$ will be called *dividing cycles*. A dividing cycle is thus one which is homologous to a border cycle.

The border cycles form a free group with q basis elements b_i, one for each contour. If the surface is orientable we can and will choose the b_i so that $\sum b_i = \sum \partial s_j^2$; in the nonorientable case there is no preferred direction of the contours. In the orientable case $\sum b_i \sim 0$, but except for multiples of this homology relation there are no other homologies between the b_i. If K is nonorientable the b_i are homologically independent.

It follows that $H_1B(K)$ is a free group of rank $q-1$ or q, depending on the orientability. In the orientable case we can choose the homology classes of $b_2, \cdots, b_q$ as basis elements, and this system can be extended to a basis of $H_1(K)$. In the nonorientable case we must separate the torsion group which consists of 0 and the homology class of a certain cycle z_0, but the result is similar.

Theorem. *Let K be a finite bordered polyhedron. If K is orientable there exist cycles $z_1, \cdots, z_m$ such that every cycle on K satisfies a unique homology relation of the form*

$$z \sim x_2 b_2 + \cdots + x_q b_q + t_1 z_1 + \cdots + t_m z_m.$$

If K is nonorientable, the corresponding representation is

$$z \sim x_1 b_1 + \cdots + x_q b_q + t_0 z_0 + t_1 z_1 + \cdots + t_m z_m$$

with $t_0 = 0$ or 1. The cycle z is dividing if and only if all the t_i are 0.

On comparing the ranks with the known value of p_1 (26C) we obtain $m = \rho - q + 2$ in the orientable case and $m = \rho - q + 1$ in the nonorientable case.

Observe that we can regard the z_i as generators of the quotient group $H_1(K)/H_1 B(K)$. This quotient group is called the *homology group modulo dividing cycles*. It consists of cosets of absolute cycles and should not be confused with $H_1(K/B)$.

27D. We proceed to the study of $H_1(K/B) = Z_1(K/B)/B_1(K/B)$. The elements of $Z_1(K/B)$ are cosets $x + C_1(B)$ with $\partial x \in C_0(B)$, where $C_0(B)$ and $C_1(B)$ denote the groups of 0- and 1-dimensional border chains. Let us choose a fixed vertex s_i^0 on each b_i, and chains c_i with $\partial c_i = s_i^0 - s_1^0$. Given x with $\partial x \in C_0(B)$ we let y_i be the sum of the coefficients of ∂x at the vertices that lie on b_i. It is clear that $\partial x - \sum y_i s_i^0$ is the boundary of a chain u on B. Because $\sum y_i = 0$ we conclude that $x - y_2 c_2 - \cdots - y_q c_q - u$ is an absolute cycle z, and we find that $x + C_1(B)$ can be represented in the form $z + y_2 c_2 + \cdots + y_q c_q + C_1(B)$. This representation is unique, except for border cycles that can be added to z. Indeed, if $z + y_2 c_2 + \cdots + y_q c_q \in C_1(B)$, then $y_2(s_2^0 - s_1^0) + \cdots + y_q(s_q^0 - s_1^0)$ is the boundary of a border chain. In such a boundary the coefficients at vertices on b_i must add up to 0, from which it follows that $y_i = 0$.

A coset $x + C_1(B)$ belongs to $B_1(K/B)$ if there exists a 2-dimensional chain v such that $x - \partial v \in C_1(B)$. We have thus $x + C_1(B) = \partial v + C_1(B)$, and by the uniqueness of the representation $z + y_2 c_2 + \cdots + y_q c_q + C_1(B)$ it follows that all y_i are 0 while z differs from ∂v by a border cycle. The latter condition means that z is a dividing cycle.

From these considerations we conclude that $H_1(K/B)$ is isomorphic with the direct sum of $H_1(K)/H_1 B(K)$ and a free group with $q-1$ generators. By virtue of Theorem 27C the result can be expressed as follows:

Theorem. *On a finite orientable K every relative cycle z satisfies a unique relative homology relation of the form*

$$z \sim t_1 z_1 + \cdots + t_m z_m + y_2 c_2 + \cdots + y_q c_q \pmod{B}.$$

In the nonorientable case a term $t_0 z_0$ with $t_0 = 0$ or 1 must be added.

We see that $H_1(K)$ and $H_1(K/B)$ are isomorphic in the orientable case, but not in the nonorientable case. It is more significant that the quotient group $H_1(K)/H_1B(K)$ can always be identified with the subgroup of $H_1(K/B)$ that consists of the relative homology classes of absolute cycles.

28. Subdivisions

28A. It is time to introduce the very simple idea of a subdivision or refinement of a triangulation. The process of subdivision consists in repeating a single step, referred to as *elementary subdivision*. Its geometric nature is illustrated in Fig. 1.

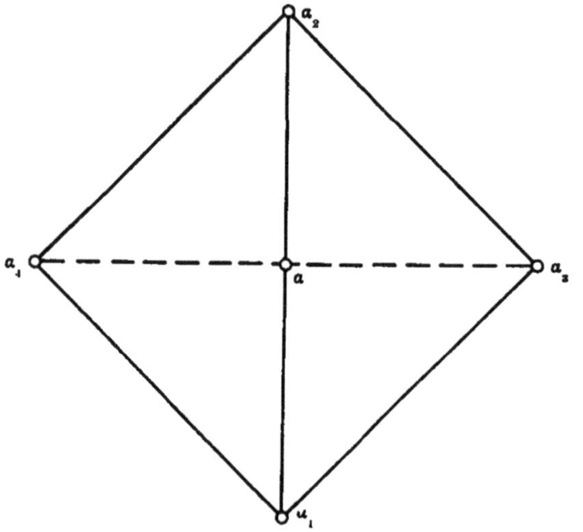

FIG. 1

Formally, the process can be described as follows. Let a be a 1-simplex in a polyhedron K. We write $a=(\alpha_1\alpha_2)$ and assume first that a belongs to two 2-simplices $A_1=(\alpha_1\alpha_2\alpha_3)$, $A_2=(\alpha_1\alpha_2\alpha_4)$. We introduce a new 0-simplex α and replace a, A_1, A_2 by four 1-simplices $(\alpha\alpha_1)$, $(\alpha\alpha_2)$, $(\alpha\alpha_3)$, $(\alpha\alpha_4)$ and four 2-simplices $(\alpha\alpha_1\alpha_3)$, $(\alpha\alpha_2\alpha_3)$, $(\alpha\alpha_1\alpha_4)$, $(\alpha\alpha_2\alpha_4)$. All other simplices remain untouched. If a belongs to only one 2-simplex the corresponding simpler process is defined in a similar manner.

The complex K' obtained by this construction is said to result from K by elementary subdivision, applied to a. It is obvious that K' is again a polyhedron, and that the geometric complexes K_g and K'_g are homeomorphic. In fact, we can even effect the homeomorphism by means of a piecewise affine mapping which is uniquely determined if we let α correspond to the midpoint of $(\alpha_1\alpha_2)$.

The process of elementary subdivision can be applied simultaneously to any finite or infinite number of simplices a, provided that no two of these 1-simplices belong to the same 2-simplex. Any such simultaneous application of elementary subdivisions is said to constitute a *simple subdivision.*

Finally, we can apply a finite number of simple subdivisions in succession. The resulting polyhedron K' is called a *subdivision* of K. It is again obvious that K_g' is homeomorphic to K_g by means of a piecewise affine mapping. In this way, a triangulation of a surface F or $\bar{F}$ (22C), together with a homeomorphic mapping of K_g, becomes a triangulation by means of K_g'. We call it a *refinement* of the original triangulation.

28B. A slightly more general notion is that of *equivalent* polyhedrons. We say that K_1 and K_2 are equivalent if it is possible to pass from one to the other by way of a sequence $K_1 = K^1, K^2, \cdots, K^n = K_2$ in which either K^{i+1} is a subdivision of K^i, or K^i is a subdivision of K^{i+1}. Equivalent polyhedrons have homeomorphic geometric complexes. Furthermore, they agree in the following properties:

Theorem. *Equivalent polyhedrons have isomorphic homology groups. They have the same character of orientability, the same number of contours, and, if finite, the same characteristic.*

Proof is needed only for the case that K' is a simple subdivision of K. The group $C_1(K)$ can be mapped isomorphically into $C_1(K')$ by letting $a = (\alpha_1\alpha_2)$ be carried into $(\alpha_1\alpha) + (\alpha\alpha_2)$. One verifies that cycles correspond to cycles and boundaries to boundaries. Hence $Z_1(K)$ and $B_1(K)$ can be thought of as subgroups of $Z_1(K')$ and $B_1(K')$ respectively. Further explicit verification of the most elementary kind shows that $Z_1(K) \cap B_1(K') = B_1(K)$. For this reason two cycles on K are homologous in K' if and only if they are homologous in K. Finally, every cycle on K' is homologous (with respect to K') to a cycle on K. In fact, a cycle must have equally many 1-simplices beginning and ending at α. On the other hand, a chain of the type $(\alpha_1\alpha) + (\alpha\alpha_3)$ can be replaced by $(\alpha_1\alpha_3)$ in the sense of homology.

These considerations show that there is an isomorphic correspondence between $H_1(K)$ and $H_1(K')$.

28C. The verification of the assertions with respect to orientability and number of contours are trivial. The isomorphism of the 2-dimensional homology groups is a consequence, by virtue of 25C.

The invariance of the characteristic follows if we make use of the result in 26C, but the direct verification is equally easy. If an elementary subdivision is applied to an inner 1-simplex we find that n_0, n_1, n_2 are increased by 1, 3 and 2 respectively. In the case of a border 1-simplex the increments are 1, 2, 1. In both cases $\rho = -n_0 + n_1 - n_2$ remains invariant.

28D. We speak of a *barycentric subdivision* if each 2-simplex is divided as indicated in Fig. 2. It is obvious how to construct the barycentric subdivision by repeated simple subdivisions. The new vertices that are introduced on each 1- and 2-simplex are called *barycenters*. The process of passing to the barycentric subdivision of a complex has the advantage that it can be repeated indefinitely in exactly the same way.

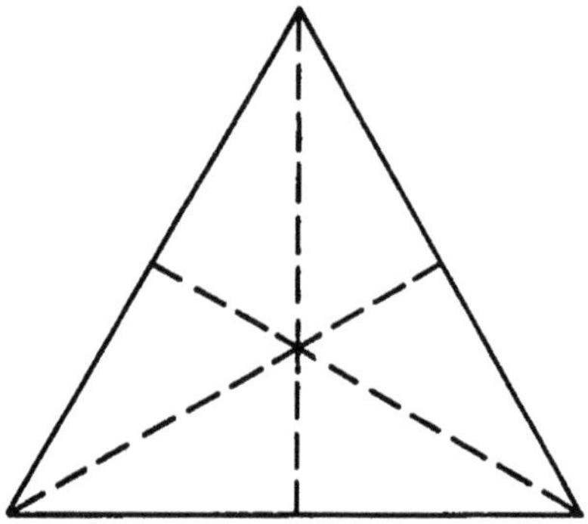

FIG. 2

If K_g is a polyhedron we denote by K' the complex obtained from K by barycentric subdivision, and by K'_g the corresponding geometric complex. There exists a unique piecewise affine mapping of K'_g onto K_g which carries each vertex into the barycenter (center of gravity) of the corresponding 0-, 1- or 2-simplex on K_g. We shall identify K'_g with its image under this mapping. The identification makes it possible to regard the simplices of K'_g as subsets of K_g.

We recall the notion of star that was introduced in 25D. The star $S(\alpha)$ of a vertex α is the union of all geometric simplices which contain α. The interior of $S(\alpha)$, denoted by $S_0(\alpha)$, is referred to as the open star of α. If α is a vertex of the polyhedron K_g we form its star $S'(\alpha)$ on K'_g; this is the *barycentric star* of α. It is easily seen that each $S'(\alpha)$ is homeomorphic with a closed disk, a property that is not necessarily shared by $S(\alpha)$. Moreover, two or three barycentric stars intersect if and only if their vertices belong to a common simplex.

28E. The process of barycentric subdivision can be iterated. We denote the iterated subdivisions by $K^{(n)}$ and the corresponding geometric complexes by $K_g^{(n)}$. The same rule as before permits us to regard the simplices of $K_g^{(n)}$ as subsets of K_g. The following theorem shows that the barycentric subdivisions can be repeated until the simplices are arbitrarily small:

Theorem. *Given an open covering of a finite polyhedron K_g by sets V, the barycentric subdivisions can be repeated until every star on $K_g^{(n)}$ is contained in a V.*

Since K_g is imbedded in a Euclidean space there is a natural concept of length on K_g. It is easily seen that the diameters of the stars tend to zero under repeated barycentric subdivisions. On the other hand, a familiar compactness argument shows the existence of a number $\delta > 0$ such that every set of diameter $< \delta$ is contained in a V. It is thus sufficient to continue until the diameters of all stars lie under this bound.

29. Exhaustions

29A. In order to study an infinite polyhedron K it is necessary to consider approximations of K by finite subcomplexes. The approximating subcomplexes are required to satisfy certain special conditions. We show in this section that these requirements can be met, if not for K, then at least for a suitable subdivision of K. It would not be difficult to treat the most general case, but for reasons of brevity we prefer to assume that K is open.

Consider in K a finite subcomplex P, which is itself a polyhedron. The complement of P with respect to K becomes a complex if the border simplices of P are added. For convenience, when we speak of the complement $K-P$ we shall always mean this complex, rather than the set-theoretical difference. The connected components of $K-P$ are bordered polyhedrons which may be finite or infinite; their contours are identical with the contours of P. We shall call P a *canonical subcomplex* if it is a polyhedron, if all components of $K-P$ are infinite, and if each component of $K-P$ has a single contour.

A sequence of subcomplexes $P_1, P_2, \cdots, P_n, \cdots$ is an *exhaustion* of K if the following conditions are fulfilled: (1) each P_n is a polyhedron, (2) P_n is a subcomplex of P_{n+1}, (3) the border simplices of P_{n+1} are not in P_n, (4) every simplex in K belongs to a P_n. We speak of a *canonical exhaustion* if all the P_n are canonical.

Our aim is to prove:

Theorem. *Every open polyhedron has a subdivision which permits a canonical exhaustion.*

29B. The proof requires several steps. Let A be a finite subcomplex of the given polyhedron K. As before, A' and K' denote the barycentric subdivisions of A and K. We show first that K' has a subcomplex P which is a polyhedron and contains A'.

We may assume, without loss of generality, that A is connected. We take P to be the union of the barycentric stars (28D) of the vertices of A.

P is a subcomplex of K', and it evidently contains A'. The connectedness of A implies that P is connected. It remains to show that P is a polyhedron.

Let α be a vertex of P. According to the definition of barycentric star α is either a vertex of K, the barycenter of a 1-simplex of K, or the barycenter of a 2-simplex of K. In the first case α is in A, and it is surrounded by 1- and 2-simplices of K' which all belong to P. In the second case α belongs to four 2-simplices of K'. Since the 2-simplices that lie in the same barycentric star are either all in P or all not in P, the 2-simplices of P that contain α will form either a sequence of two or a cycle of four adjacent ones. In the third case α belongs to six 2-simplices of K', and of these six either one, two, or three pairs of adjacent ones will lie in P. In any event, the 2-simplices of P that contain α form a connected sequence or a full cycle. We have proved that P is a polyhedron.

29C. We show next that there exists an iterated barycentric subdivision $K^{(n)}$ which contains a *canonical* polyhedron that in turn contains $A^{(n)}$. For this purpose, consider all n and all polyhedrons which are subcomplexes of $K^{(n)}$ and contain $A^{(n)}$. The previous reasoning has shown that this set of polyhedrons is not empty. Among all polyhedrons in the set, let P be one with the minimum number of contours. We claim that P is canonical.

For simplicity, if P is a subcomplex of $K^{(n)}$, we write K for $K^{(n)}$ and A for $A^{(n)}$. Then P is a subcomplex of K. It is clear that all components of $K-P$ are infinite. Indeed, a finite component could be added to P, and we would obtain a polyhedron with fewer contours.

Suppose now that two contours b_1 and b_2 of P belong to the same complementary component Q. Choose vertices on b_1, b_2 and join them by a simple polygon σ whose sides are 1-simplices in Q. It is permissible to assume that none of these 1-simplices lies on the border of Q; indeed, we can follow σ to the last vertex on b_1, and from there to the first vertex on another contour, which, if it is not b_2, can be used in place of b_2. We regard the union $L=\sigma \cup b_1 \cup b_2$ as a 1-dimensional complex. The total border of P is denoted by B.

Repeat the construction in 29B with K' in the place of K, $P' \cup L'$ in the place of A. The resulting polyhedron, a subcomplex of K'', will be denoted by P_0. We will show that P_0 has fewer contours than P.

A look at Fig. 3 is more instructive than a detailed proof. The figure suggests a way to associate with L, and with each of the other contours, a cyclically ordered sequence of 1-simplices, the outer perimeter in the figure. We need to show that all these 1-simplices are distinct. Assume that, in the figure, $a_1=a_2$. The centers α_1, α_2 of the corresponding barycentric stars belong to $B' \cup L'$. Because the barycentric stars have a common side, $(\alpha_1\alpha_2)$ is a 1-simplex on K'. Then one of the end points, say

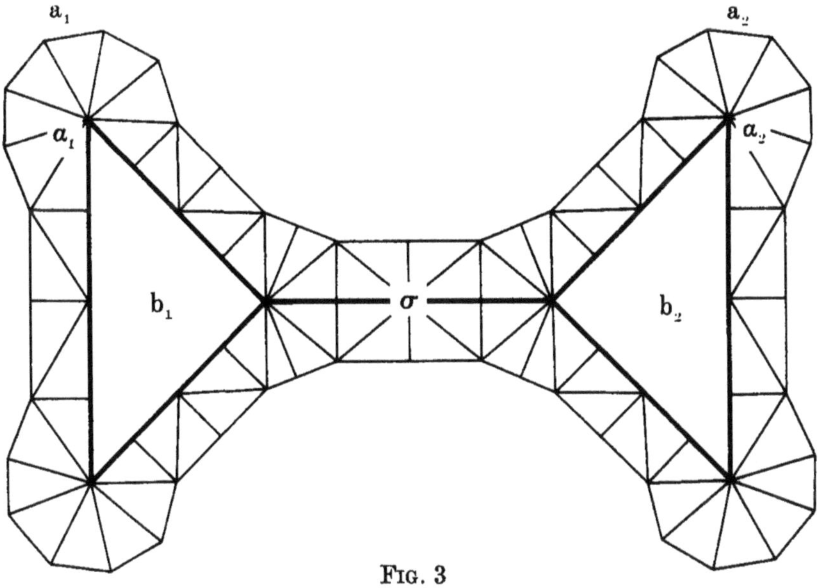

FIG. 3

α_1, does not belong to K, by the construction of K'. Hence α_1 is the barycenter of a 1-simplex on $B \cup L$, and we see that α_1, α_2 are neighboring vertices on L' (or on the same contour). This would mean that the common side $a_1 = a_2$ of the stars has an end point on $B'' \cup L''$, contrary to the construction of the perimeter.

We conclude that P_0 has indeed fewer contours than P, and since this is impossible P must be canonical.

29D. Returning to the original notation we let K be an arbitrary open polyhedron. It is possible to find a sequence of triples $\{A_i, n_i, P_i\}$ with the following properties:

(a) A_i is a finite connected subcomplex of K,

(b) n_i is a nonnegative integer,

(c) P_i is a canonical polyhedron on $K^{(n_i)}$,

(d) $A_i^{(n_i)} \subset P_i$,

(e) $P_i \subset A_{i+1}^{(n_i)}$,

(f) the border of P_{i+1} is not in P_i.

Indeed, we can begin by choosing A_1 as a single vertex. As soon as A_i has been found we can determine n_i and P_i by the method described in 29C. Next, let A_{i+1} comprise all 2-simplices of K whose barycentric subdivision of order n_i contains at least one vertex of P_i. It is obvious that (e) is fulfilled. As for (f), the statement needs further clarification. What we mean is that in a common subdivision the border simplices of P_{i+1} are not in P_i. The construction shows that the vertices of P_i are completely surrounded by simplices in A_{i+1}, and hence by simplices in P_{i+1}. We conclude that all conditions are satisfied.

29E. If the numbers n_i were bounded we could pass to a common barycentric subdivision of all the P_i, and the subdivided polyhedrons would constitute a canonical exhaustion.

In the general case of unbounded n_i we may as well assume that $n_i < n_{i+1}$. Under this condition we can consider $P_i^{(n_{i+1}-n_i)}$ as a subcomplex of P_{i+1}. Construct the complement $P_{i+1} - P_i^{(n_{i+1}-n_i)}$ which by definition includes the subdivided border of P_i; for $i=0$ the complement is replaced by P_1. The subdivisions of adjacent complements do not agree on the common borders. An easy adjustment takes care of this difficulty. Consider all 2-simplices of P_{i+1} which have at least one side on the border of P_{i+1}. Join the barycenter to the three vertices of the 2-simplex, and in addition to all new vertices on the sides that lie on the border. When this construction is carried out for all i we obtain a subdivision of the original K, and the P_i become canonical subcomplexes which define a canonical exhaustion. The proof of Theorem 29A is now complete.

30. Homologies on open polyhedrons

30A. We have not yet determined the homology group $H_1(K)$ of an infinite polyhedron. This will now be done, although under the slight restriction that K is orientable and open. The general case offers practically no added interest.

The result will be that $H_1(K)$ is a free group with a countable basis. If this were all we wanted to show the proof could be made very short. It is essential, however, that we introduce a particularly simple homology basis.

It has already been observed that the homology groups do not change if we pass to a subdivision. On evoking Theorem 29A we can therefore assume from the start that K possesses a canonical exhaustion $\{P_n\}$.

30B. With respect to a given canonical exhaustion we agree on a set of useful notations. Let the contours of P_1 be denoted by b_i, $i=1,\cdots,q$. At the same time the b_i will be generators of the cycles that lie on the various contours; moreover, they shall be positively directed with respect to the orientation of P_1 which in turn is induced by a fixed orientation of K. The orientation makes it possible to regard P_1 as a 2-dimensional chain, and the notation is such that $\partial P_1 = \sum b_i$.

Consider the complement $P_2 - P_1$, always in the sense that the border is included in the complement. Because the exhaustion is canonical there will be exactly q components, and we denote by Q_i the component which has the contour b_i in common with P_1. The remaining contours of Q_i will be denoted by b_{ij}, $1 \leqq j \leqq q_i$, and we choose the notation so that $\partial Q_i = \sum_j b_{ij} - b_i$. Next, the complement $P_3 - P_2$ consists of components Q_{ij} where Q_i and Q_{ij} have the common contour b_{ij}.

When we continue in this way we obtain a symbol $b_{i_1 \cdots i_n}$ for each contour of P_n, with the subscript i_n running from 1 to a number $q_{i_1 \cdots i_{n-1}}$. The components of $P_{n+1} - P_n$ have matching names $Q_{i_1 \cdots i_n}$. We agree further that $\partial Q_{i_1 \cdots i_n} = \sum_j b_{i_1 \cdots i_n j} - b_{i_1 \cdots i_n}$. For the sake of conformity, Q will be another name for P_1, and b will be 0.

30C. Let z be a 1-dimensional cycle on K, and suppose that z lies in P_n. It is clear that z has a decomposition $z = z' + \sum z_{i_1 \cdots i_{n-1}}$ where z' is in P_{n-1} and $z_{i_1 \cdots i_{n-1}}$ belongs to $Q_{i_1 \cdots i_{n-1}}$. In this decomposition, which ordinarily is not unique, z' and $z_{i_1 \cdots i_{n-1}}$ are chains, but not necessarily cycles. However, by the condition $\partial z = 0$ it follows that $\partial z_{i_1 \cdots i_{n-1}}$ lies at once on the border of $Q_{i_1 \cdots i_{n-1}}$ and on the border of its complement in P_n. Hence $\partial z_{i_1 \cdots i_{n-1}}$ lies on $b_{i_1 \cdots i_{n-1}}$, and since it is a boundary the sum of its coefficients is 0. It follows that $\partial z_{i_1 \cdots i_{n-1}}$ is the boundary of a chain on $b_{i_1 \cdots i_{n-1}}$, and we conclude easily that all the $z_{i_1 \cdots i_{n-1}}$ can be chosen as cycles.

According to Theorem 27C it is possible to introduce a homology basis on $Q_{i_1 \cdots i_{n-1}}$ which consists of the border cycle $b_{i_1 \cdots i_{n-1}}$, the border cycles $b_{i_1 \cdots i_n}$ with $i_n > 1$, and a finite number of cycles which we denote by $z_{i_1 \cdots i_{n-1}, j}$. The latter form a relative homology basis.

We are now able to prove:

Theorem. *A homology basis for the open orientable polyhedron K is formed by the combined system of all cycles $b_{i_1 \cdots i_n}$ with $i_n > 1$ and all cycles $z_{i_1 \cdots i_n, j}$.*

We must show that each cycle is homologous to a linear combination of the cycles mentioned in the theorem. This is manifestly so for cycles in P_1, and we assume that the assertion has been proved for cycles in P_{n-1}. If z is in P_n we use the representation $z = z' + \sum z_{i_1 \cdots i_{n-1}}$. Each $z_{i_1 \cdots i_{n-1}}$ is homologous to a combination of $b_{i_1 \cdots i_{n-1}}$, cycles $b_{i_1 \cdots i_n}$ with $i_n > 1$, and cycles $z_{i_1 \cdots i_{n-1}, j}$. Of these, $b_{i_1 \cdots i_{n-1}}$ may not be of the desired type, namely if $i_{n-1} = 1$. However, $b_{i_1 \cdots i_{n-1}}$ can always be incorporated in z', and since z' has a representation of the desired kind we have proved the assertion for cycles in P_n, and consequently for all cycles.

It remains to verify that the cycles are linearly independent in the sense of homology. Suppose that a certain linear combination, which we denote by z, is homologous to 0 on K. Then z is also homologous to 0 on some P_n. We set $z = \partial y$ and write $z = z' + \sum z_{i_1 \cdots i_{n-1}}, y = y' + \sum y_{i_1 \cdots i_{n-1}}$ where z', y' are in P_{n-1} and $z_{i_1 \cdots i_{n-1}}$, $y_{i_1 \cdots i_{n-1}}$ in $Q_{i_1 \cdots i_{n-1}}$; the $z_{i_1 \cdots i_{n-1}}$ are chosen as cycles. We see as on the similar previous occasion that $z_{i_1 \cdots i_{n-1}} - \partial y_{i_1 \cdots i_{n-1}}$ must lie on $b_{i_1 \cdots i_{n-1}}$. This is a homology relation on $Q_{i_1 \cdots i_{n-1}}$ between elements of the homology basis. It can be satisfied only so that $z_{i_1 \cdots i_{n-1}}$ is a multiple of $b_{i_1 \cdots i_{n-1}}$, and since there are no 2-dimensional cycles it follows in addition that $y_{i_1 \cdots i_{n-1}} = 0$. Consequently, z lies in P_{n-1} and is homologous to 0 on P_{n-1}. Since we could have assumed that n is the smallest index for which $z \sim 0$ on P_n, we conclude that z must be 0, and we have proved the linear independence.

30D. A cycle z on K is called a *dividing cycle* if, for any finite subcomplex L of K, z is homologous to a cycle outside of L (compare 27C). Another suggestive terminology is to say that a dividing cycle is *homologous to* 0 *modulo the idea boundary*. It is clear that the homology classes of the dividing cycles form a subgroup of $H_1(K)$ which in analogy with the finite case will be denoted by $H_1\beta(K)$. The quotient group $H_1(K)/H_1\beta(K)$ is the homology group modulo dividing cycles. It can also be called the relative homology group with respect to the ideal boundary.

Theorem. *The cycles $b_{i_1 \cdots i_n}$, $i_n > 1$, determine a basis for $H_1\beta(K)$, and the cycles $z_{i_1 \cdots i_n, j}$ are elements of a homology basis modulo dividing cycles. In other words, each cycle z satisfies a unique homology relation*

$$z \sim \sum_{i_n > 1} x_{i_1 \cdots i_n} b_{i_1 \cdots i_n} + \sum t_{i_1 \cdots i_n, j} z_{i_1 \cdots i_n, j}$$

where the sums are finite, and z is a dividing cycle if and only if all $t_{i_1 \cdots i_n, j}$ are 0.

Every $b_{i_1 \cdots i_n}$ is homologous to the sum of all $b_{i_1 \cdots i_n i_{n+1}}$ for variable i_{n+1}. Each $b_{i_1 \cdots i_{n+1}}$ is in turn homologous to a sum of contours in the next generation, and so on. It follows that $b_{i_1 \cdots i_n}$ is a dividing cycle. Conversely, suppose that z is a dividing cycle. We know that z lies in a P_n and is homologous to a cycle z' in the complement of P_n. We set $z = z' + \partial y + \partial y'$ where y is in P_n and y' in the complement. It follows that $z - \partial y$ lies on the border of P_n. Thus z is homologous to a linear combination of contours $b_{i_1 \cdots i_n}$. Those with $i_n = 1$ can be replaced by contours of the preceding generation, and ultimately z will be expressed in terms of the basis elements, without use of the $z_{i_1 \cdots i_n, j}$. This proves our theorem.

It can happen that all cycles are dividing cycles. Then there are no basis elements of the form $z_{i_1 \cdots i_n, j}$, and the polyhedron K is said to be *planar*. It can also happen that there are no dividing cycles other than those which are homologous to 0. In that case all the P_n have a single contour.

Finally, we mention that every sequence of the form $b_{i_1}, b_{i_1 i_2}, \cdots, b_{i_1 \cdots i_n}, \cdots$ is said to determine a *boundary component*. To say that $H_1\beta(K)=0$ is equivalent to saying that there is only one boundary component. This property is evidently independent of the choice of exhaustion.

31. Intersection theory

31A. We have to define the *intersection number* $x \times y$ of two cycles x, y in a way which formalizes the intuitive idea of the number of crossings of two oriented polygons. The somewhat roundabout method that must be used has its origin in the difficulties that occur when the cycles have common sides.

Let K be an orientable polyhedron, and consider its barycentric sub-

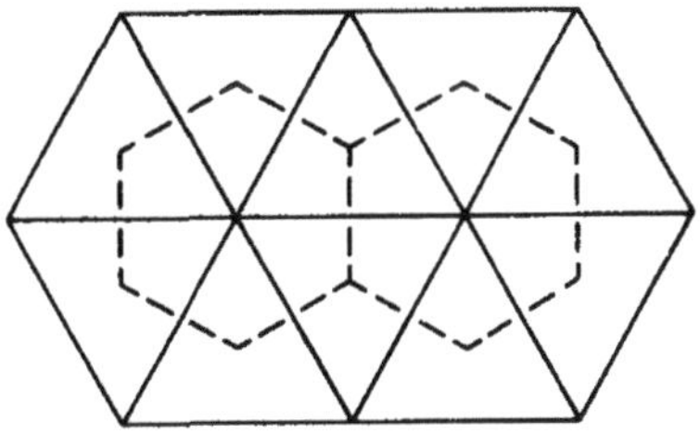

FIG. 4

division K'. We denote by a_i the oriented 1-simplices of K. In the subdivision K' they are replaced by chains $a_i = a_i' + a_i''$. Similarly, we let $b_i = b_i' + b_i''$ denote the chain on K' which leads from the barycenter of the 2-simplex on the left of a_i to the barycenter of the one on the right, left and right being determined by a fixed orientation of K. If a_i is a border simplex there is no corresponding b_i, but there is a b_i' or a b_i'', depending on the orientation.

A chain x on K' is called a K-chain if it can be expressed in the form $x = \sum x_i a_i$, and a chain y^* in K' is said to be a K^*-chain if it is of the form $y^* = \sum y_i b_i$. Note carefully that all chains under consideration are chains on K'; we are merely attaching the name of K-chain or K^*-chain to chains on K' whose coefficients satisfy certain special conditions. As shown in Fig. 4 we may think of K^* as a "polyhedron" whose faces are barycentric stars. It is called the *dual* of K.

We begin by setting $a_i \times b_i = 1$ and $a_i \times b_j = 0$ for $i \neq j$. The intersection number of an arbitrary K-chain and an arbitrary K^*-chain can then be defined by distributivity. With the previous notations we obtain $x \times y^* = \sum x_i y_i$. The intersection number is thus determined by a bilinear form. It depends on the orientation of K, but not on the orientations of the 1-simplices a_i.

31B. The intersection number is of special interest when x and y^* are cycles, for then it depends only on the homology classes. The notation is intended to imply that x is a K-cycle and y^* a K^*-cycle. Moreover, all homologies are with respect to K'.

Lemma. *The intersection number $x \times y^*$ of two cycles does not change when x and y^* are replaced by K- and K^*-cycles which are homologous to x and y^* respectively.*

We need only show that $x \sim 0$ or $y^* \sim 0$ implies $x \times y^* = 0$. As far as x is concerned it is sufficient to consider $x = \partial s_j^2$. The condition $\partial s_j^2 \times y^* = 0$ has a simple interpretation. It means that y^* contains equally many b_i which begin and end at the barycenter of s_j^2, and it is fulfilled because y^* is a cycle. The same reasoning applies to y^*, except that s_j^2 is replaced by a barycentric star.

31C. *The intersection numbers remain unchanged if we pass to a subdivision of K.* To see this, it is sufficient to investigate the effect of an elementary subdivision. With the notations of Fig. 1, 28A, let a be the simplex $(\alpha_1\alpha_2)$. Then b leads from the barycenter of $(\alpha\alpha_4)$ to that of $(\alpha\alpha_3)$, and it is homologous to a chain which passes in turn through the barycenters of $(\alpha\alpha_4)$, $(\alpha\alpha_2\alpha_4)$, $(\alpha\alpha_2)$, $(\alpha\alpha_2\alpha_3)$, $(\alpha\alpha_3)$. This chain has intersection 1 with $a = (\alpha_1\alpha) + (\alpha\alpha_2)$, and we conclude that no change takes place.

31D. It follows by the considerations in 28B, or by a simple direct proof, that any homology class on K' contains a K-cycle. Let us prove, moreover, that every K-cycle is homologous to a K^*-cycle. To see this, let x be a given K-cycle. We can write $x = \sum A_i$ where A_i is the part of x that falls within $S(\alpha_i)$; the sum is finite. Because $\partial x = 0$ we obtain $\partial A_j = -\sum_{i \neq j} \partial A_i$, and it becomes evident that ∂A_j lies on the border of $S(\alpha_j)$. From this we conclude that A_j is homologous to a chain B_j on the border of $S(\alpha_j)$. We find the representation $x \sim \sum B_i$ where the sum on the right is a K^*-cycle.

In view of Lemma 31B this result permits us to extend the definition of the intersection number $x \times y$ to any pair of cycles on K'. In particular, it becomes possible to compare $x \times y$ and $y \times x$. In this respect we prove:

Lemma. *The intersection number is skew-symmetric: $x \times y = -y \times x$.*

We may assume that x and y are K-cycles. In order to prove the skew symmetry it is sufficient to show that all K-cycles satisfy $x \times x = 0$ for then

$0=(x+y)\times(x+y)=x\times y+y\times x$. In other words, if x is homologous to the K^*-cycle x^* we are merely required to show that $x\times x^*=0$. For this purpose we are free to choose x^* arbitrarily within its homology class. Our method will be to fix a specific x^* and then determine $x\times x^*$ by explicit computation.

We recall the notations $a_i=a_i'+a_i''$, $b_i=b_i'+b_i''$ introduced in 31A. When we compute $x\times x^*$ it will be convenient to represent x as a linear combination of the a_i and x^* as a linear combination of the b_i' and b_i''. The result will be correct if we introduce the conventional values $a_i\times b_i'=a_i\times b_i''=\frac{1}{2}$; all other intersection numbers are set equal to 0.

We use the same representation $x=\sum A_i$, $x^*=\sum B_i$ with $A_i\sim B_i$ as above, but this time the B_i will be chosen in a specific manner. It will be sufficient to show that $x\times B_i=0$ for all i.

In order to avoid double indices we let α represent any α_j. The 1-simplices of K that contain α will be denoted by $a_1, \cdots, a_n$, and we assume that this is the positive cyclic or linear order induced by the orientation of K. We may assume, without loss of generality, that α is the initial point of all the a_i. The corresponding A is of the form $A=t_1a_1'+\cdots+t_na_n'$ with $t_1+\cdots+t_n=0$. On setting $u_1=t_1+\cdots+t_i$ we get $A=u_1(a_1'-a_2')+u_2(a_2'-a_3')+\cdots+u_{n-1}(a_{n-1}'-a_n')$. But $a_i'-a_{i+1}'$ is homologous to $b_i'+b_{i+1}''$. Therefore we can choose $B=u_1(b_1'+b_2'')+u_2(b_2'+b_3'')+\cdots+u_{n-1}(b_{n+1}'+b_n'')$. In forming $x\times B$ the only nonzero contributions come from the terms t_ia_i in x. We obtain

$$\begin{aligned} x\times B &= \tfrac{1}{2}t_1u_1+\tfrac{1}{2}t_2(u_1+u_2)+\cdots+\tfrac{1}{2}t_n(u_{n-1}+u_n) \\ &= \tfrac{1}{2}u_1^2+\tfrac{1}{2}(u_2^2-u_1^2)+\cdots+\tfrac{1}{2}(u_n^2-u_{n-1}^2) \\ &= \tfrac{1}{2}u_n^2=0. \end{aligned}$$

The theorem follows.

31E. According to Lemma 31B a cycle x which is homologous to 0 is such that $x\times y=0$ for all cycles y. The same is true if x is homologous to a border cycle, for a border cycle has trivially the intersection 0 with every K^*-cycle. Finally, if K is open the same conclusion can be drawn for a cycle x which is homologous to 0 modulo the ideal boundary, for x can be replaced by a cycle which does not meet a given y^*.

We show now that the converse is also true:

Lemma. *If a cycle x is such that $x\times y=0$ for all cycles y, then $x\sim 0$ if K is closed, $x\sim 0$ modulo the border if K is finite and bordered, $x\sim 0$ modulo the ideal boundary if K is open.*

We suppose first that K is finite. Set $x=\sum \xi_i a_i$, and let β_j denote the barycenter of s_j^2. For each j there exists a K^*-chain y_j^* with $\partial y_j^*=\beta_j-\beta_1$. Although y_j^* is not unique, the intersection number $x\times y_j^*$ is completely determined, for the difference of two y_j^* that correspond to the same j is a cycle and has thus the intersection 0 with x. We set $\eta_j = x \times y_j^*$ and consider the boundary $u = \sum \eta_j \partial s_j^2$ where the s_j^2 are coherently oriented by the same orientation that was used to define the relationship between a_i and b_i (see 31A). Let a_i be an interior 1-simplex on K, and suppose that $\partial b_i = \beta_k - \beta_l$. The coefficient of a_i in u is then $\eta_k - \eta_l$. But $\eta_l^* + b_i - y_k^*$ is a cycle, and on forming the intersection with x we obtain $x \times (y^* + b_i - y_k^*) = \eta_l + \xi_i - \eta_k = 0$, or $\xi_i = \eta_k - \eta_l$. Thus a_i has the same coefficient in u as in x. We conclude that $x - u$ lies on the border, and since u is a boundary we have proved the lemma for finite K.

If K is infinite we must choose u as a finite partial sum of $\sum \eta_j \partial s_j^2$. Since the partial sum can be made to include any given finite set of 2-simplices it follows readily by the same reasoning that x is homologous to a cycle that lies outside of any finite subcomplex, and hence $x \sim 0$ modulo the ideal boundary.

31F. If a cycle x is homologous to a multiple mx', $m > 1$, then it is clear that all intersection numbers $x \times y$ are divisible by m. Again, the converse is true, and we state it as follows:

Lemma. *If a cycle x is not homologous to a genuine multiple of another cycle modulo the border or the ideal boundary, then there exists a cycle y with $x \times y = 1$.*

All intersection numbers $x \times y$ form a module, and are thus the multiples of an integer m. If $m = 0$ we have already seen that x is homologous to 0 modulo the border in the case of a finite polyhedron, and modulo the ideal boundary in the case of an open polyhedron. If $m \geq 1$ we can repeat the proof in 31E, with the difference that all coefficients are reduced to their remainders modulo m. In the finite case it follows that x is homologous modulo the border to a cycle whose coefficients are divisible by m. This contradicts the hypothesis, unless $m = 1$. Hence there exists a cycle y with $x \times y = 1$.

In the infinite case a more detailed reasoning is needed. Suppose that x is contained in a finite subcomplex K_0 of K. The complement of K_0 has a finite number of components, and we choose a fixed barycenter β_{j_n} in each component. We must again choose chains y_j^* with $y_j^* = \beta_j - \beta_1$. If β_j is contained in K_0 no other condition is imposed, and it follows that $u_j = x \times y_j^*$ is determined modulo m, where m has the same meaning as above. The chains $y_{j_n}^*$ will also be arbitrary, but for a β_j that is contained in the component of β_{j_n} we specify that y_j^* shall be the sum of $y_{j_n}^*$ and a chain which lies outside of K_0 and has the boundary $\beta_j - \beta_{j_n}$. This has the effect that u_j is constant in each component. The same reasoning as in 31E will then show that x is homologous to a cycle of the form $mx' + z$ where x' is a fixed cycle in K_0, and z can be chosen to lie outside of any finite subcomplex. In other words, $x \sim mx'$ modulo the ideal boundary, and by hypothesis this implies $m=1$. We conclude again that there exists a cycle y with $x \times y = 1$.

31G. We use the notion of intersection number to introduce what will be called a *canonical homology basis*. A finite or infinite sequence of cycles, labelled in pairs by A_i, B_i, is said to be canonical if $A_i \times A_j = B_i \times B_j = 0$, $A_i \times B_i = 1$ and $A_i \times B_j = 0$ for $i \neq j$. The elements of a canonical sequence are always linearly independent in the sense of absolute or relative homology, for from a relation $\sum (x_i A_i + y_i B_i) \sim 0$ we obtain, by forming the intersection with A_j and B_j, $y_j = 0$ and $x_j = 0$ for all j. A canonical sequence which is also a basis is called a canonical homology basis.

We treat closed polyhedrons, finite bordered polyhedrons, and open polyhedrons simultaneously. Accordingly, homologies will mean either absolute homologies or homologies relatively to the border or the ideal boundary. The case of infinite bordered polyhedrons is left aside, but it could easily be treated in the same manner.

Theorem. *Every orientable closed polyhedron has a canonical homology basis, and every orientable finite bordered or open polyhedron has a canonical homology basis modulo the border or the ideal boundary.*

We treat first the finite case. Assume, by way of induction, that we have found a canonical sequence $A_1, B_1, \cdots, A_k, B_k$; the initial step $k=0$ is included in the general reasoning. Denote by H_1 the absolute homology group or the homology group modulo the border, as the case may be, and let G be the subgroup of H_1 which is generated by the homology classes of $A_1, B_1, \cdots, A_k, B_k$. The nonzero elements of H_1/G are of infinite order. In fact, suppose that $mz \sim x_1 A_1 + y_1 B_1 + \cdots + x_k A_k + y_k B_k$. Then $x_i = m(z \times B_i)$, $y_i = -m(z \times A_i)$, and since the elements of H_1 are of infinite order it follows that $z \in G$. Furthermore, H_1/G is finitely generated. Theorem 24C permits us to conclude that it has a basis $\{\bar{z}_n\}$; the bars indicate cosets of G.

If $H_1 = G$ we have found a canonical basis. In the contrary case we choose an element z_1 of the coset $\bar{z}_1$, and construct $A_{k+1} = z_1 - \sum_{i=1}^{k} (z_1 \times B_i) A_i + \sum_{i=1}^{k} (z_1 \times A_i) B_i$. It is found that A_{k+1} has intersection 0 with $A_1, B_1, \cdots, A_k, B_k$. Moreover, A_{k+1} cannot be homologous to a genuine multiple of a cycle. Indeed, $A_{k+1} \sim my$ would imply $\bar{z}_1 = m\bar{y} = m(a_1 \bar{z}_1 + \cdots + a_r \bar{z}_r)$, and hence $ma_1 = 1$, $m = \pm 1$. By Lemma 31F we can find y_1 so that $A_{k+1} \times y_1 = 1$. On setting $B_{k+1} = y_1 - \sum_i (y_1 \times B_i) A_i + \sum_i (y_1 \times A_i) B_i$ we obtain a new canonical sequence $A_1, B_1, \cdots, A_{k+1}, B_{k+1}$.

The process breaks off in a finite number of steps, and determines a canonical basis $A_1, B_1, \cdots, A_p, B_p$. The number p is known as the *genus* of the polyhedron. For a closed polyhedron we conclude that the 1-dimentional Betti number is $2p$, and hence necessarily even. For a polyhedron with q contours the Betti number is $2p+q-1$, and it is connected with the Euler characteristic through $\rho=2p+q-2$.

By use of Theorem 30D the finite case immediately leads to a proof for the infinite case. Indeed, the generators $z_{i_1\cdots i_n, j}$ of the relative homology group of $Q_{i_1\cdots i_n}$ can be replaced by canonically intersecting generators $A_{i_1\cdots i_n, j}, B_{i_1\cdots i_n, j}$. Since the generators in different $Q_{i_1\cdots i_n}$ cannot intersect the combined system of all these generators is canonical.

We emphasize that the canonical basis $\{A_i, B_i\}$ obtained in this manner has two special properties: (1) there is a finite subsequence $A_1, B_1, \cdots, A_{k(n)}, B_{k(n)}$ which is a canonical basis of P_n, (2) the cycles A_k, B_k with $k>k(n)$ lie outside of P_n.

32. Infinite cycles

32A. For the study of open polyhedrons it is indispensable to introduce infinite chains. It presents no difficulty to consider formal sums $c=\sum x_i s_i^{(n)}$, $n=0, 1, 2$, with integral coefficients that are not subject to the earlier condition that only a finite number be different from 0. It is clear how to add such chains, and since there are only a finite number of simplices that contain a given 0- or 1-simplex it is also possible to form the boundary ∂c. We shall say that c is a *relative cycle* if and only if $\partial c=0$. According to this terminology the group of relative cycles contains the subgroup of cycles (or finite cycles).

There are two kinds of homology. We say that two relative cycles z_1 and z_2 are *weakly homologous* if the difference z_1-z_2 is the boundary of a finite or infinite chain, and they are *strongly homologous* if z_1-z_2 is the boundary of a finite chain. Since the latter condition requires z_1-z_2 to be finite, strong homology is too restrictive to be useful. The group formed by the weak homology classes is called the *weak homology group*.

We consider only orientable open polyhedrons. Simple considerations which are quite similar to the ones in 25C show that the 2-dimensional weak homology group is an infinite cyclic group. Because it is possible to construct an infinite polygon with a single end point the 0-dimensional weak homology group reduces to 0. Thus only the 1-dimensional case is of interest.

32B. We shall define the intersection number $x\times y$ of two relative cycles only if at least one is a finite cycle. For definiteness, we agree that x is always finite. The meaning of the intersection number is clear if x belongs to K and y to the dual K^*. One verifies that it depends only on the homology class of x and the weak homology class of y. The definition can therefore be extended, exactly as in 31B, to the case where x and y are both on K, or on a subdivision of K. The relation of skew symmetry, $x \times y = -y \times x$, has meaning only if x and y are both finite.

32C. Weak homology is related to the notion of dividing cycle through the following proposition:

Theorem. *A finite cycle is weakly homologous to 0 if and only if it is dividing.*

If y is weakly homologous to 0 we have $x \times y = 0$ for all finite cycles x. For finite y we conclude by Lemma 31E that y is a dividing cycle.

To prove the converse we use a canonical exhaustion $\{P_n\}$ and the notations $b_{i_1 \cdots i_n}$, $Q_{i_1 \cdots i_n}$ introduced in 30B. We recall that $b_{i_1 \cdots i_n}$ is the only common contour of $Q_{i_1 \cdots i_{n-1}}$ and $Q_{i_1 \cdots i_n}$.

If y is a dividing cycle we know by Theorem 30D that y is homologous to a finite linear combination of cycles $b_{i_1 \cdots i_n}$. But $b_{i_1 \cdots i_n}$ is the boundary of the infinite chain

$$-\sum Q_{i_1 \cdots i_n j_{n+1} \cdots j_{n+p}}$$

where the sum is over all permissible subscripts that begin with $i_1 \cdots i_n$. Hence $b_{i_1 \cdots i_n}$ is weakly homologous to 0, and the theorem is proved.

32D. The following lemma will be needed:

Lemma. *If a relative cycle z has zero intersection with all contours of a finite polyhedron $P \subset K$, then z can be written as the sum of a cycle on P and a relative cycle on $K-P$.*

We may regard z as a relative cycle on the barycentric subdivision K', and as such it is weakly homologous to a relative cycle z^* on the dual K^*. We set $z^* = z_1^* + z_2^*$ where z_1^* lies on P' (the barycentric subdivision of P) and z_2^* on $K'-P'$. The fact that z^* has zero intersection with the contours of P implies that the coefficients of ∂z_1^* which correspond to vertices on a contour have the sum 0. It follows that ∂z_1^* is the boundary of a 1-dimensional chain c on the border of P'. We obtain a representation $z^* = (z_1^* - c) + (z_2^* + c)$ by a cycle on P' and a relative cycle on $K'-P'$. But $z_1^* - c$ is homologous to a K-cycle z_1 on P, and $z_2^* + c$ is weakly homologous to a relative K-cycle z_2 on $K-P$. We have $z \sim z_1 + z_2$ on K', and because all three cycles belong to K this weak homology relation is also true on K. Hence $z = z_1 + z_2 + \partial a$ where a is a 2-dimensional chain on K. On setting $a = a_1 + a_2$ where a_1 is the part on P we find that $z = (z_1 + \partial a_1) + (z_2 + \partial a_2)$. This representation satisfies the requirements.

32E. It is often possible to form infinite sums of chains. Indeed, if a sequence $\{c_i\}$ of 1-dimensional finite or infinite chains has the property that each 1-simplex has a nonzero coefficient in at most finitely many c_i, then it is clear how to interpret an equation $c = \sum c_i$. We say, when this equation is meaningful, that c is represented as a *convergent sum*.

Theorem. *A relative cycle z has a representation $\sum z_i$ as a convergent sum of finite cycles if and only if z has intersection number* 0 *with all finite dividing cycles.*

The necessity is almost trivial. Indeed, if $z = \sum z_i$ and b is any finite cycle we obtain $b \times z = \sum (b \times z_i)$, where only a finite number of terms $b \times z_i$ can be different from 0. If b is dividing all $b \times z_i$ vanish, and we find that $b \times z = 0$.

To prove the sufficiency we use again a canonical exhaustion $\{P_n\}$ and the notations $b_{i_1 \cdots i_n}$, $Q_{i_1 \cdots i_n}$. If we assume that z has zero intersection with all dividing cycles it follows by Lemma 32D that there exists a decomposition $z = x_{i_1 \cdots i_n} + y_{i_1 \cdots i_n}$ where $x_{i_1 \cdots i_n}$ is a cycle on $Q_{i_1 \cdots i_n}$ and $y_{i_1 \cdots i_n}$ a relative cycle on $K - Q_{i_1 \cdots i_n}$.

We form $\sum x_{i_1 \cdots i_n}$ where the sum is over all subscripts for which $Q_{i_1 \cdots i_n}$ is defined; this includes the empty subscript which, by convention, corresponds to $Q = P_1$. The difference $z - \sum x_{i_1 \cdots i_n}$ lies in the complement of each $Q_{i_1 \cdots i_n}$, and is therefore a sum of multiples of the contours $b_{i_1 \cdots i_n}$. We obtain a representation of z as a sum of cycles $x_{i_1 \cdots i_n}$ and multiples of $b_{i_1 \cdots i_n}$ which is obviously convergent.

32F. The preceding theorem shows that a relative cycle z is determined up to a convergent sum of finite cycles by its intersection numbers with all dividing cycles, and hence by the intersection numbers $b_{i_1 \cdots i_n} \times z$, $i_n > 1$, for a given canonical exhaustion $\{P_n\}$. We show now that these intersection numbers can be chosen arbitrarily.

Consider a vertex on $b_{i_1 \cdots i_n}$, $i_n > 1$. It can be joined by a polygon in $Q_{i_1 \cdots i_n}$ to a vertex on $b_{i_1 \cdots i_n 1}$. This vertex can in turn be joined to a vertex on $b_{i_1 \cdots i_n 11}$ by a polygon in $Q_{i_1 \cdots i_n 1}$, and so on. In the opposite direction we can pass from the vertex on $b_{i_1 \cdots i_n}$ through $Q_{i_1 \cdots i_{n-1}}$ to a vertex on $b_{i_1 \cdots i_{n-1} 1}$, then through $Q_{i_1 \cdots i_{n-1} 1}$ to a vertex on $b_{i_1 \cdots i_{n-1} 11}$, etc. The sum of the polygons between consecutive vertices is a relative cycle which we denote by $c_{i_1 \cdots i_n}$. Its direction can be fixed so that $b_{i_1 \cdots i_n} \times c_{i_1 \cdots i_n} = 1$. Then $b_{i_1 \cdots i_n 1 \cdots 1} \times c_{i_1 \cdots i_n} = 1$, $b_{i_1 \cdots i_{n-1} 1 \cdots 1} \times c_{i_1 \cdots i_n} = -1$, and all other intersection numbers $b_{j_1 \cdots j_r} \times c_{i_1 \cdots i_n}$ are 0.

With arbitrary integers $t_{i_1 \cdots i_n}$ we form

$$z = \sum_{i_n > 1} t_{i_1 \cdots i_n} c_{i_1 \cdots i_n}.$$

The sum is obviously convergent, and $b_{i_1 \cdots i_n} \times z = t_{i_1 \cdots i_n}$, $i_n > 1$. The most general relative cycle with these intersection numbers is of the form $z + y$ where y is a convergent sum of finite cycles.

32G. If we are interested merely in weak homology a convergent sum of finite cycles can of course be expressed through the canonical generators A_i, B_i introduced in 31G. The theorem that we obtain is an analog of Theorem 27D.

Theorem. *The weak homology classes are represented, in a unique manner, by infinite linear combinations of the cycles A_i, B_i and the relative cycles $c_{i_1 \cdots i_n}$, $i_n > 1$.*

Because we are allowing infinite sums the generating elements do not form a basis in the conventional sense.

§5. SINGULAR HOMOLOGY

The simplicial homology theory has two weaknesses. In the first place, it applies only to polyhedrons, and thus to surfaces which permit a triangulation. Secondly, it is far from evident that two triangulations of the same surface have isomorphic homology groups. On the other hand, it has the great advantage that the homology groups can be computed explicitly.

In this section we introduce the notion of singular homology. It is in some ways more elementary than simplicial homology, and applies with equal facility to any surface or bordered surface. For this reason we shall make a point of using singular homology whenever possible. It is difficult, however, to gain detailed information about the singular homology group, except in the presence of a triangulation. For polyhedrons it is shown that the singular and simplicial homology groups are isomorphic. Since the singular homology groups are defined in an invariant way with respect to homeomorphisms it is an automatic consequence that two triangulations of the same or topologically equivalent surfaces have isomorphic homology groups.

33. Definitions and relation to the fundamental group

33A. In what follows F will denote a surface or a bordered surface. The earlier practice of using a special notation $\bar{F}$ for a bordered surface is temporarily abandoned.

In 7A an arc on F was defined as a continuous mapping $f: t \to f(t)$ of the closed unit interval $0 \leqq t \leqq 1$ into F. In the present connection an arc will be called a *singular* 1-*simplex*. Note that different parametrizations of the same arc are regarded as different singular 1-simplices.

By analogy, a *singular 2-simplex* will be defined as a continuous mapping of a fixed geometric 2-simplex into F. For definiteness, we let the basic 2-simplex Δ be defined as the set of all (t, u) with $0 \leq u \leq t \leq 1$. A singular 2-simplex is thus defined by a mapping $f:(t, u) \rightarrow f(t, u)$ of Δ into F. Finally, a *singular 0-simplex* is just a point on F. To guard against misunderstandings, we emphasize that the singular 2-simplex must not be confused with the image $f(\Delta)$.

We use the generic notation σ^n for singular n-simplices. A singular n-chain is a formal sum $\sum x_i \sigma_i^n$ with a finite number of nonzero integral coefficients. In other words, the group $C_n(F)$ of singular n-chains is defined as the free Abelian group generated by the singular n-simplices on F. The notations $C_n(F)$ and $C_n(K)$ are unlikely to cause confusion since F refers to a surface and K to a complex.

33B. The next step is to define the boundary of a singular simplex. The boundary of a singular 1-simplex, defined by f, is the 0-chain $f(1) - f(0)$. The definition is extended to 1-chains by setting $\partial(\sum x_i \sigma_i^1) = \sum x_i \, \partial \sigma_i^1$, and it becomes clear what is meant by a singular 1-cycle. Clearly, the notion of singular 1-cycle is a direct generalization of the notion of closed curve.

The boundary of a 2-simplex σ^2, given by the mapping f of Δ, is defined as follows. Let e_1 denote the mapping $t \rightarrow (t, 0)$, e_2 the mapping $t \rightarrow (1, 1-t)$, and e_3 the mapping $t \rightarrow (1-t, 1-t)$. Let σ_i^1, $i = 1, 2, 3$, be the 1-simplices defined by the mappings $f \circ e_i$. With these notations we set $\partial \sigma^2 = \sigma_1^1 - \sigma_2^1 + \sigma_3^1$. The operation ∂ is extended to arbitrary 2-chains by linearity. Simple verification shows that $\partial \partial \sigma^2 = 0$, and hence that every boundary is a cycle.

The groups $Z_1(F)$ and $B_1(F)$ of singular cycles and boundaries are now well defined, and we can introduce the singular homology group $H_1(F) = Z_1(F)/B_1(F)$. The 0- and 2-dimensional singular homology groups are of no interest to us and will not be considered.

33C. Contrary to the procedure in 23B, a singular chain $-\sigma^n$, $n = 0, 1, 2$, is never identified with a singular simplex. In this way we avoid all difficulties that would arise whenever $-\sigma^n$ would be identified with σ^n. Actually, no identifications are needed as long as we are interested only in homologies.

It is important, however, to point out certain standard homologies which are used all the time. In the first place, let σ^1 be a point-arc, defined by a constant f. We denote by σ^2 the singular 2-simplex that is defined by means of the same constant. Applying the definition of boundary we find that $\partial \sigma^2 = \sigma^1 - \sigma^1 + \sigma^1 = \sigma^1$, and hence every point-arc is homologous to 0. Secondly, let γ and γ^{-1} be opposite arcs, defined by $t \rightarrow f(t)$ and $t \rightarrow f(1-t)$.

We consider the 2-simplex given by $(t, u) \to f(t)$. Its boundary is seen to be $\gamma - f(1) + \gamma^{-1}$, where $f(1)$ is a point-arc. Hence $\gamma + \gamma^{-1} \sim 0$.

The last relation shows that if σ^2 is defined by a mapping f of Δ, then $\partial\sigma^2$ is obtained, within homology, by restricting f to the positively oriented sides of Δ, each side being referred to the unit interval by means of a linear mapping. This interpretation should be thought of as underlying the notion of singular homology.

33D. It is evident from the definition that the singular homology group is a topological invariant. This fact is further emphasized by proving the following connection between the fundamental group and the singular homology group.

Theorem. *The singular homology group $H_1(F)$ is isomorphic with the quotient group $\mathscr{F}/\mathscr{F}^{(1)}$ of the fundamental group $\mathscr{F}(F)$ with respect to its commutator subgroup* (see 19D).

We represent $\mathscr{F}$ as $\mathscr{F}_O$ with a fixed origin. Each closed curve from O can be considered as a singular 1-simplex which is a cycle. Homotop curves determine homologous 1-simplices; this fact needs verification, but the verification is quite automatic and will be omitted. A similar verification shows that $\alpha\beta$ corresponds to a 1-simplex which is homologous to $\alpha + \beta$. Therefore, the correspondence defines a homomorphic mapping of $\mathscr{F}_O$ into $H_1(F)$.

The kernel of the homomorphism contains the commutator subgroup, for $\alpha\beta\alpha^{-1}\beta^{-1}$ corresponds to $\alpha + \beta + \alpha^{-1} + \beta^{-1}$ which we know is homologous to 0. To prove, conversely, that the kernel is contained in the commutator subgroup, let γ be a closed curve from O, and suppose that $\gamma \sim 0$ when regarded as a singular cycle. We set $\gamma = \sum x_i \, \partial\sigma_i^2$ and write $\partial\sigma_i^2 = \alpha_{i1} - \alpha_{i2} + \alpha_{i3}$ where the α_{ij} are 1-simplices. The relation $\gamma = \sum x_i(\alpha_{i1} - \alpha_{i2} + \alpha_{i3})$ must be a formal identity. Hence all terms in the sum must cancel except for a single remaining term α_{rs} which is identical with γ.

To continue, set $\partial\alpha_{ij} = \epsilon_{ij2} - \epsilon_{ij1}$. For each ϵ_{ijk}, let η_{ijk} be an arc which joins O to ϵ_{ijk}, and suppose that we choose identical arcs for coinciding points. The arc $\beta_{ij} = \eta_{ij1}\alpha_{ij}\eta_{ij2}^{-1}$ is a closed curve from the origin. Moreover, one proves without difficulty that $\beta_{i1}\beta_{i2}^{-1}\beta_{i3}$ is homotop to 1. We consider now the product $\Pi(\beta_{i1}\beta_{i2}^{-1}\beta_{i3})^{x_i}$, where the order of the factors is immaterial. On one hand it is homotop to 1. On the other hand, if we let the factors β_{ij} commute they must cancel against each other in exactly the same way as the terms in $\sum x_i(\alpha_{i1} - \alpha_{i2} + \alpha_{i3})$. It follows that the commutator subgroup contains $\beta_{rs} - \eta_{rs1}\gamma\eta_{rs2}^{-1}$. Here η_{rs1} and η_{rs2} are identical closed curves from O, and we conclude that γ is itself in the commutator subgroup. We have thus proved that the kernel is identical with $\mathscr{F}^{(1)}$.

It remains to prove that the image is all of $H_1(F)$. Let $x = \sum x_i \sigma_i^1$ be a singular cycle. We replace σ_i^1 by $\beta_i = \eta_{i1} + \sigma_i^1 - \eta_{i2}$, where η_{i1} and η_{i2} are arcs from O to the initial and terminal points of σ_i^1. These arcs are to be chosen so that they depend only on the end point. Since x is a cycle, it follows that $x = \sum x_i \beta_i$. Each β_i may be regarded as a closed curve from O, and we see that x is the image of $\Pi \beta_i^{x_i}$.

34. Simplicial approximation

34A. We suppose now that F permits a triangulation. In other words, we take F to be a geometric polyhedron K_g, and prove:

Theorem. *The singular homology group $H_1(K_g)$ is isomorphic with $H_1(K)$.*

Each oriented 1-simplex on K can be considered as a singular 1-simplex arising from a linear mapping of the unit interval on the corresponding geometric 1-simplex. Cycles correspond to cycles, and a simplex boundary ∂s^2 becomes, within homology, a singular boundary $\partial \sigma^2$, where σ^2 is defined by an affine mapping of Δ. We obtain in this way a homomorphic mapping i of $H_1(K)$ into $H_1(K_g)$.

To obtain a mapping in the opposite direction we are going to use the

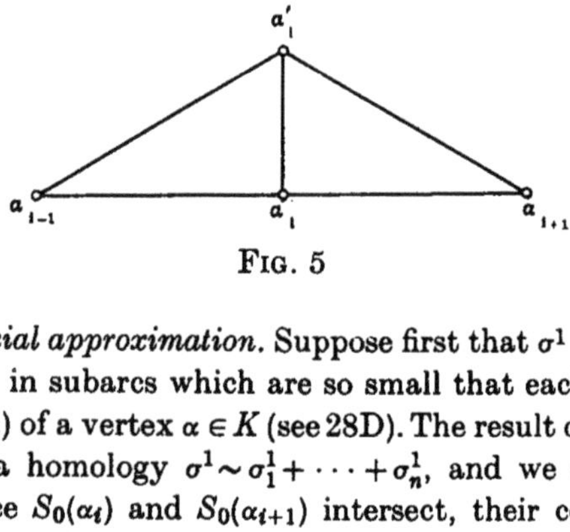

FIG. 5

method of *simplicial approximation*. Suppose first that σ^1 is a closed curve. We can divide σ^1 in subarcs which are so small that each is contained in the open star $S_0(\alpha)$ of a vertex $\alpha \in K$ (see 28D). The result of the subdivision is expressed by a homology $\sigma^1 \sim \sigma_1^1 + \cdots + \sigma_n^1$, and we suppose that σ_i^1 lies in $S_0(\alpha_i)$. Since $S_0(\alpha_i)$ and $S_0(\alpha_{i+1})$ intersect, their centers are either identical or they are the end points of a 1-simplex $(\alpha_i \alpha_{i+1})$. We assign to σ^1 the cycle $h(\sigma^1)$ formed by all $(\alpha_i \alpha_{i+1})$ with $\alpha_i \neq \alpha_{i+1}$. If we agree that $(\alpha_i \alpha_{i+1}) = 0$ if $\alpha_i = \alpha_{i+1}$ we can write $h(\sigma^1) = (\alpha_0 \alpha_1) + (\alpha_1 \alpha_2) + \cdots + (\alpha_{n-1} \alpha_n)$ with $\alpha_n = \alpha_0$. We call $h(\sigma^1)$ a simplicial approximation of σ^1.

The construction is not unique, but we can show that the homology class of $h(\sigma^1)$ is completely determined. The first ambiguity lies in the choice of α_i. It is sufficient to consider the replacement of one α_i by another vertex α'_i. Since $S_0(\alpha_{i-1})$, $S_0(\alpha_i)$ and $S_0(\alpha'_i)$ intersect, α_{i-1}, α_i and α'_i must belong to a common 2-simplex, and the same is true of α_{i+1}, α_i, α'_i. The nondegenerate case is shown in Fig. 5, and we see at once that $(\alpha_{i-1}\alpha_i) + (\alpha_i \alpha_{i+1}) \sim (\alpha_{i-1}\alpha'_i) + (\alpha'_i \alpha_{i+1})$. In the case of coinciding vertices the verification is even simpler.

The second source of ambiguity is in the choice of subdivision. Two subdivisions can be superimposed, and for this reason it is sufficient to consider the case of a subarc σ_i^1 being divided in two. The smaller arcs are both contained in $S_0(\alpha_i)$, and we can choose identically the same simplicial approximation as before. It follows that any two subdivisions yield homologous simplicial approximations.

34B. The construction can be generalized to an arbitrary singular cycle $x = \sum x_i \sigma_i^1$. We subdivide each arc σ_i^1 and approximate it in the manner that has just been described. The only precaution that we must take is that a fixed vertex shall be assigned to each end point, the same for all arcs that begin or end at that point. When this is done the simplicial approximation will be a cycle $h(x)$, and the homology class of $h(x)$ is uniquely determined.

We must prove that $x \sim y$ implies $h(x) \sim h(y)$. For the proof we need only show that $h(\partial\sigma^2) \sim 0$. This is obvious if σ^2 is contained in a star. In the general case we apply repeated barycentric subdivisions to Δ, until the image of each subsimplex is contained in a star. If we examine one elementary subdivision at a time it becomes evident that there is a natural way to attach a singular simplex σ_i^2 to each 2-dimensional subsimplex of Δ. Moreover, explicit verification in the case of an elementary subdivision shows that $\sum \partial\sigma_i^2$ will represent a subdivision of $\partial\sigma^2$. It follows that $h(\partial\sigma^2) \sim 0$ as asserted.

34C. We have found that h may be interpreted as a homomorphic mapping of $H_1(K_g)$ into $H_1(K)$. Trivial considerations show that a cycle on K may be regarded as its own simplicial approximation. In other words, $h \circ i$ is the identity. In order to conclude that $i \circ h$ is also the identity we must show that σ^1 is homologous to $h(\sigma^1)$ when the latter is regarded as a singular cycle.

The case of a closed curve σ^1 is sufficiently general to illustrate the reasoning. We can draw straight line segments from the initial point of the subarc σ_i^1 to the vertices α_{i-1} and α_i, as shown in Fig. 6. It is possible

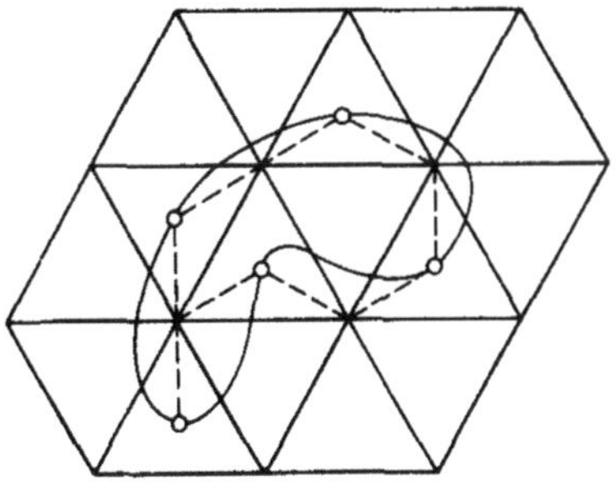

FIG. 6

to define a singular 2-simplex whose boundary consists of σ_i^1 and the line segments from a_i to its end points, taken with proper orientations and signs. In fact, if σ_i^1 is defined by $t \to f_i(t)$, $t \in [0, 1]$, we map Δ by setting $f(t, 0) = f_i(t)$, $f(1, 1) = a_i$ and declaring that the mapping shall be linear on every straight line through the vertex $(1, 1)$. On the other hand, the triangle whose vertices are at a_{i-1}, a_i and the initial point of σ_i^1 can also be regarded as a singular simplex obtained by an affine mapping of Δ. On forming the boundaries the auxiliary line segments drop out, and we find that $\sigma^1 \sim h(\sigma^1)$. Thus h is indeed the inverse of i, and we have proved that $H_1(K_g)$ and $H_1(K)$ are isomorphic.

As a corollary, topologically equivalent polyhedrons have isomorphic homology groups. We have actually proved a little more, for we have shown that a given homeomorphism determines, by the method of simplicial approximation, a definite isomorphic correspondence between the two homology groups.

35. Relative singular homology

35A. We discuss the topics connected with relative singular homology only for the case of an open orientable surface F. The definitions are similar to the ones for open polyhedrons, except that finite subcomplexes are replaced by compact sets.

Accordingly, a singular cycle is said to be a dividing cycle, or homologous to 0 modulo the ideal boundary, if it is homologous to a singular cycle that lies outside of any given compact set. The group formed by the homology classes of dividing singular cycles is denoted by $H_1\beta(F)$. If F is an open polyhedron with triangulation K it is shown that $H_1\beta(F)$ is isomorphic with $H_1\beta(K)$.

Finite polyhedrons are replaced by regular regions (see 13G). We say that F has an exhaustion if every compact set is contained in a regular region. If F is countable this property can be used to construct an exhausting sequence of regular regions. However, there is no need to use such sequences. It follows from the existence of an exhaustion that the union of any two regular regions is contained in a third. It is this property that is needed in practically all applications. We express it by saying that the regular regions form a *directed set* under inclusion.

A regular region is called *canonical* if each component of the complement has a single contour. We speak of a canonical exhaustion in the corresponding sense.

35B. We assume now that F is a polyhedron. This means that F has a triangulation by means of a complex K. It will be shown in 46H that it is possible to choose K so that a given regular region on F is triangulated by a subcomplex of K. This subcomplex is then automatically a finite bordered polyhedron, and since we assume F to be orientable it is also orientable.

This result will be anticipated and used for proving:

Theorem. *An open orientable polyhedron F has the following properties:*

(a) *The positively oriented boundary of a regular region, considered as a singular cycle, is homologous to* 0.

(b) *F has a canonical exhaustion.*

(c) *If Ω is a canonical region on F, then a singular cycle on Ω is homologous to* 0 *modulo the ideal boundary if and only if it is homologous to a finite linear combination of the contours of Ω, considered as singular cycles.*

Essentially, point (a) has been taken for granted. Point (b) follows by the fact that a canonical polyhedron is a canonical region. As for (c), if we choose K so that $\bar{\Omega}$ is a canonical polyhedron it is immediate, by the equivalence of singular and simplicial homology, that each contour of Ω is dividing in the sense of singular homology. Suppose now that a singular cycle γ in Ω is homologous to a singular cycle γ' that lies outside of a large polyhedron. We replace γ and γ' by their simplicial approximations in a sufficiently fine subdivision. These approximations are simplicially homologous, and it follows that γ is homologous to a cycle on the boundary of Ω.

§6. COMPACTIFICATION

In the Alexandroff compactification (4F) a single point represents the whole ideal boundary. On the other hand, for a bordered surface we obtained a realization of the ideal boundary which associates a continuum of points with each contour (13C). For many purposes an intermediate approach is the most adequate. In this section we introduce a realization of the ideal boundary which in the case of a bordered surface represents all contours as separate single points. The ideal points will thus generalize the notion of contour.

36. Boundary components

36A. We say that a surface F is imbedded in a topological space M if there is given a topological mapping of F into M. We may, and this is frequently convenient, identify F with its topological image, so that F appears as a subset of M with the induced topology. The imbedding is called a *compactification* if (1) M is compact, (2) F (or its image) is dense in M, that is, $\bar{F}=M$, (3) F is an open subset of M. We shall regard $\beta=M-F$ as a realization of the ideal boundary. The points of β are called ideal boundary points of F.

Two compactifications M and M', both containing F, are equivalent if there exists a topological mapping of M onto M' which reduces to the identity on F. Equivalent compactifications can be considered as identical, and all uniqueness should be interpreted in this sense.

36B. There are, of course, many different compactifications, and the choice of a compactification depends on the use that will be made of it. As already indicated, we wish to construct a compactification that serves to generalize the notion of contour. With a suitable formulation of the desired properties it turns out that there is one and only one compactification which meets the requirements.

To simplify the formulation we introduce the following auxiliary definition:

Definition. *A set $\beta \subset M$ is said to be nonseparating on M if every region $G \subset M$ remains connected after removal of all points in $G \cap \beta$.*

We state the basic existence and uniqueness theorem:

Theorem. *There exists a unique compactification M with the following properties:*

(B1) *M is a locally connected Hausdorff space.*

(B2) *β is totally disconnected.*

(B3) *β is nonseparating on M.*

The proof will be based on an explicit construction of the set β and a topology on $M = F \cup \beta$. The points of β are called *boundary components*. In case F is an open geometric polyhedron K_g it will be shown that the boundary components of K_g can be identified with the boundary components of K, the latter being defined as in 30D.

36C. For convenience we agree to denote by Q, with or without subscripts, any region on F which is not relatively compact on F, but has a compact boundary. Thus, whenever we speak of a set Q we shall always mean a region with these characteristic properties. Note that Q is never void.

We choose a formal definition which identifies each boundary component with a set of subsets of F:

Definition. *A boundary component is a nonvoid collection q of sets Q which satisfies these conditions:*

(C1) *If $Q_0 \in q$ and $Q \supset Q_0$, then $Q \in q$.*

(C2) *If $Q_1, Q_2 \in q$ there exists a $Q_3 \subset Q_1 \cap Q_2$ which belongs to q.*

(C3) *The intersection of all closures $\bar{Q}$, $Q \in q$, is empty.*

It is an easy consequence of (C2) and (C3) that there exists a $Q \in q$ which fails to meet any given compact set. Indeed, for each point of the given set there exists, by (C3), a $Q \in q$ which does not meet a certain neighborhood of that point. The compact set can be covered by a finite number of these neighborhoods, and is hence disjoint from a finite intersection $Q_1 \cap \cdots \cap Q_N$ of sets in q. By (C2) this intersection contains a $Q \in q$.

36D. Let β be the set of all boundary components of F. For a given Q we denote by $\beta(Q)$ the set of all $q \in \beta$ such that $Q \in q$; in other words, the statements $q \in \beta(Q)$ and $Q \in q$ shall be equivalent. We remark that $Q_1 \cap Q_2 = 0$ implies $\beta(Q_1) \cap \beta(Q_2) = 0$. Indeed $q \in \beta(Q_1) \cap \beta(Q_2)$ would mean that $Q_1, Q_2 \in q$, and because of (C2) the sets Q_1 and Q_2 could not be disjoint.

To extend the topology of F we agree that all sets $Q \cup \beta(Q)$ shall be open. More precisely, we introduce a basis for the open sets on M which consists of all open subsets of F and all sets of the form $Q \cup \beta(Q)$. It must be verified that the intersection of any two such sets is a union of sets in the basis. The only case that is not completely trivial occurs for a point $q \in \beta(Q_1) \cap \beta(Q_2)$. By (C2) there exists a $Q_3 \subset Q_1 \cap Q_2$ such that $q \in \beta(Q_3)$, and (C1) implies that $\beta(Q_3) \subset \beta(Q_1) \cap \beta(Q_2)$. Thus $q \in Q_3 \cup \beta(Q_3) \subset [Q_1 \cup \beta(Q_1)] \cap [Q_2 \cup \beta(Q_2)]$ which is precisely what we need to know.

Let us show that M is a Hausdorff space. For $p_1, p_2 \in F$ the existence of disjoint neighborhoods is clear because F is a Hausdorff space, and for $p \in F$, $q \in \beta$ disjoint neighborhoods exist because of (C3). Suppose now that q_1, q_2 are distinct boundary components. There exists, for instance, a $Q_1 \in q_1$ that does not belong to q_2. We can find a $Q_2 \in q_2$ that does not meet the boundary of Q_1. Since Q_1 is connected, either $Q_2 \subset Q_1$ or $Q_1 \cap Q_2 = 0$. The first alternative would imply $Q_1 \in q_2$, by (C1), which is contrary to the assumption. Hence Q_1, Q_2 are disjoint, and the same is true of $Q_1 \cup \beta(Q_1)$ and $Q_2 \cup \beta(Q_2)$. Also, M is locally connected, for each $Q \cup \beta(Q)$, being contained between the connected set Q and its closure, is itself connected.

F is open on M, by definition, and $\bar{F} = M$, for every open set on M contains points of F. We prove next that $\beta(Q)$, besides being relatively open with respect to β, is also closed. Assume that q_0 is not in $\beta(Q)$ and determine a $Q_0 \in q_0$ which does not meet the boundary of Q. It follows by the same reasoning as above that Q and Q_0 are disjoint, and hence that $Q_0 \cup \beta(Q_0)$ does not meet $\beta(Q)$. Consequently, $\beta(Q)$ is closed. We can now conclude that β is totally disconnected. In fact, if q_1 and q_2 are distinct points of β there exists a $\beta(Q)$ which contains q_1 but not q_2. A component of β that contains q_1 and q_2 would have to meet the relative boundary with respect to β of $\beta(Q)$. But $\beta(Q)$ has an empty relative boundary. Hence β is totally disconnected.

To prove (B3), let G be any region on M, and suppose that $G \cap F = O_1 \cup O_2$ is a decomposition into open sets. For each $q \in G \cap \beta$ there exists a $Q \in q$ which is contained in G. But every $Q \subset G$ is contained in O_1 or O_2. The union of O_1 and all sets $Q \cup \beta(Q)$ such that $Q \subset O_1$ is open, and so is the union of O_2 and all $Q \cup \beta(Q)$, $Q \subset O_2$. The two unions would constitute a decomposition of G into open sets. Hence either O_1 or O_2 is void, and $G \cap F$ is connected.

37. The compactness proof

37A. It remains to show that M is compact. In the most general case the proof uses the uncountable version of the axiom of choice. For this reason we give a separate and perhaps more perspicuous proof in the case that F is an open geometric polyhedron K_g. At the same time this separate treatment serves the purpose of showing that the two notions of boundary component that we have introduced are essentially identical (see the remark at the end of 36B).

We recall that K (the underlying complex of K_g), or at least a sufficiently fine barycentric subdivision of K, has a canonical exhaustion $\{P_n\}$. We assume that K itself can be so exhausted. We use again the notations $b_{i_1\cdots i_n}$ for the contours of P_n, and we let $R_{i_1\cdots i_n}$ be the component of $K-P_n$ with the contour $b_{i_1\cdots i_n}$ (in other words, $R_{i_1\cdots i_n}$ contains $Q_{i_1\cdots i_n}$). For simplicity we agree, in this connection, that $R_{i_1\cdots i_n}$ will actually stand for the geometric polyhedron $(R_{i_1\cdots i_n})_g$. Also, we shall write F for K_g.

Suppose that M is covered by open sets O_α. If M were not compact there would exist a component R_{i_1} which does not have a finite subcovering. Next we could find an $R_{i_1 i_2} \subset R_{i_1}$ with the same property, and so on, until we obtain a sequence $R_{i_1} \supset R_{i_1 i_2} \supset \cdots \supset R_{i_1\cdots i_n} \supset \cdots$ of components that have no finite subcovering. This sequence defines a boundary component (in the sense of Definition 36C), namely the collection q of all sets Q which contain an $R_{i_1\cdots i_n}$ from the sequence. Indeed, conditions (C1)–(C3) are all easily seen to be satisfied. By assumption, q lies in an O_α, and this O_α contains a $Q \in q$. There is an $R_{i_1\cdots i_n} \subset Q$, and this $R_{i_1\cdots i_n}$ would have a subcovering which consists of a single set, contrary to the construction. The contradiction proves that M is compact.

In the course of the reasoning we have associated a q with every sequence $R_{i_1} \supset R_{i_1 i_2} \supset \cdots \supset R_{i_1\cdots i_n} \supset \cdots$. Conversely, if q is given there exists, first of all, a $Q_1 \in q$ which does not meet P_1. Because Q_1 is connected it must be contained in an R_{i_1}, and it follows by (C1) that $R_{i_1} \in q$. Next there exists a $Q_2 \in q$ which does not meet P_2. Because Q_2 must meet Q_1 it is necessarily contained in an $R_{i_1 i_2}$, and we find as before that $R_{i_1 i_2} \in q$. In this way we obtain an infinite nested sequence of components $R_{i_1\cdots i_n} \in q$. For an arbitrary $Q \in q$ it happens for a certain n that $R_{i_1\cdots i_n}$ does not meet the boundary of Q. This is possible only so that $R_{i_1\cdots i_n} \subset Q$. We have shown that the correspondence between boundary components q and nested sequences $\{R_{i_1\cdots i_n}\}$ is one to one. But the nested sequences are nothing else than the boundary components according to our earlier definition (30D), and we have thus shown that the two definitions are equivalent.

In view of this equivalence Definition 36C may seem like a luxury. It should not be forgotten, however, that it is the equivalence of the two definitions which shows at a glance that the definition by sequences is independent of the triangulation, and that Definition 36C does not even require the existence of a triangulation.

37B. We shall now give the general proof. It may be omitted by readers who are not interested in extreme generality.

A family of sets is said to be *totally ordered*, by inclusion, if for every two sets in the family one is contained in the other. We shall use *Zorn's lemma*, in the following form:

If a family of sets contains the union of the sets in any totally ordered subfamily, then it contains a maximal set, i.e., one that is not properly contained in any other set in the family.

Let Φ_0 be a collection of closed sets on M with the finite intersection property, that is, any finite number of sets in Φ_0 have a common point. Consider the family of all collections Φ of closed sets which contain Φ_0 and retain the finite intersection property. This family satisfies the hypothesis of Zorn's lemma, for if we choose a finite number of closed sets from the union of the collections Φ in a totally ordered subfamily, then, by the total ordering, these sets are already contained in a single Φ, and hence they have a common point. The lemma is applicable, and we conclude that there exists a collection Φ_1 which cannot be enlarged without violating the finite intersection property. If we prove that all sets in Φ_1 have a common point, then the same is true of Φ_0, and it follows by Theorem 4B that M is compact.

If one of the sets $C_0 \in \Phi_1$ is known to be compact there is nothing left to prove, for then the sets $C_0 \cap C$, $C \in \Phi_1$, have the finite intersection property and would have a common point. We may therefore assume that no set $C \in \Phi_1$ is a compact subset of F. We remark further that by the maximality of Φ_1 the intersection of any finite number of sets in Φ_1 must belong to Φ_1, and that a closed set which intersects all $C \in \Phi_1$ must itself be a member of Φ_1.

Let us define q as the collection of all Q such that $Q \cup \beta(Q)$ contains a $C \in \Phi_1$. If we can prove that q is a boundary component, then it is a common point of all C. Indeed, if q were not in $C' \in \Phi_1$ there would exist a neighborhood $Q \cup \beta(Q)$ which does not meet C' and hence cannot contain any $C \in \Phi_1$.

Condition (C1) is trivially fulfilled. In order to prove (C2), suppose that $C_1, C_2 \in \Phi_1$ and $C_1 \subset Q_1 \cup \beta(Q_1)$, $C_2 \subset Q_2 \cup \beta(Q_2)$. The boundary of $Q_1 \cap Q_2$ with respect to F is a compact set. It can therefore be enclosed in a regular subregion Ω of F. The complement $F - \Omega$ has only a finite number of components. Those that are compact may be added to Ω. We can therefore suppose that the components of $F-\bar{\Omega}$, denoted by $Q_1', \cdots, Q_n'$, are not relatively compact on F. It is clear that $\beta = \cup \beta(Q_i')$. Indeed, if $q_0 \in \beta$, there exists a $Q_0 \in q_0$ which does not meet the boundary of Ω. It follows that Q_0 is contained in a Q_i', and hence that $Q_i' \in q_0$, or $q_0 \in \beta(Q_i')$.

Suppose that Q_i' does not belong to q. Then, by the definition of q, $M-[Q_i' \cup \beta(Q_i')]$ meets all $C \in \Phi_1$ and is therefore, by an earlier remark, contained in Φ_1. If this were true for $i=1, \cdots, n$ the intersection of all the sets $M-[Q_i' \cup \beta(Q_i')]$, i.e., the set $M-\cup Q_i' - \cup \beta(Q_i') = F - \cup Q_i' = \bar{\Omega}$, would also be in Φ_1. But $\bar{\Omega}$ is a compact set on F, and we have assumed that there are no such sets in F. Hence there is a $Q_i' \in q$. Because Q_i' is connected and does not meet the boundary of $Q_1 \cap Q_2$ it is either contained in or disjoint from $Q_1 \cap Q_2$. The latter is impossible, for then a set $C' \subset Q_i' \cup \beta(Q_i')$ could not meet $C_1 \cap C_2$. Hence $Q_i' \subset Q_1 \cap Q_2$, and we have proved (C2).

The same reasoning shows that condition (C3) is satisfied. Indeed, we can choose Ω so that it contains a point given in advance, and we have constructed a $Q_i' \in q$ whose closure does not meet Ω. In other words, the intersection of all closures $\bar{Q}$ for $Q \in q$ is empty.

We have proved that M is compact. The compactness implies, in particular, that β is not empty.

Remark. Boundary components for open surfaces were introduced by B. Kerékjártó [1] and studied in greater detail by S. Stoïlow [14]. Our treatment is more in line with that of H. Freudenthal [1] who showed, in the countable case, that boundary components can be defined for any locally compact and locally connected Hausdorff space.

37C. The uniqueness assertion in Theorem 36B is an essential feature of the theory of boundary components. To prove it, let M_0 be another compactification of F which satisfies (B1)–(B3). We have to construct a topological mapping of M_0 onto M which is the identity on F.

In the following we use bars to indicate closure with respect to M_0. Similarly, the boundary of a set will always be the boundary on M_0. We write $\beta_0 = M_0 - F$.

Lemma. *Every point $q_0 \in \beta_0$ is contained in an arbitrarily small region V whose boundary lies in F.*

Let U be any open neighborhood of q_0. Because β_0 is totally disconnected it is possible, by Theorem 4G and Lemma 4G, to separate q_0 from $\beta_0 \cap (M_0 - U)$. In other words, there exists a decomposition $\beta_0 = A \cup B$ where A and B are closed, disjoint, and such that $q_0 \in A$, $\beta_0 \cap (M_0 - U) \subset B$. The sets A and $B \cup (M_0 - U)$ can be enclosed in disjoint open sets V_1 and V_2 respectively. We have $q_0 \in V_1 \subset U$. The points of A are interior points of V_1 and the points of B are exterior points.

Hence V_1 has no boundary points on β_0. The component of V_1 that contains q_0 is a region V with the desired properties.

For a given q_0, let V run through all regions with these properties, and set $Q = V \cap F$. We claim that the sets Q define a boundary component q of F. First, Q is a region, by condition (B3). Second, the relative boundary of Q with respect to F is compact. Indeed, the relative boundary is $(\bar{V} \cap F) - V = (\bar{V} - V) \cap F = \bar{V} - V$, the boundary of V, a compact set. Third, the relative closure $\bar{Q} \cap F = \bar{V} \cap F$ is not compact, for then it would be closed on M_0, and $\bar{V}$, in spite of being connected, would have a nontrivial decomposition $\bar{V} = (\bar{V} \cap F) \cup (\bar{V} \cap \beta_0)$ into closed sets. These facts justify the notation Q.

To verify (C1), assume that $Q_0 = V_0 \cap F$ and $Q \supset Q_0$. Set $V = Q \cup (\bar{Q} \cup \beta_0)$. From $Q \subset V \subset \bar{Q}$ we conclude that V is connected and $\bar{V} = \bar{Q}$. It is evident that V contains q_0. To prove that V is open and has its boundary in F we must show that every point $q_1 \in \bar{Q} \cap \beta_0$ has a neighborhood in V. Let V_1 be an open connected neighborhood of q_1 that does not meet the relative boundary of Q in F. Because $V_1 \cap F$ is connected and contains points of Q we have necessarily $V_1 \cap F \subset Q$. Since F is dense in V_1 it follows that $V_1 \subset \bar{Q}$. Hence $V_1 = (V_1 \cap F) \cup (V_1 \cap \beta_0) \subset Q \cup (\bar{Q} \cap \beta_0) = V$, which is what we wanted to show. Finally, $V \cap F = Q \cap F = Q$ so that $Q \in q$.

Conditions (C2) and (C3) are immediate consequences of the lemma.

37D. Define a mapping f of M_0 into M by $f(p) = p$ for $p \in F$, $f(q_0) = q$ for $q_0 \in \beta_0$. The mapping is one to one, for distinct points q_0', q_0'' have disjoint neighborhoods V', V''. It takes open sets into open sets, for if $Q = V \cap F$, then $f(V) = Q \cup \beta(Q)$. It is also continuous, for every neighborhood of $q = f(q_0)$ contains a $Q \cup \beta(Q)$, and we can repeat the reasoning in 37C (the proof of (C1)) to show that q_0 has a neighborhood V which corresponds to Q.

Because $f(M_0)$ is at once open and closed it must coincide with M. We have proved that f is a topological mapping of M_0 onto M, and the uniqueness proof is complete. All parts of Theorem 36B have now been established.

38. Partitions of the ideal boundary

38A. In the applications we shall frequently find cause to consider partitions of the ideal boundary. Such partitions have a natural relation to the contours of regular subregions.

For each regular subregion Ω of the open surface F we consider the finite set whose elements are the components of $F - \Omega$, all noncompact. Let P denote a partition of this set into mutually disjoint subsets. We can also regard P as a partition of the set of contours of Ω, but with this interpretation it must be postulated that all contours of a complementary component belong to the same part.

The *identity partition* I consists of a single part, and the *canonical partition*, denoted by Q, is obtained by letting each component of $F-\Omega$ be a part by itself.

38B. Suppose now that there is given a whole system $\{P\}$ of partitions $P=P(\Omega)$, one for each regular region Ω. We shall say that the system is *consistent* if, whenever $\bar{\Omega}\subset\Omega'$, the following is true: any two components of $F-\Omega'$ which belong to the same part in $P(\Omega')$ are contained in components of $F-\Omega$ which belong to the same part in $P(\Omega)$.

Another way of describing the same relationship is as follows: the partition $P(\Omega)$ induces, in a natural way, a partition of the set of components of $F-\Omega'$; it is required that $P(\Omega')$ is a refinement of the induced partition.

The systems $\{I\}$ and $\{Q\}$ of identity partitions and canonical partitions are evidently consistent.

38C. We show that a consistent system $\{P\}$ induces a partition P, or $P(F)$, of the ideal boundary β, considered as a set of boundary components. We define it in the following way:

Given a boundary component q and a regular region there is a unique component $R=R(q,\Omega)$ of $F-\Omega$ such that, in the terminology of 36D, $q\in\beta(R)$. We agree that q_1, q_2 will belong to the same part in the partition $P(F)$ if and only if, for every Ω, $R(q_1,\Omega)$ and $R(q_2,\Omega)$ belong to the same part in $P(\Omega)$.

It must be verified that the relation is transitive. This is trivial, however, for if $R(q_1,\Omega)$, $R(q_2,\Omega)$ belong to the same part and $R(q_2,\Omega)$, $R(q_3,\Omega)$ belong to the same part, then $R(q_1,\Omega)$ and $R(q_3,\Omega)$ belong to the same part in $P(\Omega)$.

Note that $\{Q\}$ induces the canonical partition Q in which every part is a single boundary component.

38D. The induced partition of β is not arbitrary. On the contrary, it is found to satisfy a rather strong condition of regularity.

In order to describe this condition we identify points of β that belong to the same part; we denote by $\bar{q}$ the part that contains q. The resulting space, suitably denoted by β/P, has an induced topology (see 2F). We recall that a set $E\subset\beta/P$ is open if and only if the union of all sets $\bar{q}\in E$ is open on β.

We contend that β/P is a totally disconnected Hausdorff space. A partition which satisfies this condition will be called a *regular* partition.

To prove that P is regular, consider $\bar{q}_1\neq\bar{q}_2$ and choose $q_1\in\bar{q}_1$, $q_2\in\bar{q}_2$.

There exists an Ω such that $R_1 = R(q_1, \Omega)$ and $R_2 = R(q_2, \Omega)$ belong to different parts in $P(\Omega)$. Let $\bar{R}_1, \bar{R}_2$ be the parts that contain R_1, R_2 respectively. We form the sets $U_1 = \cup\, \beta(R)$, $R \in \bar{R}_1$, and $U_2 = \cup\, \beta(R)$, $R \in \bar{R}_2$. Then U_1 and U_2 are open and closed subsets which are disjoint and contain q_1 and q_2. Moreover, the definition of P implies that they are unions of sets $\bar{q}$. Therefore U_1/P and U_2/P are disjoint open and closed neighborhoods of $\bar{q}_1, \bar{q}_2$. It follows that β/P is a Hausdorff space, and that the component of $\bar{q}_1$ does not contain $\bar{q}_2$. In other words, each component reduces to a point, and the space is totally disconnected.

38E. We prove conversely that every regular partition of β is induced by a consistent system of partitions. To summarize:

Theorem. *Every consistent system $\{P\}$ of partitions induces a regular partition of β, and every regular partition of β is induced by such a system.*

Given a regular partition P of β we let $M(P)$ denote the space obtained from $M = F \cup \beta$ by identifying points of β which belong to the same part. For each regular region Ω we consider the components of $M(P) - \Omega$ and define $P(\Omega)$ by letting two components of $F - \Omega$ belong to the same part if and only if they are contained in the same component of $M(P) - \Omega$. The partitions $P(\Omega)$ are consistent, for if $\bar{\Omega} \subset \Omega'$, then each component of $M(P) - \Omega'$ is contained in a component of $M(P) - \Omega$.

It remains to show that the system $\{P(\Omega)\}$ induces P. Suppose first that q_1 and q_2 are in the same part with respect to P, that is to say, $\bar{q}_1 = \bar{q}_2$. We set $R_1 = R(q_1, \Omega)$, $R_2 = R(q_2, \Omega)$. Then $q_1 \in \beta(R_1)$, $q_2 \in \beta(R_2)$, and we see that $\bar{q}_1$ belongs to the closure of R_1 as well as the closure of R_2 with respect to the topology of $M(P)$. Hence R_1 and R_2 are in the same component of $M(P) - \Omega$, and we have shown that q_1, q_2 belong to the same part in the partition induced by $\{P(\Omega)\}$.

Suppose now that $\bar{q}_1 \neq \bar{q}_2$. Because β/P is totally disconnected we can find open and closed disjoint neighborhoods of $\bar{q}_1, \bar{q}_2$ (see 4G). These neighborhoods are of the form U_1/P, U_2/P where U_1, U_2 are disjoint open and closed unions of sets $\bar{q}$. Because U_1, U_2 are open they are unions of sets $\beta(Q)$. Because they are compact they are finite unions of such sets: $U_1 = \beta(Q_1) \cup \cdots \cup \beta(Q_m)$, $U_2 = \beta(Q_{m+1}) \cup \cdots \cup \beta(Q_{m+p})$.

Choose Ω so large that it contains the boundaries of $Q_1, \cdots, Q_{m+p}$. Then, for each component R of $F - \Omega$ and each Q_i, either $R \subset Q_i$ or $R \cap Q_i = 0$. It follows that $\beta(R) \subset U_1$, $\beta(R) \subset U_2$, or $\beta(R) \subset \beta - U_1 - U_2$. Let $\bar{R}_1, \bar{R}_2$ be the sets of components R with $\beta(R) \subset U_1$, $\beta(R) \subset U_2$ respectively. We have $U_1 = \cup\, \beta(R)$, $R \in \bar{R}_1$, and $U_2 = \cup\, \beta(R)$, $R \in \bar{R}_2$. Our contention that q_1, q_2 belong to different parts in the partition induced by $\{P(\Omega)\}$ will be proved if we show $R_1 \in \bar{R}_1$ and $R_2 \in \bar{R}_2$ cannot be contained in the same component of $M(P) - \Omega$.

For this purpose set $S_1 = \cup\, (R \cup \beta(R))$, $R \in \bar{R}_1$. This is an open and closed set on $M-\Omega$, and $S_1 \cap \beta = U_1$ is a union of sets $\bar{q}$. Therefore the corresponding set on $M(P)-\Omega$, obtained by identification, is also open and closed, and hence a union of components. It contains R_1 and does not meet R_2. Hence R_1 and R_2 cannot be contained in the same component, and our proof is complete.

§7. THE CLASSIFICATION OF POLYHEDRONS

The classification theory depends on deeper combinatorial considerations than those which occur in homology theory. This is evidenced by the fact that homology theory can be formulated just as easily in any number of dimensions, while not even the 3-dimensional analogue of the classification theory is known.

The introduction of models makes the theory of surfaces much more concrete, and it is the only known way to determine the fundamental group in an explicit manner.

39. Cell complexes

39A. Our immediate aim is to obtain a complete classification of all topologically different finite polyhedrons. The method, briefly described, will be as follows: We imbed the complexes of finite polyhedrons in a larger class of objects, called *cell complexes*. In this larger class we define an equivalence relation, similar to the one introduced in 28. This new equivalence relation, when restricted to polyhedrons, will actually be identical with our earlier notion of equivalence, but it will not be necessary to prove this identity as long as we can show that the new equivalence will possess the fundamental property that equivalent geometric polyhedrons are homeomorphic. The next step is to exhibit a complete array of model polyhedrons, one from each equivalence class. It turns out that the models are topologically different. As a consequence two finite polyhedrons are homeomorphic if and only if they are equivalent (in the new sense), and we are led to a complete solution of the topological classification problem.

39B. Our definition of a cell complex is purely abstract. As indicated above, certain cell complexes will later be identified with polyhedrons, and will thus receive a geometric interpretation. This interpretation forms the motivation for the terminology and is of course tacitly present from the start.

A cell complex K consists of two finite sets of elements a and A, referred to as the *sides* and *faces* of K. Each element a and A has an inverse a^{-1}, A^{-1}, which is again a side or face, and these inversions are supposed to be involutory mappings in the sense that $(a^{-1})^{-1}=a$, $(A^{-1})^{-1}=A$. It is assumed that $a \neq a^{-1}$ and $A \neq A^{-1}$.

To each face A there is assigned a cyclically ordered set $a_1a_2\cdots a_n$ of sides, the "boundary" of A, in such a way that the inverse cycle $a_n^{-1}\cdots a_2^{-1}a_1^{-1}$ is assigned to A^{-1}. The notation is misleading for typographical reasons: a_1 is meant to be the successor of a_n, and there is no distinguished first element. If $n>1$ the cycle $a_n\cdots a_1$ is different from $a_1\cdots a_n$.

In addition to these basic assumptions we assume that K satisfies the following special conditions:

(B1) Each side a occurs either once or twice as an element of a boundary.

(B2) K is connected.

Condition (B1) is to be interpreted in the sense that a can appear twice in the same boundary, but then it does not appear in any other boundary. The connectedness means that K is not the union of two disjoint non-empty systems which satisfy all other requirements.

By way of clarifying the definition we agree that K is not allowed to be empty. However, K may consist of a single pair A, A^{-1} with empty boundaries.

39C. The notion of *vertex* can be described in terms of sides and faces. A successor of a is a side which appears immediately after a in a boundary. If a appears in two places it has two successors, and we call them a pair of successors even if they should happen to be identical.

A cyclically ordered set $\alpha=(a_1, a_2, \cdots, a_n)$ is called an *inner vertex* if a_i has a_{i-1}^{-1} and a_{i+1}^{-1} as its pairs of successors. If the same condition holds for $1<i<n$, while a_1 and a_n have only the successors a_2^{-1} and a_{n-1}^{-1} respectively, then α, regarded as a linearly ordered set, is called a *border vertex*. In the definition of vertices we make no distinction between $(a_1, \cdots, a_n)$ and $(a_n, \cdots, a_1)$.

With these conventions every side belongs to exactly one vertex. In fact, starting from an arbitrary a we find its neighbors in a vertex by taking the inverses of its successors. This can be continued in both directions until the cycle closes, or until we hit sides with only one successor. Using a geometric language it is convenient to say that each side leads to a vertex, its terminal point.

There is a natural way in which any finite polyhedron can be regarded as a cell complex. Conversely, a cell complex can be regarded as a polyhedron if the following conditions are fulfilled:

(C1) If a, b are distinct and lead to the same vertex, then a^{-1}, b^{-1} lead to distinct vertices.

(C2) The boundary of each face has the form abc.

(C3) Different faces have different boundaries.

It is a consequence of (C1) that a and a^{-1} cannot lead to the same vertex. Furthermore, the sides a, b, c in (C2) lead to different vertices and are hence distinct. In fact, since abc is a boundary a and b^{-1} determine the same vertex, and by the preceding remark b cannot belong to that vertex.

39D. By the use of cell complexes an elementary subdivision of a polyhedron (28A) can be decomposed in even simpler steps. For this purpose, we define two elementary operations on cell complexes:

P1. The first operation consists in replacing a pair of sides a, a^{-1} by bc and $c^{-1}b^{-1}$ in all boundaries.

P2. In the second operation we replace a face A with the boundary $a_1 \cdots a_p a_{p+1} \cdots a_n$ by two faces A', A'' with the boundaries $a_1 \cdots a_p d$ and $d^{-1} a_{p+1} \cdots a_n$. The corresponding decomposition is applied to A^{-1}.

With the help of these operations we define the notions of *refinement* and *equivalent cell complexes* exactly as in 28.

39E. Since every polyhedron can be regarded as a cell complex, we have, at least formally, introduced two different equivalence relations between polyhedrons. On comparing the operation of elementary subdivision with the operations P1 and P2 it is evident that two polyhedrons which are equivalent as polyhedrons, i.e., by the definition in 28, are also equivalent as cell complexes. To prove the opposite conclusion by a combinatorial argument is discouragingly complicated. Fortunately, as remarked in 39A, it is sufficient to establish the weaker result that polyhedrons which are equivalent as cell complexes are a fortiori topologically equivalent.

For this purpose we associate a topological space $\bar{K}$ with every cell complex K. Provided that no face A has an empty boundary, the construction is as follows: For each pair A, A^{-1} we choose one face, say A, and represent it as a closed circular disk. If the boundary of A is $a_1 a_2 \cdots a_n$ we divide the corresponding circumference into n arcs. These arcs, described in positive direction with respect to the disk, are named $a_1, a_2, \cdots, a_n$ and the opposite arcs $a_1^{-1}, a_2^{-1}, \cdots, a_n^{-1}$. Whenever, in the complete system, two arcs have the same name, they will be mapped onto each other in such a way that arc lengths are proportional to each other, and so that initial and terminal points correspond. The space $\bar{K}$ is obtained by identifying points that are mapped on each other. The end points of the arcs a must of course be identified in a transitive manner.

With this definition of $\bar{K}$ it is easy to verify that equivalent cell complexes K_1, K_2 lead to homeomorphic spaces $\bar{K}_1, \bar{K}_2$. Moreover, if K is a polyhedron, i.e., if (C1)–(C3) are satisfied, then $\bar{K}$ is homeomorphic to K_g.

To make the definition complete we must still consider the case where K has a face A with empty boundary. It follows by (B2) that A and A^{-1}

are the only faces. Application of P2 permits us to replace A by A', A'' with the boundaries d and d^{-1} respectively. The corresponding topological space will be the double of a disk. Hence, if we want $\bar{K}$ to be topologically invariant under P2 we must choose $\bar{K}$ as a sphere.

Our assertion was:

If two polyhedrons K_1, K_2 are equivalent as cell complexes, then $(K_1)_g$ is topologically equivalent to $(K_2)_g$.

The proof is immediate, for $(K_1)_g$ is homeomorphic to $\bar{K}_1$, $(K_2)_g$ is homeomorphic to $\bar{K}_2$, and $\bar{K}_1$, $\bar{K}_2$ are topologically equivalent by reason of the equivalence.

39F. We wish to show that every cell complex has a refinement which is a polyhedron. This will permit us to conclude, without further proof, that every $\bar{K}$ is a surface.

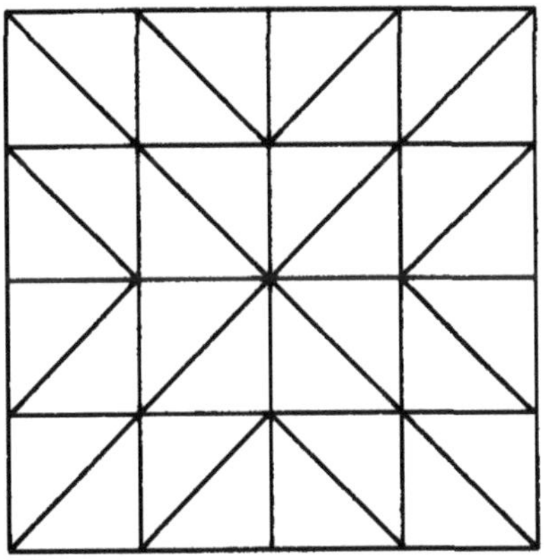

FIG. 7

The construction requires several steps. We may assume that no A has an empty boundary, for we have seen in 39E that every cell complex has a refinement with this property. The first step consists in applying P1 to all sides. If a is replaced by bc, a new vertex (b, c^{-1}) is introduced. It follows that b and b^{-1} lead to different vertices, and the same is true of c and c^{-1}.

For the second step, suppose that A has the boundary $a_1a_2\cdots a_n$. It is known that $n \geqq 2$, for otherwise a_1 and a_1^{-1} would lead to the same vertex. Application of P2 replaces the boundary by a_1d and $d^{-1}a_2\cdots a_n$. We use P1 to replace d by $d_1d_n^{-1}$. The boundaries can then be written as $d_n^{-1}a_1d_1$ and $d_1^{-1}a_2\cdots a_nd_n$, and by repeated application of P2 the latter boundary can be decomposed into $d_1^{-1}a_2d_2$, $d_2^{-1}a_3d_3$, $\cdots$, $d_{n-1}^{-1}a_nd_n$. When this transformation is applied to all faces, the resulting cell complex satisfies (C2) and (C3) but not necessarily (C1). It is important to notice that the result accomplished by the second step remains in force. In fact, for a given A all the d_i form a single vertex $(d_1, d_2, \cdots, d_n)$.

As a third step we consider a boundary $a_1a_2a_3$ and use P1 to replace it by $b_1c_1b_2c_2b_3c_3$. By three applications of P2 this boundary can be decomposed into $c_1b_2d_3$, $c_2b_3d_1$, $c_3b_1d_2$ and $d_1^{-1}d_2^{-1}d_3^{-1}$. It is an easy matter to verify that the cell complex obtained by this final step is a polyhedron.

The process which we have described is unique up to isomorphisms. Its application in the case of a square is illustrated in Fig. 7. Note that the construction produces a triangulation even if the sides of the square were originally identified in the manner of a torus.

With every cell complex K we have associated a polyhedron K_s which will be called *the simplicial subdivision* of K. The construction shows that K is equivalent to K_s, and $\bar{K}$ homeomorphic with $(K_s)_g$.

40. Canonical forms

40A. We have already derived, in 28B, three necessary conditions for the equivalence of polyhedrons. It was shown that two equivalent polyhedrons must agree in orientability, number of contours, and Euler characteristic. These conditions can be reformulated in terms of cell complexes.

We begin with the most basic distinction which is that between the orientable and nonorientable case. The faces of a cell complex K are divided into pairs A, A^{-1}. A choice of notation whereby one out of each pair of inverse faces is denoted by A, the other by A^{-1}, constitutes an orientation of K. The orientation is *coherent* if each side a appears at most once in the boundaries of the faces A. In other words, if a appears twice in the whole system of boundaries, then it appears in the boundaries of a face A_1 and of a face A_2^{-1}.

A cell complex K is called orientable if it has a coherent orientation. One verifies at once that any refinement of an orientable K is orientable. Conversely, if the refinement is orientable, so is K. We conclude that the cell complexes in an equivalence class are either all orientable or all nonorientable. For polyhedrons, the new definition of orientability coincides with the old.

40B. The contours of a cell complex are formed by the sides that appear in only one boundary. An individual contour is a cyclically ordered set $\langle a_1, \cdots, a_n \rangle$ of such sides, with the property that a_i and a_{i+1}^{-1} lead to the same vertex. When we count the number of contours we do not distinguish between $\langle a_1, \cdots, a_n \rangle$ and $\langle a_n^{-1}, \cdots, a_1^{-1} \rangle$.

It is clear that neither P1 nor P2 changes the number of contours. Therefore, equivalent cell complexes have the same number of contours.

40C. The Euler characteristic of a cell complex is found by counting the vertices, the pairs of sides a, a^{-1}, and the pairs of faces A, A^{-1}. In terms of these numbers n_0, n_1, n_2 the characteristic is $\rho = -n_0 + n_1 - n_2$.

The operation P1 increases the number of sides by one and introduces one new vertex while n_2 remains unchanged. Similarly, P2 increases n_1 and n_2 by one and leaves n_0 unchanged. Hence equivalent cell complexes have equal Euler characteristics.

There is an important exception. If K consists of a pair A, A^{-1} with empty boundaries, the application of P2 produces boundaries d, d^{-1} and hence a vertex (d, d^{-1}). In order to remove the discrepancy we must agree that the original system K possesses an empty vertex.

40D. The classification of cell complexes will be achieved by showing that each cell complex is equivalent to one and only one canonical cell complex. The canonical cell complexes are completely described through the following explicit requirements:

Each canonical cell complex contains a single pair of faces A, A^{-1}, and the boundary of A is of the form (I) or (II) listed below.

(I) $a_1b_1a_1^{-1}b_1^{-1}\cdots a_pb_pa_p^{-1}b_p^{-1}c_1h_1c_1^{-1}\cdots c_qh_qc_q^{-1}$ $(p \geqq 0, q \geqq 0)$.

(II) $a_1a_1\cdots a_pa_pc_1h_1c_1^{-1}\cdots c_qh_qc_q^{-1}$ $(p \geqq 1, q \geqq 0)$.

40E. The canonical cell complexes are all inequivalent. In the first place, the systems of type (I) are orientable, and the systems of type (II) are nonorientable. This is obvious since there are only two orientations to choose from.

There are q border vertices, defined by the linear sequences (h_i, c_i, h_i^{-1}). Since h_i and h_i^{-1} lead to the same vertex, it follows that $\langle h_i \rangle$ is a contour all by itself, and there are q contours. We conclude that two canonical complexes cannot be equivalent unless they have the same q.

In order to compute the characteristics we note that the system (I) has one inner vertex, namely $(a_1^{-1}, b_1, a_1, b_1^{-1}, \cdots, a_p^{-1}, b_p, a_p, b_p^{-1}, c_1^{-1}, \cdots, c_q^{-1})$. The total number of vertices is thus $q+1$, the number of sides is $2p+2q$, and there is only one face. This gives $\rho = -(q+1)+2(p+q)-1 = 2p+q-2$.

In the system (II) the inner vertex is $(a_1^{-1}, a_1, \cdots, a_p^{-1}, a_p, c_1^{-1}, \cdots, c_q^{-1})$, and we find that $\rho = -(q+1)+(p+2q)-1 = p+q-2$.

Since equivalent canonical cell complexes have the same ρ and q, they must also have the same p. We have shown that equivalent canonical complexes must be identical up to isomorphism.

It is to be observed that the computation of ρ agrees with our special convention in the case $p=q=0$. In fact, we postulate the existence of an empty vertex and obtain $\rho = -2$, in accordance with the formula $\rho = 2p+q-2$.

41. Reduction to canonical forms

41A. The inverse operations of P1 and P2 will be denoted as $(P1)^{-1}$ and $(P2)^{-1}$. They are applicable under circumstances that are easy to recognize. Clearly, $(P1)^{-1}$ can be applied to a pair of sides b, c provided that (b, c^{-1}) is a vertex, and the operation eliminates this vertex. Similarly, $(P2)^{-1}$ can be used to eliminate a side d provided that there are two faces A_1 and A_2 which are neither identical nor inverse to each other, and whose boundaries contain d and d^{-1} respectively.

We will now introduce a series of systematic reductions by which a given cell complex can be brought to one of the canonical forms.

41B. *Cancellation.* Suppose that one of the boundaries has the form $aa^{-1}(1)$, where the symbol (1) stands for an unspecified sequence of sides, a sequence that could even be empty. By use of P2 the boundary can be replaced by ad and $d^{-1}a^{-1}(1)$. $(P1)^{-1}$ allows us to write $ad=c$, and c can be eliminated by $(P2)^{-1}$. The net result is that aa^{-1} can be omitted.

41C. *Vertex reduction.* We suppose that all cancellations have been carried out. If the remaining system has no sides it must be of type (I) with $p=q=0$, and we have reached our goal. In the opposite case there remains a side, and therefore at least one vertex.

Consider an inner vertex $\alpha=(b_1, \cdots, b_m)$, and assume, for instance, that b_1^{-1} does not lead to α. Then $m \geqq 2$, for otherwise one of the boundaries would contain $b_1b_1^{-1}$. We locate the succession $b_1b_2^{-1}$ and use P2 to introduce a boundary of the form $b_1b_2^{-1}c$. Since b_2 is different from b_1, b_1^{-1}, c, c^{-1} it is possible to eliminate b_2 by $(P2)^{-1}$. The effect is to shorten α. In fact, c leads to the same vertex as b_1^{-1}, and c^{-1} leads to the same vertex as b_2^{-1}. Hence, if b_2^{-1} does not belong to α, nothing is added to α and b_2 is dropped; if b_2^{-1} does belong to α, c^{-1} is added and b_2, b_2^{-1} are both dropped.

This can be repeated until α consists of b_1 alone, and then b_1 can be cancelled, causing α to disappear. The eliminating process can be applied to any inner vertex, provided that it is not formed by pairs of sides which are inverse to each other. On the other hand, if the b_i in $\alpha=(b_1, \cdots, b_m)$ are inverse in pairs, then every b_i has elements b_j and b_k for successors, and it follows easily by the connectedness of K that α is the only vertex. The reduction can thus be continued until there are no inner vertices, or until a single inner vertex remains.

In the presence of a border we prefer to leave one inner vertex, even though it could be eliminated. If there is no inner vertex from the beginning, one can be introduced by application of P2 and P1. Let the remaining inner vertex be α_0. There must exist a border vertex $\beta=(h, c_1, \cdots, c_m, k)$ such that a c_i^{-1} leads to α_0, for otherwise the system would not be connected. The same method as above permits us to eliminate c_{i+1} and c_{i-1}, until the remaining vertex has the form (h, c_i, k). If $k \neq h^{-1}$ it is even possible to eliminate c_i, and the remaining vertex (h, k) can be removed by $(\text{P1})^{-1}$. We can thus continue until we come to a vertex of the form (h, c, h^{-1}) with c^{-1} leading to α_0.

Any remaining vertex can be reduced in the same way without interfering with the reductions that have already been carried out. This is easily seen if we observe that a vertex (h, c, h^{-1}) entails the presence of a succession chc^{-1} in the boundaries. Such a succession, which we call a *loop*, remains unaffected by any operation which does not involve h or c. Because c, h and h^{-1} lead to the vertex that has already been reduced, while c^{-1} leads to α_0, we see that h and c are not involved in any further reductions. We conclude that the whole process leads to a cell complex with one inner vertex and border vertices of the form (h_i, c_i, h_i^{-1}).

41D. *Cross-caps.* The resulting system may consist of several pairs of faces A, A^{-1}. It can always be reduced to a single pair by applying $(\text{P2})^{-1}$ as many times as possible. This does not introduce any new vertices, and the loops are not affected.

Suppose that the boundary of the remaining A contains two identical sides; we indicate this by writing the boundary in the form $a(1)a(2)$. It can be replaced by $a(1)b, b^{-1}a(2)$, and if a is eliminated we obtain the boundary $bb(2)^{-1}(1)$. The succession bb is called a *cross-cap*. When the process is repeated the cross-caps and loops that are already present will not be destroyed. We can therefore continue until all pairs of identical sides appear as cross-caps.

We note the formal rule $a(1)a(2) \sim aa(2)^{-1}(1)$ (read: equivalent to), or $aa(1)(2) \sim a(2)a(1)^{-1}$.

41E. *Handles.* A second formal rule is obtained if we consider a boundary of the form $a(1)(2)a^{-1}(3)$. We break it into $a(1)b, b^{-1}(2)a^{-1}(3)$ and eliminate a. The result is a boundary $b^{-1}(2)(1)b(3)$, and we can write $a(1)(2)a^{-1}(3) \sim a(2)(1)a^{-1}(3)$.

We apply this rule repeatedly to a boundary $a(1)b(2)a^{-1}(3)b^{-1}(4)$ and find that it is equivalent to $ab(2)(1)a^{-1}(3)b^{-1}(4) \sim aba^{-1}(3)(2)(1)b^{-1}(4) \sim aba^{-1}b^{-1}(4)(3)(2)(1)$. The succession $aba^{-1}b^{-1}$ is called a *handle*. Repetition does not destroy the handles, cross-caps, and loops which are already present.

As a further simplification we show that a handle and a cross-cap can be replaced by three cross-caps. This follows by the first rule, for we obtain $aa(1)bcb^{-1}c^{-1}(2) \sim ab^{-1}c^{-1}(2)ac^{-1}b^{-1}(1)^{-1} \sim b^{-1}b^{-1}a^{-1}(1)c^{-1}(2)ac^{-1} \sim c^{-1}c^{-1}(1)^{-1}abb(2)a \sim aa(1)ccbb(2)$.

41F. When these reductions have been carried out, we claim that the remaining boundary is completely decomposed in loops and cross-caps or handles. Clearly, we must show that there does not remain any pair c, c^{-1} which is not part of a loop or handle. If such a pair exists we can write the boundary of A in the form $c(1)c^{-1}(2)$, where no side in (1) is equal or inverse to a side in (2).

We note that the result of the vertex reduction is still in force, for the subsequent reductions did not make use of P1 or its inverse. Since c and c^{-1} are not part of a loop, they must both lead to the inner vertex α_0. On the other hand, it is clear from our assumption that both successors of c are in (1), and that both successors of c are in (1) and that each successor of a side in (1) is either itself in (1), or else identical with c^{-1}. It follows that c^{-1} is not in the vertex determined by c, and we have reached a contradiction.

41G. In order to arrive at the canonical forms we must still show that the loops can be placed in succession. This is easily accomplished, for by our second formal rule we obtain $chc^{-1}(1)dkd^{-1}(2) \sim chc^{-1}dkd^{-1}(2)(1)$. Thus any two loops can be brought next to each other without breaking up previous successions.

Our proof is now complete, and we have shown that every cell complex is equivalent to one and only one canonical complex.

42. Topological classification

42A. We know now that any polyhedron K is equivalent, as a cell complex, to a canonical cell complex. If two polyhedrons K and K' are equivalent to the same canonical cell complex, then K_g and K'_g are homeomorphic (39E). We have indicated in (39A) that this sufficient condition for topological equivalence is also necessary.

To prove the necessity it is sufficient to show that the simplicial subdivisions (see 39F) of the different canonical cell complexes determine topologically different geometric polyhedrons. But this is an immediate consequence of the fact that the canonical forms differ either in orientability, in number of contours, or in Euler characteristic. Indeed, homeomorphic surfaces must have the same character of orientability, and the same number of contours. In addition, they must have isomorphic homology groups, and we have shown that the homology group determines the Euler characteristic.

We have thus obtained a complete proof of the fundamental theorem in classification theory:

Theorem. *Two triangulated compact surfaces or bordered surfaces are topologically equivalent if and only if they agree in character of orientability, number of contours, and Euler characteristic.*

42B. An important consequence of this is that we can construct simple model surfaces corresponding to the canonical forms. Consider, for instance, a circular disk in which we make p circular holes. The double of this region is a closed orientable surface, and in various simple ways we can construct an explicit triangulation whose characteristic turns out to be $2p-2$. Hence it represents a surface of type (I) with $q=0$ and the given p. It is visualized as a sphere with p handles. Removal of q disks transforms it into a surface of the same type with arbitrary q.

Starting from a disk with p circular holes another closed surface can be constructed by identification of diametrically opposite points on each contour (but no doubling). The resulting surface is nonorientable, and it has a triangulation with the characteristic $p-2$. Therefore, it is a model of type (II) with $q=0$ and given p. A general surface of type (II) can again be obtained by removing q disks. The term cross-cap refers to a concrete way of realizing the identification of opposite points.

43. Determination of the fundamental group

43A. Even more important than the classification is the fact that we can now compute, explicitly, the fundamental group of a finite polyhedron. For this purpose we are going to rely once more on the technique of simplicial approximation which was introduced in 34A to determine the singular homology group of K_g.

We begin by defining, in combinatorial terms, the fundamental group of a cell complex. A *polygon* on a cell complex K is a succession $a_1a_2\cdots a_n$ of sides with the property that a_i and a_{i+1}^{-1} determine the same vertex. The initial point is the vertex determined by a_1^{-1}, and terminal point is the vertex determined by a_n. The polygon is closed if they coincide.

The product of $a_1\cdots a_n$ and $a_{n+1}\cdots a_{n+p}$ is defined only if the terminal point of the first polygon coincides with the initial point of the second; it is taken to be the polygon $a_1\cdots a_{n+p}$.

We define two elementary deformations of polygons. The deformation D1 consists in omitting a succession of the form aa^{-1}, and D2 consists in omitting the boundary of a face. Two polygons are said to be homotop if one can be reduced to the other by means of a finite number of deformations D1, D2 and their inverse operations. The closed polygons from a fixed vertex α_0 fall into homotopy classes, and multiplication of polygons induces a multiplication of these classes. With this law of composition the homotopy classes form a group $\mathscr{F}(K)$, the fundamental group of K. The unit element is the homotopy class of the empty polygon.

It is readily shown that $\mathscr{F}(K)$ does not depend, as an abstract group, on the choice of the vertex α_0. Furthermore, it does not change under the operations P1 and P2. Hence equivalent cell complexes have isomorphic fundamental groups.

43B. We can determine the fundamental group of an arbitrary cell complex by considering the corresponding canonical system. In the case $q=0$ there is only one vertex, and all polygons are closed. For a system of type (I) the fundamental group is generated by $a_1, b_1, \cdots, a_p, b_p$, and the generators are connected by the relation $a_1b_1a_1^{-1}b_1^{-1}\cdots a_pb_pa_p^{-1}b_p^{-1}=1$. The special case $p=0$ is included, for then the only polygon is the empty polygon, and the fundamental group reduces to its unit element. Similarly, the fundamental group of a canonical cell complex of type (II) is generated by $a_1, \cdots, a_p$, subject to the relation $a_1a_1\cdots a_pa_p=1$.

If $q \geqq 1$ we take α_0 to be the inner vertex. It is easy to see that the closed polygons are products of a_i, b_i and $e_i=c_ih_ic_i^{-1}$. These generators satisfy relations $a_1b_1a_1^{-1}b_1^{-1}\cdots a_pb_pa_p^{-1}b_p^{-1}e_1\cdots e_q=1$ or $a_1a_1\cdots a_pa_pe_1\cdots e_q=1$. With the aid of these relations e_q can be expressed in terms of the other generators. Thus the fundamental group is a free group with $2p+q-1$ generators in the orientable case and $p+q-1$ generators in the nonorientable case.

43C. If K is a polyhedron we show that $\mathscr{F}(K)$ is isomorphic to $\mathscr{F}(K_g)$, the fundamental group of the surface or bordered surface K_g. First of all, the vertex α_0 may be identified with a point O on K_g, and we represent $\mathscr{F}(K_g)$ as $\mathscr{F}_0(K_g)$ with respect to O as origin. Each closed polygon from α_0 can be considered as a closed curve from O. We find that homotop polygons determine homotop curves. Therefore, the correspondence defines a homomorphic mapping of $\mathscr{F}(K)$ into $\mathscr{F}(K_g)$.

Suppose now that γ is a closed curve from O. Exactly as in 34A we determine a simplicial approximation of γ, but this time the approximation is interpreted as a polygon rather than as a cycle. First, γ is divided into subarcs, $\gamma=\gamma_1\gamma_2\cdots\gamma_n$, such that each γ_i is contained in a star $S(\alpha_i)$. The centers α_i and α_{i+1} are either identical, or the initial and terminal points of a side a_i; in the case of identical centers we set $a_i=1$. We make sure to choose α_1 and α_n equal to the given vertex α_0, and construct the polygon $a_1a_2\cdots a_{n-1}$ as simplicial approximation of γ. It is not unique, but a very simple argument shows that its homotopy class is uniquely determined.

We must show that homotop curves have homotop simplicial approximations. This requires a slightly more detailed argument than in 34B where only homology was involved. Suppose that γ' and γ'' can be deformed into each other by means of the mapping $(t, u)\to f(t, u)$. We can divide the unit square into rectangles $[t_i, t_{i+1}]\times[u_j, u_{j+1}]$ such that the image of each rectangle under the mapping f is contained in a star. As a consequence, it is possible to choose identical simplicial approximations of the curves $t\to f(t, u_j)$ and $t\to f(t, u_{j+1})$. It follows that any two simplicial approximations of γ' and γ'' are homotop, as asserted.

By simplicial approximation we can thus define a homomorphic mapping of $\mathscr{F}(K_g)$ into $\mathscr{F}(K)$. We must show that it is inverse to the homomorphism in the opposite direction which we obtained by interpreting each polygon as a closed curve. If the polygon P gives rise to the curve π, it is quite clear that P is in turn a simplicial approximation of π. Conversely, it must be shown that a closed curve $\gamma=\gamma_1\gamma_2\cdots\gamma_n$ is homotop to its simplicial approximation $P=a_1a_2\cdots a_{n-1}$ when the latter is considered as a curve. This becomes clear when we note that the intersection of the stars $S(\alpha_i)$ and $S(\alpha_{i+1})$ is simply connected. As a first step we join α_i to the initial point of γ_i by a line segment σ_i in $S(\alpha_i)$. Then γ is homotop to $(\gamma_1\sigma_2^{-1})(\sigma_2\gamma_2\sigma_3^{-1})\cdots(\sigma_n\gamma_n)$. Here $\sigma_i\gamma_i\sigma_{i+1}^{-1}$ can be deformed into a_i within $S(\alpha_i)\cap S(\alpha_{i+1})$, and it follows that γ is homotop to P.

We have now proved that the fundamental groups of K and K_g are isomorphic, and we are thus in a position to compute the fundamental group of any compact polyhedron.

43D. On examination of the different fundamental groups, enumerated under 43B, we find that the fundamental group is trivial in two cases, namely for canonical complexes of type (I) with $p=0$, $q=0$ and $p=0$, $q=1$. This means that the only simply connected finite polyhedrons are the *sphere* and the *closed disk*.

It is also of interest to determine all finite polyhedrons whose fundamental group is Abelian. In addition to the cases already considered this happens for a *torus*, type (I), $p=1, q=0$, for an *annulus*, (I), $p=0, q=2$, the *projective plane*, (II), $p=1, q=0$, and the *Möbius band*, (II), $p=1$, $q=1$.

43E. Having determined the fundamental group of a finite polyhedron we can automatically write down the homology group by application of Theorems 33D and 34A. All that is needed is to make the fundamental group Abelian and write it additively. The same generators will now satisfy the relation $e_1+\cdots+e_q=0$ in the orientable case and $2a_1+\cdots+2a_p+e_1+\cdots+e_q=0$ in the nonorientable case. If $q>0$ these relations serve to eliminate e_q, and the homology group is thus a free Abelian group with $2p+q-1$ generators in case (I) and $p+q-1$ generators in case (II). If $q=0$ the homology group is still a free Abelian group in case (I), now with $2p$ generators, but in case (II) the relation $2(a_1+\cdots+a_p)=0$ shows that there exists an element of order 2. Since there is no other relation the homology group is the direct product of a group of order 2 and a free Abelian group with $p-1$ generators. All this merely reconfirms our earlier results which were obtained without use of the classification theory. In particular, for orientable polyhedrons the number p coincides with the genus, as defined in 31G.

With the help of the canonical cell complexes we can even construct a homology basis in an explicit way. We know that a finite polyhedron K_g is homeomorphic to the geometric complex that corresponds to the simplical subdivision of some canonical cell complex. If we use this triangulation there is an obvious way of associating definite cycles with a_i, b_i and e_i (or, preferably, with the contours h_i). These cycles form a concrete homology basis. What is more, in the orientable case we can introduce the dual subdivision, and find by easy considerations that either $a_1, b_1, \cdots, a_p, b_p$ or $b_1, a_1, \cdots, b_p, a_p$ constitute a canonical basis (31G).

44. Topological properties of open polyhedrons

44A. No complete classification of open polyhedrons is known, but we can obtain partial results which throw light on the situation. We make use of canonical exhaustions as introduced in 29. Our first aim is to prove:

Theorem. *The fundamental group of an open polyhedron is a free group.*

Given an open polyhedron K we were able to construct, for a suitable subdivision, an exhaustion P_n with the property that the complementary components of P_n in P_{n+1} have each a single contour in common with P_n. The contours of P_n were named $b_{i_1 \cdots i_n}$ and the complementary components in P_{n+1} had matching names $Q_{i_1 \cdots i_n}$. The proof in 43C is valid for infinite as well as for finite polyhedrons. Thus we need only determine the group $\mathscr{F}(K)$ formed by the homotopy classes of closed polygons from a fixed vertex.

For the determination of $\mathscr{F}(K)$ we can replace K by an equivalent infinite cell complex, and we choose this cell complex in a definite way. We replace P_1 by a cell complex which has an inner vertex α_0 and a single vertex α_i on each contour b_i. Similarly, we let Q_i have the vertex α_i on b_i and a single vertex α_{ij} on each b_{ij}, and so on.

The vertex α_0 is chosen as initial point of the polygons. Our method will be to determine $\mathscr{F}(P_{n+1})$ when $\mathscr{F}(P_n)$ is given, and this passage can be accomplished by adding one $Q_{i_1 \cdots i_n}$ at a time. Therefore, the passage from P_1 to $P_1 \cup Q_1$ is completely typical.

For the purpose of comparing $\mathscr{F}(P_1)$ and $\mathscr{F}(P_1 \cup Q_1)$ we can temporarily choose α_1 as origin. The group $\mathscr{F}(P_1)$ is free, and its generators will be denoted by $g_1, \cdots, g_r$. The group $\mathscr{F}(Q_1)$ is likewise free, and we can choose its generators $h_1, \cdots, h_s$ so that $h_1 = b_1$. Since α_1 is the only vertex on b_1 it is clear that any closed polygon on $P_1 \cup Q_1$ which begins at α_1 can be decomposed into a product of closed polygons which lie alternately in P_1 and in Q_1. Each polygon on P_1 is homotop to a product of generators g_i, and each polygon on Q_1 can be expressed as a product of the h_j. Since h_1 belongs also to P_1 it follows that $\mathscr{F}(P_1 \cup Q_1)$ is generated by $g_1, \cdots, g_r, h_2, \cdots, h_s$.

We contend that these generators are independent. Any relation that they satisfy could be written in the form $p_1q_1p_2q_2\cdots p_kq_k \approx 1$ where each p_i is a product of factors $g_1, \cdots, g_r$ and each q_i is a product of factors $h_2, \cdots, h_s$. A deformation D1 or D2 (43A) brought to bear on $p_1q_1\cdots p_kq_k$ can be thought of as acting on an individual factor, and the same is true of the inverse operations, except that factors h_1 may be introduced in the q_i. The only way in which a q_i can disappear is therefore that it be reducible to a power of h_1. But this is impossible since $h_1, \cdots, h_s$ are independent. The relation must hence be of the form $p_1 \approx 1$, and since the g_i are independent this must be the identity relation. We have proved that $\mathscr{F}(P_1 \cup Q_1)$ is the free group generated by $g_1, \cdots, g_r, h_2, \cdots, h_s$.

If we return to the origin α_0 the generators are replaced by polygons of the form $c_1g_ic_1^{-1}$, $c_1h_jc_1^{-1}$. It is clear that the construction can be repeated indefinitely, and we find that each $\mathscr{F}(P_n)$ is a free group.

A closed polygon on K is homotop to 1 if and only if it is homotop to 1 on some P_n. Therefore the generators remain independent on K, and in their totality they generate $\mathscr{F}(K)$, which is consequently a free group.

44B. We denote by $\rho_{i_1\cdots i_n}$ the characteristic, by $p_{i_1\cdots i_n}$ the genus, and by $q_{i_1\cdots i_n}$ the number of outer contours of $Q_{i_1\cdots i_n}$; the outer contours are those different from $b_{i_1\cdots i_n}$. For a consistent notation we agree again that Q with an empty index is another name for P_1, while ρ, p and q are the characteristic, genus and number of contours of Q. Finally, we denote the corresponding numbers for P_n by $\bar{\rho}_n$, $\bar{p}_n$, $\bar{q}_n$.

It is clear, first of all, that

$$\bar{q}_{n+1} = \sum q_{i_1\cdots i_n}$$

where the subscripts are subject to the condition $i_m \leqq q_{i_1\cdots i_{m-1}}$. Next, the characteristics are additive, so that

$$\bar{\rho}_{n+1} = \sum_{m \leqq n} \rho_{i_1\cdots i_m}.$$

We consider only the orientable case and have thus $\rho = 2p+q-2$ and $\rho_{i_1\cdots i_m} - 1 = 2p_{i_1\cdots i_m} + q_{i_1\cdots i_m} - 1$ for $m > 0$. When these values are substituted in the formula for $\bar{\rho}_{n+1}$ the term -1 occurs once for each $Q_{i_1\cdots i_m}$, $0 < m \leqq n$. The total number of these terms is

$$\sum_{m<n} q_{i_1\cdots i_m}.$$

Therefore we obtain

$$2\bar{p}_{n+1} + \bar{q}_{n+1} - 2 = 2\sum_{m \leqq n} p_{i_1\cdots i_m} + \sum q_{i_1\cdots i_n} - 2,$$

and hence

$$\bar{p}_{n+1} = \sum_{m \leqq n} p_{i_1\cdots i_m}.$$

We see that $\bar{p}_n$ is nondecreasing. The limit $\bar{p}$ as $n \to \infty$ is called the genus of K, while $\bar{q} = \lim \bar{q}_n$ is the number of boundary components. The number of generators of the fundamental group $\mathscr{F}(K)$ is obtained by following the construction in 44A. The number of generators in $Q_{i_1 \cdots i_n}$ is $1 + \rho_{i_1 \cdots i_n}$, but one is dropped each time a component is adjoined. The total number is therefore $1 + \lim \rho_n = 2\bar{p} + \bar{q} - 1$. This is also the number of generators of the homology group $H_1(K)$, a result that could also have been derived from Theorem 30C. Finally, by Theorem 30D the group $H_1\beta(K)$ has $\bar{q}-1$ generators, and the quotient group $H_1(K)/H_1\beta(K)$ has $2\bar{p}$ generators.

44C. The results mentioned in the last paragraph are of special interest if $\bar{p}$ or $\bar{q}$ is finite; otherwise, they show merely that the groups in question have a countable infinity of generators. We focus our attention on the case where $\bar{p}$ is finite, and prove:

Theorem. *An orientable open polyhedron K_g of finite genus $\bar{p}$ is homeomorphic to an open subset of a closed polyhedron of genus $\bar{p}$.*

If $\bar{p}$ is finite there exists an m such that $\bar{p}_n = \bar{p}$ for $n \geqq m$. This means that every $Q_{i_1 \cdots i_n}$ with $n \geqq m$ is of genus zero. We begin by representing $(P_m)_g$ as a closed polyhedron of genus $\bar{p}$ from which a finite number of closed disks have been removed, one for each contour (42B). From each disk we remove the requisite number of smaller disks, in such a way that $(Q_{i_1 \cdots i_m})_g$ can be mapped topologically on the remainder. The mappings can be fitted together so that the two maps of the common contour $b_{i_1 \cdots i_m}$ agree with each other. This process can be repeated indefinitely and leads to a mapping of the desired nature.

Corollary. *A planar polyhedron is homeomorphic to a plane region.*

This is the case $\bar{p} = 0$ of the theorem. We obtain a mapping into the extended plane, and if the polyhedron is not a sphere we can assume that the point at infinity does not belong to the image.

44D. If $\bar{p}$ and $\bar{q}$ are both finite the situation is even simpler. For all sufficiently large n there are exactly $\bar{q}$ complements $Q_{i_1 \cdots i_n}$, each of genus 0 and with two contours. From each disk only one smaller disk needs to be removed, and in the limit we find that we have removed $\bar{q}$ closed disks or points. We conclude that the open polyhedron can be mapped on the interior of an orientable polyhedron of genus $\bar{p}$ with $\bar{q}$ contours, or, if we prefer, on a closed polyhedron of genus $\bar{p}$ which has been punctured at $\bar{q}$ points.

In particular, a simply connected open polyhedron is homeomorphic to an open disk, for the fundamental group is trivial only if $\bar{p} = 0$, $\bar{q} = 1$.

§8. EXISTENCE OF TRIANGULATIONS

In this section we prove that every countable surface permits a triangulation. The combinatorial theory which has been developed for polyhedrons has thus a very wide range of applicability.

The existence proof is in two parts. First, we prove that a surface can be triangulated if there exists an open covering with certain desirable properties. In the second part such a covering is constructed. This disposition of the reasoning is motivated by the fact that the existence of a covering with the desired properties is often known from the start, as for instance in the theory of Riemann surfaces.

The proofs in this section make essential use of the Jordan curve theorem, which is taken for granted. It may seem slightly inconsistent that we have not included a proof of the Jordan curve theorem in a presentation that is otherwise self-contained. The reason is of course that a proof would cast no light on the questions with which we are primarily concerned.

45. Coverings of finite character

45A. We know that every surface F has an open covering by Jordan regions. In what follows the *closed* Jordan regions will be denoted by J, their boundaries by γ. The covering is said to be of *finite character* if the following is true:

(A1) Each J meets at most a finite number of others.

(A2) The intersection of any two boundaries γ consists of at most a finite number of points or arcs.

It is an immediate consequence of (A1) together with the connectedness of F that a covering of finite character consists of a countable number of regions J. They can consequently be ordered in a sequence $\{J_n\}$. The boundary of J_n is denoted by γ_n.

If γ_n is not contained in J_m it intersects the interior of J_m along a finite number of *cross-cuts*. Precisely, a cross-cut of J_m is the interior of an arc in J_m whose end points, and only the end points, lie on the boundary. Naturally, the set of cross-cuts that lie on γ_n can be empty.

45B. For the proof that follows we are going to need the *Jordan curve theorem*, in the sharp form due to Schoenflies:

If γ is a Jordan curve in the plane there exists a topological mapping of the plane onto itself which transforms γ into a circle.

It follows that γ divides the extended plane in two Jordan regions with common boundary. The Jordan-Schoenflies theorem has the following corollary:

A cross-cut of a Jordan region divides it into two Jordan regions, and the boundary of each subregion consists of the cross-cut and one of the boundary arcs between its endpoints.

For the proof we denote the region by Δ, the cross-cut by σ, and the two boundary arcs by γ_1, γ_2. Let Δ_1, Δ_2 be the Jordan regions in the plane which are bounded by $\gamma_1 \cup \sigma$ and $\gamma_2 \cup \sigma$ respectively. Since Δ_1 has boundary points in Δ, while Δ has no boundary points in Δ_1, it follows by a connectedness argument that $\Delta_1 \subset \Delta$, and the same applies to Δ_2. Since Δ_1 and Δ_2 have different boundaries, and none has a boundary point in the other, it is seen in the same way that Δ_1 and Δ_2 are disjoint. In order to see that these are the only complementary regions of σ we identify all boundary points of Δ. Then $\bar{\Delta}$ becomes a sphere, and σ together with the boundary becomes a Jordan curve. It follows that σ divides Δ into two regions only, and they must be identical with Δ_1 and Δ_2.

45C. We are now ready to prove:

Theorem. *A surface is a polyhedron if it possesses an open covering by Jordan regions which is of finite character.*

From the covering $\{J_n\}$ we discard all those J_n which are contained in a J_m, $m \neq n$. The remaining J will still form a covering, for otherwise a point would be contained in an infinite nested sequence $J_{n_1} \subset J_{n_2} \subset J_{n_3} \subset \cdots$, in clear violation of property (A1). The new covering will again be denoted as $\{J_n\}$.

It is possible that $\gamma_n \subset J_m$ for some $m \neq n$. Since J_n is not contained in J_m, the Jordan region which γ_n encloses in J_m must be complementary to J_n. When this is so, the surface F is a sphere, and a trivial triangulation can be found. This case will now be left aside.

A fixed J_m is intersected by each γ_n along a finite number of cross-cuts. We begin by considering the cross-cuts on γ_1, if any. The first cross-cut divides J_m into two Jordan regions. One of these subregions is divided in the same way by the second cross-cut, and so on. We find that γ_1 divides J_m into a finite number of Jordan regions. Each of these either does not meet γ_2, or is in turn divided by γ_2 into a finite number of Jordan regions. When continued, the process ends in a finite number of steps, for J_m meets only finitely many γ_n.

We denote by J_{mi} the closed subregions of J_m obtained by this construction. Any two regions J_{mi} and J_{nj} are either identical or have disjoint interiors, for neither has a boundary point which lies in the interior of the other.

At the same time we consider the arcs γ_{mi} into which γ_m is divided by the points on a γ_n or the end points of arcs that are common to γ_m and γ_n. Again, γ_{mi} and γ_{nj} are either identical or have at most end points in common. It is evident from the construction that the boundary of each J_{mi} is a union of arcs γ_{nj}.

45D. In order to construct a triangulation we introduce a complex K whose vertices are the end points of the arcs γ_{mi} and, in addition, an interior point of each γ_{mi} and each J_{nj}. The interior point of J_{nj} can be joined to the vertices on the boundary of that closed region by Jordan arcs which have only the initial point in common. They divide J_{nj} into triangular closed regions. We have thus a natural way of defining the 1- and 2-simplices of K, and at the same time a set $\sigma(s)$ is associated with each simplex. The conditions (C1)–(C4) enumerated in 22C are all fulfilled in a trivial manner.

We have proved that a triangulation exists. By a previous result (Theorem 22D) F is homeomorphic with K_g and is thus a polyhedron.

46. The existence proof

46A. Every polyhedron has a countable basis for the open sets. Therefore, a surface F cannot be a polyhedron unless it is countable. We are going to show that this necessary condition is also sufficient.

Theorem. *A surface is a polyhedron if and only if it has a countable basis for the open sets.*

According to Theorem 45C we need only prove that a countable surface has a covering of finite character.

46B. As a preliminary step we prove the following lemma:

Lemma. *On a countable surface F it is possible to find two finite or infinite sequences $\{V_n\}$ and $\{W_n\}$ of Jordan regions with these properties:*

(B1) $\bar{V}_n \subset W_n$,

(B2) $\cup V_n = F$,

(B3) No point belongs to infinitely many $\bar{W}_n$.

There exists a basis in the form of a sequence $\{U_i\}$ of Jordan regions (see 8E). Each U_i is a countable union of Jordan regions U_{ij} with $\bar{U}_{ij} \subset U_i$. Similarly, U_{ij} is a union of Jordan regions U_{ijk} with $\bar{U}_{ijk} \subset U_{ij}$. We rearrange the U_{ijk} in a sequence $\{V_n\}$, and if $V_n = U_{ijk}$ we set $W_n = U_{ij}$. Then every open set O is a union of sets V_n with $\bar{V}_n \subset W_n \subset \bar{W}_n \subset O$.

A finite or infinite sequence of integers n_k is determined by the following condition: n_1 is 1 and n_k, $k > 1$, is the least integer for which

$$\bar{V}_1 \cup \cdots \cup \bar{V}_{n_{k-1}} \subset V_1 \cup \cdots \cup V_{n_k}.$$

Such an n_k will always exist, for $\bar{V}_1 \cup \cdots \cup \bar{V}_{n_{k-1}}$ is compact and the V_n form an open covering. It is greater than n_{k-1}, except if $V_1 \cup \cdots \cup V_{n_{k-1}} = F$. In fact, $n_k \leqq n_{k-1}$ would mean that

$$\bar{V}_1 \cup \cdots \cup \bar{V}_{n_{k-1}} \subset V_1 \cup \cdots \cup V_{n_{k-1}} \subset \bar{V}_1 \cup \cdots \cup \bar{V}_{n_{k-1}}.$$

Hence $V_1 \cup \cdots \cup V_{n_{k-1}}$ would be open and closed, and consequently equal to F. This can and will happen only if F is compact, and then $V_1, \cdots, V_{n_{k-1}}$ and $W_1, \cdots, W_{n_{k-1}}$ are finite sequences with the desired properties.

In the noncompact case we set $G_k = V_1 \cup \cdots \cup V_{n_k}$. The defining property implies $\bar{G}_{k-1} \subset G_k$. Each $\bar{G}_k$ is compact, and the union of all G_k is F. Write $H_k = \bar{G}_{k+1} - G_k \subset G_{k+2} - \bar{G}_{k-1}$. The set $G_{k+2} - \bar{G}_{k-1}$ is open and can therefore be represented as a union of sets $V_n \subset \bar{W}_n \subset G_{k+2} - \bar{G}_{k-1}$. Since H_k is compact it is covered by a finite number of these sets. We denote the sets in this finite covering by V_{kl}, and the corresponding W_n by W_{kl}.

If $i \leqq k-3$ the set $\bar{W}_{ij}$ cannot meet a set $\bar{W}_{kl}$, for $\bar{W}_{ij} \subset G_{i+2} \subset G_{k-1}$ while $\bar{W}_{kl} \cap G_{k-1} = 0$. Hence every $\bar{W}_{ij}$ meets only a finite number of $\bar{W}_{kl}$. The double sequences $\{V_{kl}\}$ and $\{W_{kl}\}$, suitably rearranged, satisfy the requirements of the lemma.

46C. A set Γ of Jordan arcs on F is said to be *discrete* if every point on F has a neighborhood which meets at most a finite number of arcs $\gamma \in \Gamma$. The next lemma is the key to the construction of a triangulation.

Lemma. *If Γ is a discrete set of Jordan arcs on F, then any two points in a region $G \subset F$ can be joined by a Jordan arc $\sigma \subset G$ whose intersection with the arcs $\gamma \in \Gamma$ consists of a finite number of points and subarcs of σ.*

We remark first that if p_0 can be joined to p_1 and p_1 can be joined to p_2 in the sense of the lemma, then p_0 can be joined to p_2.

For fixed $p_0 \in G$, let E consist of p_0 and all points $p \in G$ which can be joined to p_0 in the desired manner. If $p \in E$ does not lie on any $\gamma \in \Gamma$ there exists an open connected neighborhood $V(p) \subset G$ which does not intersect any $\gamma \in \Gamma$. The initial remark shows that $V(p) \subset E$.

Suppose now that $p \in E$ lies on at least one $\gamma \in \Gamma$. Then we can find an open connected neighborhood $V(p) \subset G$ which intersects only those $\gamma \in \Gamma$ on which p lies. Any $q \in V(p)$ can be joined to p by a Jordan arc in $V(p)$. We follow this arc from q to the first intersection with a $\gamma \in \Gamma$, and continue along this γ to p. The construction makes it clear that $V(p) \subset E$. Hence E is open, and the same reasoning can be repeated to show that $G - E$ is open. Since G is connected, it follows that $E = G$ as asserted.

46D. If G is a Jordan region the result of Lemma 45C can be extended to the case of two boundary points p_1, p_2, provided that they do not lie on any $\gamma \in \Gamma$. In fact, we can first join p_1 and p_2 by a Jordan arc σ_0 which lies in G save for its end points. Near p_1 and p_2 we can find subarcs σ_1, σ_2 which do not meet any $\gamma \in \Gamma$. Their end points in G can be joined by an arc σ' in the sense of the lemma. It is clear that a connecting Jordan arc can be extracted from the union $\sigma_1 \cup \sigma' \cup \sigma_2$.

On the other hand, if $p_1 \in \gamma$, say, it could easily happen that any arc leading to p_1 must cross γ infinitely many times, and the construction would not be possible.

46E. Consider the sequences $\{V_n\}$ and $\{W_n\}$ introduced by Lemma 46B. We will show that there exist closed Jordan regions J_n, such that $V_n \subset J_n \subset W_n$, whose boundaries γ_n have only a finite number of common points or arcs. They will then form a covering of finite character.

We take $J_1 = \bar{V}_1$ and assume that $J_1, \cdots, J_{n-1}$ have been constructed. We must find J_n so that γ_n intersects $\gamma_1 \cup \cdots \cup \gamma_{n-1}$ along a finite number of points and arcs. For the construction we represent $\bar{W}_n$ homeomorphically as a closed disk whose center corresponds to a point of V_n.

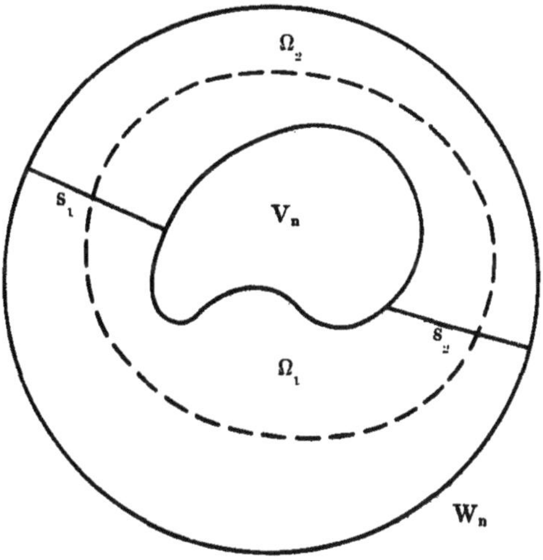

FIG. 8

It is a consequence of the Jordan curve theorem that a Jordan curve has an empty interior. For this reason it is possible to find two points p_1, p_2 in W_n on distinct radii, which do not lie on $\gamma_1 \cup \cdots \cup \gamma_{n-1}$. In addition they can be chosen so close to the circumference that the radial segments between p_1, p_2 and the boundary circle do not intersect $\bar{V}_n$. These segments, extended from the circle toward the center to the first intersection with $\bar{V}_n$, will be denoted by s_1, s_2 (see Fig. 8).

We apply the cross-cut theorem twice, first to the cross-cut formed by s_1, s_2 and one of the boundary arcs of V_n between p_1, p_2, then to the second boundary arc. It follows that s_1 and s_2 divide $W_n - \bar{V}_n$ into two Jordan regions Ω_1, Ω_2 whose boundaries are as indicated by the figure. We apply Lemma 46C, or rather its extension in 46D, to join p_1 and p_2

by cross-cuts σ_1, σ_2 of Ω_1, Ω_2 which meet $\gamma_1 \cup \cdots \cup \gamma_{n-1}$ in a finite number of points and arcs. The cross-cuts combine to a Jordan curve γ_n which bounds a closed Jordan region $J_n \subset W_n$.

It remains to prove that $V_n \subset J_n$. Since J_n has no boundary points in V_n, either $V_n \subset J_n$ or else V_n lies in the complement of J_n. In the latter case we see that the whole boundary of Ω_1 would belong to the outside region determined by γ_n, with the exception of p_1, p_2 that lie on its boundary. Thus Ω_1 would have no boundary points in the interior of J_n, and this would imply $\Omega_1 \cap J_n = 0$, contrary to the fact that σ_1 belongs at once to Ω_1 and to J_n. It follows that $V_n \subset J_n$, and the proof of Theorem 46A is complete.

46F. Obvious modifications of the proof make it applicable to bordered surfaces. If $\bar{F}$ is a bordered surface we construct its double $\hat{F}$. For $\hat{F}$ we construct a covering of finite character with the additional property that each γ_n intersects the border B in a finite number of points and arcs. This can be done by including the contours in the system Γ to which Lemma 46C is applied.

Finally, the proof of Theorem 45C needs a slight modification which consists in subdividing each J_m not only by cross-cuts on the curve γ_n, but also by all cross-cuts on B.

46G. Our method makes it easy to show that any two triangulations of a surface or bordered surface are equivalent.

Let two triangulations of F with the complexes K_1, K_2 be given. The Jordan arcs which correspond to their 1-simplices form a discrete set. Therefore we can construct a new triangulation whose 1-simplices intersect those of K_1, K_2 in a finite number of points or arcs. Observe that Lemma 46C must be used twice, once to construct a covering of finite character, and a second time in the proof of Theorem 45C when we subdivide the regions J_{mi}. If K is the new complex it is a simple matter to show that K, K_1 on one side and K, K_2 on the other have common refinements. It follows that K_1 and K_2 are equivalent.

Note the connection with the two definitions of equivalence that were compared with each other in 39. The result that we have just proved, together with the italicized statement in 39E, shows that two finite polyhedrons are simultaneously equivalent as polyhedrons and as cell complexes. Because a combinatorial proof would be quite involved we did not, in 39–42, make use of the identity of these two notions of equivalence.

46H. Let Ω be a regular region on F. Its contours form a discrete set of Jordan arcs. We can therefore find a triangulation whose 1-simplices intersect the boundary of Ω only in a finite number of points (or arcs). On passing to a suitable subdivision we obtain a simultaneous triangulation of F and $\bar{\Omega}$; more precisely, $\bar{\Omega}$ is triangulated by a subcomplex of the complex K which triangulates F. The existence of a complex K with this property was needed in 35B.

CHAPTER II
Riemann Surfaces

In current terminology *Riemann surfaces* are the domains of most general type which can be used to replace the complex plane in the theory of analytic functions of one complex variable. This is in strict accordance with the spirit of Riemann's own work, for Riemann was the first to recognize that plane regions are not sufficiently general to give a complete picture of the ideas that dominate function theory, even when restricted to a single variable.

The present chapter is of a preparatory nature, being devoted mainly to definitions and basic properties. In §1 we pay special attention to the construction of Riemann surfaces. This is followed in §2 by a discussion of some elementary aspects of function theory on Riemann surfaces, including some properties of subharmonic functions. The latter are used in §3 to construct the solution of Dirichlet's problem by Perron's method.

§1. DEFINITIONS AND CONSTRUCTIONS

We choose a definition of Riemann surface which emphasizes that we are dealing with manifolds that possess a certain structure. This approach makes it possible to use more general structures for comparison.

We are also stressing the methods of effective construction. These methods play an important role in the more advanced theory, for instance for the classification of Riemann surfaces (Ch. IV).

1. Conformal structure

1A. The notion of *Riemann surface* can be defined in many different ways. It is possible to base the theory on conformal mappings, or one may wish to single out the class of analytic functions, or, alternatively, the class of harmonic functions. All such definitions are essentially equivalent and the choice is a matter of taste.

The approach through conformal mappings is perhaps the most natural because it does justice to the geometric aspects of the theory.

1B. In what follows W will denote a connected Hausdorff space. We impose very strong conditions which will imply that W is a surface. The conditions take the form of relations between certain local homeomorphisms. They will be divided into three sets of postulates: the first set defines a *family* Φ *of local homeomorphisms*, the second defines a *structure class* Ψ, and the third describes the connection between Φ and Ψ.

As far as Φ is concerned the postulates are as follows:

(B1) Each $h \in \Phi$ is a topological mapping of an open set $V \subset W$ onto an open set in the complex plane.

(B2) If $h \in \Phi$ has domain V, then the restriction of h to any open set $V' \subset V$ is also in Φ.

(B3) Let h be a topological mapping of an open set $V \subset W$ onto an open set in the complex plane, and suppose that V is covered by open subsets V'. If the restriction of h to each V' is in Φ, then the same shall be true of h with domain V.

(B4) The domains of all $h \in \Phi$ form a covering of W.

Condition (B3) is the most restrictive. It means that the mappings in Φ are defined by local properties alone.

1C. Let W_0 be the complex plane. A family Ψ of local homeomorphisms with domain and range on W_0 is called a *structure class* if the following is true:

(C1) The identity mapping belongs to Ψ.

(C2) If $g \in \Psi$, then the inverse mapping g^{-1} is also in Ψ.

(C3) If $g_1, g_2 \in \Psi$, then the composite mapping $g_1 \circ g_2$ is in Ψ, provided that it is defined.

We recall that the composite mapping is defined if the range of g_2 is contained in the domain of g_1.

A trivial example is the class C^0 of all homeomorphisms with open domain and range. More restrictive is the class C^k of homeomorphisms g such that g and g^{-1} have continuous partial derivatives up to the order k. The most important example for our purposes is the class of directly conformal mappings. Other examples are the class C^∞ of homeomorphisms with continuous partial derivatives of all orders, and the class C^ω of homeomorphisms that are analytic in the real sense.

1D. We return to the case of an arbitrary W and introduce a connection between Φ and a structure class Ψ. We shall say that Φ defines a *structure of class* Ψ if it satisfies these conditions:

(D1) If $h_1, h_2 \in \Phi$ have the same domain V, then $h_1 \circ h_2^{-1} \in \Psi$.

(D2) If $h \in \Phi$ and $g \in \Psi$, then $g \circ h \in \Phi$ provided that it is defined.

Observe that (D1) is meaningful for any two mappings whose domains overlap, for by (B2) we can consider the restrictions to the common part. Similarly, (D2) has a meaning as soon as the domain of g overlaps the range of h.

To abbreviate, we speak frequently of the structure Φ rather than the structure defined by the family Φ.

1E. The structure class which is formed by all directly conformal mappings will be denoted by A. A structure of class A will also be called a *conformal structure*. We are now ready to formulate the definition of a Riemann surface.

Definition. *A Riemann surface is a connected Hausdorff space W together with a conformal structure defined by a family Φ of local homeomorphisms on W.*

The correct notation for a Riemann surface is (W, Φ), but it is frequently abbreviated to W. It is clear from (B1) and (B4) that W, as a topological space, is a surface. Moreover, since all mappings of the form $h_1 \circ h_2^{-1}$ are sense-preserving, it is an orientable surface.

The class A of directly conformal mappings can be replaced by the structure class $\hat{A}$ of directly or indirectly conformal mappings. A space with a structure of class $\hat{A}$ may be referred to as a *Riemann surface without orientation*. The theory of such surfaces differs only in obvious ways from the theory of ordinary Riemann surfaces and we shall refer to them only occasionally.

1F. A structure Φ of a given class Ψ is completely determined by a sufficiently inclusive subset of Φ. We call a set B a basis for a structure of class Ψ if it satisfies the following requirements:

(F1) Each $h \in B$ is a topological mapping of an open set $V \subset W$ onto an open set in the complex plane.

(F2) The composite mappings $h_1 \circ h_2^{-1}$ with $h_1, h_2 \in B$ are of class Ψ in their domain of definition.

(F3) The domains of all $h \in B$ form a covering of W.

If B satisfies these conditions we obtain a structure Φ of class Ψ by including all topological mappings h' with open domain and range which are such that $h' \circ h^{-1} \in \Psi$ for all $h \in B$. (B2) and (B3) are trivially fulfilled, and it follows from (F2) that $B \subset \Phi$. Next, $h \circ h'^{-1} \in \Psi$ by virtue of (C2), and if $h', h'' \in \Phi$ we find by (C3) that $h' \circ h''^{-1} = (h' \circ h^{-1}) \circ (h'' \circ h^{-1})^{-1} \in \Psi$ locally, and therefore in its whole domain. Finally, (D2) is likewise an immediate consequence of (C3), for $(g \circ h') \circ h^{-1} = g \circ (h' \circ h^{-1}) \in \Psi$. We conclude that Φ is a structure of class Ψ. Conversely, it is the only such structure which contains B.

Suppose that $\Psi_1 \subset \Psi_2$ are two structure classes, one contained in the other. Then any structure of class Ψ_1 is a basis for a structure of class Ψ_2. In this sense any structure determines a structure class C^0, which is just the topological structure, and a conformal structure determines a structure of arbitrary class C^k.

1G. For the complex plane, or any subregion of the complex plane, the identity mapping by itself can be considered as a basis for a structure of any class, and in particular for a conformal structure. Unless the contrary is mentioned we shall always have this conformal structure of the plane in mind.

In the case of the sphere we use a basis consisting of two mappings, the identity mapping defined on the finite plane, and the mapping $z \to 1/z$ defined for $z \neq 0$ and extended to the point at infinity by $\infty \to 0$. It is clear that these mappings determine a conformal structure, and the sphere with this structure is called the *Riemann sphere.*

We remark that any subregion of a Riemann surface becomes a Riemann surface by retaining only those mappings in Φ whose domain is contained in the subregion. Whenever we consider a subregion it will be endowed with this induced structure.

2. Analytic mappings

2A. We wish to define *analytic mappings* from one Riemann surface to another. We begin by considering the case of an *analytic function* which is a mapping from a Riemann surface to the complex plane. We need only consider open connected domains, and it is thus no restriction to assume that the domain is the whole surface.

Definition. *A complex-valued function f is said to be analytic on the Riemann surface (W, Φ) if and only if $f \circ h^{-1}$ is analytic on $h(V)$, in the classical sense, for every $h \in \Phi$ with domain V.*

It is sufficient to assume that $f \circ h^{-1}$ is analytic for the mappings h that belong to a basis B. In fact, for arbitrary $h' \in \Phi$ with domain V' we have $f \circ h'^{-1} = (f \circ h^{-1}) \circ (h \circ h'^{-1})$, the latter function being defined and analytic on $h'(V \cap V')$. Since the domains V form an open covering it follows that $f \circ h'^{-1}$ is analytic on all of $h'(V')$.

2B. In the case of two Riemann surfaces the corresponding definition is as follows:

Definition. *A continuous mapping f of one Riemann surface (W_1, Φ_1) into another Riemann surface (W_2, Φ_2) is called analytic if every function $h_2 \circ f \circ h_1^{-1}$ with $h_1 \in \Phi_1$, $h_2 \in \Phi_2$ is analytic in its domain of definition.*

If h_1, h_2 are defined in V_1, V_2 respectively, the composite function $h_2 \circ f \circ h_1^{-1}$ is defined on $h_1(V_1 \cap f^{-1}(V_2))$. The continuity of f guarantees that this domain is open. It is again easy to see that it is sufficient to consider mappings h_1, h_2 that belong to bases of Φ_1, Φ_2 respectively. This remark makes it clear that the two definitions agree when W_2 is the complex plane, for then we may take h_2 to be the identity mapping.

2C. If f is one to one we speak of a *conformal mapping*. In this case the range of each mapping $h_2 \circ f \circ h_1^{-1}$ is open, and the inverse $h_1 \circ f^{-1} \circ h_2^{-1}$

is analytic. Thus $f(V_1 \cap f^{-1}(V_2)) = f(V_1) \cap V_2$ is open, and since the V_2 form a covering, $f(V_1)$ is open. Hence f is a homeomorphism, and f^{-1} is analytic.

Two Riemann surfaces which can be mapped conformally onto each other are said to be *conformally equivalent*. One of the central problems in the theory of Riemann surfaces is to determine explicit conditions for conformal equivalence.

2D. Although our main concern is with Riemann surfaces it is indispensable that we consider more general structures. In particular, it is necessary to define differentiable functions and differentiable mappings.

In 1C a structure class was defined as a family of local homeomorphisms. We must now consider extended structure classes which contain other functions as well. An extended structure class Ψ is still required to contain a subclass formed by a family of local homeomorphisms that satisfy (C1)–(C3). In addition, Ψ may contain functions g that have no inverse, or whose inverse is not in Ψ. These functions shall have open domains, and the basic conditions (B2), (B3) remain in force. Moreover, we require that they satisfy (C3), i.e. Ψ shall be closed under composition.

It is clear that each class C^k has a natural extension of this sort, namely to the class of all complex functions whose real and imaginary parts have continuous partial derivatives up to order k. Similarly, the extension of the class A of directly conformal mappings consists of all analytic functions. We retain the names C^k and A for the extended classes.

2E. A structure Φ of class Ψ will still be defined in terms of the homeomorphisms of class Ψ, that is to say by homeomorphisms g with g and $g^{-1} \in \Psi$. It is only when we define mappings of class Ψ from one surface to another that the extended class is needed.

Definition. *Suppose that the structures of* (W_1, Φ_1) *and* (W_2, Φ_2) *are of class* Ψ_1, Ψ_2 *respectively with* $\Psi_1 \subset \Psi$, $\Psi_2 \subset \Psi$. *A continuous mapping of* W_1 *into* W_2 *is said to be of class* Ψ *with respect to these structures if and only if* $h_2 \circ f \circ h_1^{-1} \in \Psi$ *for all* $h_1 \in \Phi_1$, $h_2 \in \Phi_2$.

Because of the condition $\Psi_1 \cup \Psi_2 \subset \Psi$ it is sufficient to verify that $h_2 \circ f \circ h_1^{-1} \in \Psi$ when h_1, h_2 are taken from bases of Φ_1, Φ_2 respectively. Also, the condition remains valid for the structures of class Ψ which are determined by Φ_1, Φ_2 as bases. In other words, we would not have lost any generality by requiring that $\Psi_1 = \Psi_2 = \Psi$. However, we wish to emphasize that it makes sense to speak of functions of class C^k on a Riemann surface, but it makes no sense to speak of analytic functions on a surface of class C^k.

In the case of a one to one mapping we can no longer guarantee that

f^{-1} is of the same class. Therefore, we speak of a homeomorphism of class Ψ only when it is known that f and f^{-1} are both of class Ψ.

3. Bordered Riemann surfaces

3A. The notion of bordered surface was introduced in Ch. I, 13B. It has a rather obvious analogue for Riemann surfaces which will play an important role. Since nothing essentially new is added we shall point very briefly to the modifications in the previous definitions that become necessary.

The connected Hausdorff space will now be denoted by $\overline{W}$. In the definition of a family of local homeomorphisms we require that the range of each $h \in \Phi$ is a relatively open subset of the closed upper half-plane. A structure class Ψ is now a family of local homeomorphisms defined on the closed half-plane. In the case of the classes C^k this means that certain derivatives are one-sided. A function $g \in A$ is one which has an analytic extension to an open subset of the plane.

With these changes a structure Φ of class Ψ on $\overline{W}$ can be defined exactly as before. Clearly, if $\overline{W}$ carries a structure of this type it is a bordered surface, provided that the border B is not empty. If the structure is conformal, $\overline{W}$ is called a *bordered Riemann surface*. The interior W is then a Riemann surface with the induced conformal structure.

It is worth pointing out that the intrinsic characterization of border points (see Ch. I, 13C) is a much simpler matter in the presence of a conformal structure. In fact, let g be an analytic homeomorphism whose domain V is contained in the open half-plane and whose range lies in the closed half-plane. Since g is not a constant, the hypothesis $\operatorname{Im} g(z) \geqq 0$ implies $\operatorname{Im} g(z) > 0$ in V by virtue of the maximum principle for harmonic functions. In other words, $g(V)$ cannot meet the real axis, and we find quite trivially that the border points are uniquely characterized by having all their parametric images on the real axis.

Since $\overline{W}$ is orientable, the border B has a positive direction (Ch. I, 13E).

3B. We recall also the notions of 1-dimensional submanifold and regular imbedding that were introduced in Ch. I, 13. These terms were used with a purely topological meaning, but when applied to surfaces (W, Φ) of a given structure class Ψ they should be appropriately reinterpreted. The details are very simple, and only brief indications are needed.

A subset Γ of W is said to be a 1-dimensional submanifold of (W, Φ) if every $p \in \Gamma$ has an open neighborhood V which is mapped by a homeomorphism $h \in \Phi$ onto $|z| < 1$, in such a way that $V \cap \Gamma$ corresponds to the real diameter. In the case of a conformal structure Γ is also called an *analytic submanifold*, and its components are *analytic curves*.

The definition of a regularly imbedded subregion can be phrased exactly as before. We recall that Ω is regularly imbedded if Ω and its exterior have a common boundary which is a 1-dimensional submanifold. A regularly imbedded subregion of a Riemann surface is thus bounded by analytic curves.

As before, we use the name *regular region* for any regularly imbedded region with compact closure whose complement has no compact components.

3C. If $(\overline{W}_1, \Phi_1)$ and $(\overline{W}_2, \Phi_2)$ are two bordered Riemann surfaces with homeomorphic borders B_1, B_2 there is frequently reason to weld them together to a single Riemann surface by identifying the borders. Identified points must correspond to each other under a homeomorphism φ which maps B_1 onto B_2. In order that the welding be possible two conditions must be fulfilled: (1) the homeomorphism must be *analytic*, (2) it must be *direction reversing*.

To make these conditions explicit, let $p_1 \in B_1$ and $p_2 = \varphi(p_1) \in B_2$ be corresponding border points. We choose mappings $h_1 \in \Phi_1$, $h_2 \in \Phi_2$ whose domains V_1, V_2 contain p_1, p_2 respectively. The function $\omega = h_2 \circ \varphi \circ h_1^{-1}$, restricted to the part of $h_1(V_1 \cap B_1)$ where it is defined, is a real function of a real variable.

Our first condition requires ω to be analytic in the real sense at $z_1 = h_1(p_1)$. This means that ω has a complex analytic extension to a disk Δ_1 about z_1. Observe that the condition is obviously independent of the choice of h_1 and h_2. The same condition shall hold for ω^{-1}, and hence $\omega'(z_1) \neq 0$.

The second condition is expressed by the inequality $\omega'(z_1) < 0$. It is again independent of the choice of h_1, h_2, for if they are replaced by k_1, k_2 we have $k_2 \circ \varphi \circ k_1^{-1} = (k_2 \circ h_2^{-1}) \circ \omega \circ (h_1 \circ k_1^{-1})$, and the analytic functions $k_2 \circ h_2^{-1}$, $h_1 \circ k_1^{-1}$ have positive derivatives as seen by the fact that they take reals into reals and preserve the sign of the imaginary part. If Δ_1 is sufficiently small it follows from our condition that ω maps Δ_1 onto a symmetric region Δ_2 in such a way that the lower half of Δ_1 corresponds to the upper half of Δ_2.

3D. We can now proceed to define the welding. First of all, we form the topological sum of $\overline{W}_1$, $\overline{W}_2$ and identify corresponding border points p_1 and $p_2 = \varphi(p_1)$. It is easy to see that this leads to a connected Hausdorff space W. Next, we define a basis for a conformal structure on W. This basis will contain all mappings $h_1 \in \Phi_1$, $h_2 \in \Phi_2$ with domains $V_1 \subset W_1$, $V_2 \subset W_2$. In addition, we must have mappings whose domains contain the identified boundary points. We consider an identified pair p_1, p_2 and determine ω, Δ_1, Δ_2 as above; upper and lower halves will be denoted by Δ_i^+, Δ_i^-, and we suppose that $\Delta_i^+ \subset h_i(V_i)$, $i = 1, 2$. Take V to be the union of $h_1^{-1}(\Delta_1^+)$ and $h_2^{-1}(\Delta_2^+)$. In the first of these sets we choose $h = h_1$, in the second we define h by $h(p)=\omega^{-1}(h_2(p))$. These definitions agree on the border, for there $\omega^{-1}\circ h_2=h_1\circ\varphi^{-1}$. Moreover, h is a topological mapping of V onto Δ_1.

The condition for a conformal basis is trivially satisfied if one of the overlapping domains is contained in W_1 or W_2. If h and k are defined in overlapping neighborhoods of two border points we see that $h\circ k^{-1}$ is analytic off the real axis, and since it is continuous and real on the real axis it follows by the reflection principle that it is also analytic on the real axis. Therefore we have defined a basis which leads to a conformal structure on W. Observe that $\overline{W}_1$ and $\overline{W}_2$ are analytically imbedded in W and have the induced structure.

The welding can also take place along open subsets of the borders. If the complete borders are not used the resulting surface is a bordered Riemann surface.

3E. The welding process can be used to form the *double* of a bordered Riemann surface. We note first that every Riemann surface (W, Φ), bordered or not, has a conjugate surface (W, Φ^*), obtained by replacing each $h\in\Phi$ by $h^*:p\to-\bar{h}(p)$. The new structure is conformal, for $h_1^*\circ h_2^{*-1}$ has the explicit expression $z\to-\bar{\varphi}_{12}(-\bar{z})$, where $\varphi_{12}=h_1\circ h_2^{-1}$, and this is manifestly a directly conformal mapping.

In the case of a bordered Riemann surface $(\overline{W}, \Phi)$ we weld it to $(\overline{W}, \Phi^*)$ by means of the identity mapping of the border. In the neighborhood of a border point ω may be taken to be of the form $h^*\circ h^{-1}$, whence $\omega(z)=-\bar{z}=-z$ for real z. The correspondence is thus analytic and direction reversing, so that the welding is indeed possible.

4. Riemann surfaces as covering surfaces

4A. A nonconstant analytic mapping f of a Riemann surface (W_1, Φ_1) into another Riemann surface (W_2, Φ_2) defines W_1 as a covering surface (W_1, f) of W_2. To see this, let q_1 and $q_2=f(q_1)$ be corresponding points, and choose mappings $h_1\in\Phi_1$, $h_2\in\Phi_2$ defined in neighborhoods of q_1, q_2. The function $\varphi=h_2\circ f\circ h_1^{-1}$ is analytic. If the derivative $\varphi'(h_1(q_1))$ is $\neq 0$, then φ is a homeomorphism in a neighborhood of $h_1(q_1)$, and hence f is a local homeomorphism. In the general case φ' has at most an isolated zero at $h_1(q_1)$. Therefore we can find a neighborhood V_1 of q_1 with the property that V_1-q_1 is a smooth covering surface of W_2 under the mapping f. Moreover, we can choose V_1 so small that $f(q)\neq f(q_1)$ for $q\in V_1-q_1$; this permits us to regard V_1-q_1 as a smooth covering surface of W_2-q_2. By comparison with Definition 20B, Ch. I, we conclude that (W_1, f) is indeed a covering surface of W_2.

4B. The reverse situation is more interesting. We prove in this respect:

Theorem. *Suppose that (W_1, f) is a covering surface of W_2, and that W_2*

has a conformal structure Φ_2. Then there exists a unique conformal structure Φ_1 on W_1 which makes f an analytic mapping of (W_1, Φ_1) into (W_2, Φ_2).

The structure Φ_1 is said to be induced by the structure Φ_2. If an induced structure exists its mappings h_1 must be such that $\varphi = h_2 \circ f \circ h_1^{-1}$ is analytic for all $h_2 \in \Phi_2$. Let us therefore denote by Φ_1 the family of all homeomorphisms h_1 with this property. The existence and uniqueness of the induced structure will both be proved if we show that Φ_1 satisfies conditions (B1)–(B4) in 1B, and (D1)–(D2) in 1D, the latter with respect to the structure class A of directly conformal mappings.

Conditions (B1)–(B3) are trivially fulfilled. To prove (B4), let q_1 be an arbitrary point on W_1. Set $q_2 = f(q_1)$, and choose a mapping $h_2 \in \Phi_2$ whose domain contains q_2; we may assume that $h_2(q_2) = 0$. As we showed in Ch. I, 20E, there exists a homeomorphism h_1, defined in a neighborhood of q_1, which satisfies an equation $h_2(f(p_1)) = h_1(p_1)^n$; n is the multiplicity of the branch point at q_1. The equation may be written in the form $h_2 \circ f \circ h_1^{-1} = \varphi$ where $\varphi(z_1) = z_1^n$. If k_2 is another mapping in Φ_2 we obtain $k_2 \circ f \circ h_1^{-1} = (k_2 \circ h_2^{-1}) \circ \varphi$, and since $k_2 \circ h_2^{-1} \in A$ the composed function is analytic whenever it is defined. In other words, we have shown that $h_1 \in \Phi_1$, and hence that the domains of mappings in Φ_1 form a covering of W_1.

As for (D1), suppose that $h_1, k_1 \in \Phi_1$ have the same domain V_1, and write $h_2 \circ f \circ h_1^{-1} = \varphi$, $h_2 \circ f \circ k_1^{-1} = \psi$; we assume that the domain of $h_2 \in \Phi_2$ contains $f(V_1)$. If f is topological in V_1 we obtain $h_1 \circ k_1^{-1} = (\varphi^{-1} \circ h_2 \circ f) \circ (f^{-1} \circ h_2^{-1} \circ \psi) = \varphi^{-1} \circ \psi$, and this is an analytic function. If f is not known to be topological the same conclusion holds for a neighborhood of any point in $k_1(V_1)$ which does not correspond to a branch point. Consequently $h_1 \circ k_1^{-1}$ is analytic in $k_1(V_1)$, except for isolated singularities. But because $h_1 \circ k_1^{-1}$ is known to be continuous these singularities are removable. We have verified that condition (D1) is fulfilled.

Finally, if g is complex analytic and $h_1 \in \Phi_1$ it is almost trivial that $g \circ h_1 \in \Phi_1$, for $h_2 \circ f \circ (g \circ h_1)^{-1} = (h_2 \circ f \circ h_1^{-1}) \circ g^{-1}$, and the function on the right is obtained by composing two analytic functions. Thus (D2) is also satisfied, and we have proved that Φ_1 is a conformal structure. The fact that f becomes an analytic mapping from (W_1, Φ_1) to (W_2, Φ_2) is an immediate consequence of the definition of Φ_1.

4C. In the preceding theorem we have supposed that Φ_2 is given, and the problem was to find Φ_1. One can also ask whether there exists a Φ_2 which induces a given Φ_1. In this respect there is an almost obvious, but very important, necessary condition. Suppose that (W_1, f), regarded as a covering surface of W_2, has a cover transformation φ, that is to say a mapping of W_1 onto itself which satisfies $f \circ \varphi = f$. Consider a point $q_1 \in W_1$ which is not a branch point. There exists a neighborhood V of q_1 which is mapped topologically by f. Then $\varphi(V)$ is a neighborhood of $\varphi(q_1)$ which is also mapped topologically. If f_1, f_2 denote the restrictions of f to V and $\varphi(V)$ respectively, we can write $\varphi = f_2^{-1} \circ f_1$ in V. Granted the existence of Φ_2 it would follow that φ is at least locally obtained by composition of two analytic mappings, and would therefore be analytic. In other words, it is necessary that Φ_1 be compatible with φ in the sense that φ is analytic with respect to the structure Φ_1.

The next theorem describes a situation in which this necessary condition is also sufficient:

Theorem. *Let $\mathcal{G}$ be a group of one to one conformal mappings φ of a Riemann surface (W_1, Φ_1) onto itself. We suppose that the following condition is satisfied:*

(**C**) *Given any two compact sets A, $B \subset W_1$ there are at most a finite number of mappings $\varphi \in \mathcal{G}$ such that $\varphi(A)$ meets B.*

Then it is possible to find a Riemann surface (W_2, Φ_2) and an analytic mapping f of (W_1, Φ_1) onto (W_2, Φ_2) which is such that the group of cover transformations of (W_1, f) is exactly $\mathcal{G}$.

The construction is unique up to conformal mappings of (W_2, Φ_2).

In standard terminology, condition (C) means that $\mathcal{G}$ is *properly discontinuous.*

4D. Before proving the theorem we discuss a familiar application. Let ω_1, ω_2 be complex numbers whose ratio ω_1/ω_2 is nonreal. In the complex plane the transformations $z \to z + n_1\omega_1 + n_2\omega_2$ with integral n_1, n_2 form a group of conformal self-mappings. It is elementary to show that condition (C) is fulfilled. If the self-mappings are to become cover transformations, the underlying surface must have one point for each class of points that correspond to each other under the mappings. We have thus to identify points which are equivalent in this sense. The resulting surface is a torus, and the theorem asserts that this torus has a unique conformal structure which induces the conformal structure of the plane.

4E. The example suggests the method of construction. Let W_2 be the topological space obtained by identifying all points on W_1 which correspond to each other under a mapping $\varphi \in \mathcal{G}$. We denote by f the mapping which carries each $p_1 \in W_1$ into its equivalence class, and we recall that a set on W_2 is open if and only if its inverse image is open (Ch. I, 2F). It follows that f is continuous and carries open sets into open sets. The precise proof of the latter property is as follows: Let O be an open set on W_1. Then all images $\varphi(O)$ are open, and hence $f^{-1}[f(O)] = \cup\ \varphi\ (O)$ is open; this means, by definition, that $f(O)$ is open.

We have to prove that W_2 is a Hausdorff space. Consider two points $a, b \in W_1$ with $f(a) \neq f(b)$. By (C) there are only a finite number of images $\varphi(a)$ in a compact neighborhood of b, and for this reason we can find a compact neighborhood B of b which does not contain any $\varphi(a)$. Similarly, only a finite number of $\varphi(B)$ meet a given compact neighborhood of a, and therefore there exists a compact neighborhood A of a which does not meet any $\varphi(B)$. The images of A are then disjoint from the images of B, for if $\varphi_1(A)$ meets $\varphi_2(B)$, then $\varphi_1^{-1}(\varphi_2(B))$ would meet A. This proves that $f(A)$ and $f(B)$ are disjoint. On the other hand, $f(A)$ and $f(B)$ are neighborhoods of $f(a)$ and $f(b)$ respectively. Indeed, we can find an open set O such that $a \in O \subset A$. This implies $f(a) \in f(O) \subset f(A)$, and since $f(O)$ is open $f(A)$ is a neighborhood. We conclude that W_2 is a Hausdorff space.

4F. We prove next that W_2 is a surface, and simultaneously that (W_1, f) is a covering surface of W_2. A conformal mapping which is not the identity has isolated fixed points, and it follows from (C) that there are only a finite number of $\varphi \in \mathcal{G}$ with a given fixed point. Let $\varphi_1, \cdots, \varphi_n$ be the mappings with the fixed point q_1; they form a subgroup of $\mathcal{G}$. For a given compact neighborhood A of q_1 there are only a finite number of $\varphi(A)$ which meet A. If any of these φ is such that $\varphi(a) \neq a$ we can replace A by a smaller neighborhood A' for which $\varphi(A') \cap A' = 0$. In a finite number of steps we can thus construct an open neighborhood U_1 which intersects $\varphi(U_1)$ only is φ is one of the mappings $\varphi_1, \cdots, \varphi_n$; we may also assume that U_1 does not contain any fixed points other than q_1. The set $V_1 = \varphi_1(U_1) \cap \cdots \cap \varphi_n(U_1)$ is mapped onto itself by $\varphi_1, \cdots, \varphi_n$, and $V_1 \cap \varphi(V_1) = 0$ for all other $\varphi \in \mathcal{G}$. Every $p_1 \neq q_1$ in V_1 has thus exactly n equivalent points in V_1.

We can choose U_1 arbitrarily small. For this reason we can assume the existence of a mapping $h_1 \in \Phi_1$ which is defined on V_1, and we may normalize the mapping so that $h_1(q_1) = 0$. The product $\prod_{i=1}^{n} h_1(\varphi_i(p_1))$ has the same value at equivalent points in V_1. Therefore it represents a function on $V_2 = f(V_1)$, that is to say there exists a function h_2 on V_2 such that $h_2(f(p_1)) = \prod_{i=1}^{n} h_1(\varphi_i(p_1))$ for all $p_1 \in V_1$. The function $g = h_2 \circ f \circ h_1^{-1} = \prod_{i=1}^{n} (h_1 \circ \varphi_i \circ h_1^{-1})$ is analytic in $h_1(V_1)$ and has a zero of order n at the origin, for each factor is analytic with a simple zero.

Because g has a zero of order n we can find a disk $\Delta_1 \subset h_1(V_1)$, centered at the origin, which is so small that g takes no value more than n times in Δ_1. Without changing the notation we restrict U_1 so that it is contained in $h_1^{-1}(\Delta_1)$. Then $V_1 \subset U_1$ is also contained in $h_1^{-1}(\Delta_1)$, and the functions h_2 and g remain unchanged except for being restricted to smaller domains. The restriction has the effect that h_2 will now be one to one. In fact, for any $p_2 \neq f(q_1)$ in V_2 there are n distinct points $p_{11}, \cdots, p_{1n} \in V_1$ with

$f(p_{1i})=p_2$. By the identity $g \circ h_1 = h_2 \circ f$ it follows that the equation $g(z)=h_2(p_2)$ has the roots $h_1(p_{11}), \cdots, h_1(p_{1n})$, all distinct and in Δ_1. Since there can be no other roots it follows that the p_{1i}, and hence p_2, are completely determined by the value $h_2(p_2)$. The special value 0 is taken only for $p_2=f(q_1)$. We conclude that h_2 is one to one.

To prove that h_2 is a homeomorphism we investigate the direct and inverse images of open sets. Let O_2 be an open subset of $V_2=f(V_1)$. Because the range of f includes V_2 we have $f(f^{-1}(O_2))=O_2$. The identity $g \circ h_1 = h_2 \circ f$ yields $h_2(O_2)= h_2[f(f^{-1}(O_2))]=g[h_1(f^{-1}(O_2))]$. But f^{-1}, h_1 and g transform open sets into open sets. Hence $h_2(O_2)$, and in particular $h_2(V_2)$, are open. Similarly, let O_2' be an open subset of $h_2(V_2)$. Then $h_2^{-1}(O_2')=h_2^{-1}[g(g^{-1}(O_2'))]=f[h_1^{-1}(g^{-1}(O_2'))]$ is open. We have shown that h_2 maps V_2 topologically on an open set in the plane. Hence W_2 is a surface.

4G. To continue the proof, we replace V_1 by the connected component which contains q_1. Let it be V_1', and denote its image by $V_2'=f(V_1')$. Since (Δ_1, g) is a covering surface of the plane, we see that $(V_1', g \circ h_1) =(V_1', h_2 \circ f)$ is a covering surface of its projection $h_2(V_2')$, and hence (V_1', f) is a covering surface of V_2'. This local result implies that (W_1, f) is a covering surface of W_2. The branch points occur at points with $n>1$, that is to say at the fixed points of transformations $\varphi \in \mathscr{G}$ other than the identity.

The mappings h_2 form a basis for a conformal structure Φ_2 on W_2, and f is analytic from (W_1, Φ_1) to (W_2, Φ_2). It is clear that every $\varphi \in \mathscr{G}$ satisfies $f \circ \varphi = f$ and is thus a cover transformation. Conversely, suppose that ψ is a cover transformation, i.e. a homeomorphism which satisfies $f \circ \psi = f$. At any given point q_1 this condition implies that $\psi(q_1)=\varphi(q_1)$ for some $\varphi \in \mathscr{G}$. If q_1, and hence $\varphi(q_1)$, is not a branch point, f is a local homeomorphism near $\varphi(q_1)$, and from $f \circ \psi = f \circ \varphi$ we are able to conclude that $\psi(p_1)=\varphi(p_1)$ in a neighborhood of q_1. When the branch points are removed W_1 remains connected, and we find in the familiar way that $\psi=\varphi$ everywhere. In other words, every cover transformation belongs to $\mathscr{G}$.

For the uniqueness proof we assume that f' maps (W_1, Φ_1) in a similar manner onto a Riemann surface (W_2', Φ_2'). Since equivalent points have coinciding images there exists a one to one mapping g of W_2 onto W_2' which satisfies $g \circ f = f'$. If $O' \subset W_2'$ is open, $g^{-1}(O')=f[f'^{-1}(O')]$ is open. Hence g is continuous, and the same reasoning applies to g^{-1}, showing that g is a homeomorphism. It is clear that g is analytic at any point which is not the projection of a branch point. Because of its topological character it is then analytic everywhere, and it is consequently a conformal mapping. This concludes the proof.

4H. As a further application of Theorem 4C we consider the case of a *Fuchsian group*. The group $\mathcal{G}$ is now a group of linear transformations which map the unit disk $|z|<1$ on itself; we assume expressly that the group is properly discontinuous. The surface W_1 is the unit disk, and the theorem asserts that we obtain a Riemann surface W_2 when equivalent points are identified. If $\mathcal{G}$ contains no elliptic transformations there are no fixed points inside the unit circle, and hence W_1 will be a regular covering surface of W_2. Moreover, since W_1 is simply connected it will be the universal covering surface of W_2, and we see that $\mathcal{G}$ can be identified with the fundamental group of W_2.

It will later be shown that with few exceptions the universal covering surface of a Riemann surface can always be represented as the unit disk, in which case the fundamental group appears as a Fuchsian group without fixed points. From this point of view the theory of Riemann surfaces and the theory of Fuchsian groups are thus almost equivalent. Some formal advantages can be gained from this equivalence, but it is usually preferable to study the Riemann surface by more direct methods.

5. Metric structures

5A. In classical differential geometry one considers surfaces in three dimensional space with a given equation $F(x_1, x_2, x_3)=0$. It is slightly more general to assume that the equation is given in parametric form: $x_i=x_i(u, v)$, $i=1, 2, 3$. In this case the surface is usually thought of as a point set which is covered by parametric regions. The latter are in homeomorphic correspondence with open sets in the plane, and the parametric equations are the explicit expressions for these homeomorphisms.

The geometry of the surface is determined by the fundamental form

$$ds^2 = \sum_{i=1}^{3}\left(\frac{\partial x_i}{\partial u}du+\frac{\partial x_i}{\partial v}dv\right)^2,$$

which is invariant under a change of the parameters. In particular, the angle between two curves is calculated by means of the fundamental form. This makes it possible to speak of conformal mappings, and we must show that this geometric notion of conformality can be used to introduce a conformal structure in the sense of our earlier definitions.

5B. We can take a more general point of view and assume that there is given an abstract surface W which is endowed with a *Riemannian metric*, determined by a fundamental form

$$ds^2 = E\,dx^2+2F\,dxdy+G\,dy^2.$$

This form is supposed to be positive definite, and invariant under transformations of the parameters, now denoted by x, y.

The invariance condition needs perhaps a few words of explanation. It is assumed, first of all, that W carries a structure Φ of class C^1. To every mapping $h \in \Phi$ with domain V correspond three real-valued functions E, F, G in $h(V)$, satisfying the conditions $E > 0$, $EG - F^2 > 0$. If h, h_1 have overlapping domains we denote the mapping $h_1 \circ h^{-1}$ as $(x, y) \to (x_1, y_1)$, and the functions associated with h_1 by E_1, F_1, G_1. These functions, or more precisely the composite functions $E_1 \circ h_1 \circ h^{-1}$ etc., shall be related to E, F, G by identities which are most conveniently expressed in the form

$$E_1\, dx_1^2 + 2F_1\, dx_1 dy_1 + G_1\, dy_1^2 = E\, dx^2 + 2F\, dxdy + G\, dy^2$$

with the accepted meaning of this equation.

Since we are interested only in angles we can go one step further and replace the above relation by one of the form

$$E_1\, dx_1^2 + 2F_1\, dx_1 dy_1 + G_1\, dy_1^2 = \rho(E\, dx^2 + 2F\, dxdy + G\, dy^2)$$

where ρ is an arbitrary positive factor. This means that only the ratio $E : F : G$ is determined, while the scale remains arbitrary. Since questions of conformality are independent of the scale we say in these circumstances that (W, Φ) carries a *conformal metric.*

5C. An *isothermal* parameter system is a mapping $h \in \Phi$ whose associated coefficients E, F, G satisfy $E = G$, $F = 0$. If h_1 and h_2 are both isothermal, then it is well known from elementary differential geometry that $h_1 \circ h_2^{-1}$ is directly or indirectly conformal. Conversely, if g is conformal, then $g \circ h$ is isothermal together with h. We assume henceforth that Φ is an oriented structure. Then all mappings $h_1 \circ h_2^{-1}$ are directly conformal, and in order to conclude that the subset $\Phi_0 \subset \Phi$ formed by all isothermal parameter systems defines a conformal structure on W, we need merely show that every point is in the domain of an isothermal parameter system. When this is so, the conformal structure Φ_0 is said to be induced by the given conformal metric.

5D. The remaining existence proof cannot be carried out without additional regularity assumptions. We shall consider only the classical case of an analytic conformal metric, that is to say we suppose that Φ belongs to the structure class C^ω (see 1C), and that the given ratios $E : F : G$ are analytic in the real sense.

Theorem. *Every analytic conformal metric on an orientable surface induces a conformal structure.*

Suppose that q lies in the domain of $h \in \Phi$. The problem is to find an isothermal mapping h_0 whose domain includes q. If we denote $h_0 \circ h^{-1}$ as $(x, y) \to z_0$ the condition reads

$$|dz_0|^2 = \rho(E\, dx^2 + 2F\, dxdy + G\, dy^2),$$

whence

$$\left|\frac{\partial z_0}{\partial x}\right|^2 = \rho E, \quad \operatorname{Re}\left\{\frac{\partial z_0}{\partial x}\frac{\partial \bar{z}_0}{\partial y}\right\} = \rho F, \quad \left|\frac{\partial z_0}{\partial y}\right|^2 = \rho G.$$

We set $\frac{\partial z_0}{\partial x} = \lambda \frac{\partial z_0}{\partial y}$ and eliminate ρ. Then $|\lambda|^2 = E/G$, $\operatorname{Re} \lambda = F/G$, and consequently

$$\lambda = \frac{F - i\sqrt{EG-F^2}}{G},$$

except perhaps for the sign of the square root. But the Jacobian of the mapping turns out to be $-\left|\frac{\partial z_0}{\partial y}\right|^2 \operatorname{Im} \lambda$. Since it must be positive, we conclude that we have made the correct choice of sign.

We have now to integrate the partial differential equation

$$\frac{\partial z_0}{\partial x} = \lambda \frac{\partial z_0}{\partial y},$$

where λ is a known analytic function. The classical method is to consider the associated ordinary differential equation

$$\frac{dy}{dx} = -\lambda\,(x, y).$$

We assume for simplicity that $h(g) = (0, 0)$. Then λ can be extended by power series development to a complex neighborhood of the origin, and the associated equation has a unique complex solution $y = Y(x, \eta)$ with initial value η for $x = 0$, provided that $|x|$ and $|\eta|$ are sufficiently small. It is known that Y will be analytic in both variables. Since $Y(0, \eta) = \eta$, the partial derivative $\partial Y/\partial\eta$ is 1 for $x = 0$. Hence we can solve for η, obtaining $\eta = H(x, y)$ with analytic H. The identity

$$Y(x, H(x, y)) = y$$

holds in a complex neighborhood of (0, 0). On differentiating with respect to x and y we obtain

$$-\lambda + Y_\eta H_x = 0$$
$$Y_\eta H_y = 1,$$

and hence $H_x = \lambda H_y$. This remains true when x, y are restricted to real values, and our problem is solved by taking $z_0 = H(x, y)$. The mapping is a local homeomorphism, for $H_y = 1$ at the origin, and the Jacobian is $-|H_y|^2 \operatorname{Im} \lambda$.

We have proved the existence of an isothermal mapping whose domain contains an arbitrary point q. The other conditions for a conformal structure are trivially satisfied.

Remark. The classical problem was solved already by Gauss. The best result, due to L. Lichtenstein [1] and A. Korn [3], is that the problem has a solution whenever λ satisfies a Hölder condition of the form $|\lambda(x, y)-\lambda(x_1, y_1)| \leqq M(|x-x_1|^\alpha+|y-y_1|^\alpha)$, $\alpha>0$. For relatively simple proofs the reader is referred to S. Chern [1] and L. Ahlfors [23].

5E. Very often the given metric has singularities, and it may nevertheless be possible to find a corresponding conformal structure. For instance the following important theorem is almost trivial:

Theorem. *Every orientable polyhedron can be endowed with a conformal structure.*

We let the polyhedron be represented as a geometric complex K_g. On each triangle we introduce the conformal structure which is determined by a sense-preserving congruence mapping h into the plane. For adjacent triangles these mappings can be chosen so that they agree on the common side. Consider now a vertex α, and suppose that the sum of the angles at α is $2\pi\omega$. It is clear that we can find congruence mappings h of the triangles in the star $S(\alpha)$ in such a way that $h^{1/\omega}$ defines a topological mapping of the whole open star. The mapping $z \rightarrow z^{1/\omega}$ is directly conformal for $z \neq 0$. By virtue of the fact that no two open stars contain a common vertex it follows that the mappings form a basis for a conformal structure.

§2. ELEMENTARY THEORY OF FUNCTIONS ON RIEMANN SURFACES

On Riemann surfaces it is possible to consider the categories of analytic functions, harmonic functions, and subharmonic functions. In a way, analytic functions are in the center of interest, and we obtain harmonic functions by forming the real and imaginary parts of analytic functions. However, a harmonic function does not necessarily have a single-valued conjugate. In view of our deliberate restriction to single-valued functions the theory of harmonic functions is thus more general. Subharmonic functions are used as an important tool for the study of harmonic functions.

Most theorems in classical function theory have obvious analogues on Riemann surfaces, but there are a few significant differences which have to be pointed out. This concerns above all the integration formulas which are used in potential-theoretic reasonings. The general discussion of integrals is postponed, but for immediate purposes a preliminary treatment of the Dirichlet integral is included in this section.

An important side result is that every Riemann surface is countable.

6. Harmonic functions

6A. It is time that we simplify our notations. For a given Riemann surface (W, Φ) we consider a mapping $h \in \Phi$ with domain V. A point

$p \in V$ is uniquely determined by the corresponding parameter value $z = h(p)$. Different z determine different $p \in V$, and this gives us the right to speak of z as a *local variable*. It is also permissible to think of the complex value z as a name for the corresponding point p, and with a slight "abus de langage" we can even speak of the point $z \in W$. If used with caution this practice simplifies the formal part of many reasonings and serves to emphasize the similarity between function theory on Riemann surfaces and in plane regions.

There would be little gain if it were always necessary to specify the mapping h. However, all intrinsic properties of the Riemann surface must be independent of the choice of h. In other words, if expressed in terms of a local variable z they must be *invariant* under directly conformal mappings. We shall be careful to use the notation z for a point on W only when this invariance condition is fulfilled.

We have already defined an analytic function f on W as one which makes $f \circ h^{-1}$ analytic as a function of the local variable z. This definition is manifestly invariant under conformal mappings. True to our convention we shall therefore use the notation $z \to f(z)$. With the slightly ambiguous simplification that is customary in function theory we shall even allow ourselves to speak of the analytic function $f(z)$.

6B. We introduce the class of harmonic functions on Riemann surfaces:

Definition. *A real-valued function u on a Riemann surface (W, Φ) is said to be harmonic if for every $h \in \Phi$ the composite function $u \circ h^{-1}$ is harmonic on the range $h(V)$.*

Because of the conformal invariance a harmonic function can be denoted by $u(z)$. The real and imaginary parts of an analytic function are harmonic, and the imaginary part is called a conjugate harmonic function of the real part. It is convenient to denote a conjugate harmonic function of u, if one exists, by u^*. The conjugate is determined up to an additive constant. An arbitrary harmonic u has a conjugate in a neighborhood of each point, but in general these local conjugates cannot be pieced together to a conjugate on W. However, if W is simply connected it follows by the monodromy theorem that a conjugate exists.

6C. Analytic and harmonic functions on Riemann surfaces have the same local properties as in the plane. Therefore, all theorems in classical function theory which have a purely local character remain valid. A typical example is the maximum principle for analytic and harmonic functions. Such theorems will not be listed, but will be used freely when the need arises.

Theorems which concern the convergence of sequences are not completely local. Frequently the proof depends on a countability argument, for instance in the theory of normal families. In such cases the theorems remain valid for Riemann surfaces with a countable basis. It will be shown (in 12) that this is no restriction, but for the present these theorems cannot be used indiscriminately.

6D. A function is analytic or harmonic on a bordered Riemann surface $\overline{W}$ only if it has an analytic or harmonic extension to an open set on the double $\hat{W}$ (Ch. I, 13H and Ch. II, 3E). To prove that this is the case it is frequently necessary to appeal to the *reflection principle*.

Suppose that u is harmonic in the interior W, continuous on $\overline{W}$, and zero on the whole border B. Then u can be continued to the double by requiring that it takes opposite values at symmetric points: $u(p^*) = -u(p)$. Indeed, the extension is harmonic on W^*, and by the classical form of the reflection principle the extended function is also harmonic at each point of B. In particular, the hypotheses are thus sufficient to conclude that u is harmonic on $\overline{W}$.

In the same circumstances, if u has a conjugate harmonic function u^* on W, then we can extend the definition of u^* to W^* by setting $u^*(p^*) = u^*(p)$. If we represent a neighborhood of a border point by a plane disk the classical reflection principle shows that u^* has a continuous extension to B and gives rise to a function which is harmonic on the whole double $\hat{W}$. Hence $f = u + iu^*$ is analytic on $\hat{W}$, and its extension satisfies the symmetry relation $f(p^*) = -\bar{f}(p)$.

6E. There is no invariant meaning attached to the derivative $f'(z)$ of an analytic function. In fact, suppose that z_1 is another local variable which is related to z by $z_1 = \varphi(z)$. In terms of z_1 the function f is expressed by $f_1(z_1)$, where $f(z) = f_1(\varphi(z))$. We obtain $f'(z) = f_1'(\varphi(z))\varphi'(z) = f_1'(z_1)\, dz_1/dz$. It is convenient to express this relationship by writing, formally, $f'(z)\, dz = f_1'(z_1)\, dz_1$, and we say that f has an invariantly defined *differential* $df = f'(z)\, dz$.

Differentials will be studied extensively in Ch. V. At present we make merely some preliminary remarks. We note that an arbitrary function g of class C^1 has a differential $dg = \frac{\partial g}{\partial x}\, dx + \frac{dg}{\partial y}\, dy$ where the coefficients are now *covariant*. This means that two representations $a\, dx + b\, dy$ and $a_1\, dx_1 + b_1\, dy_1$ of the same differential are connected by the relations $a = a_1 \frac{\partial x_1}{\partial x} + b_1 \frac{\partial y_1}{\partial x}$, $b = a_1 \frac{\partial x_1}{\partial y} + b_1 \frac{\partial y_1}{\partial y}$.

The conjugate differential of dg is defined as $dg^* = -\frac{\partial g}{\partial y}\, dx + \frac{\partial g}{\partial x}\, dy$. It has invariant meaning, because of the Cauchy-Riemann differential equations, but in general it is not the differential of a function. However, if u is harmonic and has a conjugate harmonic function u^*, then $du^* = d(u^*)$, and the notation is free from ambiguity. Even if u^* is defined only locally we can use both interpretations of the symbol du^*.

Whenever u is harmonic we refer to du and du^* as *harmonic differentials*. The linear combination $du + i\,du^* = \left(\frac{\partial u}{\partial x} - i\,\frac{\partial u}{\partial y}\right) dz$ is an *analytic differential*, although it need not be the differential of an analytic function.

6F. A covariant differential $a\,dx + b\,dy$ has more than formal meaning when used as integrand in a line integral. An arc $t \to z(t)$ in the plane is said to be piecewise differentiable if the parameter interval can be divided into a finite number of subintervals with the property that $z(t)$ is of class C^1 in each. This notion is invariant, and we can thus speak of piecewise differentiable arcs on a Riemann surface. Let γ be such an arc, and assume first that it is contained in the domain of a local variable z. Then

$$\int_{\gamma} a\,dx + b\,dy$$

can be evaluated as a line integral in the z-plane, and its value is independent of the choice of local variable. An arbitrary piecewise differentiable arc can be divided into subarcs with this property, and we can define the integral as the sum of the integrals over the subarcs. It is clear that the result is independent of the subdivision.

In the future, whenever we consider a line integral it will be tacitly understood that the path of integration is piecewise differentiable. The value of the integral does not depend on the parametrization, and in the case of a closed path it does not depend on the choice of initial point.

7. The Dirichlet integral

7A. Suppose that u is real and of class C^1 on a plane region W. The integral

$$D_W(u) = \iint_W \left[\left(\frac{\partial u}{\partial x}\right)^2 + \left(\frac{\partial u}{\partial y}\right)^2\right] dxdy$$

is known as the *Dirichlet integral* of u over the region W. Its value is finite or $+\infty$, and it is 0 if and only if u is a constant.

An essential property of the Dirichlet integral is its invariance with respect to conformal mappings. More precisely, if φ is a conformal mapping of W onto W_1, and if $u = u_1 \circ \varphi$, then $D_W(u) = D_{W_1}(u_1)$.

If $u, v \in C^1$ and $D_W(u)$, $D_W(v)$ are both finite, then we can also consider the mixed Dirichlet integral

$$D_W(u, v) = \iint_W \left[\frac{\partial u}{\partial x}\frac{\partial v}{\partial x} + \frac{\partial u}{\partial y}\frac{\partial v}{\partial y}\right] dxdy.$$

It is likewise invariant, as seen by the identity $4D_W(u, v) = D_W(u+v) - D_W(u-v)$. The important inequality $D_W(u, v)^2 \leqq D_W(u)D_W(v)$ is constantly used. It follows from the fact that $D_W(u+tv) = D_W(u) + 2tD_W(u, v) + t^2D_W(v)$ is a positive quadratic form.

7B. We must extend this definition to the case where W is a Riemann surface. If the integral is restricted to a domain which is contained in a parametric region, then a unique value can be assigned by virtue of the conformal invariance. It would be natural to treat the general case by dividing the surface into small subregions. Clearly, this requires something in the nature of a triangulation with sufficiently smooth sides. If W is known to have a countable basis, this plan can be carried out, but the details require careful attention. Fortunately, there is a much better way.

We remark first that the conformal invariance applies more generally to integrals of the form

$$\iint_W \left[\left(\frac{\partial u}{\partial x}\right)^2 + \left(\frac{\partial u}{\partial y}\right)^2\right] g\, dxdy$$

where g is any continuous function on W. Therefore, even if W is a Riemann surface, this integral can be defined as soon as g is identically zero outside of a parametric region V. Suppose now that it is possible to write $g = \sum_i g_i$ where each g_i is identically zero outside of a parametric region V_i. Then it is natural to set

$$\iint_W \left[\left(\frac{\partial u}{\partial x}\right)^2 + \left(\frac{\partial u}{\partial y}\right)^2\right] g\, dxdy = \sum_i \iint_W \left[\left(\frac{\partial u}{\partial x}\right)^2 + \left(\frac{\partial u}{\partial y}\right)^2\right] g_i\, dxdy,$$

at least if the convergence is assured. In particular, we can define the Dirichlet integral by means of nonnegative functions e_i with $\sum_i e_i = 1$. Functions with this property are said to constitute a *partition of unity*.

The definition will be independent of the partition of unity that is used, for if $\{e_i\}$ and $\{e_j'\}$ are two partitions, we have clearly

$$\sum_i \iint_W [\quad] e_i\, dxdy = \sum_{i,j} \iint_W [\quad] e_i e_j'\, dxdy = \sum_j \iint_W [\quad] e_j'\, dxdy.$$

7C. We prove the existence of a partition of unity for the case where W is a closed surface or the interior of a compact bordered surface. It is important for later applications that we can choose the e_i to be for instance of class C^2.

For a Riemann surface (W, Φ) the notion of *parametric disk* is redefined as follows: V is a parametric disk if there exists a mapping $h \in \Phi$ with domain V', such that $\bar{V} \subset V'$ and $h(V)$ is the disk $|z|<1$. For a bordered Riemann surface a parametric half-disk is defined in the corresponding manner. It is clear that a compact Riemann surface or bordered surface has a finite covering by parametric disks or half-disks V_i with corresponding mappings h_i.

On the disk $|z|<1$ we define g by $g(z)=(1-|z|^2)^3$. If the definition is extended to the whole complex plane by setting $g(z)=0$ for $|z| \geqq 1$, it is easily verified that g is of class C^2. The function g_i which is equal to $g \circ h_i$ on V_i and vanishes identically outside of V_i is of class C^2 on W. A partition of unity is obtained by setting

$$e_i(p) = g_i(p)(\sum_j g_j(p))^{-1},$$

and this partition can be used to define the Dirichlet integrals $D_W(u)$ and $D_W(u, v)$ for functions u, v that are harmonic on $\bar{W}$.

The definition of $D_W(u)$ and $D_W(u, v)$ for arbitrary open W will be given in 13.

8. Green's formula

8A. Let $\bar{W}_0$ be the upper half-plane $y \geqq 0$, and let B_0 denote the x-axis, traced in the positive direction. The formulas

$$\iint_{W_0} \frac{\partial u}{\partial x}\, dxdy = 0$$

$$\iint_{W_0} \frac{\partial u}{\partial y}\, dxdy = -\int_{B_0} u\, dx$$

are trivial for functions u of class C^1 which vanish identically in a neighborhood of infinity. Replace u by $v\dfrac{\partial u}{\partial x}$ in the first formula and by $v\dfrac{\partial u}{\partial y}$ in the second, where u is now of class C^2 and $v \in C^1$ vanishes outside of a compact set. On adding the formulas we obtain

$$\iint_{W_0} \left(\frac{\partial u}{\partial x}\frac{\partial v}{\partial x}+\frac{\partial u}{\partial y}\frac{\partial v}{\partial y}\right) dxdy + \iint_{W_0} v\Delta u\, dxdy = -\int_{B_0} v\frac{\partial u}{\partial y}\, dx$$

where Δu is the Laplacian. For harmonic u we have thus

$$D_{W_0}(u, v) = \int_{B_0} v\, du^*.$$

If $\bar{W}_0$ is replaced by the whole plane we find in the same way that the mixed Dirichlet integral is zero. If v is known to vanish for $|z| \geqq 1$ the integral can be restricted to the unit disk or half-disk, and in the latter case B_0 can be replaced by the real diameter.

8B. Suppose now that u and v are given on a compact bordered Riemann surface $\overline{W}$ with border B; u is again supposed to be harmonic while v is an arbitrary function of class C^1. We use a partition of unity $\{e_i\}$, and obtain

$$D_{\overline{W}}(u, e_i v) = \int_B e_i v \, du^*,$$

where the integral on the right can be thought of as a line integral extended over the positively oriented border.

The mixed Dirichlet integral is bilinear. Therefore, the sum of all $D_{\overline{W}}(u, e_i v)$ is $D_{\overline{W}}(u, v)$, and we find

$$D_{\overline{W}}(u, v) = \int_B v \, du^*.$$

This relation will be referred to as *Green's formula*. Naturally, it is also valid for a closed Riemann surface W, and in that case the mixed Dirichlet integral is zero.

Observe that only u need be harmonic. It is also useful to note the fact that the formula remains valid if u is defined only locally.

8C. If u and v are both harmonic and globally defined we obtain, by the symmetry of the Dirichlet integral,

$$\int_B v \, du^* - u \, dv^* = 0.$$

The special case $v = 1$ yields

$$\int_B du^* = 0.$$

These particular consequences of Green's formula are used very frequently.

When integrating over a given arc it is often convenient to replace a conjugate differential du^* by the expression $\frac{\partial u}{\partial n} ds$, where the derivative is in the direction of the right normal. With this notation the first formula reads

$$\int_B \left(v \frac{\partial u}{\partial n} - u \frac{\partial v}{\partial n} \right) ds = 0.$$

9. Harnack's principle

9A. The classical principle of Harnack states that a monotone sequence of harmonic functions converges to a harmonic limit function unless it diverges everywhere. The proof uses only local considerations and can thus be applied to harmonic functions on a Riemann surface.

We shall need a more general result which we formulate as follows:

Theorem. *Suppose that a family $\mathscr{U}$ of harmonic functions on a Riemann surface W satisfies the following condition:*

(**A**) *To any $u_1, u_2 \in \mathscr{U}$ there exists a $u \in \mathscr{U}$ with $u \geqq \max(u_1, u_2)$ on W.*

Then the function

$$U(z) = \sup_{u \in \mathscr{U}} u(z)$$

is either harmonic or constantly equal to $+\infty$.

It is obvious that the classical Harnack principle is a special case.

9B. For the proof, consider an arbitrary point $z_0 \in W$. There exists a sequence of functions $u_n \in \mathscr{U}$ with

$$\lim_{n\to\infty} u_n(z_0) = U(z_0).$$

Set $\bar{u}_1 = u_1$ and choose, by induction, $\bar{u}_n \in \mathscr{U}$ such that $\bar{u}_n \geqq \max(u_n, \bar{u}_{n-1})$. Then it is also true that

$$\lim_{n\to\infty} \bar{u}_n(z_0) = U(z_0),$$

and the functions $\bar{u}_n$ form a nondecreasing sequence. By the original form of Harnack's principle,

$$U_0(z) = \lim_{n\to\infty} \bar{u}_n(z)$$

is either harmonic or identically $+\infty$. It satisfies $U_0(z_0) = U(z_0)$.

We repeat the construction for another point z_0'. This time, starting from a sequence of functions $u_n' \in \mathscr{U}$ with

$$\lim_{n\to\infty} u_n'(z_0') = U(z_0')$$

we determine $\bar{u}_n' \in \mathscr{U}$ according to the condition $\bar{u}_n' \geqq \max(u_n', \bar{u}_n, \bar{u}_{n-1}')$. The corresponding limit function $U_0' = \lim \bar{u}_n'$ will then satisfy $U_0' \geqq U_0$ as well as $U_0'(z_0) = U_0(z_0)$ and $U_0'(z_0') = U(z_0')$.

If U_0 and U_0' are finite we conclude that $U_0 - U_0'$ has a maximum equal to 0 at z_0, and by the maximum principle this implies $U_0 = U_0'$. In particular, $U(z_0') = U_0(z_0')$, and since z_0' is arbitrary we have proved that U is identical with the harmonic function U_0.

If $U_0 = +\infty$ we have $U_0'(z_0) = +\infty$, and consequently $U(z_0') = U_0'(z_0') = +\infty$. In this case U is identically equal to $+\infty$.

10. Subharmonic functions

10A. We recall that a real-valued function v is said to be *subharmonic* in a plane region W if it satisfies the following requirements:

(**A1**) *v is upper semicontinuous* ($=$u.s.c.) *in W, i.e.,* $v(z) \geqq \overline{\lim}_{z' \to z} v(z')$.

(**A2**) *For any function u which is harmonic in a region $W' \subset W$, the difference $v-u$ is either constant or fails to have a maximum in W'.*

An u.s.c. function is conventionally allowed to take the value $-\infty$, but not the value $+\infty$. It has a finite maximum on any compact set on which it is not identically $-\infty$. We note that a function is u.s.c. if and only if it can be represented as the limit of a nonincreasing sequence of continuous functions.

We say that v is *superharmonic* if $-v$ is subharmonic. A harmonic function is simultaneously subharmonic and superharmonic. The converse of this statement is true, but requires proof.

10B. The definition of subharmonicity is of local character. In other words, if a function is subharmonic in a neighborhood of every point on W, then it is subharmonic on W.

The subharmonic character is invariant under conformal mapping. If φ maps W conformally onto W_1 this statement means that $v = v_1 \circ \varphi$ is subharmonic together with v_1.

The definition given under 10A can be applied without change to functions on an arbitrary Riemann surface W. The two properties that we have just mentioned show that a function is subharmonic on W if and only if it is subharmonic when expressed in terms of a local variable.

10C. If v_1 and v_2 are both subharmonic, then $v = \max(v_1, v_2)$ is also subharmonic. Condition (A1) is practically obvious. To prove (A2), let u be harmonic in W', and suppose that $v-u$ has a maximum at $z_0 \in W'$. We assume that $v(z_0) = v_1(z_0)$. Then

$$v_1(z) - u(z) \leqq v(z) - u(z) \leqq v(z_0) - u(z_0) = v_1(z_0) - u(z_0)$$

for $z \in W'$, showing that $v_1 - u$ has a maximum. Hence $v_1 - u$ is constant, and the above inequalities show that $v-u$ is likewise constant.

Naturally, the result can be extended to the maximum of any finite number of subharmonic functions. As for an infinite family, the same would hold, except that we can no longer prove that the supremum is u.s.c.

10D. In order to derive further properties of subharmonic functions we use the Poisson integral. Let v be defined and continuous on the circumference of a disk Δ with radius ρ and center z_0. The Poisson integral of v with respect to this disk is defined as

$$(1) \qquad P_v(z) = \frac{1}{2\pi}\int_0^{2\pi} \frac{\rho^2-|z-z_0|^2}{|\zeta-z|^2}\, v(\zeta)\, d\Theta$$

where $\zeta = z_0 + \rho e^{i\Theta}$. It represents a harmonic function in Δ with the property

$$(2) \qquad \lim_{z\to\zeta} P_v(z) = v(\zeta).$$

The definition can be extended to u.s.c. v, for instance by interpreting (1) as a Lebesgue integral. The same purpose is achieved, more directly, by setting $P_v(z) = \inf P_W(z)$ where w ranges over all continuous majorants of v. For u.s.c. v, (2) is replaced by

$$(3) \qquad \overline{\lim_{z\to\zeta}}\, P_v(z) \leqq v(\zeta).$$

Hence the function which is defined as P_v in Δ and as v on the boundary is u.s.c. in the closed disk. Observe that P_v is either harmonic or identically $-\infty$ in Δ, as seen by applying Theorem 9A to the functions P_w.

10E. In order to emphasize the elementary nature of the Poisson integral we prove the formula

$$(4) \qquad P_{v_1+v_2} = P_{v_1} + P_{v_2}$$

without use of the Lebesgue integral. The relation is evident for continuous v_1, v_2. In the case of u.s.c. functions, suppose that w_1, w_2 are continuous majorants of v_1, v_2. Then $P_{v_1+v_2} \leqq P_{w_1+w_2} = P_{w_1} + P_{w_2}$, and we obtain $P_{v_1+v_2} \leqq P_{v_1} + P_{v_2}$. For the opposite inequality, let w be a continuous majorant of $v_1 + v_2$. We find that

$$\overline{\lim_{z\to\zeta}}\, (P_{v_1} + P_{v_2}) \leqq v_1(\zeta) + v_2(\zeta) \leqq w(\zeta) = \lim_{z\to\zeta} P_w,$$

and the maximum principle permits us to conclude that $P_{v_1} + P_{v_2} \leq P_w$. Hence $P_{v_1} + P_{v_2} \leqq P_{v_1+v_2}$, and (4) is proved.

10F. We remark that $v \leqq 0$ implies $P_v \leqq 0$, and that the strict inequality $P_v < 0$ holds everywhere in Δ unless v is identically zero on the boundary. In fact, if there is a single value $v(\zeta) < 0$, then (3) shows that P_v cannot be identically zero, and by the maximum principle P_v cannot vanish anywhere.

10G. Returning to the theory of subharmonic functions we prove:

Theorem. *An u.s.c. function v is subharmonic in a plane region W if and only if*

$$(5) \qquad v(z) \leqq P_v(z)$$

in all disks Δ with $\bar{\Delta} \subset W$.

Suppose first that v is subharmonic, and consider a continuous majorant w on the boundary of Δ. Then, by the semicontinuity,

$$\overline{\lim_{z\to\zeta}}\, v(z) \leqq v(\zeta) \leqq w(\zeta) = \lim_{z\to\zeta} P_w(z)$$

for each ζ on the boundary of Δ. Since $v-P_w$ cannot have a maximum without being constant, it follows by the usual compactness argument that $v \leqq P_w$ in Δ, and hence that $v \leqq P_v$.

Conversely, if (5) is satisfied, let u be harmonic in $W' \subset W$. Suppose that $v-u$ has a maximum at $z_0 \in W'$. Since (5) implies $v-u \leqq P_{v-u}$ it means no loss of generality to take $u \equiv 0$, and we may even assume that the maximum of v at z_0 is zero. For a sufficiently small disk, centered at z_0, we have then $v(\zeta) \leqq 0$ and $0 = v(z_0) \leqq P_v(z_0) \leqq 0$. But we remarked in 10F that $P_v(z_0) = 0$ implies $v(\zeta) = 0$ on the whole circumference. It follows that $v = 0$ in a neighborhood of z_0, and by the connectedness of W' we conclude that v (or $v-u$) is constant in W'. This is what we set out to prove.

10H. For continuous functions condition (5) leads to the familiar mean value property

$$v(z_0) \leqq \frac{1}{2\pi}\int_0^{2\pi} v(z_0+\rho e^{i\Theta})\, d\Theta,$$

and if the Riemann integral is replaced by the Lebesgue integral the inequality holds for arbitrary subharmonic v. If $v \in C^2$ it is an easy consequence that v is subharmonic if and only if $\Delta v \geqq 0$.

A function which is at once subharmonic and superharmonic is continuous and satisfies

$$v(z_0) = \frac{1}{2\pi}\int_0^{2\pi} v(z_0+\rho e^{i\Theta})\, d\Theta.$$

Therefore, such a function is harmonic.

If Theorem 10G is combined with (4) it is seen that the sum of two subharmonic functions is subharmonic.

These results are of a local character. Therefore they hold not only in a plane region, but on an arbitrary Riemann surface.

10I. Let v be subharmonic on a Riemann surface W, and consider a parametric disk Δ. We denote by P_v the Poisson integral of v in Δ which is formed by means of a specific conformal mapping of Δ onto a circular disk.

Theorem. *The function v_0 which is equal to P_v in Δ and equal to v on the complement of Δ is subharmonic on W.*

We have already shown (10D) that v_0 is u.s.c. To prove (A2), let u be harmonic in $W' \subset W$. Suppose that v_0-u has a maximum at $z_0 \in W'$. If

$z_0 \in \Delta$, then $v_0 - u$ is constant in a component of $W' \cap \Delta$. Either this component is identical with W', or else Δ has a boundary point in W', and since $v_0 - u$ is u.s.c. the maximum is attained at that boundary point. If z_0 belongs to the exterior of Δ the same reasoning applies, and we find that we need only consider the case of a z_0 which lies on the boundary of Δ. We have then

$$v(z) - u(z) \leqq v_0(z) - u(z) \leqq v(z_0) - u(z_0)$$

for all $z \in W'$. Thus $v - u$ has a maximum in W' and must be constant. The double inequality shows that $v_0 - u$ is likewise constant, and we have proved (A2).

§3. THE DIRICHLET PROBLEM AND APPLICATIONS

We need the solution of Dirichlet's problem for comparatively simple regions as a basic tool. In the case of a circular disk the solution is given explicitly by Poisson's integral, and we are required to pass from a single disk to a finite union of disks. This can be accomplished by the alternating method, but not without tedious attention to detail. We prefer to solve the problem in a more general setting by a method which requires very little detailed analysis.

11. The Dirichlet problem

11A. In the following G denotes a *relatively compact* subregion of a Riemann surface W. Its boundary, which is supposed to be nonvoid, is denoted by Γ. The Dirichlet problem deals with the construction of harmonic functions in G with given values on Γ.

Quite precisely, there is given a continuous real-valued function f on Γ. We are required to construct a continuous function u on $\bar{G} = G \cup \Gamma$ which coincides with f on Γ and is harmonic in G.

It follows immediately from the maximum principle that such a function, if it exists, is necessarily uniquely determined.

11B. Very simple examples show that the Dirichlet problem does not always have a solution. For instance, let G be the punctured disk $0 < |z| < 1$, and choose $f(0) = 0, f(\zeta) = 1$ for $|\zeta| = 1$. A solution u to this Dirichlet problem would be bounded, and hence the singularity at the origin would be removable. But then the fact that $u(0) = 0$ while u tends to 1 when z tends to the circumference would violate the maximum principle. It follows that no solution exists.

11C. We are going to attack the Dirichlet problem by a method which was introduced by O. Perron [1]. It is a remarkably simple and direct method. When analyzed, it is found to depend on essentially the same arguments as most classical methods, with the difference that these arguments appear in their purest form.

One of the advantages of Perron's method is that it associates a function u which is either harmonic or completely degenerate with any function f, continuous or not, and regardless of whether the Dirichlet problem has a solution. The function u is obtained by a universal construction, and the Dirichlet problem is replaced by a study of the boundary behaviour of u.

11D. We denote by $\mathscr{V}(f)$ the class of all subharmonic functions v in G which satisfy the condition

$$\overline{\lim_{z\to\zeta}}\, v(z) \leqq f(\zeta)$$

for all $\zeta \in \Gamma$. Except for being real, the function f is not subject to any restrictions; in particular, it may take the values $+\infty$ and $-\infty$.

Theorem. *The function u, defined by*

$$u(z) = \sup_{v\in\mathscr{V}(f)} v(z),$$

is either harmonic, identically $+\infty$, or identically $-\infty$ in G.

The class $\mathscr{V}(f)$ is never empty, but it may contain only the function $v = -\infty$. In that case u is identically $-\infty$.

Consider a parametric disk Δ with closure in G. For any $v \in \mathscr{V}(f)$ we form the corresponding function v_0 which is equal to P_v in Δ and equal to v outside of Δ (cf. 10I). We conclude by Theorem 10I that $v_0 \in \mathscr{V}(f)$, and by Theorem 10G that $v \leqq v_0$. It follows that $u(z) = \sup v_0(z)$ in Δ.

The functions v_0 are either harmonic or identically $-\infty$ in Δ. If they are not all infinite, we consider the class of all those which are harmonic in Δ. This class satisfies condition (A) of Theorem 9A. Indeed, suppose that v_0', v_0'' are formed from $v', v'' \in \mathscr{V}(f)$. Then $v = \max(v', v'')$ belongs to $\mathscr{V}(f)$, and the corresponding v_0 is a common majorant of v_0' and v_0''. We conclude that $u = \sup v_0$ is either harmonic or identically $+\infty$ in Δ.

We have shown that u is either harmonic, identically $+\infty$, or identically $-\infty$ in each parametric disk. Because of the connectedness one of these alternatives must hold throughout G, and the theorem is proved.

11E. It remains to consider the boundary behavior of u. We shall be content to study the case where f is bounded, $|f| \leqq M$.

A function β in G will be called a *barrier* at $\zeta_0 \in \Gamma$ if it satisfies these conditions:

(E1) β is subharmonic in G,

(E2) $\lim\limits_{z\to\zeta_0} \beta(z) = 0$,

(E3) $\overline{\lim\limits_{z\to\zeta}}\, \beta(z) < 0$ for all $\zeta \neq \zeta_0$, $\zeta \in \Gamma$.

A boundary point ζ_0 will be called *regular* if and only if there exists a barrier at ζ_0. Consider a neighborhood V of ζ_0 and a barrier β. The barrier is strictly negative, and because of (E3) it has a negative upper bound $-m$ outside of V. The function $\beta_V = \max(\beta/m, -1)$ is a new barrier with the additional property that $\beta_V = -1$ outside of V. We call β_V a normalized barrier with respect to V.

It is clear that the normalized barrier β_V can be used as a barrier for any region G' with $G' \cap V = G \cap V$, provided that it is defined as -1 at all points of G' outside of V. It follows that the existence of a barrier is a local property which depends only on the geometric nature of G in the immediate vicinity of ζ_0.

11F. We are now ready to prove the following central theorem:

Theorem. *At a regular point ζ_0 the function u, introduced in* 11D, *satisfies*

$$\underline{\lim}_{\zeta\to\zeta_0} f(\zeta) \leqq \underline{\lim}_{z\to\zeta_0} u(z) \leqq \overline{\lim}_{z\to\zeta_0} u(z) \leqq \overline{\lim}_{\zeta\to\zeta_0} f(\zeta), \tag{6}$$

provided that f is bounded.

Set $A = \overline{\lim}_{\zeta\to\zeta_0} f(\zeta)$, and determine a closed neighborhood V of ζ_0 in which $f(\zeta) < A+\epsilon$, $\epsilon > 0$. For any $v \in \mathscr{V}(f)$ the function

$$\varphi = v - A + (M-A)\beta_V$$

is subharmonic and has the property

$$\overline{\lim}_{z\to\zeta} \varphi(z) < \epsilon$$

for all $\zeta \in \Gamma$, whether in or outside of V. In fact, for $\zeta \in V$ we have $\overline{\lim}\, v(z) \leqq f(\zeta) < A+\epsilon$ and $\overline{\lim}\, \beta_V(z) \leqq 0$. For ζ not in V we have $\overline{\lim}\, v(z) \leqq M$ and, because V is closed, $\lim \beta_V(z) = -1$. We conclude that $\varphi(z) < \epsilon$ in G, and since this is true for all $v \in \mathscr{V}(f)$ we obtain

$$u(z) \leqq A - (M-A)\beta_V(z) + \epsilon.$$

Letting z tend to ζ_0 we find

$$\overline{\lim}_{z\to\zeta_0} u(z) \leqq A + \epsilon.$$

This proves the last part of (6).

We consider now the function

$$\psi = (B+M)\beta_V + B - \epsilon$$

where $B = \underline{\lim}_{\zeta\to\zeta_0} f(\zeta)$ and $f(\zeta) > B - \epsilon$ in V, a closed neighborhood of ζ_0. It is subharmonic and satisfies

$$\overline{\lim}_{z\to\zeta} \psi(z) \leqq B - \epsilon < f(\zeta)$$

for $\zeta \in V$, while

$$\overline{\lim_{z\to\zeta}}\ \psi(z) = -M-\epsilon < f(\zeta)$$

for any ζ outside of V. It follows that $\psi \in \mathscr{V}(f)$, and hence $u(z) \geqq \psi(z)$. On letting z tend to ζ_0 we obtain

$$\underline{\lim}_{z\to\zeta_0}\ u(z) \geqq B-\epsilon$$

and the first part of (6) is proved.

11G. As a corollary of Theorem 11F we find that the Dirichlet problem with continuous boundary values f has a unique solution for any region G with only regular points. Conversely, if the Dirichlet problem has a solution for arbitrary continuous f, then every boundary point is regular. Indeed, it is possible to find a continuous f which is zero at ζ_0 and strictly negative at all other boundary points. The solution of Dirichlet's problem for this function is evidently a barrier at ζ_0.

11H. Explicit necessary and sufficient conditions for the regularity of a boundary point are known, but they cannot be expressed in a form that would make them immediately useful. For many applications the following elementary result is sufficiently general.

Theorem. *The point ζ_0 is a regular boundary point of G whenever the component of the boundary Γ which contains ζ_0 does not reduce to a point.*

We have already remarked that regularity is a local property. For this reason it is sufficient to consider the case of a plane region G which we regard as a subset of the Riemann sphere. By assumption, ζ_0 belongs to a component E of the complement of G which contains a point $\zeta_1 \neq \zeta_0$. An auxiliary linear transformation permits us to choose $\zeta_0 = \infty$, $\zeta_1 = 0$.

Since the complement of E is simply connected we can define a single-valued branch of $s = \sigma + i\tau = \log z$ in G. It maps G onto a region G' whose intersection with any line $\sigma = \sigma_0$ consists of segments with a total length $\leqq 2\pi$. For a fixed σ_0 we denote the end points of such a segment by s_i', s_i'' with $\operatorname{Im} s_i'' > \operatorname{Im} s_i'$. For $\sigma \geqq \sigma_0$ we define

$$\omega_i(s) = \arg \frac{s_i'-s}{s_i''-s}, \quad 0 \leq \omega_i \leqq \pi.$$

The function

$$\alpha(s) = -\frac{1}{\pi}\sum_i \omega_i(s)$$

is harmonic and satisfies the double inequality

$$-\frac{2}{\pi}\arc\tan\frac{\pi}{\sigma-\sigma_0} \leqq \alpha(s) \leqq 0.$$

It is equal to -1 on the segments (s_i', s_i''). Hence α becomes subharmonic in G' if defined as -1 for $\sigma < \sigma_0$.

Although $\alpha(\log z)$ is subharmonic, negative, and has the limit 0 for $z \to \infty$, it is not yet known to be a barrier, for it could happen that α tends to 0 at a finite boundary point. Let $\{\sigma_n\}$ be a sequence of real numbers, tending to $+\infty$. If σ_0 is replaced by σ_n in the definition of α, the resulting function is denoted by α_n. We consider the function β defined by

$$\beta(z) = \sum_{n=0}^{\infty} 2^{-n}\alpha_n(\log z).$$

The series converges uniformly in G. Therefore, β is subharmonic and tends to 0 for $z \to \infty$. In a neighborhood of a finite boundary point the functions α_n will be identically -1 from a certain n on. It follows that the upper limit of β is strictly negative, and we have shown that β is a barrier.

11I. We have restricted our treatment of the Dirichlet problem to the case of a relatively compact region G on a Riemann surface W. However, the method can be used under more general circumstances. In fact, suppose that we imbed G in a compact Hausdorff space, for instance by Alexandroff compactification of W (Ch. I, 4C). Then G has a compact boundary Γ with respect to the enlarged space, and Γ is endowed with a Hausdorff topology. A close examination of our proofs reveals that these are the only properties of Γ that have been used.

As a typical application we prove the following:

Theorem. *Let G be the complement of a closed parametric disk with respect to an arbitrary Riemann surface W. Then there exists a nonconstant harmonic function on G.*

If W is noncompact we replace it by its Alexandroff compactification. Then Γ will consist of α, the boundary of the disk, and a single point β, the ideal boundary of W. We choose for f any continuous nonconstant function of α and let $f(\beta)$ be an arbitrary finite number. Perron's method yields a harmonic function u on G which tends to $f(\zeta)$ as $z \to \zeta \in \alpha$. Because of this property u is not constant.

12. Existence of a countable basis

12A. It has already been mentioned (see 6C) that every Riemann surface is countable. We are now in a position to give a very short proof of this fact. The proof is based on Theorem 11I together with the following lemma:

Lemma. *The existence of a nonconstant harmonic function u on a Riemann surface W implies that W has a countable basis.*

The shortest proof runs as follows: u has a single-valued conjugate harmonic function u^* on the universal covering surface W_∞ of W. The analytic function $f=u+iu^*$ defines W_∞ as a covering surface of the plane. We proved in Ch. I, 21D, that every covering surface of a countable surface is itself countable. Therefore W_∞ is countable, and the same is true of its projection W.

12B. There is an alternative proof which is more elementary in that it does not make use of covering surfaces and their properties. If u is a nonconstant harmonic function we can introduce a metric on W with the distance function

$$d(z_1, z_2) = \inf \int_\gamma |du + idu^*|,$$

where the infimum is with respect to all piecewise differentiable arcs from z_1 to z_2. It must be verified that any two distinct points have a positive distance, but this is evident by local considerations.

The set $U(z_0, \rho)$ of all points z with $d(z, z_0) < \rho$ is open. We denote by $\rho(z_0)$ the least upper bound of all ρ for which $U(z_0, \rho)$ is relatively compact. It is >0, and if $\rho(z_0) = \infty$ for a single z_0 there is nothing to prove, for then W is the union of the sets $U(z_0, n)$ and can thus be covered by a sequence of compact sets. We may therefore assume that $\rho(z)$ is finite. It is continuous as a function of z, for it satisfies the obvious inequality $|\rho(z_1) - \rho(z_2)| \leqq d(z_1, z_2)$.

We choose a fixed point $z_0 \in W$ and denote by G_n the set of all points z which can be reached from z_0 by means of a chain $z_0, z_1, \cdots, z_n = z$ with $d(z_{i-1}, z_i) \leqq \frac{1}{2}\rho(z_{i-1})$. It is clear that G_n is closed and contained in the interior of G_{n+1}. Because of the latter property, $\bigcup_{n=1}^{\infty} G_n$ is an open set.

The complement of $\cup G_n$ is likewise open. To see this, let ζ be a point in the complement, and consider the set $U(\zeta, \frac{1}{3}\rho(\zeta))$, which is a neighborhood of ζ. Suppose that this neighborhood contains a point $z_n \in G_n$. Then $d(\zeta, z_n) < \frac{1}{3}\rho(\zeta)$ and $\rho(z_n) \geqq \rho(\zeta) - d(\zeta, z_n) > \frac{2}{3}\rho(\zeta)$. These inequalities imply $d(\zeta, z_n) < \frac{1}{2}\rho(z_n)$, and hence ζ would belong to G_{n+1}, contrary to the assumption. We have thus proved the existence of a neighborhood of ζ which is contained in the complement of $\cup G_n$. In other words, the complement is open.

It follows that $W = \bigcup_{1}^{\infty} G_n$, and the lemma will be proved as soon as we show that each G_n is compact. It is trivial that G_1 is compact. Let us therefore suppose that G_{n-1} is known to be compact. It has an open covering by sets $U(z, \frac{1}{4}\rho(z))$, $z \in G_{n-1}$. We select a finite subcovering by the sets $U(z_i, \frac{1}{4}\rho(z_i))$.

Consider a point $\zeta \in G_n$. There exists a $\zeta_{n-1} \in G_{n-1}$ with $d(\zeta, \zeta_{n-1}) \leqq \frac{1}{2}\rho(\zeta_{n-1})$, and a z_i with $d(\zeta_{n-1}, z_i) < \frac{1}{4}\rho(z_i)$. These inequalities imply

$$\rho(\zeta_{n-1}) \leqq d(\zeta_{n-1}, z_i) + \rho(z_i) < \tfrac{5}{4}\rho(z_i)$$

and

$$d(\zeta, z_i) \leqq d(\zeta, \zeta_{n-1}) + d(\zeta_{n-1}, z_i) < \tfrac{1}{2}\rho(\zeta_{n-1}) + \tfrac{1}{4}\rho(z_i) < \tfrac{7}{8}\rho(z_i).$$

Therefore, G_n is completely covered by the sets $U(z_i, \frac{7}{8}\rho(z_i))$. Each of these sets is contained in a compact set, and it follows that G_n is compact. The lemma is proved.

12C. The proof of the main theorem is now quite trivial. We remark that the theorem is due to T. Radó [3] who proved it by an entirely different method.

Theorem. *Every Riemann surface has a countable basis.*

We remove a parametric disk. By Theorem 11I there exists a nonconstant harmonic function on the complement. Hence the complement has a countable basis, and the same is then true of the original surface.

12D. This is a convenient connection to discuss exhaustions of Riemann surfaces. Since we know now that a Riemann surface is countable we can conclude by Theorem 46A, Ch. I that it is a polyhedron. It is worth pointing out that the proof can be made a little simpler than in the general case. Indeed, since all parametric disks have analytic boundaries, a covering by parametric disks is automatically of finite character (Ch. I, 45A). Therefore, the proofs in Ch. I, 46 can be dispensed with. Even so, the Jordan curve theorem is not completely eliminated, for it was used also in Ch. I, 45.

As a result of the triangulability there exists a canonical exhaustion in the combinatorial sense (Ch. I, 29), and hence a countable exhaustion by regions P_n which are regular in the topological sense (Ch. I, 13G).

This approach is unsatisfactory on two accounts: first, it depends on the cumbersome apparatus of triangulation, and second, it yields only an exhaustion by topologically regular regions, rather than by regions which are regular with respect to the conformal structure (3B). The proof that follows below is more elementary and leads to a stronger result.

Theorem. *On every open Riemann surface W there exists a sequence of regular subregions W_n such that $\overline{W}_n \subset W_{n+1}$ and $W = \bigcup_1^\infty W_n$.*

The first part of the proof is identical with an argument used in Ch. I, 46B, but we repeat it for the convenience of the reader. Let there be given a countable covering by parametric disks V_n. Take $n_1 = 1$ and define n_k, recursively, as the smallest integer with the property that

$$\overline{V}_1 \cup \cdots \cup \overline{V}_{n_{k-1}} \subset V_1 \cup \cdots \cup V_{n_k}.$$

Such an integer exists because the set on the left hand side is compact, and $n_k > n_{k-1}$ because W is open. We set $G_k = V_1 \cup \cdots \cup V_{n_k}$. Then $\bar{G}_k$ is compact, $\bar{G}_k \subset G_{k+1}$, and $\bigcup_1^\infty G_k = W$.

Denote the boundary of V_i by β_i. Because the β_i are analytic, and may be supposed distinct, $\beta_1, \cdots, \beta_{n_k}$ intersect only at a finite number of points. Let β_{ij} be the subarcs of β_i between consecutive points of intersection; we introduce an additional pair of points on any β_i that remains undivided. With respect to a V_m, $m \neq i$, $m \leqq n_k$, the interior of any arc β_{ij} is contained either in V_m or in its exterior. It follows easily that the total boundary Γ_k of G_k consists of finitely many arcs β_{ij}, and perhaps a finite number of isolated points. We modify G_k by including the isolated boundary points; for the sake of simplicity we refrain from changing the notation.

The set $G_{k+1} - \bar{G}_k$ is bounded by $\Gamma_k \cup \Gamma_{k+1}$, and no component of the boundary reduces to a point. We can therefore solve the Dirichlet problem with the boundary values 0 on Γ_k and 1 on Γ_{k+1}, separately for each component of $G_{k+1} - \bar{G}_k$. The solution u_k will be completed by setting $u_k = 0$ on G_k. For any constant η_k, $0 < \eta_k < 1$, the set W_k on which $u_k < \eta_k$ is open, and $\bar{G}_k \subset W_k \subset G_{k+1}$. It follows that $\bar{W}_k \subset W_{k+1}$ and $\bigcup_1^\infty W_k = W$.

We contend that each component of W_k is regularly imbedded, provided that the numbers η_k are suitably chosen. Let Ω be a component of $G_{k+1} - \bar{G}_k$ which has part of its boundary on Γ_k and part on Γ_{k+1}, so that u_k is not constant in Ω. The equation $\frac{\partial u_k}{\partial x} - i\,\frac{\partial u_k}{\partial y} = 0$ has invariant meaning and is fulfilled only at isolated points. We can therefore choose η_k so that the level curves $u_k = \eta_k$ do not pass through any of these points. Let z_0 be a point on the level curve $u_k = \eta_k$, choose a conjugate function u_k^* in a neighborhood of z_0, and set $f(z) = -i(u_k - \eta_k) + u_k^*$. Then $f(z)$ is analytic with $f'(z_0) \neq 0$. There is hence a neighborhood V of z_0 in which $f(z)$ is one to one and such that $\operatorname{Im} f(z) = 0$ if and only if $z \in V \cap \Gamma_k$. By the definition in 3B this means that W_k is regularly imbedded.

The W_k are not necessarily connected. It is clear, however, that we get an exhaustion by regularly imbedded regions if we replace W_k by the component that contains V_1, and an exhaustion by regular regions if we include the compact components of $W - W_k$.

Corollary. *Every compact set $A \subset W$ is contained in a regular region.*

In fact, the W_n form an open covering of A, and the corollary is an immediate consequence of the compactness of A. We wish to point out, however, that the corollary can be easily proved without using the countability. To do so, we enclose A in a relatively compact region G_1

formed by a finite number of parametric disks, and next we enclose $\bar{G}_1$ in a similar region G_2. The same construction as above yields a regular region W_1 such that $G_1 \subset W_1 \subset G_2$.

13. Convergence in Dirichlet norm

13A. We recall that the Dirichlet integral $D_W(u)$ of a function $u \in C^1$ was defined in 7 only under the condition that W is closed or bordered and compact. It is now clear how to remove this restriction, for using the exhaustion in Theorem 12D we can evidently set

$$D_W(u) = \lim_{n \to \infty} D_{W_n}(u).$$

It is necessary to verify that the limit is independent of the sequence $\{W_n\}$. The verification offers no difficulty, but the very need for a verification shows that this approach is somewhat artificial. A better method will be given below.

13B. Consider an open Riemann surface W, and let Ω be a generic notation for a regular subregion of W (see 3B). Very often there is given a numerically valued set function $F(\Omega)$, defined for all such regions Ω. We wish to define the limit of $F(\Omega)$ as Ω tends to W. This can be done without recourse to Theorem 12D if we make use of the notion of *directed limit*.

By definition, we say that

$$\lim_{\Omega \to W} F(\Omega) = a$$

if and only if there exists, to every $\epsilon > 0$, a compact set A with the property that $|F(\Omega) - a| < \epsilon$ for all Ω which contain A. The limit, if it exists, is uniquely determined. For suppose that a and a' are two limits. Given ϵ, we determine corresponding sets A, A'. By the Corollary in 12D there exists an Ω which contains $A \cup A'$. It satisfies $|F(\Omega) - a| < \epsilon$ and $|F(\Omega) - a'| < \epsilon$. The resulting inequality $|a - a'| < 2\epsilon$ proves that $a = a'$. For the existence of a limit it is necessary and sufficient that Cauchy's condition is fulfilled: to any $\epsilon > 0$ there exists a compact set A such that $|F(\Omega) - F(\Omega')| < \epsilon$ whenever $A \subset \Omega \cap \Omega'$.

Obvious modifications are needed to define infinite limits and the notions of *limes inferior* and *limes superior*.

If F is increasing, that is to say if $F(\Omega) \leqq F(\Omega')$ for $\Omega \subset \Omega'$, the existence of a finite or infinite limit is assured. In fact, the limit is then equal to $\sup F(\Omega)$, as seen by completely trivial considerations. In particular, we are now able to define the Dirichlet integral by

$$D_W(u) = \lim_{\Omega \to W} D_\Omega(u).$$

13C. Let G be an arbitrary subregion of W, and suppose that a harmonic function u_Ω, defined in G, is assigned to every sufficiently large Ω. The notion of uniform convergence has an obvious interpretation, and if $u_\Omega(z)$ converges uniformly to $u(z)$ on every compact subset of G, then $u(z)$ is harmonic in G.

We shall have to deduce the existence of a limit function when it is known that the Dirichlet integrals $D_G(u_\Omega - u_{\Omega'})$ tend to zero as Ω and Ω' approach W. The theorem can be formulated as follows:

Theorem. *Suppose that $D_G(u_\Omega - u_{\Omega'}) \to 0$ as Ω and Ω' tend to W. Then there exists a harmonic function u in G with these properties:*

(**C1**) $D_G(u_\Omega - u) \to 0$

(**C2**) $D_G(u_\Omega) \to D_G(u)$

(**C3**) $u_\Omega(z) - u_\Omega(z_0) \to u(z)$.

In (C3), z_0 *is an arbitrary point in G, fixed in advance, and the convergence is uniform on every compact subset of G.*

13D. For the proof, let Δ be a parametric disk in G, represented by $|z| < 1$. If v is harmonic in Δ, then $f(z) = \dfrac{\partial v}{\partial x} - i\dfrac{\partial v}{\partial y}$ is analytic. Since the real and imaginary parts of $f(z)^2$ are harmonic we have

$$f(z)^2 = \frac{1}{\pi\rho^2}\int_0^\rho\int_0^{2\pi} f(z+re^{i\Theta})^2\, r\, dr\, d\Theta$$

with $\rho = 1 - |z|$. It follows that

$$|f(z)|^2 \leqq \pi^{-1}(1-|z|)^{-2}\iint_\Delta |f|^2\, dx\, dy = \pi^{-1}(1-|z|)^{-2} D_\Delta(v).$$

For $|z| \leqq \frac{1}{2}$, say, we have thus

$$|f(z)| \leqq (4\pi^{-1} D_\Delta(v))^{\frac{1}{2}},$$

and if $|z_1|, |z_2| \leqq \frac{1}{2}$ we find

$$|v(z_1) - v(z_2)| \leqq \int_{z_1}^{z_2} |f|\,|dz| \leqq (4\pi^{-1} D_\Delta(v))^{\frac{1}{2}}. \tag{7}$$

For a given Δ we denote by Δ' the concentric disk that corresponds to $|z| < \frac{1}{2}$. A given compact set $A \subset G$ can be covered by a finite number of Δ_i'. A point in each Δ_i' can be joined to z_0 by an arc in G, and this arc can again be covered by a finite number of disks Δ'. In this way we get a connected covering by disks $\Delta_1', \cdots, \Delta_N'$, and on applying (7) to $v = u_\Omega - u_{\Omega'}$ we obtain

$$|u_\Omega(z) - u_{\Omega'}(z) - u_\Omega(z_0) + u_{\Omega'}(z_0)| \leqq N(4\pi^{-1} D_G(u_\Omega - u_{\Omega'}))^{\frac{1}{2}}$$

for all $z \in A$. We conclude that $u_\Omega(z) - u_\Omega(z_0)$ converges uniformly to a limit function $u(z)$ which is evidently harmonic.

CHAPTER III

Harmonic Functions on Riemann Surfaces

The solution of Dirichlet's problem does not yet yield the classical existence theorems for harmonic functions with given singularities. Although Perron's method could be extended to somewhat more general situations, it remains essentially a technique for local problems. Especially in the case of open Riemann surfaces it becomes mandatory to introduce new ways of constructing harmonic functions.

The problem of constructing harmonic functions with given singularities on a closed Riemann surface was first solved by H. A. Schwarz who used an iteration process which he called the alternating method. The initial method applies only to compact surfaces. We develop in this chapter a method of greater scope and flexibility. It depends on the use of a whole class of linear operators which we call normal operators. They are equally well adapted to isolated singularities as to prescribed modes of behavior near the ideal boundary. In this way we gain a simple and unified approach to a wide variety of existence theorems.

The theory centers around the notion of a principal function.

In §1 the method is developed in full generality; the section culminates with Theorem 3A, the main existence theorem. The next step is to adapt the method to more specific problems. In this task, which occupies §§2 and 3, attention is drawn to a number of important extremal problems. The most general of these results is Theorem 9E. In §4 we specialize further to the case of planar surfaces, and §5 deals with the so-called capacity functions.

§1. NORMAL OPERATORS

In this section we formulate and prove the main existence theorem in very general terms. The result is illustrated by applications to the classical case of harmonic functions with prescribed singularities on compact surfaces and relatively compact subregions of open surfaces.

1. Linear classes of harmonic functions

1A. The harmonic functions on a Riemann surface form a *linear space* or *vector space*. In other words, if u_1 and u_2 are harmonic, so is every linear combination $c_1u_1+c_2u_2$ with constant real coefficients.

This simple fact is of utmost importance for the whole study of Riemann surfaces. It indicates that linear methods must find fruitful applications in the theory, and that the concepts of *linear subspace* and *linear operator* will play a central role.

1B. The space of all harmonic functions on a Riemann surface W will be denoted by $H(W)$. A linear subspace $H_0(W)$ is any subset with the property that $u_1 \in H_0(W)$, $u_2 \in H_0(W)$ imply $c_1u_1+c_2u_2 \in H_0(W)$.

Numerous examples of linear subspaces can be given. For instance the bounded harmonic functions on W form a linear subspace which we shall designate by $HB(W)$. The constant functions form a very trivial subspace which we denote by C, and the space that contains only the constant 0 is itself denoted by 0.

Let W' be an open subset of W. There is an obvious way in which $H(W)$ can be identified with a linear subspace of $H(W')$.

1C. If a function u is defined and harmonic outside of a compact set on an open Riemann surface W we will say that it belongs to $H(\beta)$ where β stands for the ideal boundary of W (see Ch. I, 13A). It is understood that the compact set may depend on the function u. Nevertheless, $H(\beta)$ is a linear space with respect to the obvious definition of vector sum and scalar multiple. The relation of a function v to different subspaces of $H(\beta)$ will serve to describe the behavior of v in a neighborhood of the ideal boundary.

1D. Let H be a linear space of harmonic functions, and let H_0 be a linear subspace of H. Algebraically, H is an Abelian group with operators, and H_0 is a subgroup of H. A coset S of H_0 is formed by all elements of the form $s+u$ where s is a fixed element of H and u runs through all elements of H_0.

S is an *affine subspace*. By this we mean that $c_1v_1+c_2v_2 \in S$ whenever $v_1, v_2 \in S$ and $c_1+c_2=1$. The equation $c_1(s+u_1)+c_2(s+u_2)=s+(c_1u_1+c_2u_2)$ shows that S has this property.

Conversely, any affine subspace S is a coset of a subgroup. We choose an arbitrary $s \in S$. Then the differences $v-s$, $v \in S$ form a linear subspace H_0, for $c(v-s)=cv+(1-c)s-s$ and $(v_1-s)+(v_2-s)=\frac{1}{2}(2v_1-s)+\frac{1}{2}(2v_2-s)-s$. It is clear that S is the coset of H_0 which contains s.

1E. Consider the space $H(\beta)$ defined in 1C. Let S be an affine subspace of $H(\beta)$, and denote the corresponding linear subspace by H_0. We shall find it an important problem to determine the intersection $S \cap H(W)$, which is evidently affine. We are particularly interested in finding conditions under which the intersection is nonvoid and reduces to a coset of the linear subspace C of all constants. When this is so there exists a function in S which has a harmonic extension p to all of W, and this function is unique except for additive constants.

In more explicit formulation, we are given a function $s \in H(\beta)$ and a linear subspace $H_0 \subset H(\beta)$. We wish to construct a function p which is harmonic on all of W and such that $p - s \in H_0$. Conditions will be formulated which guarantee the existence and essential uniqueness of such a function.

It is a natural interpretation to say that p has a prescribed behavior near the ideal boundary in the sense determined by the subspace H_0. Briefly, p has the behavior s modulo H_0.

2. Linear operators

2A. To illustrate the general problem raised in 1E we consider first a classical application, the determination of harmonic functions with given singularities on a closed Riemann surface W_0. It follows from the maximum principle that $H(W_0)$ coincides with C, the space of constants. This explains why singularities must be introduced in order to yield a significant problem.

Let E be a finite point set on W_0, and set $W = W_0 - E$. The problem of finding a function on W_0 with given singularities at the points of E is more accurately formulated as a problem which requires the construction of a function on W with a prescribed behavior near the ideal boundary. Accordingly, we introduce the class $H(\beta)$ of functions which are defined and harmonic on W outside of a compact subset. A function in $H(\beta)$ is thus harmonic in a punctured neighborhood of each point $z_i \in E$. A linear subspace H_0 is formed by those functions which can be continued to the full neighborhoods, in other words the functions with removable singularities.

In terms of a local variable the most general harmonic function with an isolated singularity at z_i can be written in the form

$$\operatorname{Re} \sum_{n=0}^{\infty} a_n^{(i)} (z - z_i)^n + \operatorname{Re} \sum_{n=1}^{\infty} b_n^{(i)} (z - z_i)^{-n} + c^{(i)} \log |z - z_i|$$

where the two infinite series are supposed to converge, the first for sufficiently small $|z - z_i|$ and the second for all $z \neq z_i$; the coefficients $a_n^{(i)}, b_n^{(i)}$ are complex and $c^{(i)}$ is real. A singularity function $s \in H(\beta)$ is given by the singular parts

$$\operatorname{Re} \sum_{n=1}^{\infty} b_n^{(i)} (z - z_i)^{-n} + c^{(i)} \log |z - z_i|,$$

and the corresponding coset S consists of all functions with these singular parts. The intersection $H(W) \cap S$ is the class of all functions which are harmonic on W_0 except for isolated singularities with the given singular parts at the points z_i. We will prove that such functions exist if and only if $\sum_i c^{(i)} = 0$. It is trivial that any two functions in $H(W) \cap S$ differ by a constant.

2B. We return to the general problem with the purpose of imposing suitable restrictions on the subspace H_0. In order to achieve uniqueness we must make sure that $H(W) \cap H_0$ contains only the constants. In the above example this was a consequence of the maximum principle. It is natural to introduce a restriction which in the general case permits use of the same reasoning.

Let $W' \subset W$ be complementary to a compact set, and denote the boundary of W' by α. We require that any $u \in H_0$ which is defined on the closure of W' takes its maximum on α. In other words, since $-u$ is also in H_0, we postulate the double inequality

$$\min_\alpha u \leqq u(z) \leqq \max_\alpha u$$

for all $z \in W'$. When this is so we say, briefly, that the functions in H_0 satisfy the maximum principle.

Under this condition it is readily seen that any function in $H(W) \cap H_0$ has a maximum on W. It must consequently reduce to a constant.

2C. The next condition requires H_0 to be sufficiently inclusive. Specifically, we want the boundary value problem to be solvable. We consider again a set W' with a compact complement and the boundary α. If f is any continuous function on α we require that there exist a $u \in H_0$, defined in W', with a continuous extension to $\overline{W}'$ which equals f on α. It follows from the maximum principle that u is uniquely determined.

We remark that the solvability of the boundary value problem and the maximum principle need to be postulated only for a single fixed region W'. In all applications, however, these properties will stay in force for arbitrary regions W' with compact complement, and we can profit from this fact to choose, without loss of generality, a W' that is regularly imbedded. An assumption to this effect will be embodied in the hypotheses of Theorem 3A.

2D. By the corollary to Theorem 12D, Ch. II it is possible to enclose the complement of W' in a regular region G with positively oriented boundary $\beta(G)$. For a harmonic function u in W' the integral

$$\int_{\beta(G)} du^*$$

is called the *flux* over $\beta(G)$.

Suppose that G is replaced by another region G_1 with the same properties and the boundary $\beta(G_1)$. There exists a third region G_2 which contains $\bar{G} \cup \bar{G}_1$. We apply Green's formula (Ch. II, 8C) to $\bar{G}_2 - G$ and $\bar{G}_2 - G_1$, regarded as bordered surfaces, and obtain

$$\int_{\beta(G)} du^* = \int_{\beta(G_2)} du^* = \int_{\beta(G_1)} du^*.$$

The flux is thus independent of G and may be thought of as a quantity attached to the function u and the ideal boundary β. Accordingly, we agree to write the flux as a symbolic integral over β.

We introduce the crucial assumption that the flux vanishes,

$$\int_{\beta} du^* = 0, \tag{1}$$

for all functions $u \in H_0$. The reason for this strong restriction is primarily its effectiveness for the existence proof. But one can also present the heuristic argument that a nonvanishing flux indicates the presence of a source or sink. Therefore, if we want the functions $u \in H_0$ to be regular at the boundary in some reasonable sense, it is natural to require that the flux vanishes.

Condition (1) implies that all functions s in a coset S of H_0 have the same flux

$$\int_{\beta} ds^*. \tag{2}$$

On the other hand, a function in $H(W)$ has of course the flux zero. Hence $H(W) \cap S$ is void unless the common value of the integrals (2) is zero. We see that if (1) is fulfilled, a necessary condition for the existence of a harmonic function on W with the singularity s modulo H_0 is that the flux of s vanishes. Subject to the earlier assumptions concerning H_0 we intend to show that this condition is also sufficient.

2E. The hypotheses which we have introduced can be assembled and reformulated in a more condensed notation. According to 2C there exists a function $u \in H_0$ with given boundary values on α, and by 2B it is uniquely determined. We may therefore write $u = Lf$, where L is a functional whose domain of definition consists of all continuous real-valued functions on α. The basic assumption is that Lf is continuous on $\bar{W}'$ and harmonic in W'. In addition, L has the following properties:

(E1) $Lf = f$ on α,

(E2) $L(c_1 f_1 + c_2 f_2) = c_1 Lf_1 + c_2 Lf_2$,

(E3) $L1 = 1$,

(E4) $Lf \geqq 0$ if $f \geqq 0$,

(E5) $\int_{\beta} (dLf)^* = 0$.

An operator which satisfies (E1)–(E5) will be called a *normal operator*. The first condition states that Lf solves the boundary value problem. The second property shows that L is a linear operator. Conditions (E3) and (E4) are equivalent with the validity of the maximum principle. Indeed, they are obvious consequences of the maximum principle, and if they are fulfilled we see at once that $m \leqq f \leqq M$ implies $m \leqq Lf \leqq M$. Finally, (E5) is a restatement of the hypothesis introduced in 2D.

If v is any function defined on a set containing α we use the notation Lv to indicate the result of operating with L on the restriction of v to α. With this interpretation the operator L is idempotent: $L^2f = Lf$.

2F. We have considered the linear operator L as determined by the subspace H_0. Conversely, we can start from a given operator L. The corresponding subspace H_0 is then formed by all functions u, harmonic in W' and continuous in $\overline{W}'$, which satisfy the condition $u = Lu$. If L satisfies (E1)–(E5) it is evident that H_0 has the desired properties, and L is the linear operator associated with H_0.

In terms of the operator L the problem of finding a harmonic function with a given behavior s modulo H_0 amounts to constructing a harmonic function p on W which satisfies the nonhomogeneous linear equation

$$p-s = L(p-s).$$

2G. Several examples of normal operators will be considered. We show here that the classical problem introduced in 2A leads to a normal operator.

As before, E is a finite set on the closed surface W_0, and W is the open Riemann surface $W_0 - E$. We choose W' as a union of disjoint punctured parametric disks about the points of E. The subspace H_0 is the class of harmonic functions in W' with removable singularities at E. For these functions the maximum principle holds, and we know that the boundary value problem has a solution. Finally, condition (E5) is satisfied because each component of β (more precisely, $\beta(G)$) is contained in a simply connected subregion of W_0 so that the conjugate periods of any function in H_0 must vanish. Hence the corresponding operator L is indeed a normal operator.

For a singularity function s with the logarithmic terms $c^{(i)} \log |z-z_i|$ we find

$$\int_\beta ds^* = -2\pi \sum_i c^{(i)}$$

by Green's formula and the usual method of excluding the singularities by small circles. The vanishing of the flux is thus equivalent to the vanishing of $\sum c^{(i)}$.

If we anticipate the existence theorem we can consequently conclude that the problem of finding a harmonic function with a finite number of prescribed singularities has a solution if and only if the sum of the coefficients of the logarithmic terms vanishes.

3. The main existence theorem

3A. In this section we bring forth the proof of the existence theorem that has already been foreshadowed:

Theorem. *Let W be an open Riemann surface, and L a normal operator in the sense of 2E, defined with respect to a regularly imbedded open set $W' \subset W$ with compact complement.*

Given s, harmonic in W' and continuous on $\overline{W}'$, a necessary and sufficient condition that there exist a harmonic function p on W which satisfies

$$p - s = L(p - s) \tag{3}$$

in W' is that

$$\int_{\beta} ds^* = 0 \tag{4}$$

where β represents the ideal boundary.

The function p is uniquely determined up to an additive constant, and p is constant if and only if $s = Ls$.

The necessity and uniqueness have already been proved, and the final remark is a trivial consequence of the unicity. Indeed, if $s = Ls$ then $p = 0$ is a solution of (3), and if $p = 0$ is a solution, then (3) implies $s = Ls$.

The existence of p under condition (4) remains to be proved.

3B. We begin with the proof of a simple lemma:

Lemma. *Let A be a compact set on a Riemann surface W. There exists a positive constant $k < 1$, depending only on A and W, such that*

$$\max_A |u| \leqq k \sup_W |u|$$

for all functions u which are harmonic on W and not of constant sign on A.

If $\sup_W |u| = 0$ or ∞ there is nothing to prove. In all other cases u can be normalized by the condition $\sup_W |u| = 1$. If the lemma were not true we could find a sequence of normalized harmonic functions u_n for which $\max_A |u_n|$ tends to 1. Since W is known to have a countable basis it would be possible to select a subsequence which converges to a harmonic limit function u, and this function would attain either its maximum 1 or its minimum -1 on A. Hence it would be constantly 1 or -1. But this is not possible, for by assumption $\min_A |u_n| \leqq 0 \leqq \max_A |u_n|$, and this property must be shared by the limit function. The contradiction proves the lemma.

In this form the proof uses the countability of W. However, A can be enclosed in a relatively compact region $W_1 \subset W$. The lemma can then be proved for W_1 and follows trivially for W since $\sup_{W_1}|u| \leqq \sup_W |u|$.

3C. We recall the notations used in 2D. G is a regular region which contains the complement of W'. It has the boundary $\beta(G)$, and $G \cap W'$ is bounded by α and $\beta(G)$. We have oriented $\beta(G)$ positively with respect to G, and hence with respect to $G \cap W'$. Since W' is regularly imbedded we can also orient α, and we elect to let α denote the negatively oriented boundary of W'. The complete positively oriented boundary of $G \cap W'$ is thus $\beta(G) - \alpha$.

Lf represents a harmonic function in W' with the boundary values f on α. Similarly, if φ is continuous on $\beta(G)$ we denote by $K\varphi$ the unique harmonic function in G with the boundary values φ on $\beta(G)$.

To solve equation (3) it is sufficient to find functions f on α and φ on $\beta(G)$ which satisfy

$$\begin{aligned} \varphi - s &= L(f-s) \quad \text{on} \quad \beta(G), \\ f &= K\varphi \qquad\quad \text{on} \quad \alpha. \end{aligned} \tag{5}$$

In fact, if these conditions are fulfilled we can set $p = K\varphi$ in G, $p = Lf + s - Ls$ in W'. The two definitions coincide in $G \cap W'$, for $K\varphi = f = Lf + s - Ls$ on α, and $Lf + s - Ls = \varphi = K\varphi$ on $\beta(G)$. Furthermore, from $p - s = L(f-s)$ on $\overline{W}'$ it follows that $L(p-s) = L^2(f-s) = L(f-s) = p - s$. In other words, p is a solution to our problem.

It is clear that the system (5) is equivalent to the single equation

$$\varphi - LK\varphi = s - Ls \quad \text{on} \quad \beta(G). \tag{6}$$

We set $s - Ls = s_0$ and note that (6) has a formal solution represented by the Neumann series

$$\varphi = \sum_{n=0}^{\infty} (LK)^n s_0, \tag{7}$$

where $(LK)^n$ is the nth iterate of LK. We must show that the series converges and that φ satisfies (6).

3D. In order to make use of Lemma 3B we show that the functions Ks_0 and $K(LK)^n s_0$ are not of constant sign on α.

Let ω be the harmonic function in $G \cap W'$ which has the boundary values 0 on α and 1 on $\beta(G)$. It follows from the reflexion principle that ω is still harmonic on α and $\beta(G)$. If v is any function which is likewise harmonic on the closure of $G \cap W'$ we obtain by Green's formula

$$\int_{\beta(G)} v\, d\omega^* - \omega\, dv^* = \int_{\alpha} v\, d\omega^* - \omega\, dv^*.$$

In case v has vanishing flux this reduces to

$$\int_{\beta(G)} v\,d\omega^* = \int_{\alpha} v\,d\omega^*. \tag{8}$$

We apply (8) to $v = Ks_0 - s_0$. It has vanishing flux, for $s_0 = s - Ls$ and the flux of s vanishes by assumption. Since v is identically zero on $\beta(G)$ it remains harmonic on $\beta(G)$. The same applies to s_0 on α, and we see that v is harmonic on the closure of $G \cap W'$. By (8) we obtain

$$\int_{\alpha} (Ks_0 - s_0)\,d\omega^* = \int_{\beta(G)} (Ks_0 - s_0)\,d\omega^* = 0,$$

and thus

$$\int_{\alpha} Ks_0\,d\omega^* = \int_{\alpha} s_0\,d\omega^* = 0.$$

Let us now choose $v = K(LK)^n s_0 - (LK)^n s_0$ with $n > 0$. Again, v vanishes on $\beta(G)$, and $(LK)^n s_0 = K(LK)^{n-1} s_0$ on α. This is sufficient to conclude that v is harmonic on the closure of $G \cap W'$. Since the flux vanishes we obtain by (8)

$$\int_{\alpha} K(LK)^n s_0\,d\omega^* = \int_{\alpha} (LK)^n s_0\,d\omega^* = \int_{\alpha} K(LK)^{n-1} s_0\,d\omega^*.$$

Together with the previous result we have thus

$$\int_{\alpha} K(LK)^n s_0\,d\omega^* = 0$$

for all $n \geqq 0$.

This suffices to conclude that $K(LK)^n s_0$ changes sign on α, unless it is constantly zero. In fact, any local conjugate ω^* is increasing in the positive direction of α. Hence $v \geqq 0$ on α would imply $\int_{\alpha} v\,d\omega^* \geqq 0$, and $v \leqq 0$ would give the opposite inequality. Moreover, since ω^* cannot be constant on any open arc, the inequalities would be strict unless v is identically zero. The desired conclusion follows on taking $v = K(LK)^n s_0$.

3E. We are now ready for the convergence proof. Lemma 3B is applicable to G, the set α, and the function $K(LK)^{n-1} s_0$. It yields

$$\max_{\alpha} |K(LK)^{n-1} s_0| \leqq k \max_{\beta(G)} |(LK)^{n-1} s_0|.$$

On the other hand the maximum principle, which we have postulated for the operator L, leads to the inequality

$$\max_{\beta(G)} |(LK)^n s_0| \leqq \max_{\alpha} |K(LK)^{n-1} s_0|.$$

These estimates imply

$$\max_{\beta(G)} |(LK)^n s_0| \leqq k^n \max_{\beta(G)} |s_0|,$$

and we find that Neumann's series

$$\varphi = \sum_{n=0}^{\infty} (LK)^n s_0$$

converges uniformly on $\beta(G)$.

By use of the maximum principle together with uniform convergence we obtain

$$LK\varphi = \sum_{n=1}^{\infty} (LK)^n s_0 = \varphi - s_0,$$

and this is the relation that was needed to complete the proof of Theorem 3A.

Remark. The notion of normal operator was introduced by L. Sario in [4, 11, 20]. These papers contain proofs of Theorems 3A and 9E.

4. Examples of normal operators

4A. The generality of our approach would be pointless if we did not exhibit other examples of normal operators than the relatively trivial one which was used to construct harmonic functions with given singularities on closed Riemann surfaces.

We begin with a general remark which shows how different principles of construction can be combined with each other. As before, W is an arbitrary Riemann surface, and $W' \subset W$ is a regularly imbedded set with compact complement A and boundary α. In general, W' is not connected. We can therefore consider representations $W' = \cup W_i'$ as finite unions of disjoint open sets. Each W_i' consists of a finite number of components of W', and its boundary α_i is formed by certain components of α.

For any function f on α we let f_i denote its restriction to α_i. Suppose that a normal operator $L^{(i)}$ is defined for each W_i'. Then we can define a linear operator L on W' by setting $Lf = L^{(i)}f_i$ in W_i'. It is trivial that L satisfies conditions (E1)–(E5) in 2E. We say that L is the direct sum of the operators $L^{(i)}$.

This construction has the advantage that we can solve problems of a mixed type by choosing the $L^{(i)}$ according to different principles. On the other hand, we must be well aware that a normal operator L cannot always be decomposed into normal components $L^{(i)}$. In fact, the normality of $L^{(i)}$, defined as the restriction of L to W_i', requires that the flux of Lf vanishes when restricted to W_i', while the normality of L requires merely that the sum of the fluxes is zero.

4B. As a first example we construct the Green's function of a relatively compact subregion of a Riemann surface. The subregion is denoted by W, and we assume that W is contained in a larger Riemann surface W_0. By hypothesis, W shall be relatively compact and regular for the Dirichlet problem. The Green's function of W is defined as the harmonic function on W with singularity $-\log|z-\zeta|$ at a point $\zeta \in W$ which vanishes continuously on the boundary β of W (in this connection β means merely the point set theoretical boundary of W). Its uniqueness is evident.

Naturally, our problem concerns the punctured surface $W-\zeta$. We surround ζ with two concentric parametric disks contained in W, and we let W_1' denote the punctured smaller disk while W_2' will be the complement of the closed larger disk. We take W' to be the union $W_1' \cup W_2'$, and we define L by means of its components $L^{(1)}$ and $L^{(2)}$. Here $L^{(1)}$ is the operator associated with the class of harmonic functions with a removable singularity at ζ. For the construction of $L^{(2)}$ some preliminary considerations are needed.

4C. The region W_2' is regular for the Dirichlet problem. We can therefore construct a harmonic function Uf which has given boundary values f on α_2 (the circumference of the larger disk) and vanishes on β (the boundary of W). It is clear that Uf depends linearly on f.

If f is positive, the flux of Uf is negative. This would be obvious if W were regularly imbedded so that the flux could be determined by integration along β. In the general case we must make use of the harmonic function ω which is 0 on α_2 and 1 on β. If β_t denotes the level curve $\omega=t$ for $0<t<1$, positively oriented with respect to $\omega<t$, the flux of ω is

$$m = \int_{\beta_t} d\omega^*.$$

We have $m>0$, for the local conjugate ω^* increases along β_t.

By Green's formula the value of

$$\int_{\beta_t} Uf\, d\omega^* - \omega(dUf)^*$$

is independent of t. Thus

$$\int_{\beta_t} Uf\, d\omega^* - t\int_{\beta_t} (dUf)^*$$

is a constant which we can determine by letting t tend to 1. We obtain in this way

$$\int_{\beta_t} Uf\, d\omega^* = (t-1)\, m_f \tag{9}$$

where m_f denotes the flux of Uf. If $f \geqq 0$, so is Uf, and (9) shows that $m_f \leqq 0$ as asserted. Moreover, $m_f = 0$ if and only if f is identically zero.

4D. We define the operator $L^{(2)}$ by the equation

$$L^{(2)}f = Uf - \frac{m_f}{m}\,\omega. \tag{10}$$

$L^{(2)}$ is linear, $L^{(2)}1 = (1-\omega)+\omega = 1$, $f \geqq 0$ implies $L^{(2)}f \geqq 0$, and the flux of $L^{(2)}f$ is zero. Hence $L^{(2)}$ is a normal operator.

We apply Theorem 3A with $s = -\log|z-\zeta|$ in W_1' and $s = -\dfrac{2\pi}{m}\,\omega$ in W_2'. This function has the total flux zero. Hence there exists a harmonic function p which is such that $p + \log|z-\zeta|$ is regular at ζ and

$$p - s = L^{(2)}\,(p-s)$$

in W_2'. Since s and $L^{(2)}\,(p-s)$ have constant boundary values on β it follows that p is constant on β. If this constant is subtracted from p we obtain the Green's function $g(z, \zeta)$.

4E. It is clear that the above construction can be generalized to the case of arbitrary prescribed singularities at a finite number of points. We are thus able to construct a harmonic function in W which vanishes on β and possesses a finite number of isolated singularities with given singular parts.

In all problems with isolated singularities the operator $L^{(1)}$ is automatically given, and in the applications of Theorem 3A we need pay attention only to $L^{(2)}$. Accordingly, we can drop the superscript and revert to the notation L. The singularity function s is given near the singular points, and near the boundary we can choose s arbitrarily, provided we take care that the total flux vanishes. Without formulating any explicit theorem we can summarize our results by stating that the singularity problem has been effectively solved for any normal operator L on any surface. It remains to study the different types of normal operators.

§2. PRINCIPAL OPERATORS

The conditions (E1)–(E5) which a normal operator has to satisfy are not very restrictive, and there is thus a rich variety of problems to which the preceding method would apply. It must be recognized, however, that all these problems are not fundamental, and for this reason we want to distinguish a subclass of normal operators, called principal operators, which enter naturally in some problems of conformal mapping and are therefore worthy of a more detailed study.

It turns out that the principal operators have interesting extremal properties which in turn can be used to shed light on the Riemann surfaces on which the operators are defined.

5. Operators on compact regions

5A. We shall define two principal operators, L_0 and L_1, of which the first is unique, while there remains a certain amount of freedom in the definition of the second.

As far as L_0 is concerned it is essential that we limit ourselves, in this preliminary discussion, to a surface W which is given as the interior of a compact bordered Riemann surface $\overline{W}$. We choose a regularly imbedded boundary neighborhood $W' \subset W$ with compact complement and boundary α. For a given continuous f on α we construct $L_0 f$ as the harmonic function in W' which has the boundary values f on α and whose normal derivative vanishes on the border $\beta = \beta(W)$. The existence and uniqueness are practically trivial, for it suffices to pass to the double and solve the Dirichlet problem with boundary values f on α as well as on its reflection α^*.

It is clear that L_0 is a normal operator. It serves to construct a harmonic function with given isolated singularities whose normal derivative on the border vanishes. It is necessary to assume that the coefficients of the logarithmic terms have zero sum.

In this form the use of L_0 is not essential, for we could obtain the same result just by considering the singularities on the double $\hat{W}$. The usefulness lies on one side in the possibility of combining L_0 with other operators, and on the other side in the fact that the definition of L_0 can later be extended to open surfaces.

5B. The second principal operator, L_1, is nothing but the operator $L^{(2)}$ which was used in 4D to construct the Green's function. We have seen that it can be defined as soon as W is regular for the Dirichlet problem, but since we are going to push the generality much further it is sufficient at this stage to consider the compact bordered case. According to our definition $L_1 f$ is constant on β, and the constant is adjusted so that the flux vanishes.

This definition can be generalized. We consider an arbitrary partition P of the contours of W. In other words, we assume that $\beta = \cup \beta_i$ where the β_i are disjoint unions of contours. In terms of cycles, the same relationship is expressed by $\beta = \sum \beta_i$. It is then possible to define a normal operator $(P)L_1$ with the property that $(P)L_1 f$ is constant and has a vanishing flux separately along each β_i.

The existence of the function $(P)L_1 f$ is proved by means of Theorem 3A, applied to W'. First, we construct disjoint neighborhoods W_0' of α and

W_i' of β_i with boundaries $\alpha_0 \cup \alpha$ and $\alpha_i \cup \beta_i$ respectively. We define L to be the operator whose components in W_0' and W_i' are the principal operators L_1 for those regions acting on functions on α_0 and α_i. In W_0' we choose the singularity function s to be any harmonic function with the boundary values f on α. Let ω_0 be the harmonic function in W_0' which is 0 on α and 1 on α_0. By subtracting a suitable multiple of ω_0 we can adjust the definition of s so that its flux along α vanishes. To complete, we set $s=0$ in all W_i'. Then the total flux of s is zero, and we can solve the equation $p-s=L(p-s)$. It is clear that p has the boundary values f on α except for an added constant. We subtract this constant and denote the resulting function by $(P)L_1 f$.

$(P)L_1 f$ has a constant value c_i on each β_i, and the flux along β_i vanishes. To prove that $(P)L_1$ is a normal operator we must show that $f \geqq 0$ implies $c_i \geqq 0$. This can be done by induction on the number of parts β_i.

For a fixed β_i we construct, by the same method as above, an auxiliary harmonic function ω_i which is 0 on α, 1 on β_i, constant on each $\beta_j \neq \beta_i$, and such that its flux along each $\beta_j \neq \beta_i$ vanishes. Application of Green's formula yields

$$c_i \int_{\beta_i} d\omega_i^* = \int_{\alpha} f\, d\omega_i^*.$$

The induction hypothesis permits us to conclude that $\omega_i > 0$. Therefore, ω_i^* increases along α. We conclude that $f \geqq 0$ implies $c_i \geqq 0$, and that $c_i = 1$ if $f=1$. Hence $(P)L_1$ is a normal operator.

We call $(P)L_1$ the second principal operator for the partition P. If P is the identity partition (Ch. I, 38A) it reduces to L_1. As soon as the partition is agreed upon and no confusion can arise we shall feel free to simplify the notation and write L_1 for $(P)L_1$.

6. Extremal harmonic functions on compact regions

6A. In order to pass from compact bordered surfaces to open Riemann surfaces it is necessary to study the compact case in greater detail. In particular, we need a better knowledge of the Dirichlet integral, especially in its relation to the extremal properties of certain harmonic functions. These properties are interesting in themselves, and they retain their validity in the noncompact case. The purpose of this section is therefore twofold: to pave the way for the study of open surfaces, and to detect extremal properties which can later be formulated with full generality.

In this discussion $\overline{W}$ will be a compact bordered surface with more than one contour. The contours are divided into two disjoint nonempty sets α and β, called the inner and outer border. It is convenient to orient the inner contours negatively and the outer contours positively with respect to $\overline{W}$, in such a manner that the complete positively directed border of

$\overline{W}$ is represented by the cycle $\beta-\alpha$. For the orientation of the reader we remark that W takes the place of the subregion W' considered in 5, β corresponds to the border, and α represents the relative boundary of W'.

6B. In this section the Dirichlet integral $D_W(u)$ will be denoted by $D(u)$. Suppose that u is harmonic on $\overline{W}$. Then its Dirichlet integral is expressed by

$$D(u) = \int_\beta u\,du^* - \int_\alpha u\,du^*. \tag{11}$$

We separate the two integrals, and write

$$A(u) = \int_\alpha u\,du^*, \quad B(u) = \int_\beta u\,du^*.$$

The corresponding bilinear functionals are denoted by

$$A(u, v) = \int_\alpha u\,dv^*, \quad B(u, v) = \int_\beta u\,dv^*.$$

Green's formula, applied to two harmonic functions u and v, can then be expressed as

$$A(u, v) - A(v, u) = B(u, v) - B(v, u). \tag{12}$$

All the results that we are going to derive in this subsection are consequences of the identity (12), together with the inequality $D(u) \geqq 0$ and the quadratic character of $A(u)$ and $B(u)$.

6C. In the following u will be a fixed harmonic function on $\overline{W}$ which satisfies the condition $B(u)=0$. We are going to compare u with the harmonic functions v in $\overline{W}$ which satisfy either the condition $B(v, u)=0$ or $B(u, v)=0$.

Accordingly, for a given u we denote by $H_0(u)$ the class of all harmonic v with $B(v, u)=0$, and by $H_1(u)$ the class with $B(u, v)=0$. The condition $B(u)=0$ implies that $u \in H_0(u) \cap H_1(u)$.

For $v \in H_0(u)$ we obtain

$$B(v-u) = B(v) - B(u, v),$$

and (12) reduces to

$$B(u, v) = A(u, v) - A(v, u).$$

Hence

$$B(v-u) = B(v) - A(u, v) + A(v, u),$$

and when the identity

$$A(v-u) = A(v) - A(v, u) - A(u, v) + A(u)$$

is subtracted we obtain by (11)

$$D(v-u) = D(v)+2A(v, u)-A(u). \tag{13}$$

For $v \in H_1(u)$ a parallel reasoning leads to the relation

$$D(v-u) = D(v)+2A(u, v)-A(u). \tag{14}$$

6D. The relations (13) and (14) are the key to the considerations that follow. Their content becomes much more intuitive if they are interpreted as extremal properties of the function u.

To arrive at this interpretation we need only observe that $D(v-u)$ is always nonnegative, and that it is equal to zero if and only if $v-u$ is constant. It follows that

$$D(v)+2A(v, u) \geqq A(u)$$
$$D(v)+2A(u, v) \geqq A(u)$$

respectively, with equality for $v=u$. We can therefore express (13) and (14) as follows:

Theorem. *A harmonic function u with $B(u)=0$ minimizes the functional $D(v)+2A(v, u)$ in the class $H_0(u)$, and the functional $D(v)+2A(u, v)$ in the class $H_1(u)$.*

The deviation from the minimum is in each case given by $D(v-u)$. The value of the minimum is $A(u)$.

6E. We shall say that two harmonic functions u_0 and u_1 with $B(u_0)=B(u_1)=0$ are *associated* if $B(u_1, u_0)=0$. An important generalization of Theorem 6D is obtained by considering linear combinations $u_{hk}=hu_0+ku_1$ where h and k are real constants. We observe that $u_{hk} \in H_0(u_0) \cap H_1(u_1)$ for all h, k.

Let v be any function in $H_0(u_0) \cap H_1(u_1)$. The same computation as in 6C leads to

$$B(v-u_{hk}) = B(v)-hB(u_0, v)-kB(v, u_1)+B(u_{hk}) =$$
$$B(v)-h(A(u_0, v)-A(v, u_0))-k(A(v, u_1)-A(u_1, v))+B(u_{hk}).$$

We subtract

$$A(v-u_{hk}) = A(v)-h(A(u_0, v)+A(v, u_0))-k(A(u_1, v)+A(v, u_1))+A(u_{hk})$$

and obtain the relation

$$D(v-u_{hk}) = D(v)+2hA(v, u_0)+2kA(u_1, v)+D(u_{hk}). \tag{15}$$

With the same interpretation as above we can state

Theorem. *If u_0 and u_1 are associated functions, the linear combination $u_{hk}=hu_0+ku_1$ minimizes the functional*

$$M_{hk}(v) = D(v)+2hA(v, u_0)+2kA(u_1, v)$$

in the class $H_0(u_0) \cap H_1(u_1)$, *and the deviation from the minimum is* $D(v-u_{hk})$.

The first part of Theorem 6D is a special case, corresponding to $(h, k) = (1, 0)$ or $(0, 1)$. However, 6D is stronger, for in that theorem the minimum refers to a larger class.

6F. If considered in its full generality, Theorem 6E does not solve a genuine minimum problem, for the functional and the class of competing functions both depend on u_0 and u_1. However, all applications that we are going to make refer to the special case where all functions are required to have the same boundary values on α. With this restriction the theorem expresses a minimum property that will be helpful in extending L_0 and L_1 to open surfaces.

Suppose therefore that u_0 and u_1 are equal on α, and denote their common restriction to α by f. If we choose h, k so that $h+k=1$ it is also true that $u_{hk}=f$ on α. Let us now restrict the competition to functions $v \in H_0(u_0) \cap H_1(u_1)$ with the boundary values f on α. Then $A(v, u_0) = A(u_0)$ does not depend on v, and $A(u_1, v) = A(v)$. The following corollary of Theorem 6E results:

Theorem. *If* u_0 *and* u_1 *are associated functions with the same values* f *on* α, *and if* $h+k=1$, *then* $u_{hk} = hu_0 + ku_1$ *minimizes the functional*

$$B(v) + (k-h)A(v)$$

in the class of all $v \in H_0(u_0) \cap H_1(u_1)$ *which are equal to* f *on* α.

6G. It is clear that a common factor in h and k is of no importance. Therefore, the restriction $h+k=1$ amounts merely to a normalization, except in the case $h=-k$. The excluded case leads to the following special result:

Theorem. *The function* $u_0 - u_1$ *minimizes the expression* $D(v) - 2A(u_1, v)$ *among all* $v \in H_0(u_0) \cap H_1(u_1)$ *which vanish on* α.

7. Application to the principal operators

7A. We recall the definition of the principal operators L_0 and L_1 introduced in 5. $\bar{W}$ is again a compact bordered surface, and $W' \subset W$ is regularly imbedded with the relative boundary α. For a given f on α, $L_0 f$ solves the boundary value problem in W' with a vanishing normal derivative on $\beta(W)$, and $L_1 f$ equals f on α and is constant on $\beta(W)$, the constant being chosen so that the flux vanishes.

We apply the results in 6 to the region W' with $\beta = \beta(W)$ and $u_0 = L_0 f$, $u_1 = L_1 f$. It is immediately verified that $B(u_0) = B(u_1) = B(u_1, u_0) = 0$. In other words, u_0 and u_1 are associated harmonic functions (6E), and Theorems 6E–6G are applicable. The notion of associated functions was introduced to underline the range of validity of these theorems. Once we have verified that L_0f and L_1f satisfy all hypotheses the general concept of associated functions will no longer be needed.

The class $H_0(u_0)$ comprises all harmonic functions on $\overline{W}'$, and $H_1(u_1)$ consists of all harmonic functions with a vanishing flux. Consequently, in Theorems 6D–6G the minimizing property will be with respect to functions v with vanishing flux.

The application is automatic, except for a slight inconsistency in the assumption. We have introduced L_0f and L_1f for arbitrary continuous f, and then u_0, u_1 are not known to be harmonic on α as required in 6. The difficulty is resolved by restricting the use of the operator to functions f which have a harmonic extension to a neighborhood of α, i.e. to functions which are analytic in the real sense. This does not impede the usefulness of the theory, for in the proof of the basic existence theorem (Theorem 3A) the operators were used only in this manner.

7B. More generally, we can consider the operator $(P)L_1$ which corresponds to a partition P: $\beta = \cup \beta_i$. We recall that $(P)L_1f$ is constant and has zero flux along each β_i. With $u_0 = L_0f$, $u_1 = (P)L_1f$ we find that the conditions $B(u_1) = B(u_1, u_0) = 0$ are still satisfied. Thus u_0 and u_1 continue to be associated, but the class $H_0(u_0) \cap H_1(u_1)$ becomes, by definition, restricted to the functions v whose flux vanishes separately along each β_i.

7C. In order to lead up to the case of open surfaces we must discuss extremal properties in variable subregions. We shall now let W denote a fixed open surface.

We choose also, once and for all, a regularly imbedded subregion $W' \subset W$ with compact complement and relative boundary α, negatively oriented with respect to W'. Finally, we let Ω be a generic notation for a regular subregion which contains the complement of W'; it will be recalled that a regular subregion is one which is regularly imbedded, relatively compact, and whose complement $W - \Omega$ has no compact components (Ch. II, 3B). The positively oriented boundary of Ω is denoted by $\beta(\Omega)$.

The operator L_0, as applied to $\Omega \cap W'$ and acting on functions f on α, will be denoted by $L_{0\Omega}$. Similarly, if P is a partition $\beta(\Omega) = \cup \beta_i(\Omega)$ of the border of Ω, we consider the corresponding principal operator $(P)L_{1\Omega}$. The functions $u_{0\Omega} = L_{0\Omega}f$, $u_{1\Omega} = (P)L_{1\Omega}f$ are associated, the class $H_{0\Omega} = H_0(u_{0\Omega})$ consists of all harmonic functions in the closure of $\Omega \cap W'$, and $H_{1\Omega} = H_1(u_{1\Omega})$ is the subclass of functions with zero flux along each $\beta_i(\Omega)$. Observe that $H_{0\Omega}$ and $H_{1\Omega}$ are independent of f.

7D. If $\Omega \subset \Omega'$ it is evident that $H_{0\Omega'} \subset H_{0\Omega}$; more precisely, $H_{0\Omega'}$ can be identified with a subclass of $H_{0\Omega}$. It is essential that we make assumptions which permit the same conclusion with respect to $H_{1\Omega}$. For this reason the partitions P associated with different regions Ω must be related to each other in a suitable manner.

We recall the results derived in Ch. I, 38. The ideal boundary β of W can be realized as a set of boundary components with a uniquely defined topology. A regular partition P of β is a partition $\beta = \cup \beta_\mu$ into closed mutually exclusive subsets. It was found that P induces, for each regular region Ω, a finite partition $\beta(\Omega) = \cup \beta_i(\Omega)$ into sets of contours. The induced partitions have the following special properties:

(1) All contours of a component of the complement $W - \Omega$ belong to the same part $\beta_i(\Omega)$.

(2) If $\Omega \subset \Omega'$, then all contours that are part of a given $\beta_j(\Omega')$ belong to components of $\Omega' - \Omega$ whose contours on $\beta(\Omega)$ are all contained in the same part $\beta_i(\Omega)$.

Conversely, when a family of partitions of all $\beta(\Omega)$ satisfies these conditions, then the partitions are induced by a unique regular partition P of the ideal boundary. The extreme cases are the *canonical partition* Q in which the $\beta_i(\Omega)$ are identical with the boundaries of complementary components, and the *identity partition* I which leaves every $\beta(\Omega)$ undivided.

7E. For an open surface W the symbol P will denote a regular partition of the ideal boundary, and it is understood that we use the partition $\beta(\Omega) = \cup \beta_i(\Omega)$ induced by P for each regular region Ω. With this convention it is easy to show that $\Omega \subset \Omega'$ implies $H_{1\Omega'} \subset H_{1\Omega}$. In fact, let us consider a harmonic function that belongs to $H_{1\Omega'}$. Its flux along the parts $\beta_j(\Omega')$ is zero.

Consider the union of all components of $\Omega' - \Omega$ whose boundary on $\beta(\Omega)$ is contained in $\beta_i(\Omega)$. By condition (2) this set has a boundary on $\beta(\Omega')$ which is a union of sets $\beta_j(\Omega')$. Hence the flux vanishes over the part of the boundary on $\beta(\Omega')$, and we conclude that the flux over $\beta_i(\Omega)$ must also vanish. The given function is thus in $H_{1\Omega}$.

We summarize the results in the following statement:

If all partitions are induced by a regular partition of the ideal boundary, then $\Omega \subset \Omega'$ *implies* $H_{0\Omega'} \subset H_{0\Omega}$ *and* $H_{1\Omega'} \subset H_{1\Omega}$.

8. Extension to noncompact regions

8A. We shall now consider a fixed partition P of β, the ideal boundary of an open Riemann surface W. Accordingly, the notations $L_{1\Omega}$ and $u_{1\Omega}$ will refer to this partition. Otherwise, we use exactly the same terminology as in 7. The reader is advised to review the notations introduced at the beginning of 7C.

If $\Omega \subset \Omega'$ we know that $u_{0\Omega'} \in H_{0\Omega}$ and $u_{1\Omega'} \in H_{1\Omega}$. Hence formulas (13) and (14) are applicable, and we obtain

$$
\begin{aligned}
D_\Omega(u_{0\Omega'} - u_{0\Omega}) &= B_\Omega(u_{0\Omega'}) + A(u_{0\Omega}) - A(u_{0\Omega'}) \\
D_\Omega(u_{1\Omega'} - u_{1\Omega}) &= B_\Omega(u_{1\Omega'}) + A(u_{1\Omega'}) - A(u_{1\Omega}).
\end{aligned} \tag{16}
$$

Here $B_\Omega(u_{i\Omega'}) \leqq B_{\Omega'}(u_{i\Omega'}) = 0$, $i = 0, 1$, and we conclude that

$$A(u_{0\Omega'}) \leqq A(u_{0\Omega}), \quad A(u_{1\Omega'}) \geqq A(u_{1\Omega}).$$

Otherwise stated, $A(u_{0\Omega})$ decreases and $A(u_{1\Omega})$ increases with increasing Ω.

But it is also true that $u_{1\Omega} \in H_{0\Omega}$. If we apply (13) it follows that

$$D_\Omega(u_{1\Omega} - u_{0\Omega}) = A(u_{0\Omega}) - A(u_{1\Omega}).$$

Hence $A(u_{1\Omega}) \leqq A(u_{0\Omega})$, and we conclude that the limits

$$
\begin{aligned}
A_0 &= \lim_{\Omega \to W} A(u_{0\Omega}) \\
A_1 &= \lim_{\Omega \to W} A(u_{1\Omega})
\end{aligned}
$$

exist and are finite.

8B. From the existence of the limits A_0, A_1 together with the equations (16) it follows that $D_\Omega(u_{i\Omega'} - u_{i\Omega})$, $i = 0, 1$, tends to zero as $\Omega \to W$ and $\Omega \subset \Omega'$. For a fixed Ω it is easy to infer, by use of the triangle inequality, that

$$\lim_{\Omega', \Omega'' \to W} D_\Omega(u_{i\Omega'} - u_{i\Omega''}) = 0$$

without any restrictions as to the relative positions of Ω' and Ω''.

We are now ready to apply Theorem 13C of Ch. II. Consider a component of $\Omega \cap W'$, and reflect it across its boundary on α. For $\Omega' \supset \Omega$ the function $u_{i\Omega} - u_{i\Omega'}$ can be continued by symmetry to the double of $\Omega \cap W'$, which we denote by $\hat{\Omega}$, and it is true that

$$\lim_{\Omega', \Omega'' \to W} D_{\hat{\Omega}}(u_{i\Omega'} - u_{i\Omega''}) = 0.$$

We apply the theorem to $\hat{\Omega}$ with a fixed point z_0 on α. Since $u_{i\Omega'} - u_{i\Omega}$ vanishes at z_0 we may conclude that

$$\lim_{\Omega' \to W} (u_{i\Omega'}(z) - u_{i\Omega}(z))$$

exists, uniformly on every compact subset of $\hat{\Omega}$. Since the reasoning can be applied to any Ω and any component it follows that

$$u_i(z) = \lim_{\Omega' \to W} u_{i\Omega'}(z)$$

exists and is harmonic throughout the closure of W'. Moreover, $u_i - u_{i\Omega}$ vanishes on α and can be extended by symmetry.

8C. We can write $u_i = L_i f$, $i=0, 1$, where L_i is a normal operator. The linearity of L_i is evident. Furthermore, $f \geqq 0$ implies $L_{i\Omega} f \geqq 0$ and hence $L_i f \geqq 0$. To $f=1$ corresponds $L_i f = 1$. Finally, the flux of $L_i f$ can be evaluated along α, and since the derivatives of $L_{i\Omega} f$ converge uniformly to those of $L_i f$ the flux is zero.

Letting Ω' tend to W in (16) we obtain

$$D_\Omega(u_0 - u_{0\Omega}) = B_\Omega(u_0) + A(u_{0\Omega}) - A(u_0)$$
$$D_\Omega(u_1 - u_{1\Omega}) = B_\Omega(u_1) - A(u_{1\Omega}) + A(u_1).$$

Since the Dirichlet integrals are nonnegative and $B_\Omega(u_i) \leqq 0$ we have the two consequences

$$B(u_i) = \lim_{\Omega \to W} B_\Omega(u_i) = 0$$

and

$$\lim_{\Omega \to W} D_\Omega(u_i - u_{i\Omega}) = 0. \tag{17}$$

The operators L_0 and L_1 (or $(P)L_1$) are called principal operators on the open surface W.

8D. It is now an easy matter to generalize the theorems in 6. We assume that v lies in all classes $H_{1\Omega}$. Writing $u_{hk\Omega} = hu_{0\Omega} + ku_{1\Omega}$ we obtain by (15), 6E,

$$D_\Omega(v - u_{hk\Omega}) = D_\Omega(v) + 2hA(v, u_{0\Omega}) + 2kA(u_{1\Omega}, v) + D_\Omega(u_{hk\Omega}).$$

By use of the triangle inequality we deduce from (17) that

$$D_\Omega(u_{hk} - u_{hk\Omega}) \to 0,$$

and this implies, again by the triangle inequality,

$$D_\Omega(u_{hk\Omega}) \to D(u_{hk}), \quad D_\Omega(v - u_{hk\Omega}) \to D(v - u_{hk}).$$

In the limit we have thus

$$D(v - u_{hk}) = D(v) + 2hA(v, u_0) + 2kA(u_1, v) + D(u_{hk}), \tag{18}$$

which is exactly the same relation as before.

Since $B_\Omega(v) = D_\Omega(v) + A(v)$ increases with Ω it is proper to use the definition $B(v) = \lim_{\Omega \to W} B_\Omega(v) = D(v) + A(v)$ for arbitrary harmonic v. Of course, $B(v)$ is finite if and only if $D(v)$ is finite.

8E. The class of functions v for which (18) is valid must still be characterized in terms independent of the variable region. In case L_1 is associated with the identity partition we need merely require that v has a vanishing flux along each $\beta(\Omega)$, and this is so if and only if the flux along α vanishes. The same property is expressed by saying that v has zero flux along the ideal boundary of W.

For a general operator $(P)L_1$ the flux of v must vanish along all $\beta_i(\Omega)$. In the extreme case of the canonical partition this condition has a very simple interpretation: the flux must be zero along all *dividing cycles* (Ch. I, 30D). This terminology can be carried over to an arbitrary partition P. Accordingly, we say that a cycle is dividing with respect to a partition P if, for sufficiently large Ω, it is homologous to a linear combination of parts $\beta_i(\Omega)$.

There is no point in repeating the statements of Theorems 6D–6G. The present setting for the theorems is at once more general, because of the extension to open surfaces, and more restricted, because of the special choice of u_0 and u_1.

§3. PRINCIPAL FUNCTIONS

We continue our investigation of extremal properties on open surfaces with particular attention to extremal functions with given singularities. Although there are, on open surfaces, many harmonic functions with the same singularities, they are nevertheless subject to severe restrictions, and we wish to express these restrictions in the form of inequalities. The extremal functions are the ones for which the corresponding equality holds, and we shall show that they are uniquely determined.

9. Functions with singularities

9A. Having defined the operators L_0 and L_1 for open surfaces we can apply Theorem 3A to establish the existence of harmonic functions, with prescribed singularities, which are of the form $L_0 f$ or $L_1 f$ near the ideal boundary:

Theorem. *Let W be an open Riemann surface. At a finite number of points $\zeta_j \in W$ there are given singularities of the form*

$$\operatorname{Re} \sum_{n=1}^{\infty} b_n^{(j)} (z-\zeta_j)^{-n} + c^{(j)} \log |z-\zeta_j| \tag{19}$$

where the $c^{(j)}$ are real and subject to the condition $\sum_j c^{(j)} = 0$.

Suppose that the principal operators L_0 and L_1 are defined relatively to a W' whose compact complement contains all ζ_j in its interior. Then there exist functions p_0 and p_1, harmonic on W except for the singularities (19), *such that*

$$L_0 p_0 = p_0, \quad L_1 p_1 = p_1$$

in W'. These functions are unique and independent of W', save for an additive constant.

The functions p_0 and p_1 will be called the *principal functions* on W

with respect to the given singularities. The fact that they do not depend on W' can be verified directly, but it is also an immediate consequence of the extremal properties which we are going to derive.

9B. We are going to apply the results of 6 with p_0 and p_1 in the roles of u_0, u_1. The linear combination $p_{hk}=hp_0+kp_1$ has the same singularities as p_0 and p_1 provided that $h+k=1$, and for the moment we consider only this case. We shall use p as a generic notation for a function which has the given singularities and which also belongs to the intersection of all classes $H_{1\Omega}$. The latter condition means that the conjugate periods of p vanish along all dividing cycles relatively to the partition P (see 8E).

The Dirichlet integral of $p-p_{hk}$ over the whole surface W can be written as

$$D(p-p_{hk}) = D_{W'}(p-p_{hk})+A(p-p_{hk}).$$

As a measure for the deviation from p_{hk} this is more natural than the restricted Dirichlet integral $D_{W'}(p-p_{hk})$, and for this reason we prefer to write (15), 6E, in the form

$$M_{hk}(p)-M_{hk}(p_{hk})+A(p-p_{hk}) = D(p-p_{hk}) \tag{20}$$

with

$$M_{hk}(p) = B(p)-A(p)+2hA(p,p_0)+2kA(p_1,p).$$

We transform $M_{hk}(p)$ by substituting

$$p_0 = p_{hk}+k(p_0-p_1), \quad p_1 = p_{hk}-h(p_0-p_1).$$

After brief computation we obtain

$$M_{hk}(p) = B(p)-A(p)+2hA(p,p_{hk})+2kA(p_{hk},p)$$
$$+2hk[A(p,p_0-p_1)-A(p_0-p_1,p)].$$

The bracketed term is independent of p, for by Green's formula

$$A(p-p_{hk},p_0-p_1) = A(p_0-p_1,p-p_{hk}).$$

Therefore,

$$M_{hk}(p)-M_{hk}(p_{hk}) = B(p)-B(p_{hk})-A(p)-A(p_{hk})$$
$$+2hA(p,p_{hk})+2kA(p_{hk},p),$$

and after simplification (20) becomes

$$B(p)-B(p_{hk})+(h-k)[A(p,p_{hk})-A(p_{hk},p)] = D(p-p_{hk}). \tag{21}$$

Formula (21) can be interpreted as follows:

Theorem. *The function p_{hk} minimizes the expression*

$$B(p)+(h-k)[A(p,p_{hk})-A(p_{hk},p)]$$

and the deviation from the minimum $B(p_{hk})$ is measured by $D(p-p_{hk})$.

9C. At first sight Theorem 9B does not look like a useful minimum property, for the solution p_{hk} appears in the functional that is being minimized. This is only apparent, however, as we shall just see. By Green's formula the quantity

$$A(p, p_{hk}) - A(p_{hk}, p) \tag{22}$$

does not change if we move α to the vicinity of the points ζ_i. We may thus assume that the singularity function s, given by (19), is defined on the whole complement of W'. The expression $A(p, s) - A(s, p)$ has then a meaning, and its value differs from (22) by a quantity $A(p, s - p_{hk}) - A(s - p_{hk}, p)$ which by a new application of Green's formula is seen to be independent of p.

We conclude that p_{hk} minimizes the functional

$$B(p) + (h-k)[A(p, s) - A(s, p)] \tag{23}$$

and may thus be regarded as the solution of a genuine minimum problem. The deviation from the minimum is $D(p - p_{hk})$. The minimizing function is hence unique except for an additive constant. In particular, p_0 and p_1 are unique save for unimportant constants.

9D. The quantity $A(p, s) - A(s, p)$ is a residue, and can be evaluated in terms of the coefficients of s and p. If the α_j are circles about ζ_j we find first that

$$A(p, s) - A(s, p) = A(p-s, s) - A(s, p-s) = \sum_j \int_{\alpha_j} (p-s)\, ds^* - s d(p-s)^*.$$

Here $(p-s)^*$ is single-valued in a neighborhood of each ζ_i so that we can write

$$\int_{\alpha_j} (p-s)\, ds^* - s d(p-s)^* = \int_{\alpha_j} (p-s)\, ds^* + (p-s)^*\, ds.$$

For a more compact notation we set $dP = dp + i\, dp^*$ and $dS = ds + i\, ds^*$. Then $P - S$ is single-valued and regular at the points ζ_i, and we obtain

$$A(p, s) - A(s, p) = \sum_j \operatorname{Im} \int_{\alpha_j} (P-S)\, dS = 2\pi \operatorname{Re} \left\{ \sum_j \operatorname{Res} (P-S)\, dS \right\}.$$

To compute the residues we suppose that $P - S$ has the power series developments

$$P - S = \sum_{m=0}^{\infty} a_m^{(j)} (z - \zeta_j)^m.$$

The residue of $(P-S)\,dS$ at ζ_j is then

$$c^{(j)}a_0^{(j)} - \sum_{n=1}^{\infty} n b_n^{(j)} a_n^{(j)}$$

where the series is known to converge. The convergence follows from estimates $|a_n^{(j)}| \leqq M_1 r_1^{-n}$, $|b_n^{(j)}| \leqq M_2 r_2^n$ where it is possible to choose $r_2 < r_1$.

9E. In order to determine the value of the minimum it remains to compute $B(p_{hk})$ in terms of the coefficients. Let us first note that $A(p_{hk\Omega}) \to A(p_{hk})$ by uniform convergence, and that $D_\Omega(p_{hk\Omega}) \to D(p_{hk})$ by application of (17), 8C, and the triangle inequality (or by Theorem 13C, Ch. II). It follows that $B_\Omega(p_{hk\Omega}) \to B(p_{hk})$. For this reason we can just as well assume that we are dealing with a compact bordered surface. In that case we have trivially

$$B(p_{hk}) = hkB(p_0, p_1) = hk[B(p_0, p_1) - B(p_1, p_0)],$$

and by Green's formula the integrals on the right can be transferred from β to circles α_j about the ζ_j. The expression takes the form

$$hk[A(p_0, p_1 - p_0) - A(p_1 - p_0, p_0)],$$

and as in 9D its value is found to be

$$2\pi hk \operatorname{Re}\left\{\sum_j \operatorname{Res}(P_0 - P_1)dP_0\right\}.$$

Let us now introduce the notation

$$C(p) = 2\pi \operatorname{Re}\left\{\sum_j \left[c^{(j)}a_0^{(j)} - \sum_{n=1}^{\infty} n b_n^{(j)} a_n^{(j)}\right]\right\} \tag{24}$$

and set C_0, C_1 for $C(p_0)$, $C(p_1)$ respectively. The functional (23) can be written as $B(p) + (h-k)C(p)$ while its value for p_{hk} becomes, according to the above computation,

$$hk(C_0 - C_1) + (h-k)(hC_0 + kC_1) = h^2C_0 - k^2C_1.$$

We collect the results in the following main statement:

Theorem. *The function $p_{hk} = hp_0 + kp_1$, $h+k=1$, minimizes the expression $B(p)+(h-k)C(p)$ in the class of all functions p with the singularities* (19) *whose conjugate periods vanish over all dividing cycles associated with the partition P.*

Explicitly, the formula

$$B(p) + (h-k)\,C(p) = h^2C_0 - k^2C_1 + D(p - p_{hk}) \tag{25}$$

which is valid for such functions shows that the minimum is $h^2C_0 - k^2C_1$, and that the deviation from the minimum is measured by $D(p - p_{hk})$.

For the orientation of the reader we recall that the principal functions p_0 and p_1 are completely determined by the singularity and the partition P. The most important cases are the ones that correspond to the identity partition I and the canonical partition Q.

10. Special cases

10A. In the preceding section we proved an existence theorem of rather general character. Its significance is best appreciated if we examine some special cases. In the first place, there is rarely any need to use the theorem with more than a finite number of nonvanishing coefficients in the developments of the singularity function s. What is more, the principal functions p_0 and p_1 depend linearly on the singularities. For this reason there are really only two cases that need to be considered: the case where only one $b_n^{(j)}$ is different from zero, and the case where two coefficients $c^{(j)}$ are different from zero and opposite.

It is clear that the nonzero coefficients $b_n^{(j)}$ or $c^{(j)}$ can be normalized at will. It must be kept in mind, however, that the first normalization depends essentially on the choice of the local variable z at ζ_j. The linearity is with respect to real rather than complex numbers. With a given choice of the local variable the most general singularity with $b_n^{(j)} \neq 0$ is thus obtained as a linear combination of singularities with the coefficients 1 and i respectively. An equivalent way of specifying the second singularity is to replace z by iz while keeping the coefficient equal to 1.

10B. In the case of a simple pole we obtain the following statement:

Theorem. *Among all admissible functions p with a development*

$$p(z) = \operatorname{Re}\left\{\frac{1}{z-\zeta}+a_0+a_1(z-\zeta)+\cdots\right\}$$

at a given point ζ, the function p_{hk} minimizes $B(p)-2\pi(h-k)\operatorname{Re} a_1$.

We use the phrase "admissible functions" to indicate the class of harmonic functions whose conjugate periods vanish over the dividing cycles with respect to the partition P.

10C. For a singularity consisting of two logarithmic poles we find:

Theorem. *Among all admissible functions with the singularities* $\log|z-\zeta_1|$ *and* $-\log|z-\zeta_2|$, *the expression* $B(p)+(h-k)2\pi\operatorname{Re}(a_0^{(1)}-a_0^{(2)})$ *attains its minimum for $p=p_{hk}$.*

In order to prevent misunderstandings due to the notation, we emphasize that p_{hk} does not have the same meaning in Theorems 10B and 10C. Ambiguities of this kind seem less objectionable than crowded symbols which contain references to all conditions subject to choice.

10D. With regard to the parameters h, k our theorem has its simplest interpretation when $h=k=\frac{1}{2}$. The symbolic integral in the following statement is of course to be interpreted as a limit of the corresponding integral over $\beta(\Omega)$.

Theorem. *For any singularity function s, the minimum of*

$$B(p) = \int_\beta p\, dp^*$$

is attained for $p=\frac{1}{2}(p_0+p_1)$, and the minimum value is $\frac{1}{4}(C_0-C_1)$.

Since $B(p_0)=B(p_1)=0$, the minimum is always $\leqq 0$. Hence $C_0 \leqq C_1$.

10E. The cases $(h, k)=(1, 0)$ and $(0, 1)$ are interesting because of the vanishing of $B(p_0)$ and $B(p_1)$. If we restrict p to functions with $B(p)\leqq 0$ we find that the functional $C(p)$ is minimized by p_0 and maximized by p_1.

For the special functions s which were singled out in 10B and 10C we obtain the following result:

Theorem. *In the class of admissible functions with $B(p)\leqq 0$ and the singularities* $\operatorname{Re} \frac{1}{z-\zeta}$ *or* $\log|z-\zeta_1|$ *and* $-\log|z-\zeta_2|$ *respectively, the maximum and minimum of* $\operatorname{Re} a_1$ *or* $\operatorname{Re}(a_0^{(2)}-a_0^{(1)})$ *are attained for the principal functions p_0 and p_1 which correspond to these singularities.*

10F. In Theorem 9E and all its corollaries we have assumed that h, k are normalized by $h+k=1$. Without this normalization the theorems remain true provided that the competing functions p are required to have the singularity $(h+k)s$. If $h+k=0$ the function p_{hk} has no singularity, but the results continue to hold. We shall write $q=p_0-p_1$, and the extremal property of q is with respect to all regular harmonic functions u whose conjugate periods vanish along certain cycles. If $u+iu^*$ has the Taylor coefficients $a_n^{(j)}$ at ζ_j, the expression minimized by q is $D(u)+2C(u)$ where $C(u)$ is still determined by formula (24). The value of the minimum is $C_0-C_1=C(q)$. It should be noted that $\operatorname{Re} a_0^{(j)}=u(\zeta_j)$ and that the other coefficients $a_n^{(j)}$ can be expressed through partial derivatives.

The most important cases are covered in the following statement:

Theorem. *For regular admissible functions the expression* $D(u)+4\pi(u(\zeta_1)-u(\zeta_2))$ *attains its minimum when $u=q=p_0-p_1$. Here p_0, p_1 are the principal functions which correspond to the singularities* $\log|z-\zeta_1|$ *and* $-\log|z-\zeta_2|$. *Similarly,*

$$D(u)-4\pi\left(\frac{\partial u}{\partial x}\right)_{z=\zeta}$$

is minimized by the function $q=p_0-p_1$ which corresponds to the singularity $\operatorname{Re}\frac{1}{z-\zeta}$.

In both cases the minimum is $-D(q)$, *and the deviation from the minimum is* $D(u-q)$.

The simplest way to obtain the value of the minimum is perhaps to apply the theorem to $u=0$. Since the functional is then zero and must differ from the minimum by $D(q)$ it follows that the minimum is $-D(q)$. The first part of the theorem can hence be expressed through the identity

$$D(u)+4\pi(u(\zeta_1)-u(\zeta_2)) = D(u-q)-D(q). \tag{26}$$

For $u=q$ we obtain, in particular,

$$D(q) = 2\pi(q(\zeta_2)-q(\zeta_1)). \tag{27}$$

Similarly, in the second case,

$$D(u)-4\pi\left(\frac{\partial u}{\partial x}\right)_{z=\zeta} = D(u-q)-D(q),$$

and for $u=q$,

$$D(q) = 2\pi\left(\frac{\partial q}{\partial x}\right)_{z=\zeta}. \tag{28}$$

It may well happen that q reduces to a constant. If this is so, (26) shows that $u(\zeta_1)=u(\zeta_2)$ for all admissible functions with $D(u)<\infty$. Conversely, if it is known that $u(\zeta_1)=u(\zeta_2)$ for all such functions, then $q(\zeta_1)=q(\zeta_2)$ and it follows by (27) that q is constant.

10G. A convenient way of expressing the first half of the preceding theorem is in the following homogenous form:

Theorem. *Any admissible harmonic function* u *satisfies the inequality*

$$4\pi^2(u(\zeta_1)-u(\zeta_2))^2 \leqq D(u)D(q)$$

with equality only for $u=aq+b$, a *and* b *constants.*

If $D(u)=0$ or ∞ there is nothing to prove. If we apply (26) to ku with constant k we obtain

$$k^2D(u)+4\pi k(u(\zeta_1)-u(\zeta_2)) \geqq -D(q).$$

The desired result follows for

$$k = -2\pi\cdot\frac{u(\zeta_1)-u(\zeta_2)}{D(u)}.$$

§4. THE CLASS OF PLANAR SURFACES

Previous results are now specialized to the case of a planar surface. We obtain, above all, the fundamental result that any open planar Riemann surface is conformally equivalent to a plane region. Through use of the principal functions we gain a very simple access to the canonical slit mappings, and we can also study the mappings that correspond to linear combinations of the principal functions.

11. Slit mappings

11A. We are now going to study the case of a planar open surface W. By definition, every cycle on W is dividing. For this reason it is natural to employ the canonical partition Q when applying the results of the preceding section. Accordingly, in 11 and 12 we consider only harmonic functions p whose conjugate periods are all zero. This means that the analytic functions $P = p + ip^*$ are single-valued, and the theory that we are developing becomes a theory of analytic functions on planar surfaces.

The functions p are supposed to have a given singularity s, and if p^* is to be single-valued we must require that s has likewise a single-valued conjugate function s^*. In other words, we consider only singularities without logarithmic terms. We denote by P_0, P_1 the analytic functions which correspond to the principal functions p_0, p_1. They are determined up to additive complex constants.

For single-valued $P = p + ip^*$ it is possible to write

$$B_\Omega(p) = \int_{\beta(\Omega)} p\,dp^* = \tfrac{1}{2}\int_{\beta(\Omega)} p\,dp^* - p^*\,dp = \frac{i}{2}\int_{\beta(\Omega)} P\,d\bar{P}$$

and hence, symbolically,

$$B(p) = \frac{i}{2}\int_{\beta} P\,d\bar{P}. \tag{29}$$

This representation is very useful for formal computations.

11B. We fix our attention on the simplest kind of singularity, a simple pole with given residue. We suppose that the pole lies at ζ, and we choose a local variable z in a neighborhood of ζ. Let the singularity be given by $s = \operatorname{Re} \dfrac{e^{i\Theta}}{z-\zeta}$; the corresponding analytic functions have then the singularity $\dfrac{e^{i\Theta}}{z-\zeta}$. To indicate the dependence on Θ we shall use notations P^Θ, P_0^Θ, P_1^Θ, and we reserve the simpler notations P, P_0, P_1 for the special case $\Theta = 0$. Observe that these distinctions are introduced relatively to a fixed local variable.

The constants in P_0 and P_1 can be chosen so that

$$\begin{aligned} P_0^\Theta &= P_0 \cos\Theta + iP_1 \sin\Theta \\ P_1^\Theta &= iP_0 \sin\Theta + P_1 \cos\Theta. \end{aligned} \tag{30}$$

This is seen by noting, first of all, that the equated functions have the same singularities. Secondly, if we restrict attention to the case of a compact bordered $\overline{W}$, the function in the right member of the first equation has a real part whose normal derivative vanishes on the border. In the second equation the real part is constant on each contour. Since these properties are characteristic for P_0^Θ and P_1^Θ respectively, the equations hold provided that the constants are properly adjusted. By routine reasoning (see 11D) they remain valid for arbitrary W.

As a particular consequence of (30) we find that $P_0^\Theta = iP_1^{\Theta-\pi/2}$. Hence, if we intend to let Θ vary, it does not matter whether we study P_0^Θ or P_1^Θ.

11C. We will now investigate the mapping of W into the extended plane which is defined by the function P_0^Θ. The following theorem will be proved:

Theorem. *The mapping by P_0^Θ is one to one, and the image G of W is a horizontal slit region. This means that every component of the complement of G is either a point or a horizontal line segment. Moreover, the area of the complement is zero.*

The proof is particularly simple if W is the interior of a compact bordered Riemann surface $\overline{W}$. In fact, for this case we will show that the theorem follows immediately from the argument principle.

We recall first that the argument principle can be expressed in a form which does not exclude zeros and poles on the border. Suppose that f is meromorphic on $\overline{W}$, and that it is neither identically zero nor identically infinite. Then the classical relation

$$\frac{1}{2\pi i}\int_\beta \frac{df}{f} = n(0) - n(\infty) \tag{31}$$

holds, subject to the following interpretation: (1) if there are zeros or poles on the boundary, the integral is to be computed as a principal value; (2) interior zeros and poles are to be counted with their ordinary multiplicity, while zeros and poles on the border are counted with one half of their multiplicity.

The imaginary part of P_0^Θ is constant on each contour β_k, and the real part varies between a finite minimum and maximum. The image of β_k is thus a horizontal line segment σ_k. It is clear that every interior point of σ_k corresponds to at least two points on β_k.

Let w be any complex number. We apply (31) with $f = P_0^\Theta - w$. It is immediately recognized that $\int_{\beta_k} \frac{df}{f} = 0$ whether w lies on or outside of σ_k. Hence the variation of the argument is zero, and we find that $n(0) = n(\infty) = 1$. There are three ways in which this number of zeros can be realized, namely, by one simple zero in the interior, by two simple zeros on the border, or by one double zero on the border.

We conclude, first, that the mapping of W by P_0^Θ is one to one and that the complement of the image G is the union of the segments σ_k. Next, if w is an interior point of σ_k it must correspond to two simple zeros on β_k, and if w is an end point the corresponding point on β_k must be a double zero. There can be no other zeros, and we find that the σ_k are mutually disjoint. We have thus proved the theorem in our special case, and in addition we have gained complete insight in the correspondence between the boundaries.

11D. We pass to the general case of an arbitrary planar W. It has not yet been proved that $p_{0\Omega} \to p_0$ and $p_{1\Omega} \to p_1$ where $p_{0\Omega}, p_{1\Omega}$ are the principal functions for a regular region Ω that tends to W. The most instructive proof is by application of Theorem 9E.

Let the corresponding analytic functions $P_{0\Omega}$, $P_{1\Omega}$ have the developments

$$P_{0\Omega} = \frac{1}{z-\zeta} + a_\Omega(z-\zeta) + \cdots$$

$$P_{1\Omega} = \frac{1}{z-\zeta} + b_\Omega(z-\zeta) + \cdots,$$

and choose $\Omega' \subset \Omega''$. Then $B_{\Omega'}(p_{0\Omega''}) \leqq B_{\Omega''}(p_{0\Omega''}) = 0$. We apply Theorem 9E to Ω' with $(h, k) = (1, 0)$ and $p = p_{0\Omega''}$. By formula (24) we have $C(p_{0\Omega''}) = -2\pi \operatorname{Re} a_{\Omega''}$ and $C_0 = C(p_{0\Omega'}) = -2\pi \operatorname{Re} a_{\Omega'}$. Hence, by (25),

$$D_{\Omega'}(p_{0\Omega''} - p_{0\Omega'}) = B_\Omega(p_{0\Omega''}) - 2\pi \operatorname{Re}(a_{\Omega''} - a_{\Omega'}) \leqq 2\pi \operatorname{Re}(a_{\Omega'} - a_{\Omega''}).$$

This shows that $\operatorname{Re} a_\Omega$ decreases when Ω increases. The same reasoning, with $(h, k) = (0, 1)$, shows that $\operatorname{Re} b_\Omega$ is increasing. On the other hand, formula (28), 10F, yields

$$2\pi \operatorname{Re}(a_\Omega - b_\Omega) = D_\Omega(p_{0\Omega} - p_{1\Omega}) \geqq 0.$$

Therefore $\operatorname{Re} a_\Omega$ has a finite limit for $\Omega \to W$, and as a result we find that

$$D_{\Omega'}(p_{0\Omega''} - p_{0\Omega'}) \to 0$$

when $\Omega' \to W$ and $\Omega' \subset \Omega''$.

If Ω is fixed it is easy to infer, by use of the triangle inequality, that

$$D_\Omega(p_{0\Omega''} - p_{0\Omega'}) \to 0$$

when Ω', Ω'' both tend to W, without additional conditions. We are thus in a situation to which Theorem 13C, Ch. II, can be applied. It follows readily that $P_{0W} = \lim_{\Omega \to W} P_{0\Omega}$ exists and has the same singularity as the approximating functions. The same reasoning shows that $P_{1W} = \lim P_{1\Omega}$ exists.

It remains to prove that $P_i = P_{iW}$, $i = 0, 1$, at least up to a constant. We recall that p_i was defined by the condition $L_i p_i = p_i$. The conclusion will be reached if we show that $p_{iW} = \lim p_{i\Omega}$ satisfies the same condition. By definition $L_i p_{iW} = \lim L_{i\Omega} p_{iW}$ where the operator $L_{i\Omega}$ is applied to the values of p_{iW} on a fixed boundary α. But $|p_{iW} - p_{i\Omega}| < \epsilon$ on α for sufficiently large Ω. It follows that $|L_{i\Omega} p_{iW} - L_{i\Omega} p_{i\Omega}| < \epsilon$, and hence that $L_i p_{iW} = \lim L_{i\Omega} p_{iW} = \lim L_{i\Omega} p_{i\Omega} = \lim p_{i\Omega} = p_{iW}$, as asserted.

Because the relations (30) have been proved for compact regions we can infer the existence of $P_{iW}^{\Theta} = \lim P_{i\Omega}^{\Theta}$. The same reasoning as above shows that $P_i^{\Theta} = P_{iW}^{\Theta}$, $i = 0, 1$, and as a consequence (30) holds for arbitrary surfaces.

Since the functions $P_{0\Omega}^{\Theta}$ are univalent it follows by easy generalization of the corresponding theorem in the plane that the limit function P_0^{Θ}, which cannot reduce to a constant, is likewise univalent.

11E. We must still prove that the image G is a horizontal slit region. For this part of the proof we are going to make use of Riemann's mapping theorem for plane simply connected regions.

We observe that any univalent mapping $P = p + ip^*$ into the complex plane satisfies the condition $B(p) \leqq 0$. In fact,

$$B_\Omega(p) = \frac{i}{2} \int_{\beta(\Omega)} P\, d\bar{P}$$

represents the negative of the finite area enclosed by the image of $\beta(\Omega)$. We can thus deduce from Theorem 10E that P_0^{Θ} maximizes $\operatorname{Re} ae^{i\Theta}$ in the class of all univalent mappings of the form

$$P^{\Theta} = \frac{e^{i\Theta}}{z - \zeta} + a(z - \zeta) + \cdots.$$

Suppose that G lies in the complex w-plane, and let E be any complementary component which does not reduce to a point. Then the complement E' of E can be mapped conformally on a disk, and therefore on the complement of a horizontal line segment. The mapping function φ can be normalized so that it has a development

$$\varphi(w) = w + \frac{b}{w} + \cdots$$

at ∞.

Since E' is the interior of a compact bordered surface, it is clear that φ is the principal function P_0 for E' which corresponds to the singularity w at ∞. Because of the extremal property of P_0 we infer that $\operatorname{Re} b \geqq 0$, with equality only if $\varphi(w) = w$, for w is a competing univalent mapping.

On the other hand, the function $\varphi(P_0^\Theta(z))$ is univalent and normalized on W. If we write

$$P_0^\Theta = \frac{e^{i\Theta}}{z-\zeta}+a_\Theta(z-\zeta)+\cdots,$$

the development of $\varphi(P_0^\Theta)$ is of the form

$$\varphi(P_0^\Theta) = \frac{e^{i\Theta}}{z-\zeta}+(a_\Theta+be^{-i\Theta})(z-\zeta)+\cdots.$$

Hence, by the extremal property of P_0^Θ,

$$\operatorname{Re} a_\Theta e^{i\Theta} \geqq \operatorname{Re} (a_\Theta e^{i\Theta}+b),$$

so that $\operatorname{Re} b \leqq 0$. We conclude that $\operatorname{Re} b=0$, and it follows, as we have already pointed out, that $\varphi(w)=w$. But then E is a horizontal line segment, and since the same reasoning can be applied to any complementary component which is not a point, the proof of Theorem 11C is complete, except for the last statement.

11F. We have already remarked that $B(p) \leqq 0$ whenever $P=p+ip^*$ is univalent. More precisely, $B_\Omega(p)=-E_\Omega(p)$, where $E_\Omega(p)$ is the area of the point set that is not covered by the image of Ω. As $\Omega \to W$, $E_\Omega(p)$ tends to $E(p)$, the area of the complement of the whole image of W. It follows that $B(p)=-E(p)$, and in particular $E(p_0)=-B(p_0)=0$. In other words, the total area of the slits is zero.

For greater clarity we restate the extremal property enjoyed by P_0^Θ. In order to simplify the normalization we write $P_{(\Theta)}=e^{i\Theta}P_0^{-\Theta}$ for the function that maps W on a slit region with slits parallel to the direction Θ. We compare $P_{(\Theta)}$ with functions

$$P = \frac{1}{z-\zeta}+a(z-\zeta)+\cdots.$$

The following results are obtained:

Theorem. *The function* $P_{(\Theta)}$ *minimizes the functional* $B(p)-2\pi \operatorname{Re}(e^{-2i\Theta}a)$ *in the class of all single-valued* P, *and it maximizes the expression* $E(p)+2\pi \operatorname{Re}(e^{-2i\Theta}a)$ *in the class of all univalent* P.

A weaker version of the second property is that $P_{(\Theta)}$ maximizes $\operatorname{Re}(e^{-2i\Theta}a)$ in the class of univalent P. For special values of Θ we find that P_0 maximizes $2\pi \operatorname{Re} a+E(p)$ and P_1 minimizes $2\pi \operatorname{Re} a-E(p)$, or, in the weaker version, that P_0 maximizes and P_1 minimizes $\operatorname{Re} a$.

11G. We cannot leave this topic without pointing out that Theorem 11C contains, as a very special corollary, the *uniformization theorem.* A slit region, not the whole extended plane, is simply connected if and only if its boundary is a single slit. If the slit has more than one point its complement can be mapped on the interior of a disk, and if the slit reduces to a point the region is equivalent to the whole plane. Therefore, any open simply connected Riemann surface is conformally equivalent to a disk or to the plane.

The universal covering surface of any surface is simply connected, and it is closed only in the case of a sphere. When this observation is combined with the preceding statement we obtain:

Theorem (*Uniformization theorem*). *The universal covering surface of any Riemann surface is conformally equivalent to a disk, to the plane, or to the sphere.*

The uniformization theorem was first proved, conclusively, by P. Koebe [5, 7] and H. Poincaré [5]. The existence of a slit mapping with the properties stated in Theorem 11C is also due to Koebe [21].

12. The functions P_{hk}

12A. The extremal properties of the functions $P_{hk}=hP_0+kP_1$ are known to us through Theorem 10B. However, they gain in interest if P_{hk} is univalent, for then the extremum is with respect to the class of univalent mappings, and the minimum expression involves the geometric quantity represented by the area of the complement of the image region. In this respect we prove:

Theorem. *Every function P_{hk} with $h, k \geqq 0$ determines a one to one mapping of W onto a region G in the extended plane. If W is of finite connectivity, each complementary component of G is either a point or a convex set bounded by an analytic Jordan curve.*

As before, we can normalize by the condition $h+k=1$, but in the present situation it is essential that h and k are both nonnegative.

12B. We begin by studying the simplest case which is that of the function $\frac{1}{2}(P_0+P_1)$; the passage to arbitrary h, k will be easy. We prove first the part of the theorem that refers to a surface W of finite connectivity. Since we have a horizontal slit mapping at our disposal it is then possible to regard W as the interior of a compact bordered surface. Indeed, a slit which does not reduce to a point has a neighborhood which can be mapped by an elementary function in such a way that the slit corresponds to a circle. Slits that reduce to points can be added to the surface and will correspond to points under all mappings.

Consider a contour β_j of W. We know that $\operatorname{Im} P_0$ and $\operatorname{Re} P_1$ are constant on β_j, and it follows that $\operatorname{Re}[(P_0+P_1)e^{i\Theta}]$ differs from $P_0 \cos\Theta + iP_1 \sin\Theta = P_0^\Theta$ (see (30), 11B) by a complex constant on each β_j. On the other hand, since P_0^Θ is a horizontal slit mapping, we may conclude that $\operatorname{Re}[(P_0+P_1)e^{i\Theta}]$ attains every value between its minimum and maximum exactly twice on β_j. This is true for every Θ. By definition, a closed curve is (strictly) convex if it meets every straight line at most twice (i.e. for, at most, two values of the parameter). The equation of any line in the w-plane can be written as $\operatorname{Re}(w\, e^{i\Theta})=b$, and we have just shown that the equation $\operatorname{Re}[(P_0+P_1)e^{i\Theta}]=b$ has at most two solutions on β_j. Hence the image of β_j under the mapping $w=P_0+P_1$ is a convex curve. We denote it by γ_j.

12C. Let us form the derivatives P_0' and P_1' with respect to a local variable. Each derivative has two simple zeros on β_j. Furthermore, these zeros are distinct. For suppose that $P_0'=P_1'=0$ at a point of β_j. Because P_0''/P_1'' is purely imaginary, it is possible to find Θ such that $P_0''/P_1''=-i\tan\Theta$ at that point. It would follow that the first two derivatives of $P_0^\Theta=P_0\cos\Theta+iP_1\sin\Theta$ vanish simultaneously, contrary to the fact that P_0^Θ takes its values at most with multiplicity two.

As a first consequence we infer that γ_j is analytic. In fact, P_0 and P_1 can be continued across β_j, and a singular point on γ_j would be characterized by $\frac{dP_0}{ds}+\frac{dP_1}{ds}=0$, where s denotes local arc length. But since dP_0 and dP_1 are real and imaginary, respectively, it would follow that $P_0'=P_1'=0$, and we have just seen that this is impossible.

Secondly, we consider the analytic function $F=\frac{dP_1}{dP_0}=P_1'/P_0'$. It has no zeros or poles at interior points of W, and it is equal to 1 at ζ. There are two simple poles on each β_j, and $\operatorname{Re}F=0$ on the contours, except for the poles. We can show, moreover, that every purely imaginary value $i\eta$ is taken at least twice on each contour. For if we set $\eta=\cot\Theta$, then $F=i\eta$ at the points where the real part of $P_0^\Theta=P_0\cos\Theta+iP_1\sin\Theta$ is a maximum or minimum.

We apply the argument principle to $F-i\eta$. If the curve parameter is denoted by t we know that $(F-i\eta)^{-1}\frac{dF}{dt}$ is real on β, the border of W. Hence $F-i\eta$ has as many zeros as poles, provided that the ones on the boundary are counted with half their multiplicities. We have already accounted for as many zeros on the boundary as there are poles. Consequently, there are no zeros in the interior of W. This means that $F\neq i\eta$ in W for arbitrary real η; in other words, $\operatorname{Re}F\neq 0$ in W. Since $\operatorname{Re}F(\zeta)=1$ it follows that $\operatorname{Re}F>0$ throughout W. This implies that $\operatorname{Im}F$ decreases in the positive direction of β. It represents the slope of the image curves γ_j, and we infer that the curves γ_j are traced in the direction of decreasing slope. Let w_1, w_2 be the utmost left and utmost right point on γ_j respectively. As we trace γ_j in the positive direction from w_1 to w_2 the slope must go from $+\infty$ to $-\infty$.

12D. The inside and outside of a convex curve γ can be defined without appealing to the Jordan curve theorem. Indeed, we say that a point w lies outside of γ if there exists a half-line, beginning at w, which does not meet γ. A point that is neither on γ nor outside of γ is said to be inside of γ. Draw a line L through two inside points w', w''. These points divide L into a finite segment s and two disjoint half-lines L', L''. By the definition of inside point L' and L'' meet γ in points w_1, w_2 respectively. Because γ is a convex curve w_1, w_2 are the only points in which L intersects γ. We conclude that the segment s does not meet γ.

An outside point w has the winding number $n(\gamma, w)=0$, because it can be joined to ∞ without crossing γ (Ch. I, 10E). Consider now an inside point w, and draw a line L through w. Both halves of L, divided at w, must meet γ, each at one point. The arcs of γ between these points cannot lie in the same half-plane determined by L, for then a half-line from w in the other half-plane would miss γ, and w would be an outside point. In this situation one finds, by explicit computation, that $n(\gamma, w)$ is either 1 or -1. But we have just found that any two inside points can be joined without crossing γ. Therefore $n(\gamma, w)$ has the same value, 1 or -1, for all inside points. It follows further that the whole line segment between two inside points consists of inside points; in other words, the inside is a convex set. An easy modification of the reasoning shows that the inside together with γ is also a convex set. The inside and the outside are both connected, and since they are characterized by the values of the winding number they must be unions of complementary components of γ. It follows that γ divides the plane into two regions, the inside and the outside. We leave it to the reader to show that γ is the common boundary of both regions.

In the case of the curves γ_j we can determine the winding number with respect to inside points unambiguously. Consider again the points w_1, w_2 at the extreme left and extreme right. Let L be the directed line that passes first through w_1 and then through w_2. As we trace γ_j in the positive direction from w_1 to w_2 the slope starts with the value $+\infty$. It follows that the arc lies to the left of L. For a point w_0 between w_1 and w_2 the argument of $w-w_0$ must decrease by π, and we see that the winding number is -1.

12E. The univalent nature of the mapping P_0+P_1 follows by another application of the argument principle. Let w be an arbitrary complex value, and suppose that w lies inside of m and on m' curves γ_j. The change of the argument of P_0+P_1-w along the boundary is then $-m-m'/2$ times 2π. If w is taken n times in W and n' times on the boundary, it follows that $n+n'/2-1=-m-m'/2$. It is necessary to have $n' \geqq m'$, and $m'=0$ implies $n'=0$. This leaves only three possibilities: (1) $n=1$, $n'=m=m'=0$, (2) $n=n'=m'=0$, $m=1$, (3) $n=m=0$, $n'=m'=1$.

This enumeration shows clearly that the curves γ_j are disjoint and mutually exterior to each other. The mapping is one to one, and the complement of the image G consists of the closed convex regions enclosed by the curves γ_j.

Exactly the same reasoning can be applied to any linear combination hP_0+kP_1 with nonnegative h, k, for the images of the contours β_j are then obtained by applying an affine transformation to the curves γ_j.

Finally, we can pass to arbitrary W by means of the relation $P_{hk}=\lim (hP_{0\Omega}+kP_{1\Omega})$. It follows at once that P_{hk} is univalent.

12F. Let us recall the extremal property of P_{hk}. We have denoted by E the area of the complement of G, and by a the coefficient in the development $P=1/(z-\zeta)+a(z-\zeta)+\cdots$. By Theorem 10B (and 9E) we obtain:

Theorem. *For $h+k=1$, $h, k\geqq 0$, the function P_{hk} maximizes the quantity $E+2\pi(h-k)\operatorname{Re} a$ in the class of all normalized univalent mappings. In particular, the complementary area E is a maximum for the function $\frac{1}{2}(P_0+P_1)$, in which case $E=\frac{\pi}{2}[a(P_0)-a(P_1)]$.*

In this statement $a(P_0)$ and $a(P_1)$ are the coefficients of $z-\zeta$ in the developments of P_0 and P_1 respectively. We shall see in a moment that the difference $a(P_0)-a(P_1)$, known as the *span*, is always real and nonnegative. In the theorem we have already used this information when writing down the expression for the maximal complementary area.

12G. The functions P_{hk} with h, k of opposite sign are relatively uninteresting, except for the case $h=-k$ which leads to multiples of the singularity free function $Q=P_0-P_1$. We know by Theorem 10F that Q minimizes the quantity $D(U)-4\pi \operatorname{Re} a(U)$ in the class of all analytic functions on W; here $U=u+iu^*$, and we write $D(U)$ for $D(u)$.

In order to see that $a(Q)$ is necessarily real and $\geqq 0$ we compare Q with $Qe^{i\Theta}$. It is evident that $D(Qe^{i\Theta})=D(Q)$ and $a(Qe^{i\Theta})=e^{i\Theta}a(Q)$. Consequently, the minimum property of Q yields $\operatorname{Re} a(Q)\geqq \operatorname{Re}[a(Q)e^{i\Theta}]$ for all Θ, and this is possible only if $a(Q)$ is real and nonnegative.

The value of the minimum can be read off from Theorem 9E and is found to be $-2\pi a(Q)$. In other words, we have $D(Q)-4\pi a(Q)=-2\pi a(Q)$, or $D(Q)=2\pi a(Q)$. On comparison with Theorem 12F we see that $D(P_0-P_1)$ is exactly equal to $E(P_0+P_1)$, the complementary area associated with P_0+P_1.

We collect all this information in a theorem:

Theorem. *The function $Q=P_0-P_1$ minimizes the expression $D(U)-4\pi \operatorname{Re} a(U)$ in the class of all analytic functions U on W. Moreover, $a(Q)$ is nonnegative and $2\pi a(Q)=D(Q)=E$, where E denotes the complementary area associated with the mapping P_0+P_1.*

The minimizing property is expressed through the inequality

$$D(U) - 4\pi \operatorname{Re} a(U) \geqq -D(Q)$$

which in turn implies

$$4\pi^2 |a(U)|^2 \leqq D(U)D(Q).$$

The latter inequality follows on applying the first to cU with $c = 2\pi\bar{a}(U)/D(U)$.

Remark: The functions $P_0 + P_1$ and $P_0 - P_1$ were introduced and their extremal properties studied by H. Grunsky [1]. It was M. Schiffer [2] who discovered that $P_0 + P_1$ is univalent and that the boundary curves are convex.

13. Exponential mappings

13A. We continue the study of planar surfaces and direct our attention to the case of two logarithmic singularities with opposite residues. Because of the similarity with the preceding case our exposition will be very brief.

We consider two distinct points ζ_1, ζ_2 on W, and we are interested in harmonic functions p with the singularities $\log |z - \zeta_1|$ and $-\log |z - \zeta_2|$. In addition, we require that p belongs to the class of harmonic functions associated with the canonical partition Q. This means that p shall have the flux zero along any cycle which separates ζ_1 and ζ_2 from the ideal boundary of W. More precisely, the flux shall be zero along any cycle γ which is homologous on $W - \zeta_1 - \zeta_2$ to a cycle that lies outside of an arbitrarily given compact subset of W. Since W is planar it can then be concluded that the flux along an arbitrary cycle is a multiple of 2π. It follows that we can define a single-valued function $F = e^{p + ip^*}$ which is analytic except for a simple pole at ζ_2 and whose only zero is at ζ_1.

We suppose that F_0 and F_1 correspond in this way to the principal functions p_0 and p_1. They are determined up to a multiplicative constant. We choose to normalize all F so that $F'(\zeta_1) = 1$ with respect to a fixed local variable at ζ_1. Then the development at ζ_2 is of the form

$$F = \frac{c}{z - \zeta_2} + \cdots$$

where the residue $c = c(F)$ is of course independent of the local variable at ζ_2.

13B. If W has finite connectivity it is very easy to show that F_0 maps W on a region bounded by a finite number of radial slits, and that F_1 maps it on a region whose complement consists of concentric circular arcs. In both cases the mapping is one to one.

In order to make the corresponding conclusion for arbitrary W we must first determine the extremal properties of F_0 and F_1. The integral $B(p)$ can be written as

$$\int_\beta \log |F| \, d\arg F,$$

and if F is univalent it represents the negative of the logarithmic area of the complement of the image region $F(W)$. If that logarithmic area is denoted by $E_{\log}(F)$ we find:

Theorem. *In the class of all normalized univalent F, the radial slit mapping F_0 maximizes $2\pi \log |c(F)| + E_{\log}(F)$, and the circular slit mapping F_1 minimizes $2\pi \log |c(F)| - E_{\log}(F)$.*

The proof follows by application of Theorem 10C. It is first observed that $p = \log |F|$ has vanishing flux along all contours of a region Ω that contain ζ_1 and ζ_2. Indeed, the univalent function F maps each contour onto a closed curve with winding number zero about the origin, and this means that the flux vanishes. We have already shown that $B(p) = -E_{\log}(F)$, and the normalization implies that $a_0^{(1)} = 0$, $a_0^{(2)} = \log |c(F)|$. For $(h, k) = (1, 0)$ we conclude that F_0 minimizes $-E_{\log}(F) - 2\pi \log |c(F)|$, and for $(h, k) = (0, 1)$ that F_1 minimizes $-E_{\log}(F) + 2\pi \log |c(F)|$. This is what we wanted to show.

Exactly as in 11E the extremal properties can be used to show that F_0 and F_1 are slit mappings even if W has infinite connectivity. Details are left to the reader.

§5. CAPACITIES

So far, in all applications of the main existence theorem (Theorem 3A), with one exception, we have insisted that the sum of the residues of the singularity function be zero. The exception occurred in 4 where we discussed the construction of a Green's function. In that case it was necessary to modify the singularity function by adding a suitable multiple of a certain function ω, defined in a neighborhood of the boundary.

This aspect of the theory will now be pursued further. We are led to a generalization of the classical notion of logarithmic capacity and to canonical mappings onto disks with radial or concentric slits.

14. Capacity functions

14A. We begin with the case of a compact bordered surface $\overline{W}$ which need not be planar. We wish to construct a function with a single logarithmic singularity $\log |z - \zeta|$. Such a function has the flux 2π along the border β, and we show that the whole flux can be concentrated to a given part of β. Accordingly, we suppose that β is partitioned into γ and $\beta - \gamma$ where γ is a union of contours. For greater generality we assume that $\beta - \gamma$ is further subdivided into $\beta_1, \cdots, \beta_n$. In this situation the following existence theorem holds:

Theorem. *There exists a function* p_γ, *unique save for an additive constant, which is harmonic except for the singularity* $\log |z-\zeta|$, *constant on* γ *and on each* β_i, *and whose flux is zero along each* β_i.

For the proof we construct a neighborhood W'_γ of γ whose closure on $\overline{W}$ does not contain ζ and does not meet the β_i. It is of course possible to choose W'_γ so that it is regularly imbedded, and we denote its relative boundary by γ_0. There exists a harmonic function ω in W'_γ which is 0 on γ_0 and 1 on γ. Its flux m along γ is positive.

We introduce a singularity function s which is equal to $\log |z-\zeta|$ in a neighborhood of ζ, identically zero near the contours β_i, and equal to $2\pi\omega/m$ in W'_γ. The total flux of s is zero. We consider the operator L_1 associated with the partition P of β into γ and $\beta_1, \cdots, \beta_n$ (see 5B). According to Theorem 3A there exists a harmonic function p_γ which satisfies $p_\gamma - s = L_1(p_\gamma - s)$. This function fulfills the conditions in the theorem. It is uniquely determined except for an additive constant.

For definiteness we normalize so that $p_\gamma - \log |z-\zeta|$ vanishes at ζ. The normalization depends on the choice of the local variable at ζ. The normalized function p_γ will be called the *capacity function* of the subboundary γ. It has a constant value $k(\gamma)$ on γ, and we define $e^{-k(\gamma)}$ to be the *capacity* of γ. Observe that the capacity depends on ζ, the local variable at ζ, and the partition P.

14B. We are interested in generalizing the preceding result to arbitrary W. For this reason we must first determine the extremal property of the capacity function in the compact case. We retain the notations $A(p)$ and $B(p)$ of 6B and maintain:

Theorem. *The capacity function* p_γ *minimizes the functional* $B(p)$ *in the class of all harmonic* p *with zero flux along each* β_i, *and such that* $p - \log |z-\zeta|$ *vanishes at* ζ. *The value of the minimum is* $2\pi k(\gamma)$, *and the deviation from the minimum is measured by* $D(p-p_\gamma)$.

For the proof we observe that $B(p_\gamma, p) = B(p_\gamma) = 2\pi k(\gamma)$. By Green's formula we have further

$$B(p_\gamma, p) - B(p, p_\gamma) = A(p_\gamma, p) - A(p, p_\gamma),$$

where the integrals A are extended over a small circle about ζ. On letting the radius of this circle tend to zero it becomes obvious, in view of the normalization, that $A(p_\gamma, p) = A(p, p_\gamma)$. Hence $B(p_\gamma, p) = B(p, p_\gamma)$, and we obtain

$$D(p-p_\gamma) = B(p-p_\gamma) = B(p) - B(p, p_\gamma) - B(p_\gamma, p) + B(p_\gamma) = B(p) - B(p_\gamma).$$

This identity implies the statement of the theorem.

14C. We turn now to the case of an open W. Let P be a regular partition of the ideal boundary β (Ch. I, 38), and let γ be one of the parts, and hence a closed subset of β. We call γ a *subboundary*. P induces a partition of the contours $\beta(\Omega)$ of any regular region $\Omega \subset W$. We denote the part corresponding to γ by $\gamma(\Omega)$ and the remaining parts by $\beta_i(\Omega)$.

Using these partitions we can determine the capacity function $p_{\gamma(\Omega)}$ for each sufficiently large Ω; the notation will be simplified to $p_{\gamma\Omega}$. If $\Omega \subset \Omega'$ it is easily seen that $p_{\gamma\Omega'}$ is a competing function for the extremal property in Ω. Hence

$$(32) \qquad B_{\Omega'}(p_{\gamma\Omega'}) > B_{\Omega}(p_{\gamma\Omega'}) = B_{\Omega}(p_{\gamma\Omega}) + D_{\Omega}(p_{\gamma\Omega'} - p_{\gamma\Omega}),$$

and we see that $B_{\Omega}(p_{\gamma\Omega}) = 2\pi k_{\Omega}(\gamma)$ increases with Ω. Thus $k_{\Omega}(\gamma)$ has a limit $k(\gamma)$ for $\Omega \to W$, and we define $c(\gamma) = e^{-k(\gamma)}$ to be the capacity of γ, relatively to the partition P.

If the capacity is positive we conclude from (32) that $D_{\Omega}(p_{\gamma\Omega'} - p_{\gamma\Omega}) \to 0$, and it follows in the usual way that $p_{\gamma\Omega}$ has a limit function p_{γ}. The limit function is normalized, and will be regarded as the capacity function associated with the subboundary γ and the partition P. On use of the triangle inequality it follows from

$$D_{\Omega}(p - p_{\gamma\Omega}) = B_{\Omega}(p) - B_{\Omega}(p_{\gamma\Omega})$$

that

$$D(p - p_{\gamma}) = B(p) - B(p_{\gamma}).$$

This formula generalizes Theorem 14B to open surfaces.

In the case of zero capacity it is still possible to demonstrate, for instance by use of normal families, the existence of a function p_{γ} with the given singularity and the prescribed conjugate periods. However, the unicity is lost, and for this reason we prefer to dismiss this case from further consideration.

15. The capacity of the boundary

15A. We investigate the special circumstances that occur when γ is the whole ideal boundary β, in which case P must be the identity partition. We recognize at once that $k_{\Omega}(\beta) - p_{\beta\Omega}$ is the Green's function $g_{\Omega}(z, \zeta)$ of Ω. Clearly, g_{Ω} increases with Ω and has a nondegenerate limit function g if and only if β has positive capacity. When this is the case, g is called the Green's function of W, and it is related to the capacity function by $p_{\beta} = -g + k(\beta)$. The following characterization of the Green's function is important:

Theorem. *The Green's function g, if it exists, is the smallest positive harmonic function with the singularity $-\log|z-\zeta|$. It satisfies $\inf g = 0$, and if a harmonic function with the same singularity tends to 0 as z approaches the ideal boundary, then it is identical with g.*

The first assertion means that $g(z) \leqq g_0(z)$ at all points z for any positive harmonic function g_0 with the same singularity as g. It is practically trivial, for the maximum principle implies $g_\Omega(z) \leqq g_0(z)$ for $z \in \Omega$, and the result follows on letting Ω tend to W. If $\inf g = \eta$ we can take $g_0 = g - \eta$, and it follows that $\eta = 0$. Finally, suppose that $g_0 \to 0$ for $z \to \beta$. For any given $\epsilon > 0$ we can then find an Ω so that $|g_0| < \epsilon$ outside of Ω. The maximum principle implies $|g_\Omega - g_0| < \epsilon$ in Ω. Hence $|g - g_0| \leqq \epsilon$ at all points, and since ϵ is arbitrary we obtain $g = g_0$.

The last result shows that the definition of g agrees with the earlier definition that was given in a special case (4B–4D). However, the reader must be well aware that g does not always have the limit 0 as z tends to β. It can merely be asserted that the inferior limit is 0.

15B. Another way of expressing the minimum property of g is to say that p_β *has the smallest supremum among all harmonic functions* p *which satisfy the condition* $p - \log|z-\zeta| \to 0$ *for* $z \to \zeta$. We are going to prove a theorem which generalizes this property as well as the extremal property expressed by Theorem 14B.

Let $\Phi(t)$ be a convex increasing function of the real variable t; the convexity means that $\Phi'(t)$ is increasing (for the sake of simplicity we assume that Φ' exists and is continuous). Given a harmonic function p we write

$$I_\Omega(p) = \int_{\beta(\Omega)} \Phi(p)dp^*.$$

Then $I_\Omega(p)$ increases with Ω. In fact, if $\Omega \subset \Omega'$ we find by Green's formula (Ch. II, 8B)

$$I_{\Omega'}(p) - I_\Omega(p) = D_{\Omega'-\Omega}(p, \Phi(p)),$$

and the Dirichlet integral can be expressed as

$$\iint \Phi'(p)\left[\left(\frac{\partial p}{\partial x}\right)^2 + \left(\frac{\partial p}{\partial y}\right)^2\right] dxdy.$$

Since $\Phi'(p) \geqq 0$ the integral is nonnegative, and we conclude that the functional increases. In particular, it is possible to define

$$I(p) = \int_\beta \Phi(p)dp^*$$

as the limit of $I_\Omega(p)$.

Theorem. *The capacity function* p_β *minimizes* $I(p)$ *among all harmonic* p *which satisfy* $p - \log|z-\zeta| \to 0$ *for* $z \to \zeta$.

We assume first that $\overline{W}$ is compact and bordered. We write $p = p_\beta + h$ and obtain

$$I(p) - I(p_\beta) = \int_\beta [\Phi(p_\beta + h) - \Phi(p_\beta)]\, dp_\beta^* + \int_\beta \Phi(p_\beta + h)\, dh^*.$$

In the first integral $dp^*_\beta \geqq 0$, and by the convexity $\Phi(p_\beta+h)-\Phi(p_\beta) \geqq h\Phi'(p_\beta)$. But p_β is constantly equal to $k=k(\beta)$ on β, so that the integral is at least equal to

$$\Phi'(k) \int_\beta h\, dp^*_\beta.$$

Since h has the flux zero we can write

$$\int_\beta h\, dp^*_\beta = \int_\beta h\, dp^*_\beta - p_\beta\, dh^*.$$

The integral on the right can be transferred to a small circle about ζ, and because of the normalization it is seen to equal zero. We have thus proved that

$$\int_\beta [\Phi(p_\beta+h)-\Phi(p_\beta)]\, dp^*_\beta \geqq 0.$$

Next, we use Green's formula to obtain

$$\int_\beta \Phi(p_\beta+h)\, dh^* = \int_\beta \Phi(k+h)\, dh^* = \iint_W \Phi'(k+h)\left[\left(\frac{\partial h}{\partial x}\right)^2+\left(\frac{\partial h}{\partial y}\right)^2\right] dxdy.$$

The double integral is nonnegative, and we conclude that $I(p) \geqq I(p_\beta)$ as asserted.

In the general case we obtain $I(p) \geqq I_\Omega(p_{\beta\Omega}) \geqq I_{\Omega_0}(p_{\beta\Omega})$ for $\Omega \supset \Omega_0$. Letting Ω tend to W we get $I(p) \geqq I_{\Omega_0}(p_\beta)$, and since Ω_0 is arbitrary it follows that $I(p) \geqq I(p_\beta)$. This completes the proof.

15C. As an illustration we may consider the case $\Phi(p)=e^{2p}$. The function $F=e^{p+ip^*}$ has a single-valued modulus $|F|$, and it is normalized by the conditions $F(\zeta)=0$, $|F'(\zeta)|=1$. The integral becomes

$$I(p) = 2\iint_W |F'|^2\, dxdy,$$

and we find that every normalized F with single-valued modulus satisfies the condition

$$\iint_W |F'|^2\, dxdy \geqq \pi c(\beta)^{-2},$$

where $c(\beta)$ is the capacity. The minimum is attained for the function F_β that corresponds to p_β.

16. Disk mappings

16A. For a planar surface W we are mainly interested in the case where γ is a single boundary component, and P is the canonical partition. For any regular region $\Omega \subset W$ the part $\gamma(\Omega)$ will bound a single component of $W-\Omega$. The corresponding capacity function p_γ will have the conjugate period 2π along $\gamma(\Omega)$ and conjugate periods 0 along the boundaries $\beta_i(\Omega)$ of all other components of $W-\Omega$. The function $F_\gamma = e^{p_\gamma + ip_\gamma^*}$ is consequently single-valued, and it is the mapping by this function that interests us.

In the compact case it is readily seen, by use of the argument principle, that F_γ is univalent and maps W onto a disk of radius $c(\gamma)^{-1}$ with circular slits centered at the origin. We propose to show that this remains true for arbitrary W, provided that $c(\gamma) > 0$. The proof makes use of an extremal property that we shall first derive.

16B. We begin with the case of a surface W which is the interior of a compact bordered surface $\overline{W}$ with n contours. Let F be a univalent analytic function on $\overline{W}$, normalized by the conditions $F(\zeta)=0$, $F'(\zeta)=1$. F maps W onto a plane region G with one outer and $n-1$ inner contours; the outer contour does and the inner contours do not enclose the origin. Choose γ to be the contour of W which corresponds to the outer contour of G. The other contours are denoted by $\beta_1, \cdots, \beta_{n-1}$.

If $p = \log |F|$ we can easily see that

$$\int_{\beta_i} p\, dp^* < 0$$

for all β_i. In fact, if we change to the variable $w = u + iv = F(z)$, the integral equals

$$-\iint \frac{dudv}{|w|^2}$$

extended over the area that is enclosed by the image of β_i. On the other hand,

$$\int_{\beta_i} p_\gamma\, dp_\gamma^* = 0,$$

since p_γ is constant on γ with the flux zero. Thus

$$B(p) = \int_\gamma p\, dp^* + \sum_{i=1}^{n-1} \int_{\beta_i} p\, dp^* < \int_\gamma p\, dp^*$$

$$B(p_\gamma) = \int_\gamma p_\gamma\, dp_\gamma^* = 2\pi k(\gamma),$$

and we obtain

$$\int_{\gamma} p\, dp^* > 2\pi k(\gamma).$$

It is geometrically evident, and quite easy to prove, that

$$\int_{\gamma} p\, dp^* \leqq 2\pi \max_{\gamma} p.$$

We can thus conclude that $\max p > k(\gamma)$, or $\max |F| > c(\gamma)^{-1}$.

Note carefully that γ depends on F. Therefore, if we want an inequality which is valid for all univalent normalized functions F, we must express the result in the form $\max |F| > [\max_i c(\gamma_i)]^{-1}$ where γ_i runs through all contours of $\overline{W}$. The strict inequality is due to our assumption that F remains univalent on the border.

16C. We consider now an arbitrary planar W and let F be a normalized mapping onto a plane region G. It is possible to define a boundary component of G, and hence of W, which represents the outer boundary of G. Indeed, for every $\Omega \subset W$ the image $F(\Omega)$ has an outer boundary which corresponds to a contour $\gamma(\Omega)$, and if $\Omega \subset \Omega'$, then $\gamma(\Omega')$ lies in the complementary component that is bounded by $\gamma(\Omega)$. This shows that the contours $\gamma(\Omega)$ determine a boundary component γ.

Theorem. *Any normalized univalent function F on a planar surface W satisfies* $\sup |F| \geqq c(\gamma)^{-1}$ *where γ is the boundary component determined by F. If $c(\gamma) > 0$, equality holds only for $F = F_\gamma$.*

There exists a boundary component with maximal capacity, and an arbitrary normalized F is thus known to satisfy $\sup |F| \geqq [\max_{\gamma} c(\gamma)]^{-1}$.

The first statement follows from $\sup |F| > \max_{\Omega} |F| > c_\Omega(\gamma)^{-1}$ on letting Ω tend to W. As for the uniqueness, we note that the sharp form of Theorem 14B yields

$$2\pi \sup_W p > 2\pi k_\Omega(\gamma) + D_\Omega(p - p_{\gamma\Omega}).$$

If $\sup p = k(\gamma) \neq \infty$ it follows that $D_\Omega(p - p_{\gamma\Omega}) \to 0$, and hence that $p = \lim p_{\gamma\Omega} = p_\gamma$.

For the second part, set $c = \sup c(\gamma)$ and assume that $c > 0$. We choose a sequence of boundary components γ_n with $c(\gamma_n) \to c$. The functions F_{γ_n} exist for sufficiently large n and satisfy $|F_{\gamma_n}| < c(\gamma_n)^{-1}$. Since they are bounded it is possible to extract a subsequence which converges to a limit function F. The latter satisfies $|F| < c^{-1}$, and because of the normalization F is not constant, hence univalent. If γ is the boundary component determined by F we know that $c(\gamma)^{-1} \leqq \sup |F| \leqq c^{-1}$. Hence $c(\gamma) = c$, and γ has maximal capacity.

16D. Suppose that the boundary component γ of W has positive capacity. The function F_γ is a limit of functions $F_{\gamma\Omega}$. It is consequently univalent and normalized. Since $|F_{\gamma\Omega}| < c_\Omega(\gamma)^{-1}$ and $c_\Omega(\gamma) \to c(\gamma)$ it follows further that $|F_\gamma| \leqq c(\gamma)^{-1}$. We are going to prove:

Theorem. *The function F_γ maps W onto a subregion G of the disk $|w| < c(\gamma)^{-1}$ with the property that each component of its complement with respect to the disk consists of a point or of a circular arc with center* 0.

We remark that this theorem contains the Riemann mapping theorem. Suppose, in fact, that W is a simply connected proper subregion of the complex plane. A familiar construction shows that there are bounded univalent functions on W. Therefore its only boundary component γ has positive capacity. In view of the simple connectivity the theorem implies that F_γ maps W onto a disk.

The proof of the theorem consists of two parts. In the first part we show that there is no complementary component which reaches the boundary of the disk, and for this proof we use the argument that has become classical in the proof of Riemann's mapping theorem. Suppose that the unbounded complementary component of G contains a point w_0 with $|w_0| < c(\gamma)^{-1}$. Then, setting $R = c(\gamma)^{-1}$, it is possible to define functions

$$\xi(z) = \left[\frac{R(F_\gamma(z) - w_0)}{R^2 - \bar{w}_0 F_\gamma(z)}\right]^{\frac{1}{2}}$$

and

$$\eta(z) = \frac{\xi(z) - \xi(\zeta)}{1 - \overline{\xi(\zeta)}\xi(z)},$$

both < 1 in absolute value. By direct computation or by use of Schwarz' lemma one sees that $|\eta'(\zeta)| > 1/R$. Hence the normalized function $\eta(z)/\eta'(\zeta)$ would have a bound $< c(\gamma)^{-1}$, contrary to Theorem 16C. The contradiction proves the first part.

For the second part, let E be a component of $D - G$ where D denotes the disk $|w| < c(\gamma)^{-1}$. The Riemann mapping theorem is now at our disposal. Therefore, if E is not a point, the doubly connected region $D - E$ can be regarded as the interior of a compact bordered surface, and we can form its capacity mapping φ with respect to the origin, the local variable w, and the contour $|w| = c(\gamma)^{-1}$. The identity mapping of $D - E$ is univalent and normalized. Therefore, by comparison with the capacity mapping φ, $c(\gamma)^{-1} \geqq \sup |\varphi|$. On the other hand, consideration of the univalent mapping $\varphi \circ F_\gamma$ yields $\sup |\varphi| \geqq c(\gamma)^{-1}$. This is possible only if φ is the identity mapping, and we conclude that E is a circular arc. The proof is complete.

16E. Finally, we prove an analogue of Theorem 15B for univalent mappings of a planar surface. As before, $\Phi(t)$ denotes a convex increasing function, and we consider a fixed boundary component γ of positive capacity.

Theorem. *Among all normalized univalent functions $F=e^{p+ip^*}$ which send γ into the outer contour, the function F_γ minimizes the functionals*

$$I(p) = \int_\beta \Phi(p)\, dp^* \quad \textit{and} \quad J(p) = \int_\gamma \Phi(p)\, dp^*.$$

As usual, we consider first the compact case; in particular, F is supposed to remain analytic on the border. The functional $I(p)$ differs from $J(p)$ by integrals over the contours $\beta_i \neq \gamma$. With the notation $w=u+iv=F(z)$ the integral over β_i can be written as

$$-\iint \Phi'(\log|w|)\frac{dudv}{|w|^2}$$

extended over the region enclosed by $F(\beta_i)$. Hence $I(p) \leqq J(p)$, while $I(p_\gamma)=J(p_\gamma)$. For this reason it is sufficient to prove the theorem for $I(p)$.

The proof is somewhat easier if we assume that $\Phi(t)\to 0$ for $t\to -\infty$; a later modification of the proof will serve to eliminate this additional hypothesis. Because of this assumption it is true that

$$\int_{\gamma_0} \Phi(p)\, dp^* \to 0$$

when the small curve γ_0 tends to ζ, and therefore it is possible to write

$$I(p) = \iint_W \Phi'(p)\left|\frac{F'}{F}\right|^2 dxdy = \iint_{F(W)} \Phi'(\log|w|)\frac{dudv}{|w|^2},$$

where the integral is known to converge. For our purposes it is preferable to extend the integral over the whole plane, and we can do so if we introduce the characteristic function χ of the set $F(W)$. With this notation we find

$$I(p) = \iint \chi(w)\Phi'(\log|w|)\frac{dudv}{|w|^2},$$

extended over the whole plane.

In the corresponding formulas for $I(p_\gamma)$ the characteristic function χ_γ is 1 for $|w|<e^{k(\gamma)}$, except for a finite number of concentric slits, and 0 for $|w|\geqq e^{k(\gamma)}$. With the simpler notation $k=k(\gamma)$ we have thus

$$\frac{1}{|w|^2}(\chi(w)-\chi_\gamma(w))(\Phi'(\log|w|)-\Phi'(k)) \geqq 0$$

in the whole plane, except on the slits. When this inequality is integrated over the plane one obtains

$$I(p)-I(p_\gamma) \geqq \Phi'(k) \iint (\chi(w)-\chi_\gamma(w)) \frac{dudv}{|w|^2}.$$

The integral on the right can be rewritten as a line integral

$$\int_\beta p\, dp^* - p_\gamma\, dp_\gamma^* = B(p)-B(p_\gamma),$$

and we have already proved that $B(p) \geqq B(p_\gamma)$. It follows that $I(p) \geqq I(p_\gamma)$ as asserted.

The restrictive condition on Φ can be removed in the following way. We choose a number $R > \sup |F|$ and denote by χ_0 the characteristic function of the disk $|z| < R$. It is then easy to see that

$$I(p)-2\pi\Phi(\log R) = \iint (\chi-\chi_0)\Phi'(\log |w|) \frac{dudv}{|w|^2}$$

and from there on we can proceed exactly as before.

The passage to arbitrary W is quite trivial. We need only observe that

$$I_\Omega(p) = \int_{\beta(\Omega)} \Phi(p)\, dp^*$$

increases with Ω, and that $I(p)$ must be defined as the limit of $I_\Omega(p)$. From $I_\Omega(p) \geqq I_\Omega(p_{\gamma\Omega})$ it follows that $I(p) \geqq I_{\Omega_0}(p_{\gamma\Omega})$ for $\Omega \supset \Omega_0$. On letting Ω tend to W we obtain $I(p) \geqq I_{\Omega_0}(p_\gamma)$, and hence $I(p) \geqq I(p_\gamma)$.

The most obvious application results for $\Phi(t) = e^{2t}$. Then

$$I(p) = \iint_W 2e^{2p} \left|\frac{F'}{F}\right|^2 dxdy = 2 \iint_W |F'|^2\, dxdy$$

is twice the area of $F(W)$, and our theorem asserts that the image area is a minimum for the capacity function.

Corollary. *Of all normalized univalent mapping functions F which send γ into the outer contour F_γ maps the surface W on the set of least area.*

CHAPTER IV

Classification Theory

Two conformally equivalent Riemann surfaces are said to be of the same conformal type. By the uniformization theorem there are only two types of simply connected open Riemann surfaces, namely the types represented by a disk and by the whole plane. The type problem, which is historically the forerunner of classification theory, consists in trying to distinguish between these two types on the basis of more or less explicit properties of the surface.

In classification theory the dichotomy between disk and plane is extended to arbitrary open Riemann surfaces. There are, however, many intrinsic differences between the disk and the plane, and each leads to a different classification of open surfaces. Generally speaking, the surfaces that fall in the same class as the plane exhibit a certain degeneracy which usually manifests itself in the complete lack of functions with some restrictive property, for instance boundedness. The degenerate surfaces are, as a rule, more accessible to detailed study.

The aim of classification theory is twofold: to find sufficient or necessary conditions for degeneracy of a given kind, and to compare different kinds of degeneracy with each other.

In §1 we assemble some fundamental theorems relative to the most important classes of degeneracy. This is followed, in §2, by a discussion of positive harmonic functions, a topic that is closely related to classification. In §3 we introduce an important tool, the method of extremal length. §§4–5 are devoted to various tests, to be applied in §6. We discuss in §7 the special properties of plane regions, and in §8 we introduce a series of counterexamples which definitively settle the questions of inclusion relations between the classes of degeneracy.

Remark. The question of the existence of a Green's function is known as the classical type problem and was in the center of interest in the 1930's (see, e.g., L. Ahlfors [1], P. J. Myrberg [3], R. Nevanlinna [6]). The thesis of L. Sario [1], contents of which were quoted by R. Nevanlinna [9] in 1946, introduces and analyzes the class O_{AD} (see 1B), which contains the parabolic surfaces as a subclass. The idea of a more detailed classification

(L. Sario [2], L. Ahlfors [15]) spread rapidly, and the present knowledge is due to contributions from many authors.

§1. FUNDAMENTALS OF THE CLASSIFICATION THEORY

In the preceding chapter we have dealt extensively with harmonic and analytic functions with a finite Dirichlet integral. It is therefore appropriate to begin with a study of the particular kinds of degeneracy that occur when the surface fails to carry such functions. Naturally, constant functions are always excluded.

A similar, but in many aspects fundamentally different degeneracy occurs when the surface fails to carry nonconstant bounded analytic functions. Let us point out that an analytic function with finite Dirichlet integral represents the surface as a covering surface of the plane which has finite total area. On the other hand, a bounded analytic function represents it as a covering surface whose projection is bounded. Thus, in one case we obtain a distinguishing geometric property of the surface itself, in the other case a property of its projection.

1. Basic properties of O_{HD} and O_{AD}

1A. Let W denote an arbitrary Riemann surface. If u is a real function of class C^1 on W we denote its Dirichlet integral (Ch. II, 7) by

$$D_W(u) = \iint_W \left[\left(\frac{\partial u}{\partial x}\right)^2 + \left(\frac{\partial u}{\partial y}\right)^2\right] dxdy.$$

It will be recalled that the precise definition made use of a partition of unity.

The definition applies in particular when u is a harmonic function. We designate by HD the class of harmonic functions which have a finite Dirichlet integral over the Riemann surface on which they are defined. For convenience we shall often refer to functions in HD as HD-functions.

The Dirichlet integral of an analytic function $F=u+iu^*$ is defined as $D_W(F)=D_W(u)$. It can be expressed by

$$D_W(F) = \iint_W |F'(z)|^2 \, dxdy$$

where the double integral should again be computed by means of a partition of unity. Geometrically, it represents the total area of the covering surface (W, F) of the complex plane.

An analytic function with finite Dirichlet integral is referred to as a function of class AD.

Any complex constant is automatically of class AD, just as a real constant is of class HD. It may be commented that a constant differs very radically from a nonconstant analytic function, especially in that its range is not an open set. Hence it would be quite logical, and for what follows even convenient, to exclude the constants from AD and HD. The disadvantage is that the classes AD and HD would no longer be linear, and this seems reason enough not to follow the suggested procedure.

1B. In accordance with our introductory remarks we introduce the following terminology:

Definition. *A Riemann surface W is said to be of class O_{HD} or O_{AD} if there are no nonconstant HD- or AD-functions respectively on W.*

It is trivial that $O_{HD} \subset O_{AD}$, for if F is of class AD, then Re F is of class HD. A further obvious remark is that every closed surface is of class O_{HD}, and hence also of class O_{AD}. Actually, all degenerate classes that we are going to consider will contain the closed surfaces, and this fact will not be mentioned again. It has the same effect to assume that henceforth all surfaces will be open.

1C. In order to gain further insight we shall now make essential use of the results derived in Ch. III, §3. We recall that these results depended on a regular partition P of the ideal boundary, and that we introduced a class of "admissible functions" determined by that partition. If P is the identity partition, then *all* harmonic functions are admissible. In the opposite extreme case, when P is the canonical partition, the admissible harmonic functions are those whose conjugates have zero periods along all dividing cycles.

With this in mind we refer to Theorem 10F, Ch. III. We shall first apply the theorem in the case of the identity partition, and are thus allowed to let u denote any harmonic function on the surface. The following result is an immediate consequence:

Theorem. *If W is of class O_{HD}, then the principal functions p_0 and p_1, determined for any singularity, differ by a constant. Conversely, in order to conclude that $W \in O_{HD}$, it is sufficient to know that $p_0 - p_1$ is constant either for all pairs of logarithmic singularities* $\log|z-\zeta_1|$ *and* $-\log|z-\zeta_2|$, *or else for all singularities of the form* $\mathrm{Re}\,\dfrac{1}{z-\zeta}$.

The necessity is clear, for $p_0 - p_1$ is known to have a finite Dirichlet integral. Suppose now that $q = p_0 - p_1$ is constant, for instance in the first case. The relation

$$D(u) + 4\pi(u(\zeta_1) - u(\zeta_2)) = D(u-q) - D(q)$$

which expresses the minimum property reduces to $u(\zeta_1) - u(\zeta_2) = 0$, and the conclusion follows.

Under the second hypothesis we obtain, in the same way, $\left(\frac{\partial u}{\partial x}\right)_{z=\zeta} = 0$. If the local variable z is replaced by iz it follows that $\frac{\partial u}{\partial y}$ is likewise 0 at ζ, and since ζ is arbitrary we can conclude that u is a constant.

1D. The method is not applicable to O_{AD}, for if P is the canonical partition, then the conjugate of $p_0 - p_1$ may have periods along the nondividing cycles, and thus we do not know that $p_0 - p_1$ is the real part of a single-valued analytic function.

It is reasonable, however, to introduce an intermediate class H_1D, consisting of all harmonic functions with finite Dirichlet integral whose conjugates have vanishing periods along all dividing cycles. Obviously, the corresponding O-class is such that $O_{HD} \subset O_{H_1D} \subset O_{AD}$.

For the intermediate class we obtain:

Theorem. *A surface is of class O_{H_1D} if and only if the functions $p_0 - p_1$ that correspond to the canonical partition are constant.*

The proof remains the same. We remark that this theorem can of course be used as a sufficient condition for the class O_{AD}.

2. Planar surfaces of class O_{AD}

2A. If W is a planar surface it is clear that H_1D consists of the real parts of functions in AD. This means that Theorem 1D can be used to characterize the class O_{AD}. We shall show, however, that a much stronger result is true.

We know that W can be mapped on a plane region, and we may assume that ∞ is an interior point of that region. Hence W can be realized as the complement of a closed bounded set E. This realization is of course not unique, and we are led to consider the following equivalence relation:

Two compact sets E_1 and E_2, each with a connected complement, are said to be equivalent if and only if their complements are conformally equivalent.

From this point of view we can concentrate our attention on the sets E, or rather on the equivalence classes of such sets.

2B. The first theorem we prove is the following:

Theorem. *The complement of E is of class O_{AD} if and only if every set which is equivalent to E has areal measure* 0.

It is convenient to say that a set E with this property has *absolute measure* 0. Clearly, a set can very well have measure 0 without being of absolute measure 0. This is illustrated by a line segment which has measure 0 but is equivalent to a closed disk.

For the proof we make use of Theorems 12F and 12G, Ch. III. Suppose first that E has absolute measure 0. Then the function $P_0 + P_1$, corresponding to an arbitrary simple pole, yields a univalent map on a region whose complement has zero area. By the theorems that we have just quoted this implies $P_0 = P_1$. Hence we need only apply Theorem 1D to conclude that W is of class O_{AD}.

Conversely, if $W \in O_{AD}$ we know that $P_0 = P_1$ for any choice of the pole. Hence $P_0 + P_1$ defines a mapping on the complement of a set of measure 0. But by Theorem 12F this is the largest area of the complement for univalent mappings with the same pole. This means that the complementary area is always zero, and hence that E has absolute measure 0.

2C. The distinguishing feature of Theorem 2B as opposed to 1D is that the condition refers in substance, although not in form, to only one choice of the pole. In fact, if E remains of measure 0 when transformed by any mapping that leaves the point at ∞ fixed, then the same is true for arbitrary mappings. This is so because an arbitrary mapping can be changed into one with fixed pole by means of a linear transformation which of course preserves sets of measure 0.

When this remark is utilized in the proof of Theorem 2B it becomes clear that for planar surfaces the following holds:

Theorem. *If $P_0 - P_1$ is constant for one choice of the pole, then it is constant for all choices, and this condition is necessary and sufficient for W to be of class O_{AD}.*

Another way of expressing the same condition is to say that the *span* vanishes (see Ch. III, 12F), and we have proved that the span vanishes for all or no locations of the pole.

2D. It will be recalled that Theorems 12F and 12G, Ch. III, were derived as special cases of much more general propositions, and that in all extremal problems of this nature the solution was found to be unique, although the uniqueness was not always expressly stated.

Consequently, if we know that $P_0 = P_1$ it follows not only that the complementary area which corresponds to any univalent mapping is zero, but also that the only univalent mapping, up to an additive constant, is $\frac{1}{2}(P_0 + P_1) = P_0 = P_1$. When the pole is at ∞ this is evidently the identity mapping; for an arbitrary pole it reduces to a linear mapping.

Theorem. *A plane region W is of class O_{AD} if and only if the only univalent functions on W are the linear functions.*

We can express this property by saying that W is *rigidly imbedded* in the sphere, in the sense that any one to one mapping of W into the sphere is induced by a one to one mapping of the sphere onto itself.

3. Connections between O_{AD} and O_{AB}

3A. In analogy with the notations that have already been introduced we denote by AB the class of bounded analytic functions, and by O_{AB}

the class of Riemann surfaces on which every AB-function reduces to a constant.

Suppose that a surface W can be realized as a covering surface of another surface W_0. Then, quite trivially, $W \in O_{AB}$ implies $W_0 \in O_{AB}$. Indeed, if the projection mapping is f, and if W_0 carried a bounded analytic function F, then $F \circ f$ would be a nonconstant AB-function on W.

It is more interesting that closer investigation of the same situation leads to a relation between the classes O_{AB} and O_{AD}.

3B. Because of the preceding observation it is particularly important to study the AB-character of plane regions. In this respect we prove:

Theorem. *Every plane region $W \in O_{AB}$ has a complement with zero area.*

It can be assumed that W contains the point ∞. By Theorem 11C, Ch. III, there exists a function $f(z) = z + \sum a_k z^{-k}$ which maps W on a slit region whose complement has zero area. If W has a complement with positive area the function $f(z) - z$ cannot reduce to a constant. The theorem follows, for $f(z) - z$ is evidently bounded.

3C. If W is a planar surface of class O_{AB} the preceding theorem shows that any univalent conformal mapping of W into the plane leaves a complement of measure zero. According to Theorem 2B this implies that W is of class O_{AD}. The conclusion is not limited to planar surfaces, as a very simple reasoning will show.

Theorem. $O_{AB} \subset O_{AD}$.

Suppose that W is not of class O_{AD}. A nonconstant AD-function represents W as a covering surface of the plane with finite area. It follows that the projection W_0 has finite area, and hence that the complement of W_0 has infinite area. Theorem 3B shows that W_0 is not of class O_{AB}, and by the remark in 3A its covering surface W cannot be of class O_{AB}. We have proved the asserted inclusion.

4. Removable singularities

4A. The complementary set E of a plane region W of class O_{AD} can also be characterized in terms of the classical problem of removable singularities. Let G be an open set which contains E, and suppose that the function $F(z)$ is defined and analytic on $G - E$. We say that E is a *removable singularity* for F if it is possible to find an extension of F which is analytic on all of G.

4B. The following result is very easy to prove:

Theorem. *A closed set E is a removable singularity for all functions of class AD in a neighborhood of E if and only if the complement of E (with respect to the sphere) is of class O_{AD}.*

Denote the complement of E by W, and suppose first that E is a removable singularity for all AD-functions. If $F \in AD$ on W it can be extended to the whole sphere and must consequently reduce to a constant. The necessity is thus trivial and has nothing to do with the special properties of the class AD.

The sufficiency follows by application of the results in Ch. III, §1. G is an open set containing E, and we suppose that F is of class AD on $G-E$ and continuous on the boundary of G. We may suppose that G is regular for the Dirichlet problem. The operator L determines the harmonic function Lf in G with boundary values f. It satisfies the conditions for a normal operator (see Ch. III, 2G).

We set $s = \operatorname{Re} F$. The basic condition

(1) $$\int_\beta ds^* = 0$$

of Theorem 3A, Ch. III, is fulfilled. Hence there exists a harmonic function p on W which satisfies $p-s = L(p-s)$. Since $p-s$ can thus be continued to all of G it follows by use of the triangle inequality that $D_W(p)$ is finite. By assumption, this implies that p is constant. Hence s can be continued to all of G, and the same reasoning can be applied to show that the imaginary part s^* can also be so continued. This proves the theorem.

4C. Essentially the same reasoning can be applied to many other classes of functions. For instance, for the class AB of bounded analytic functions we obtain:

Theorem. *E is a removable singularity for all AB-functions if and only if the complement of E is of class O_{AB}.*

The proof differs in only one point. If $p-s$ and s are bounded we must conclude that p is bounded. This is even more obvious than in the case of Dirichlet norms.

The corresponding theorem for HD-functions is much less immediate, for then it must be proved that (1) holds. This question will therefore be postponed until we have discussed the properties of the class O_{HD} in greater detail (see 7D).

5. Classes of harmonic functions

5A. If we fix our attention on arbitrary harmonic functions, without imposing any conditions on the conjugate functions, the most important function classes are the following:

HP = *the class of positive harmonic functions*;

HB = *the class of bounded harmonic functions*;

HD = *the class of harmonic functions with finite Dirichlet integral.*

5B. The corresponding classes of Riemann surfaces satisfy the following inclusion relations:

Theorem. $O_{HP} \subset O_{HB} \subset O_{HD}$.

The first inclusion, $O_{HP} \subset O_{HB}$, is quite trivial, for a bounded harmonic function becomes positive when a sufficiently large constant is added.

In order to prove the second relation we assume that W is not of class O_{HD}. According to Theorem 1C it is then possible to choose a singularity for which the principal functions p_0 and p_1 have a nonconstant difference. But p_0 and p_1 satisfy the maximum principle on the complement of a compact set. In fact, we know that $p_0 = L_0 p_0$, $p_1 = L_1 p_1$, and since L_0, L_1 are normal operators the maximum principle holds. Consequently, if we exclude a small disk about the pole, p_0 and p_1 are bounded outside of the disk, and the difference $p_0 - p_1$ is of course bounded on the disk. Hence $p_0 - p_1$ is a bounded nonconstant harmonic function on W. Thus W is not of class O_{HB}, and we have proved that $O_{HB} \subset O_{HD}$.

5C. The proof shows a little more than was stated in Theorem 5B. In fact, if there exists a nonconstant function of class HD on W we have shown that there is one, namely $p_0 - p_1$, which is simultaneously of class HD and HB. If we set $HB \cap HD = HBD$ we have thus:

Theorem. $O_{HD} = O_{HBD}$.

6. Parabolic and hyberbolic surfaces

6A. The general uniformization theorem (see Ch. III, 11G) led very early to the classification of all simply connected Riemann surfaces into *elliptic*, *parabolic*, and *hyperbolic* surfaces. The elliptic surfaces are the closed simply connected surfaces, which by the uniformization theorem can be mapped on the Riemann sphere. The parabolic surfaces are conformally equivalent with the complex plane (without the point at ∞) and the hyperbolic surfaces can be mapped on a finite disk.

It is quite clear that the plane belongs to all the degeneracy classes that we have considered so far, in particular to the smallest class O_{HP}, while the finite disk does not. Anyone of these properties could thus be used to extend the definition of parabolicity to arbitrary open surfaces. Traditionally, however, the name of parabolic surfaces has been reserved for surfaces that have no Green's function. We shall now analyze this property and study its relation to other kinds of degeneracy.

6B. The nonexistence of a Green's function is undoubtedly the most important characteristic of a parabolic surface, but as a definition it has the disadvantage that it refers to a fixed choice of the pole. In order to justify the definition it is necessary to prove that the existence of a Green's function does not depend on the location of the pole. We can avoid this by adopting the following definition, first suggested by M. Brelot (unpublished) and M. Ohtsuka [1].

Definition. *A Riemann surface W is said to be parabolic if there are no nonconstant negative subharmonic functions on W.*

All open surfaces which are not parabolic will be called hyperbolic. In spite of the choice of definition, the class of parabolic surfaces will be denoted by O_G.

6C. We are going to show that the property in the definition is equivalent to each of three conditions that appear frequently in applications. We formulate these conditions in a very concise form which will be amplified in the course of the proof (L. Ahlfors [18]).

Theorem. *A Riemann surface is parabolic if and only if any one of the following conditions is fulfilled:*

(1) *the maximum principle is valid;*

(2) *the harmonic measure of the ideal boundary vanishes;*

(3) *there is no Green's function.*

6D. Suppose that the function v is defined and subharmonic in a region $G \subset W$ with a nonvoid boundary. Assume moreover that v is bounded above, $v < M$, and that $\overline{\lim}_{z \to \zeta} v(z) \leqq m < M$ for every boundary point ζ of G. We say that the maximum principle is valid if these conditions imply $v \leqq m$.

Suppose first that W is parabolic. We construct the function

$$v_0 = \begin{cases} \max(v-M, m-M) \text{ in } G \\ m-M \text{ outside of } G. \end{cases}$$

It follows from the assumption that v_0 is subharmonic and negative. Indeed, there can be a doubt only at a boundary point ζ of G, and for such a point the upper semicontinuity follows from $v_0(\zeta) = m - M \geqq \overline{\lim}_{z \to \zeta} v_0(z)$, while the subharmonic character is obvious because $m-M$ is the minimum of v_0. Therefore v_0 reduces to a constant, and since G has a nonvoid complement the value of the constant is $m-M$. Hence $v-M \leqq m-M$ or $v \leqq m$ in G, which is the desired conclusion.

The converse is even simpler. Let v be subharmonic and negative on all of W. Take G to be the complement of a single point ζ. If the maximum principle is valid we conclude that $v(z) \leqq \overline{\lim}_{z \to \zeta} v(z) \leqq v(\zeta)$ for any $z \neq \zeta$. It follows that v is constant. Hence W is parabolic.

6E. The harmonic measure of the ideal boundary is defined with respect to a regular subregion Ω_0. Let Ω be another regular region which contains

$\bar{\Omega}_0$. We denote by u_Ω the harmonic function in $\bar{\Omega}-\Omega_0$ which is 0 on $\beta(\Omega_0)$ and 1 on $\beta(\Omega)$. It follows from the classical maximum principle that u_Ω decreases when Ω increases. Hence the harmonic limit function

$$u_W(z) = \lim_{\Omega \to W} u_\Omega(z) \tag{2}$$

exists in $W-\Omega_0$. It is either identically 0 or positive and <1. In the first case we say that the harmonic measure of the ideal boundary vanishes. It must be shown, of course, that this property is independent of the choice of Ω_0.

If the harmonic measure does not vanish it is clear that the limit function $u_W(z)$, defined by (2), violates the maximum principle. Therefore, if W is parabolic the harmonic measure vanishes.

Suppose now that v is a negative subharmonic function on W, and denote its maximum on $\bar{\Omega}_0$ by η. Then $v \leqq \eta(1-u_\Omega)$ on the boundary of $\Omega-\bar{\Omega}_0$, and consequently also in the interior. If $u_\Omega \to 0$ it follows that $v \leqq \eta$ on the whole surface. Hence v attains its maximum, and since v is subharmonic this is possible only if it reduces to a constant. We conclude that the vanishing of the harmonic measure implies parabolicity.

6F. For a pole ζ and a region Ω that contains ζ the Green's function $g_\Omega(z, \zeta)$ is defined as the harmonic function with singularity $-\log|z-\zeta|$ and boundary values 0 (see 15A). It is evident that g_Ω increases with Ω. Therefore,

$$g(z, \zeta) = \lim_{\Omega \to W} g_\Omega(z, \zeta)$$

is either identically equal to $+\infty$ or finite and positive for all $z \neq \zeta$. If g exists, then $-g$ is negative, subharmonic, and nonconstant. Hence there is no Green's function on a parabolic surface.

In order to prove the converse we consider a pair of circles $|z-\zeta|=r_1$ and $|z-\zeta|=r_2>r_1$ in the plane of a local variable at ζ. We assume that W is hyperbolic, so that the harmonic measure u which is 0 on $|z-\zeta|=r_1$ does not vanish identically. Let the minimum of u on $|z-\zeta|=r_2$ be denoted by u_0. We write $|z-\zeta|=r$ and construct the following function:

$$v = \begin{cases} \log \dfrac{r}{r_1} & \text{for} \quad r \leqq r_1 \\[2ex] \max\left(\log \dfrac{r}{r_1}, \dfrac{u}{u_0} \log \dfrac{r_2}{r_1}\right) & \text{for} \quad r_1 < r \leqq r_2 \\[2ex] \dfrac{u}{u_0} \log \dfrac{r_2}{r_1} & \text{outside of} \quad |z-\zeta| \leqq r_2. \end{cases}$$

Since $\log \frac{r}{r_1} = \frac{u}{u_0} \log \frac{r_2}{r_1} = 0$ for $r = r_1$ and $\log \frac{r}{r_1} \leqq \frac{u}{u_0} \log \frac{r_2}{r_1}$ for $r = r_2$, it is seen that v is subharmonic. It follows by the maximum principle that

$$g_\Omega + v \leqq \frac{1}{u_0} \log \frac{r_2}{r_1}$$

for sufficiently large Ω. Hence g_Ω is bounded, and the Green's function exists. This completes the proof of Theorem 6C.

6G. As a final remark, Theorem 5B can be complemented by the following result:

Theorem. $O_G \subset O_{HP}$.

In fact, if W carries a nonconstant positive function v, then $-v$ is nonconstant, negative, and subharmonic, from which it follows that W is not parabolic.

7. Vanishing of the flux

7A. There is a further characterization of parabolic surfaces which is closely connected with property (2) in Theorem 6C, i.e. the vanishing of the harmonic measure. We point out in passing that this is a property which depends only on the structure of the surface in the immediate vicinity of the ideal boundary. In other words, if compact sets can be removed from two surfaces so that the remaining parts are conformally equivalent, then the surfaces are either both parabolic or both hyperbolic. It is understood that the equivalence mapping must make the original ideal boundaries correspond to each other, in the obvious sense.

Parabolic surfaces have also been called surfaces with a nullboundary, and the previous remark shows that it is indeed only a property of the ideal boundary that is involved.

7B. Let us consider a compact bordered Riemann surface $\overline{W}$ whose contours are divided into two classes α and β, the inner contours and the outer contours. We denote by u_0 the harmonic function which is 0 on α and 1 on β. Its flux along β, or along any cycle homologous to β, is

$$\int_\beta du_0^* = D(u_0).$$

We are interested in finding a relation, in the form of an inequality, between the flux and the Dirichlet integral of an arbitrary harmonic function on $\overline{W}$.

Lemma. *The flux d and the Dirichlet integral $D(u)$ of any harmonic function u on W satisfy the condition*

(3) $$d^2 \leqq D(u)\, D(u_0).$$

The inequality that we wish to establish is homogeneous in u. Therefore we can assume, without loss of generality, that $d=D(u_0)$, in which case (3) reduces to $D(u_0) \leqq D(u)$. Because u and u_0 have the same flux it follows that

$$D(u_0, u-u_0) = \int_{\beta-\alpha} u_0(du^*-du_0^*) = 0.$$

Hence

$$D(u) = D(u_0)+D(u-u_0),$$

and we find indeed that $D(u_0) \leqq D(u)$.

7C. We are now ready to prove:

Theorem. *A surface W is parabolic if and only if every function which is of class HD in a neighborhood of the ideal boundary β has vanishing flux along β.*

We suppose that u is defined outside of Ω_0. As in 6E we choose $\Omega \supset \Omega_0$ and denote by u_Ω the harmonic function which is 0 on $\beta(\Omega_0)$ and 1 on $\beta(\Omega)$. The flux d of u is independent of Ω, and by (3) we obtain

$$d^2 \leqq D_{\Omega-\Omega_0}(u)\ D_{\Omega-\Omega_0}(u_\Omega).$$

If $u_\Omega \to 0$ we know that $D_{\Omega-\Omega_0}(u_\Omega)$, which is equal to the flux of u_Ω along $\beta(\Omega_0)$, tends to 0. Since $D_{\Omega-\Omega_0}(u)$ is bounded by hypothesis it follows that $d=0$.

Conversely, if the harmonic measure u_W is positive, then $D_{W-\Omega_0}(u_W)$ is finite and equal to the flux of u_W; in fact, $D_{\Omega-\Omega_0}(u_\Omega) \to D_{W-\Omega_0}(u_W)$ and the flux of u_Ω tends to that of u_W. In other words, u_W is an HD-function with nonzero flux.

7D. We can now return to the question raised in 4C. For a plane region with the complement E we find:

Theorem. *E is a removable singularity for all HD-functions if and only if the complement of E is of class O_G.*

The Green's function with pole at ∞, if it exists, is of class HD in a neighborhood of E as seen by use of approximating functions g_Ω. If it could be continued to E it would possess a minimum, and would thus reduce to a constant, which is impossible. This proves that the condition is necessary.

To prove the sufficiency we proceed as in 4B. Let s be of class HD in a neighborhood of E. Since we are now assuming that $W \in O_G$ we conclude by Theorem 7C that the flux condition (1), 4B, is fulfilled. We can thus find p on W so that $p-s=L(p-s)$. But then p is of class HD on W, and since $O_G \subset O_{HD}$ we conclude that p is a constant. Hence $s=Ls$ can be continued to E.

7E. The form of Theorem 7D lets us suspect that the classes O_G and O_{HD} may be identical when we consider only planar surfaces. We shall prove that this is indeed the case.

Theorem. *For planar surfaces the classes O_G, O_{HP}, O_{HB} and O_{HD} are identical.*

Because of Theorems 5B and 6G we need only prove that $O_{HD} \subset O_G$. Let E be a plane point set whose complement W is of class O_{HD}, and suppose, by antithesis, that W has a Green's function g with pole at ∞.

Let Ω be a regular subregion of W which contains ∞. Its boundary is made up of analytic contours $\beta_1, \cdots, \beta_n$. We denote by G_i the region enclosed by β_i, and by E_i the part of E that lies in G_i.

It is not possible that g has zero flux along all β_i. Suppose that g has nonzero fluxes

$$c_i = \int_{\beta_i} dg^*, \quad c_j = \int_{\beta_j} dg^*$$

over two contours $\beta_i \neq \beta_j$. We form a singularity function s which is defined by

$$s = \begin{cases} c_j g & \text{in} \quad G_i - E_i \\ -c_i g & \text{in} \quad G_j - E_j \\ 0 & \text{in} \quad \bigcup\limits_{k \neq i,j} (G_k - E_k). \end{cases}$$

Then s has vanishing flux along $\beta_1 + \cdots + \beta_n$. Let L be the operator which solves the Dirichlet problem for $G_1 \cup \cdots \cup G_n$. The main existence theorem (Ch. III, 3A) guarantees the existence of a harmonic function p on W such that $p - s = L(p - s)$ in each $G_k - E_k$. Since s has a finite Dirichlet integral over $\bigcup\limits_{k=1}^{n} (G_k - E_k)$ it follows that p is an HD-function, and hence that p reduces to a constant. We conclude that s can be extended to a regular function on G_i, and its flux $c_j c_i$ over β_i would be 0, contrary to the hypothesis.

We have shown that there is a single β_i, call it β_1, with a nonzero flux. We can prove, further, that g has a regular extension to each $G_i \neq G_1$. In fact, to see this it is sufficient to consider the singularity function $s = g$ in $G_i - E_i$. Since the flux of s vanishes the same reasoning can be applied, and the assertion follows.

We are thus able to extend g to the whole complement of $\bar{G}_1$. The extension, which we continue to denote by g, remains positive, as seen by applying the minimum principle to each $G_i \neq G_1$.

7F. We shall now denote the G_1 that corresponds to a given Ω by $G_1(\Omega)$. If $\Omega \subset \Omega'$ it is quite evident that $G_1(\Omega') \subset G_1(\Omega)$, since this is the only way in which the flux condition could be satisfied.

Let B be the intersection of all $\bar{G}_1(\Omega)$. We wish to show that g can be extended to the complement of B. In order to do so we shall momentarily denote the extension to the complement of $\bar{G}_1(\Omega)$ by g_Ω (note that g_Ω is not the Green's function of Ω, but an extension of g, the Green's function of W). It must be shown that $g_{\Omega_1}=g_{\Omega_2}$ in their common domain. This is seen by choosing $\Omega \supset \Omega_1 \cup \Omega_2$, for it is trivial that $g_{\Omega_1}=g_\Omega$ outside of $G_1(\Omega_1)$ and $g_{\Omega_2}=g_\Omega$ outside of $G_1(\Omega_2)$.

B is closed and nonempty. We contend that it is also connected. Suppose indeed that $B=B' \cup B''$ where B', B'' are closed, disjoint, and nonvoid. Let V' be a neighborhood of B' whose closure does not meet B''. By compactness, there exists a $\bar{G}_1(\Omega)$ that does not meet the boundary of V'. But this is impossible, for $\bar{G}_1(\Omega)$ is connected and contains interior and exterior points of V'. Hence B is connected.

Either B is a point or a continuum. If it is a point it is an isolated singularity of g, and since $g>0$ the singularity is either removable or a positive logarithmic pole. Neither is possible, for there cannot exist a harmonic function on the sphere with only positive poles.

If B is a continuum its complement can be mapped on a disk. The real part of the mapping function has a finite Dirichlet integral over W. This contradicts the hypothesis $W \in O_{HD}$.

We have shown that the existence of a Green's function leads to an inevitable contradiction. Hence $O_{HD} \subset O_G$, for planar surfaces.

It would be easy to modify the proof so that it applies to all surfaces of finite genus.

§2. PARABOLIC AND HYPERBOLIC SUBREGIONS

There is a natural classification of noncompact subregions which in many respects parallels the distinction between parabolic and hyperbolic surfaces. This classification is closely connected with the theory of positive harmonic functions, and we take the opportunity to develop that theory beyond what is actually needed for classification purposes.

An application of the results leads to new characterizations of the classes O_{HP} and O_{HB}.

Remark. Many results in this section are due to M. Parreau [6]. Others had been known earlier, but were knit more closely together in this important paper.

8. Positive harmonic functions

8A. We are now going to make more systematic use of the theory of subharmonic functions as developed in Ch. II, 9–11. We recall that the solution of Dirichlet's problem was obtained by considering the least upper bound of a suitably defined family of subharmonic functions. This is a very flexible method which we shall utilize for the construction of many harmonic functions with special properties.

We ask the reader to re-examine the proof of Theorem 11D, Ch. II. If $\mathscr{V}$ is a class of subharmonic functions on a Riemann surface W we wish to assert that

$$u(z) = \sup_{v \in \mathscr{V}} v(z)$$

is either harmonic or identically $\pm\infty$. It is found that the proof depends exclusively on two properties of $\mathscr{V}$:

(1) if $v_1, v_2 \in \mathscr{V}$, then $\max(v_1, v_2) \in \mathscr{V}$;

(2) if $v \in \mathscr{V}$, then the function v_0, obtained on replacing v by its Poisson integral in a parametric disk, is likewise in $\mathscr{V}$.

Consequently, for any class $\mathscr{V}$ which satisfies these conditions, it is true that $u = \sup v$ is harmonic or completely degenerate.

8B. Let u be any positive harmonic function on W (in this connection the word "positive" will be interpreted in the wide sense, $u \geqq 0$). We shall denote by $\mathscr{V}(u)$ the class of all subharmonic functions which are *bounded above* and $\leqq u$ on W. It is clear that this class satisfies (1) and (2). Hence

$$Bu = \sup_{v \in \mathscr{V}(u)} v$$

is harmonic and satisfies $0 \leqq Bu \leqq u$.

There are two extreme cases: if $Bu = u$ we say that u is *quasi-bounded*; if $Bu = 0$ we say that u is *singular*. Note that 0 is the only function which is at once quasi-bounded and singular.

For instance, the function $-\log|z|$ is singular in $0 < |z| < 1$. The same function is quasi-bounded in the half-disk $|z| < 1$, $\operatorname{Re} z > 0$ (approximate by harmonic functions which are equal to $-\log|z|$ on the boundary except on a small segment near 0).

8C. We prove first that the operator B is linear: $B(u_1 + u_2) = Bu_1 + Bu_2$. To begin with, if $v_1 \in \mathscr{V}(u_1)$, $v_2 \in \mathscr{V}(u_2)$ it is clear that $v_1 + v_2 \in \mathscr{V}(u_1 + u_2)$; hence $Bu_1 + Bu_2 \leqq B(u_1 + u_2)$. On the other hand, if $v \in \mathscr{V}(u_1 + u_2)$, then $v - u_1 \in \mathscr{V}(u_2)$ whence $v - u_1 \leqq Bu_2$. This implies $v - Bu_2 \leqq u_1$, so that $v - Bu_2 \leqq Bu_1$, and finally $B(u_1 + u_2) \leqq Bu_1 + Bu_2$. The linearity is proved.

Next we show:

Lemma. *B is idempotent*: $BB = B$.

From $Bu \leqq u$ it follows at once that $B^2u \leqq Bu$. But if v is subharmonic, bounded above, and $\leqq u$, then it is also $\leqq Bu$, and consequently $\leqq B^2u$. Thus $Bu \leqq B^2u$, and the idempotence is proved.

It is now almost trivial to prove:

Theorem. *Every positive harmonic function u has a unique representation $u = Bu + Su$ as the sum of a quasi-bounded and a singular function.*

From $BBu = Bu$ and $BSu = Bu - BBu = 0$ we conclude that Bu is indeed quasi-bounded and Su singular. Let $u = u_0 + u_1$ be any decomposition with the same properties. Then $Bu = Bu_0 + Bu_1 = u_0$, and the uniqueness follows.

8D. We shall now let G denote a regularly imbedded open set on W. The boundary α of G is thus composed of disjoint analytic curves. All that we shall say will be trivial if $\bar{G}$ is compact. Hence the noncompact case is the only interesting one; α can be either compact or noncompact. For convenience, G will frequently be referred to as a "region", even if it is not connected.

An operator T will be introduced whose domain consists of the positive harmonic functions on W while the range is formed by positive harmonic functions on G which vanish on α.

For any positive harmonic u on W we define Tu to be the *greatest harmonic minorant* of u in G which vanishes on α. Such a greatest minorant exists, for it can be obtained as the supremum in the class $\mathscr{V}_0(u)$ of subharmonic minorants which are $\leqq 0$ on α.

The argument in the first paragraph of 8C can be repeated to show that T is linear.

8E. In the following, whenever B is applied to a function defined only in G it is understood that B means the operator associated with G rather than with W. With this proviso we can state:

Theorem. *B and T commute*: $BT = TB$.

Let v be subharmonic, bounded above, and $\leqq Tu$ on G. We define v_0 to be 0 on $W - G$ and $\max(v, 0)$ on G. Then v_0 is subharmonic, bounded, and $\leqq u$ on W. Thus $v_0 \leqq Bu$, and it follows that $BTu \leqq Bu$ (observe that B is used in two different meanings). But BTu vanishes on α, whence $BTu \leqq TBu$.

For the opposite conclusion we let $B_n u$ denote the greatest harmonic minorant of u which is $\leqq n$; it is evident that $\lim_{n \to \infty} B_n u = Bu$. Moreover, $TB_n u$ is a bounded minorant of Tu. Hence $TB_n u \leqq BTu$, and as $n \to \infty$ we obtain $TBu \leqq BTu$.

Corollary. *The mapping T carries quasi-bounded and singular functions into functions of the same kind.*

Indeed, $Bu = u$ implies $BTu = TBu = Tu$, and $Bu = 0$ yields $BTu = TBu = 0$.

8F. It is possible to determine Tu constructively, in a manner analogous to the way in which the harmonic measure u_W was introduced (6E). Let Ω be a regular region which intersects G. We construct the harmonic function $T_\Omega u$ in $G \cap \Omega$ which is 0 on $\alpha \cap \Omega$ (α is the boundary of G) and equal to u on $\beta(\Omega) \cap G$. More exactly, we determine $T_\Omega u$ by Perron's method (Ch. II, 11) applied to the boundary values 0 on $\alpha \cap \Omega$ and u on the arcs of $\beta(\Omega)$ that belong to the boundary of $G \cap \Omega$. Then $T_\Omega u$ decreases with Ω. Indeed, suppose that $\Omega \subset \Omega'$. We have $T_\Omega u \leqq u$ in $G \cap \Omega$. The function which equals $T_\Omega u$ in $G \cap \Omega$ and u in $G \cap (\Omega' - \Omega)$ is therefore superharmonic. We conclude by the maximum principle that $T_{\Omega'} u \leqq T_\Omega u$ in $G \cap \Omega$.

The functions $T_\Omega u$ have a limit function $T_W u$. We claim that $Tu = T_W u$. First, $T_W u$ vanishes on α. Second, it is a harmonic minorant of u, and third, if v is any harmonic minorant which vanishes on α, then $v \leqq T_\Omega u$ in $G \cap \Omega$, and consequently $v \leqq T_W u$ in G. Because of these properties $T_W u$ coincides with Tu.

In view of this construction it is suggestive to say that Tu is, in a generalized sense, the harmonic function which vanishes on α and is equal to u on the part of the ideal boundary which belongs to G.

9. Properties of the operator T

9A. The mapping T transforms positive harmonic functions on W into positive harmonic functions on G which vanish on α. We wish to introduce another mapping R which under certain conditions acts as an inverse of T.

Let v be a positive harmonic function on G which vanishes on α, and suppose that v has a majorant on G which can be extended to a positive harmonic function on W; briefly, v has a positive harmonic majorant on W. Under this condition we define Rv to be the *least positive harmonic majorant* of v on W. It exists, for it can be obtained as the greatest lower bound of all superharmonic majorants of the subharmonic function which is 0 on $W - G$ and equal to v on G. Once more, the same reasoning as in 8B shows that R is linear: $R(v_1 + v_2) = Rv_1 + Rv_2$.

It is evident that $RTu \leqq u$, for u is a positive majorant of Tu. Similarly, $v \leqq TRv$, for v is a minorant of Rv that vanishes on α. Hence $Rv \leqq RTRv \leqq Rv$, and we have proved that $RTRv = Rv$. But $R(TRv - v) = 0$ means, trivially, that $TRv = v$.

Since T is linear the mapping $u \to Tu$ is one to one if and only if $Tu \equiv 0$ implies $u \equiv 0$. We have $u - RTu \geqq 0$ and $T(u - RTu) = Tu - TRTu = 0$. Hence, if T is one to one we conclude that $u = RTu$. In this case R and T are true inverses. In particular, Rv ranges over all positive harmonic u on W.

9B. In order that $Tu \equiv 0$ imply $u \equiv 0$ it is necessary that $T1 \not\equiv 0$. For this reason we distinguish two cases:

Definition. *A region G is called relatively parabolic if $T1 \equiv 0$ and relatively hyperbolic if $T1 \not\equiv 0$.*

If G is a genuine region, then $T1 \not\equiv 0$ is equivalent to $T1 > 0$, but if G is disconnected $T1$ can vanish on a component without being identically 0.

There are two circumstances under which the condition $T1 \not\equiv 0$ is also sufficient to insure that $Tu \not\equiv 0$ when $u \not\equiv 0$. One of these occurs, in a rather trivial way, when W is of class O_{HP}. Then all positive functions on W are constants c, and $Tc = cT1 \equiv 0$ only if $c = 0$.

Under the hypothesis $W \in O_{HP}$ we conclude further that every Rv is a constant c, and hence that $v = TRv = Tc = cT1$, provided that Rv exists. The existence of Rv is manifest if v is bounded, and then Rv is also bounded. When this is so the conclusion $Rv = c$ and its consequence $v = cT1$ are valid already under the weaker hypothesis $W \in O_{HB}$. We formulate this result as a theorem.

Theorem. *If G is a relatively hyperbolic region on a surface of class O_{HB}, then the only bounded positive harmonic functions on G which vanish on α are the positive multiples of $T1$.*

9C. The preceding theorem has a corollary which is of interest for the classification of Riemann surfaces.

Theorem. *A surface W is of class O_{HB} if and only if there are no two disjoint relatively hyperbolic regions on W.*

Suppose first that u is a nonconstant bounded harmonic function on W, and let c be a constant that lies between the infimum and supremum of u. Let G be the region (in the sense of the agreement in 8D) on which $u > c$. It cannot be parabolic, for $u < b$ and $T1 \equiv 0$ would imply $T(u-c) \leqq (b-c)T1 \equiv 0$, while on the other hand $T(u-c) = u-c > 0$. The same reasoning, applied to $c-u$, shows that the region on which $u < c$ is likewise hyperbolic. Hence there are two disjoint hyperbolic regions.

Conversely, let G_1 and G_2 be any two disjoint hyperbolic regions. We set $G = G_1 \cup G_2$ and denote the operators associated with G, G_1, G_2 by T, T_1, T_2. By hypothesis, $T_1 1$ and $T_2 1$ are $\not\equiv 0$. We consider on one side the function v_1 in G which is equal to $T_1 1$ in G_1 and 0 in G_2, and on the other side the function v_2 which is $T_2 1$ in G_2 and 0 in G_1. According to Theorem 9B, if W is of class O_{HB}, v_1 and v_2 must both be multiples of $T1$. That is manifestly impossible, and the theorem is proved.

9D. The second case in which $T_1 \not\equiv 0$ permits us to conclude that T is one to one occurs when the complement $W - G$ is compact. When this is so $T1$ is nothing else than the harmonic measure u_W with respect to the complement, so that $T1 \not\equiv 0$ if and only if W is hyperbolic. Consider any positive harmonic function u on W, and let c be its maximum on α, a compact set. Then $u - c \leqq 0$ on α and $u - c \leqq u$ in G, from which it follows that $u - c \leqq Tu$. If $Tu \equiv 0$ we obtain $u \leqq c$ in G. The same is true in $W - G$, by the maximum principle. Hence u attains its maximum and must reduce to the constant c, and $Tc\equiv 0$ only if $c=0$.

Theorem. *Suppose that W is hyperbolic and $W-G$ compact. Then $Tu\equiv 0$ only if $u\equiv 0$, and T is a one to one linear mapping of all positive harmonic functions on W onto the set of all positive harmonic functions in G which vanish on α and have a positive harmonic majorant on W.*

9E. For parabolic regions the following is true:

Theorem. *If G is parabolic every Tu is singular, and $Tu\equiv 0$ whenever u is quasi-bounded. If α and $W-G$ are compact, then Tu is always identically* 0.

If u is bounded, $u\leqq c$, then $Tu\leqq Tc=cT1\equiv 0$. Let u be arbitrary. With the notation B_n introduced in 8E we have $TB_nu\equiv 0$. But $B_nTu\leqq TB_nu$, for B_nTu is a minorant of B_nu and vanishes on α. Hence $B_nTu=0$, and on going to the limit we find $BTu=0$. Thus Tu is singular, and if u is quasi-bounded, $u=Bu$, then $Tu=TBu=BTu=0$.

For compact α and $W-G$, G is parabolic only if W is parabolic. Hence every u is a constant c, and $Tc=cT1\equiv 0$.

9F. If $G\supset G'$, and if T, T' are the corresponding operators, then $Tu\geqq T'u$. Indeed, the function which is 0 in $G-G'$ and $T'u$ in G' is subharmonic and must hence be $\leqq Tu$. As a consequence of this result *any subregion of a parabolic region is parabolic*, and, in particular, *every subregion of a parabolic surface is relatively parabolic.*

If G and G' differ only on a compact set, that is, if they have the same intersection with the complement of a sufficiently large compact set, then they are simultaneously parabolic or hyperbolic. We may suppose that $G\supset G'$ and that $G-G'$ is relatively compact. Let c be the maximum of $T1$ on α', the compact boundary of G' with respect to G. Then $T1-c\leqq 0$ on the whole boundary of G', and $T1-c\leqq 1$ in G'. This implies $T1-c\leqq T'1$, and if $T'1\equiv 0$ we have $T1\leqq c$ in G'. The same inequality holds in $G-G'$, by the maximum principle. Hence $T1$ must be a constant, at least in the component in which the maximum is attained. Since the constant must be 0 we find that $T1$ is identically 0, and we have proved that G is parabolic together with G'.

10. Properties connected with the Dirichlet integral

10A. To complete the preceding results we shall also study the effect of the transformation T on the Dirichlet integral. In this respect we prove:

Theorem. *If $W-G$ is compact, then $D_G(Tu)\geqq D(u)$, and both Dirichlet integrals are simultaneously finite.*

If W is parabolic there is nothing to prove, for then u is constant and $Tu\equiv 0$ so that both Dirichlet integrals vanish. We may therefore assume that W is hyperbolic.

10B. We remarked in 8F that Tu can be expressed as the limit of $T_\Omega u$. We need to know, in addition, that the Dirichlet integral of $T_\Omega u$ tends to that of Tu.

In order to make use of results that have already been proved it is convenient to connect T with the principal operator L_1 (for the identity partition) introduced in Ch. III, 8C. We recall that $L_{1\Omega}u$ has the boundary values u on α, that its flux over $\beta(\Omega)$ vanishes, and that it is constant on $\beta(\Omega)$. If the value of the constant is denoted by c_Ω we see at once that

$$T_\Omega u = u - L_{1\Omega}u + c_\Omega T_\Omega 1. \tag{4}$$

Since u and $L_{1\Omega}u$ have zero flux it follows that:

$$c_\Omega = \int_\alpha d(T_\Omega u)^* : \int_\alpha d(T_\Omega 1)^*.$$

Because we suppose W to be hyperbolic, c_Ω has a finite limit

$$c = \int_\alpha d(Tu)^* : \int_\alpha d(T1)^*,$$

and (4) implies

$$Tu = u - L_1 u + cT1. \tag{5}$$

For convenience, write $W - G = \bar{\Omega}_0$. We know that $D_{\Omega-\Omega_0}(L_{1\Omega}u - L_1 u) \to 0$ (Ch. III, 8C) and $D_{\Omega-\Omega_0}(T_\Omega 1 - T1) \to 0$. On subtracting (5) from (4) we conclude that $D_{\Omega-\Omega_0}(T_\Omega u - Tu) \to 0$, and hence that $D_{\Omega-\Omega_0}(T_\Omega u) \to D_G(Tu)$.

10C. By use of Green's formula

$$\begin{aligned} D_{\Omega-\Omega_0}(L_{1\Omega}u, u) &= -\int_\alpha u\,du^* = -D_{\Omega_0}(u), \\ D_{\Omega-\Omega_0}(T_\Omega 1, u) &= \int_{\beta(\Omega)} du^* = 0. \end{aligned} \tag{6}$$

From (4) and (6) we obtain

$$D_{\Omega-\Omega_0}(T_\Omega u) = D_{\Omega-\Omega_0}(u) + D_{\Omega-\Omega_0}(L_{1\Omega}u - c_\Omega T_\Omega 1) + 2D_{\Omega_0}(u)$$

and, on going to the limit,

$$D_G(Tu) = D(u) + D_G(L_1 u - cT1) + D_{\Omega_0}(u).$$

This proves the inequality $D_G(Tu) \geqq D(u)$, and since $D_G(L_1 u)$ and $D_G(T1)$ are finite we can also conclude that $D_G(Tu)$ is finite if $D(u)$ is finite.

10D. To emphasize the meaning of Theorems 9D and 10A, let us, in the case of compact $W-G$, denote by H_0 the class of all harmonic functions in G which vanish on the boundary α. Let H_0P, H_0B, H_0D be the subclasses of functions which are positive, bounded, or have a finite Dirichlet integral. Then, for hyperbolic W, T is an *isomorphic* mapping of HP into H_0P. It maps HPB onto H_0PB and HPD into H_0PD.

11. Functions with bounded means

11A. The theory of bounded harmonic functions can be complemented by a study of functions which are, in a sense, bounded in the mean. For maximum generality, the boundedness will be defined in terms of a positive, increasing, convex norm-function $\Phi(t)$, defined for all $t \geqq 0$. It is no restriction to assume that $\Phi(0)=0$. The simplest examples are $\Phi(t)=t^p$, $p \geqq 1$.

Let z_0 be a fixed point on W, and let Ω be a regular subregion of W which contains z_0. We denote by $g_\Omega(z)=g_\Omega(z, z_0)$ the Green's function of Ω with pole at z_0. For a given harmonic function f we form the mean value

$$m_\Omega\Phi(|f|) = -\frac{1}{2\pi}\int_{\beta(\Omega)} \Phi(|f|)\, dg_\Omega^*. \tag{7}$$

Observe that the minus sign makes the integral nonnegative.

Definition. *We say that f is Φ-bounded if $m_\Omega\Phi(|f|)$ lies under a finite constant, independent of Ω.*

It will be proved that the property in the definition is independent of z_0. The definition applies equally well to functions f that are of class H_0 with respect to a region G with compact complement.

11B. There is a much simpler characterization of Φ-boundedness which does not require us to consider any approximating regions.

Theorem. *A function $f \in H$ or H_0 is Φ-bounded if and only if $\Phi(|f|)$ has a harmonic majorant.*

First, if $\Phi(|f|)$ has the harmonic majorant u, it follows immediately by (7) that $m_\Omega\Phi(|f|) \leqq u(z_0)$, and hence that f is Φ-bounded.

Let u_Ω denote the harmonic function in Ω which is equal to $\Phi(|f|)$ on $\beta(\Omega)$. Because Φ is convex, $\Phi(|f|)$ is subharmonic: in fact, from

$$|f(z)| \leqq \frac{1}{2\pi}\int_0^{2\pi} |f(z+re^{i\theta})|\, d\theta$$

it follows by Jensen's inequality (see, e.g., G. H. Hardy, G. Pólya, J. E. Littlewood [1], p. 152) that

$$\Phi(|f(z)|) \leqq \frac{1}{2\pi}\int_0^{2\pi} \Phi(|f(z+re^{i\theta})|)\, d\theta.$$

As a consequence, $\Phi(|f|) \leqq u_\Omega$ in Ω. In case $f \in H_0$ the assertion remains true if we replace $\Phi(|f|)$ by 0 inside of $\Omega_0 = W - \bar{G}$.

It is further evident that u_Ω increases with Ω. For if $\Omega \subset \Omega'$, the function which is equal to u_Ω in Ω and $\Phi(|f|)$ in $\Omega' - \Omega$ is subharmonic, and we may conclude that $u_\Omega \leqq u_{\Omega'}$ in Ω. By Harnack's principle (Ch. II, 9) $u(z) = \lim_{\Omega \to W} u_\Omega(z)$ is either harmonic or identically equal to $+\infty$. But $u_\Omega(z_0) = m_\Omega \Phi(|f|)$, so that $u(z)$ is finite if f is Φ-bounded. Moreover, u is a majorant of each u_Ω, and hence of $\Phi(|f|)$. We have proved the theorem, and at the same time we see that the definition of Φ-boundedness is independent of z_0.

11C. It is natural to consider the class $O_{H\Phi}$ of Riemann surfaces on which every Φ-bounded harmonic function reduces to a constant. We shall find, however, that this is not a new classification, inasmuch as $O_{H\Phi}$ will reduce either to O_{HP} or to O_{HB}, depending on the function Φ.

Because $\Phi(t)$ is convex the limit

$$c = \lim_{t \to \infty} \frac{\Phi(t)}{t}$$

exists and is either positive and finite or equal to $+\infty$. We shall first investigate the case of finite c. Then

$$\Phi(t) \leq ct, \tag{8}$$

and in the opposite direction an inequality

$$\Phi(t) \geq at - b \tag{9}$$

with $a > 0$ is valid.

Theorem. *If* $\lim_{t \to \infty} \Phi(t)/t < \infty$, *then* $O_{H\Phi} = O_{HP}$.

If f is a positive harmonic function on W, then f is Φ-bounded, for (8) implies that $\Phi(f)$ has the majorant cf. We conclude that $O_{H\Phi} \subset O_{HP}$.

Suppose now that f is Φ-bounded and nonconstant. Since $\Phi(|f|)$ has a harmonic majorant, it follows from (9) that $|f|$ has a harmonic majorant which we denote by u. We can write $f = u - (u - f)$ where u and $u - f$ are both nonnegative. Since they cannot both be constant we have proved that $O_{HP} \subset O_{H\Phi}$.

By way of illustration, take $\Phi(t) = t$. The theorem asserts that there are nontrivial functions with bounded mean value

$$-\frac{1}{2\pi} \int_{\beta(\Omega)} |f| \, dg_\Omega^*$$

if and only if there are nontrivial positive harmonic functions.

11D. The case $c=\infty$ is more interesting. We will prove:

Theorem. *If* $\lim_{t\to\infty} \Phi(t)/t=\infty$, *then* $O_{H\Phi}=O_{HB}$.

The relation $O_{H\Phi}\subset O_{HB}$ is trivial, for a bounded function is eo ipso Φ-bounded. To prove the opposite inclusion we assume again that f is Φ-bounded and not constant. If a is any constant, then $\frac{1}{2}(f-a)$ is also Φ-bounded, for $\Phi(|f-a|/2)\leqq\frac{1}{2}(\Phi(|f|)+\Phi(|a|))$. On changing the notation we can therefore assume that f takes positive as well as negative values.

Let G^+ be the region in which $f>0$. Then f vanishes on the boundary of G^+, and $\Phi(f)$ has a harmonic majorant in G^+. We shall see in a moment that this situation cannot prevail unless G^+ is relatively hyperbolic. Exactly the same reasoning shows that the set G^- where $f<0$ is hyperbolic. Thus W contains two disjoint relatively hyperbolic regions, and it follows by Theorem 9C that W is not of class O_{HB}. With this conclusion Theorem 11D is proved.

11E. To complete the proof we have to establish the following lemma:

Lemma. *Suppose that* $\lim_{t\to\infty} \Phi(t)/t=\infty$. *Let G be a regularly imbedded open set whose complement need not be compact. If there exists a pair of nontrivial positive harmonic functions u and v in G, such that v vanishes on the boundary of G and $\Phi(v)\leqq u$ in G, then G is relatively hyperbolic.*

Let z_0 be a point in G, and choose Ω so large that $z_0\in\Omega\cap G$. We use mean values analogous to (7), but this time with respect to $\Omega\cap G$ rather than Ω. Explicitly, let $g_{\Omega\cap G}(z, z_0)$ be the Green's function of $\Omega\cap G$ with pole at z_0, and set

$$m_{\Omega\cap G}f = -\frac{1}{2\pi}\int f\,dg^*_{\Omega\cap G}\,,$$

where the integral is over the whole boundary of $\Omega\cap G$. For any mean value process m Jensen's inequality permits us to conclude that

$$\Phi\left(\frac{m(ff_0)}{mf_0}\right)\leqq\frac{m[\Phi(f)f_0]}{mf_0},$$

provided that $f, f_0\geqq 0$ and $mf_0>0$. Apply this inequality to $f=v$, $f_0=T_\Omega 1$ and $m=m_{\Omega\cap G}$. Then $m(ff_0)=mv=v(z_0)$, $mf_0=T_\Omega 1(z_0)$ and $m[\Phi(f)f_0]\leqq mu=u(z_0)$. We obtain

$$\Phi\left(\frac{v(z_0)}{T_\Omega 1(z_0)}\right)\leqq\frac{u(z_0)}{T_\Omega 1(z_0)}$$

or

$$\frac{T_\Omega 1(z_0)}{v(z_0)}\,\Phi\left(\frac{v(z_0)}{T_\Omega 1(z_0)}\right)\leqq\frac{u(z_0)}{v(z_0)}.$$

Because of the hypothesis $\Phi(t)/t\to\infty$ this is possible only so that $T_\Omega 1$ does not tend to 0 as $\Omega\to W$. Hence G is necessarily hyperbolic.

11F. The most interesting application of Theorem 11D is obtained for $\Phi(t)=t^2$. We suppose that $z_0 \in \Omega_0 \subset \Omega$ and use the notation m_Ω introduced in (7). For harmonic f one verifies that $\Delta f^2 = 2(f_x^2+f_y^2)$. Green's formula yields

$$m_\Omega f^2 = f(z_0)^2 + \frac{1}{\pi}\iint_\Omega g_\Omega(f_x^2+f_y^2)\,dxdy$$

$$m_{\Omega_0} f^2 = f(z_0)^2 + \frac{1}{\pi}\iint_{\Omega_0} g_{\Omega_0}(f_x^2+f_y^2)\,dxdy$$

and hence

$$m_\Omega f^2 - m_{\Omega_0} f^2 = \frac{1}{\pi}\iint_{\Omega-\Omega_0} g_\Omega(f_x^2+f_y^2)\,dxdy + \frac{1}{\pi}\iint_{\Omega_0}(g_\Omega - g_{\Omega_0})(f_x^2+f_y^2)\,dxdy$$

$$\leqq \frac{K_\Omega}{\pi} D(f),$$

where K_Ω is the maximum of g_Ω on $\beta(\Omega_0)$. If W is hyperbolic K_Ω remains bounded and we conclude that $HD \subset H\Phi$, $\Phi(t)=t^2$; if W is parabolic the same is true by default. In view of Theorem 11D this result constitutes a new proof of the relation $O_{HB} \subset O_{HD}$ (see 5B).

11G. We use the notation $O_{H_0\Phi}$ for the class of surfaces on which there are no nontrivial Φ-bounded functions of class H_0 (with respect to a given G with compact complement). As a counterpart of Theorems 11C and 11D we have a much simpler result:

Theorem. *$O_{H_0\Phi} = O_G$ for all Φ.*

The inclusion $O_{H_0\Phi} \subset O_{H_0B}$ is trivial, and it is equally evident that $O_{H_0B} = O_G$. For the converse we use the fact that $\Phi(t) \geqq at-b$ for some $a>0$. We conclude that any Φ-bounded function $v \in H_0$ satisfies $|v| \leqq u$ and hence $|v| \leqq Tu$ for some positive harmonic u on W. If W is parabolic u must be a constant and Tu is identically 0. Hence $O_G \subset O_{H_0\Phi}$, and the theorem is proved.

11H. For $\Phi(t)=t^2$ it is seen, exactly as in 11F, that $H_0D \subset H_0\Phi$. On the other hand, since $T1 \in H_0D$, $O_{H_0D} \subset O_G$. We have thus:

Theorem. *$O_{H_0B} = O_{H_0D} = O_G$.*

This result is in striking contrast to the fact that the inclusions $O_G \subset O_{HB} \subset O_{HD}$ are strict (§8).

§3. THE METHOD OF EXTREMAL LENGTH

No treatment of the theory of Riemann surfaces would be complete without a discussion of the method of extremal length. The roots of this method can be traced back to crude comparisons of area and length, and to more systematic exploitation of similar ideas, especially in the papers of H. Grötzsch [1]. The present sharp formulation of the method was first conceived by A. Beurling, in unpublished work, and further developed in joint publications with L. Ahlfors [1, 2, 3].

A complete treatment of extremal lengths would require a separate volume. In this presentation we are content to develop the theory to the point where we can use it as an important tool.

12. Extremal length

12A. Existing accounts of the method of extremal length deal only with the case of plane regions, but the generalization to Riemann surfaces is quite trivial. In this theory one considers a Riemann surface W and a family Γ of rectifiable curves on W. The object is to attach a numerical conformal invariant to this configuration, and to study its dependence on the curves and on the surface. There is no preferred metric on W, and therefore a curve in Γ has no definite length. However, by using all metrics which yield the same conformal structure, it is possible to introduce a sort of collective length, which by its very definition turns out to be conformally invariant. This collective length solves a simple extremal problem and has therefore been called the *extremal length* of Γ.

The importance of the method for the study of Riemann surfaces lies in the fact that certain extremal lengths can be computed in terms of classical conformal invariants.

12B. A *linear density* ρ on W is defined in terms of the local variables in such a manner that $\rho|dz|$ is invariant. We assume that $\rho \geqq 0$, and as far as regularity is concerned we shall require that ρ is lower semicontinuous, that is to say that $\rho(z_0) \leqq \underline{\lim}_{z \to z_0} \rho(z)$. This is a compromise, designed to simplify the proofs and at the same time permit sufficient generality.

Our assumptions imply that ρ can be integrated over any rectifiable arc γ. We define the ρ-length of γ by

$$L(\gamma, \rho) = \int_\gamma \rho|dz|.$$

The precise definition of the integral is clear, and we admit ∞ as a possible value.

More generally, the definition applies to any countable set of arcs, disjoint or not. We shall therefore allow Γ to be any nonempty collection of "curves" γ, where each γ consists of countably many arcs. It does not matter whether the end points of the arcs are included or not. The rectifiability can be dispensed with, for we can agree to set $L(\gamma, \rho) = \infty$ if γ is not rectifiable and $\rho > 0$ on γ.

Similarly, in view of the semicontinuity and invariance it is possible to define the ρ-area of W by

$$A(W, \rho) = \iint_W \rho^2 \, dxdy.$$

For the definition of the double integral one uses a partition of unity. It is a consequence of the lower semicontinuity that $A(W, \rho) > 0$ unless $\rho \equiv 0$. Indeed, as soon as $\rho > 0$ at a point we have $\rho > 0$ in a whole neighborhood.

12C. Given Γ, we determine the minimum ρ-length

$$L(\Gamma, \rho) = \inf_{\gamma \in \Gamma} L(\gamma, \rho).$$

This is to be compared with $A(W, \rho)$. Since $L(\Gamma, \rho)$ is homogeneous of degree 1 and $A(W, \rho)$ is homogeneous of degree 2 in ρ it is natural to focus our attention on $L(\Gamma, \rho)^2/A(W, \rho)$. We are led to the following definition:

Definition. *The extremal length of Γ in W is defined by*

$$\lambda_W(\Gamma) = \sup_\rho \frac{L(\Gamma, \rho)^2}{A(W, \rho)}$$

where ρ ranges over all lower semicontinuous densities which are not identically 0.

In the definition it is possible to use different normalizations, obtained by multiplying ρ with suitable constants. For instance, $\lambda_W(\Gamma)$ is equal to $\sup L(\Gamma, \rho)^2$ subject to the condition $A(W, \rho) = 1$. If $\lambda_W(\Gamma) > 0$ it is also the reciprocal of $\inf A(W, \rho)$ subject to $L(\Gamma, \rho) = 1$. Another convenient normalization is expressed by the condition $L(\Gamma, \rho) = A(W, \rho)$. Due to the different degrees of homogeneity it is always fulfilled for a suitable multiple of a given ρ, provided that $L(\Gamma, \rho) > 0$. With this normalization we can write

$$\lambda_W(\Gamma) = \sup L(\Gamma, \rho) = \sup A(W, \rho),$$

except when $\lambda_W(\Gamma) = 0$. It is this relation that justifies the name "extremal length", but it must be conceded that $\lambda_W(\Gamma)$ could with equal right be called an "extremal area".

12D. The conformal invariance of $\lambda_W(\Gamma)$ is an immediate consequence of the definition. Indeed, suppose that a conformal mapping takes W into W' and Γ into Γ'. The mapping allows us to regard any local variable z on W as a variable z' on W', and we can define a density ρ' by means of the relation $\rho|dz| = \rho'|dz'|$. We obtain $L(\gamma, \rho) = L(\gamma', \rho')$ for any pair of corresponding curves γ, γ'. Similarly, $A(W, \rho) = A(W', \rho')$, and it follows that $\lambda_W(\Gamma) = \lambda_{W'}(\Gamma')$.

In this connection we point out that $\lambda_W(\Gamma)$ depends on W only through the requirement that the curves of Γ be contained in W. In fact, let us suppose that $W' \subset W$ and that all $\gamma \in \Gamma$ are contained in W'. Let ρ be given on W, and denote the restriction of ρ to W' by ρ'. Then $L(\Gamma, \rho) = L(\Gamma, \rho')$ and $A(W, \rho) \geqq A(W', \rho')$. If $\rho' \not\equiv 0$ it follows that

$$\lambda_{W'}(\Gamma) \geqq \frac{L(\Gamma, \rho')^2}{A(W', \rho')} \geqq \frac{L(\Gamma, \rho)^2}{A(W, \rho)}.$$

The same is true if $\rho' \equiv 0$, $\rho \not\equiv 0$, for then $L(\Gamma, \rho) = 0$. We conclude that $\lambda_{W'}(\Gamma) \geqq \lambda_W(\Gamma)$. Conversely, suppose that ρ' is given on W', and set $\rho = \rho'$ on W', $\rho = 0$ on $W - W'$. Then ρ is lower semicontinuous, and $L(\Gamma, \rho) = L(\Gamma, \rho')$, $A(W, \rho) = A(W', \rho')$. This proves that $\lambda_W(\Gamma) \geqq \lambda_{W'}(\Gamma)$, and we have shown that the two extremal lengths are equal. The reference to W can hence be omitted and the notation simplified to $\lambda(\Gamma)$.

12E. We make use of the above remark in the formulation and proof of the following *comparison principle*:

Theorem. *If every $\gamma \in \Gamma$ contains a $\gamma' \in \Gamma'$, then $\lambda(\Gamma) \geqq \lambda(\Gamma')$.*

More precisely, we are comparing $\lambda_W(\Gamma)$ and $\lambda_{W'}(\Gamma')$ where W and W' are subregions of a larger Riemann surface, say W_0. Both extremal lengths can be evaluated with respect to W_0, and for any ρ on W_0 we have $L(\Gamma, \rho) \geqq L(\Gamma', \rho)$. These minimum lengths are compared with the same $A(W_0, \rho)$, and it follows that the extremal lengths satisfy the corresponding inequality.

12F. In addition to the comparison principle there are two basic *laws of composition* which we formulate as follows:

Theorem. *Let W_1 and W_2 be disjoint open sets in W. Let Γ_1, Γ_2 consist of curves in W_1, W_2 respectively, while Γ is a family of curves in W.*

(1) *If every $\gamma \in \Gamma$ contains a $\gamma_1 \in \Gamma_1$ and a $\gamma_2 \in \Gamma_2$, then*

$$\lambda(\Gamma) \geqq \lambda(\Gamma_1) + \lambda(\Gamma_2).$$

(2) *If every $\gamma_1 \in \Gamma_1$ and every $\gamma_2 \in \Gamma_2$ contains a $\gamma \in \Gamma$, then*

$$\frac{1}{\lambda(\Gamma)} \geqq \frac{1}{\lambda(\Gamma_1)} + \frac{1}{\lambda(\Gamma_2)}.$$

In (1) we may suppose that $\lambda(\Gamma_1), \lambda(\Gamma_2) > 0$, for otherwise the assertion is a special case of the comparison principle. Choose ρ_1 in W_1, ρ_2 in W_2, normalized by $L(\Gamma_1, \rho_1) = A(W_1, \rho_1)$, $L(\Gamma_2, \rho_2) = A(W_2, \rho_2)$. Define $\rho = \rho_1$ in W_1, $\rho = \rho_2$ in W_2, and $\rho = 0$ in $W - W_1 - W_2$. Then $L(\Gamma, \rho) \geqq L(\Gamma_1, \rho_1) + L(\Gamma_2, \rho_2)$ and $A(W, \rho) = A(W_1, \rho_1) + A(W_2, \rho_2) = L(\Gamma_1, \rho_1) + (\Gamma_2, \rho_2)$, and we obtain

$$\lambda(\Gamma) \geqq L(\Gamma_1, \rho_1) + L(\Gamma_2, \rho_2).$$

Here ρ_1 and ρ_2 can be chosen so that $L(\Gamma_1, \rho_1)$ and $L(\Gamma_2, \rho_2)$ are arbitrarily close to $\lambda(\Gamma_1)$ and $\lambda(\Gamma_2)$. It follows that $\lambda(\Gamma) \geqq \lambda(\Gamma_1) + \lambda(\Gamma_2)$.

As for property (2), we may assume that $\lambda(\Gamma) > 0$. Let ρ be given in W and normalized by $L(\Gamma, \rho) = 1$. Then $L(\Gamma_1, \rho) \geqq 1$, $L(\Gamma_2, \rho) \geqq 1$, and this implies

$$A(W, \rho) \geqq A(W_1, \rho) + A(W_2, \rho) \geqq \frac{1}{\lambda(\Gamma_1)} + \frac{1}{\lambda(\Gamma_2)}.$$

Hence $\lambda(\Gamma)^{-1}$, the greatest lower bound of $A(W, \rho)$, satisfies $\lambda(\Gamma)^{-1} \geqq \lambda(\Gamma_1)^{-1} + \lambda(\Gamma_2)^{-1}$.

13. Extremal distance

13A. The extremal lengths that are commonly used in function theory are of a rather special kind. In order that the invariance of the extremal length be of practical use it must be possible to determine the image of the family Γ under a conformal mapping without detailed knowledge of the mapping function. This will be the case if Γ is defined by topological conditions, for then the image of Γ will be known as soon as we know the topological nature of the mapping.

The most general problems of this kind are out of reach, but we shall treat a special case with numerous applications. A typical stituation is the following: Let E_1, E_2 be nonvoid disjoint sets on the Riemann surface W, and denote by Γ the family of connected arcs which join E_1 and E_2. We write $\lambda_W(\Gamma) = \lambda_W(E_1, E_2)$ and call this quantity the *extremal distance* of E_1 and E_2 relatively to W. This is a conformal invariant connected with the configuration formed by W and the sets E_1, E_2. An important property is its monotone dependence on this configuration. It follows indeed by the comparison principle (Theorem 12E) that $\lambda_W(E_1, E_2)$ increases if E_1, E_2 and W are shrunk.

13B. We make the special assumption that E_1, E_2 are closed. We can then modify any ρ by making it zero on $E_1 \cup E_2$. This does not change the ρ-distance between E_1, E_2, and the ρ-area can only decrease. For this reason $\lambda_W(E_1, E_2)$ depends only on $W - E_1 - E_2$, or rather on the components of $W - E_1 - E_2$ whose boundary meets both E_1 and E_2. As a consequence we could just as well have assumed from the start that E_1, E_2 lie on the boundary of W with respect to a compactification (see Ch. I, 36A).

With this motivation we adopt the following formal definition:

Definition. *Let M be a compactification of a Riemann surface W and let E_1, E_2 be two closed, disjoint, and nonvoid sets on $M - W$. The extremal distance $\lambda_W(E_1, E_2)$ is defined to be the extremal length $\lambda_W(\Gamma)$ of the family of open arcs on W which tend to E_1 in one direction and to E_2 in the other.*

For instance, E_1 and E_2 could be closed sets of boundary components.

As a slight generalization it is permissible to assume that W consists of several components, $W = W_1 \cup W_2 \cup \cdots \cup W_n$. Because the values of ρ on the different components are completely independent of each other it is easily seen that

$$\lambda_W^{-1} = \sum_{i=1}^{n} \lambda_{W_i}^{-1}$$

with the understanding that $\lambda_{W_i} = \infty$ if the boundary of W_i does not meet both E_1 and E_2.

13C. There are two special cases in which the extremal length is very easy to determine explicitly.

Theorem. *In a rectangle with sides of length a and b the extremal distance between the sides of length b is a/b. In an annulus $r_1 \leqq |z| \leqq r_2$ the extremal distance between the two contours is $\frac{1}{2\pi}\log(r_2/r_1)$.*

Let the rectangle R be given by $0 \leqq x \leqq a$, $0 \leqq y \leqq b$. If we use Euclidean length, $\rho = 1$, the distance between the vertical sides is a and the area is ab. Hence the extremal distance λ satisfies $\lambda \geqq a^2/ab = a/b$.

For arbitrary ρ, choose the normalization so that the ρ-distance between the sides is 1. Then

$$\int_0^a \rho(x, y)\,dx \geqq 1$$

for each y, and hence

$$\iint_R \rho\,dxdy \geqq b.$$

It follows that

$$\iint_R (a\rho - 1)^2\,dxdy \leqq a^2 \iint_R \rho^2\,dxdy - ab,$$

and we see that the ρ-area is at least b/a. By definition, this implies $\lambda \leqq a/b$, and we have proved the first part.

In the second case the choice $\rho = 1/r$ yields $\lambda \geqq \frac{1}{2\pi}\log(r_2/r_1)$. To prove the opposite inequality we cut the annulus along a radius so that it becomes the conformal image of a rectangle. The family of curves which join the circle $|z| = r_1$ and $|z| = r_2$ within the cut annulus is contained in the corresponding family for the full annulus. By the comparison principle the extremal length of the latter family is at most equal to the extremal length of the former. We may thus conclude, without computation, that $\lambda \leqq \frac{1}{2\pi}\log(r_2/r_1)$. The proof is complete.

13D. We will now assume that W is the interior of a compact bordered surface $\overline{W}$, and that E_1, E_2 consist of a finite number of arcs or full contours on the border. Under these conditions the construction in Ch. III, 5A can be used to prove the existence and uniqueness of a bounded harmonic function u which is 0 on E_1, 1 on E_2, and whose normal derivative vanishes on the remaining part of the border. We will prove:

Theorem. *The extremal distance between E_1 and E_2 is equal to* $1/D(u)$.

The assertion is true in the two cases that we have already considered. Indeed, for the rectangle, regarded as a bordered surface, we have $u=x/a$ and hence $D(u)=ab/a^2=b/a$. For the annulus $u=\log(r/r_1)/\log(r_2/r_1)$, and a simple calculation yields $D(u)=2\pi/\log(r_2/r_1)$.

In the general case it is easy to get the right inequality in one direction. Let us choose $\rho_0=|\text{grad}\,u|=(u_x^2+u_y^2)^{\frac{1}{2}}$, which is a linear density. The ρ_0-length of an arc γ from E_1 to E_2 is

$$\int_\gamma |\text{grad}\,u|\,|dz| \geqq \int_\gamma |du| \geqq \left|\int_\gamma du\right| = 1.$$

Hence $L(\Gamma, \rho_0)\geqq 1$, and since $A(W, \rho_0)=D(u)$ we may conclude that $\lambda(E_1, E_2)\geqq 1/D(u)$.

The opposite inequality would also be easy to prove if u had a single-valued conjugate u^*. Since this is not the case we must use different branches of u^* in different parts of W. Let us suppose that we can subdivide W into regions $W_1, \cdots, W_n$ with the following properties: (1) the W_k are disjoint and $\overline{W}=\cup\overline{W}_k$, (2) each W_k has a piecewise analytic boundary, (3) there exists a single-valued u^* in W_k such that $u+iu^*$ maps W_k on a rectangle with vertical sides on $u=0$ and $u=1$.

The extremal distance between E_1 and E_2 within W_k is $\lambda_k=1/D_k(u)$ where $D_k(u)$ denotes the Dirichlet integral over W_k. For a moment, let $\tilde{\lambda}$ denote the extremal distance of E_1 and E_2 with respect to the union of all W_k. By the comparison principle we obtain, on one hand, $\lambda\leqq\tilde{\lambda}$. On the other hand, by the second law of composition (Theorem 12F), $\tilde{\lambda}^{-1}\geqq \sum\lambda_k^{-1} = \sum D_k(u)=D(u)$. It follows that $\lambda\leqq 1/D(u)$ which is the desired conclusion.

13E. It remains to show that W can be subdivided in the desired manner. Consider any point $z_0\in W$. We define a conjugate function u^* in a neighborhood of z_0 and set $U=u+iu^*$. If $U'(z_0)\neq 0$ the mapping by U is one to one, and z_0 lies in a "rectangle", defined by inequalities $a'\leqq u\leqq a''$, $b'\leqq u^*\leqq b''$. If $U'(z_0)=0$ an examination of the local mapping shows, similarly, that z_0 is a common vertex of a finite number of such rectangles. In case z_0 lies on the border the same is true, except that z_0 will belong to a side or be a vertex of a single rectangle.

Because $\overline{W}$ is compact it can be covered by a finite number of these rectangles. By a simple construction, which we find needless to describe in detail, the covering can be refined to a pavement by nonoverlapping rectangles. They form a double array $\{Q_{ik}\}$ with the property that Q_{ik} and $Q_{i+1,k}$ have a common side which lies on a level curve $u = a_i$. For a fixed k the branches of u^* in Q_{ik} and $Q_{i+1,k}$ can be adjusted, step by step, so that they coincide on the common side. In this manner it becomes possible to define, in each $W_k = \cup_i Q_{ik}$, a single-valued u^* so that $u + iu^*$ maps W_k on a rectangle. The regions W_k are nonoverlapping and their closures fill out all of $\overline{W}$. These are the properties that were essential in the preceding proof. We have deliberately omitted many details that would have made the proof lengthy and less perspicuous.

13F. Although Theorem 13D gives the exact value of the extremal distance, it is seldom of great use for practical purposes. The following method is one which in many applications leads quickly to a good lower bound for the extremal distance.

Let ρ_0 be a given linear density on W which satisfies rather strong regularity conditions. For $t > 0$, let W_t be the set of all points in W whose ρ_0-distance from E_1 is less than t. We denote the area of W_t by $S(t)$. The relative boundary of W_t consists of points whose ρ_0-distance from E_1 is exactly t. Let us suppose that this relative boundary is made up of rectifiable curves, and denote its total ρ_0-length by $\Lambda(t)$. Under sufficiently strong smoothness conditions it is geometrically clear that $S'(t) = \Lambda(t)$. Because the method that we wish to develop is used only under extremely simple explicit circumstances there is no need to analyze the conditions that would lead to this conclusion. Instead we assume outright that it is known to hold.

Theorem. *Let t_0 be the shortest ρ_0-distance from E_1 to E_2. Then*

$$\lambda(E_1, E_2) \geqq \int_0^{t_0} \frac{dt}{\Lambda(t)}. \tag{10}$$

We denote by $t = t(z)$ the ρ_0-distance from z to E_1. Choose $\rho(z) = \rho_0(z)/\Lambda(t(z))$ if $t(z) < t_0$, $\rho(z) = 0$ if $t(z) \geqq t_0$. On a curve γ from E_1 to E_2 the distance t must pass through all values between 0 and t_0. We have $\rho_0 |dz| \geqq |dt|$ along the curve, and hence

$$\int_\gamma \rho |dz| \geqq \int_0^{t_0} \frac{dt}{\Lambda(t)}.$$

For the ρ-area we obtain, because $\Lambda(t)$ is constant along the relative boundary of W_t,

$$A(W, \rho) = \int_0^{t_0} \frac{dS(t)}{\Lambda(t)^2} = \int_0^{t_0} \frac{dt}{\Lambda(t)}.$$

It follows by the definition of λ that

$$\lambda \geqq \frac{L(\Gamma, \rho)^2}{A(W, \rho)} \geqq \int_0^{t_0} \frac{dt}{\Lambda(t)},$$

and (10) is proved.

Remark. Integrals of the type (10) were first used by L. Ahlfors [5] in discussions of the classical type problem for simply connected Riemann surfaces.

14. Conjugate extremal distance

14A. We continue to investigate the configuration that is formed by a compact bordered surface $\overline{W}$ and sets E_1, E_2 on the border, each consisting of a finite number of arcs. Let Γ^* be the class of curves that separate E_1 and E_2. More precisely, each $\gamma^* \in \Gamma^*$ is a countable union of arcs in W, and E_1, E_2 are in different components of $\overline{W}-\gamma^*$. The extremal length of the family Γ^* is called the *conjugate extremal distance* between E_1 and E_2. We denote it by $\lambda^*_W(E_1, E_2)$.

Theorem. $\lambda^*_W(E_1, E_2) = 1/\lambda_W(E_1, E_2)$.

If u is the same function as in Theorem 13D we have to show that $\lambda^*_W = D(u)$. Consider again $\rho_0 = |\text{grad } u|$. Because $\rho_0|dz| \geqq |du^*|$ it is seen that the ρ_0-length of any curve γ^* is at least equal to the flux of u along γ, and thus $\geqq D(u)$. It follows that $\lambda^* \geqq D(u)^2/D(u) = D(u)$.

To obtain the opposite inequality we use again the decomposition of $\overline{W}$ into rectangles Q_{ij} and consider (u, u^*) as Cartesian coordinates in each rectangle. Let ρ be defined with respect to the local variable $u+iu^*$, and suppose that $L(\Gamma^*, \rho)=1$. For each constant u the variation of u^* is equal to the flux $D(u)$. We obtain

$$\int_0^{D(u)} \rho\, du^* \geqq 1,$$

$$\iint_W \rho\, du\, du^* \geqq 1,$$

$$\iint_W [D(u)\,\rho-1]^2\, dudu^* \leqq D(u)^2 \iint_W \rho^2\, dudu^* - D(u),$$

and thus $A(W, \rho) \geqq 1/D(u)$. This implies the desired inequality $\lambda^* \leqq D(u)$.

14B. In some connections a slightly modified conjugate extremal distance is more appropriate. Let $\tilde{\Gamma}$ be the family of *connected* arcs or closed curves which separate E_1 and E_2. The extremal length of $\tilde{\Gamma}$, denoted by

$\tilde{\lambda}_W(E_1, E_2)$, is still another conformal invariant connected with the configuration W, E_1, E_2. Since $\tilde{\Gamma}$ is a subfamily of Γ^* we have necessarily $\tilde{\lambda} \geq \lambda^*$.

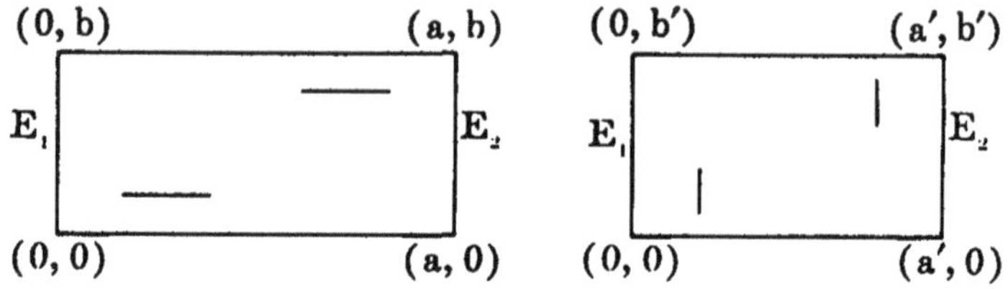

FIG. 9

We shall not try to determine $\tilde{\lambda}$ in the general case. However, Fig. 9 illustrates, in a typical situation, the relationship between $\lambda = a/b$, $\lambda^* = b/a$ and $\tilde{\lambda} = b'/a'$ (the two regions are conformally equivalent).

§4. MODULAR TESTS

When a Riemann surface is given in some concrete manner there arises the problem of determining whether it does or does not belong to a preassigned O-class, for instance O_G. It is desirable to base the decision on conditions, or tests, which are not merely theoretical, but can actually be applied to concrete cases. In this section we devise tests that depend on basic conformal invariants connected with compact bordered surfaces. It is not unreasonable to take the point of view that such invariants can be computed, or at least estimated, especially if the genus and the number of contours remain small.

15. Harmonic modules

15A. Let $\overline{W}$ be a compact bordered Riemann surface whose contours are divided into two classes α and β, sometimes referred to as inner and outer contours. We denote by $\lambda_{\alpha\beta}(\overline{W})$ the extremal distance between the inner and outer contours. According to Theorem 13D it has the value $1/D(u)$ where u is the harmonic function which is 0 on α and 1 on β. We shall call $\lambda_{\alpha\beta}(\overline{W})$ the *logarithmic module* of $\overline{W}$ between α and β. In some connections it is more convenient to consider the quantity

$$\mu_{\alpha\beta}(\overline{W}) = e^{\lambda_{\alpha\beta}(\overline{W})}$$

which will be called the *harmonic module*.

For an annulus $R_1 < |z| < R_2$ the logarithmic module is $\lambda = (1/2\pi) \log (R_2/R_1)$ and the harmonic module is $\mu = (R_2/R_1)^{1/2\pi}$.

15B. We consider an exhaustion of a given open surface W by a sequence of relatively compact G_n. It is assumed that $\overline{G}_n \subset G_{n+1}$, and the boundary of G_n is denoted by β_n. It is inconvenient for the applications to insist that the G_n be regular regions. We shall therefore allow β_n, the common boundary of G_n and $W-G_n$, to be merely piecewise analytic.

We denote by $\lambda_{\beta_1\beta_n}$ the extremal distance between β_1 and β_n with respect to $G_n-\bar{G}_1$, and by $\mu_{\beta_1\beta_n}$ the corresponding harmonic module. The modules are connected with the classification problem through the following simple theorem:

Theorem. *The surface W is parabolic if and only if* $\lim_{n\to\infty} \mu_{\beta_1\beta_n}=\infty$.

Even though the boundaries are merely piecewise analytic it is possible to regard $\bar{G}_n-G_1$ as a bordered Riemann surface. We denote by u_n the harmonic function which is 0 on β_1 and 1 on β_n. The condition $\mu_{\beta_1\beta_n}\to\infty$ is equivalent to $D(u_n)\to 0$, and hence to $u_n\to 0$. By Theorem 6C the surface is parabolic if and only if $u_n\to 0$.

15C. In most cases the theorem that we have just proved is not a practical criterion, for the regions $G_n-\bar{G}_1$ are likely to become increasingly complicated. However, we can derive more useful tests by combining the theorem with estimates of the modules.

Write $\lambda_n=\lambda_{\beta_n\beta_{n+1}}$ and $\mu_n=e^{\lambda_n}$. The first law of composition (Theorem 12F) yields $\lambda_{\beta_1\beta_n}\geqq\lambda_1+\lambda_2+\cdots+\lambda_{n-1}$. We find:

Theorem. *W is parabolic if, for some exhaustion,*

$$\prod_1^\infty \mu_n = \infty. \tag{11}$$

When this theorem is applied it is frequently the case that $G_{n+1}-\bar{G}_n$ has several components. As we have already remarked in 13B the module λ_n is then connected with the modules λ_{nj} of the components by the relation $\lambda_n^{-1}=\sum_j \lambda_{nj}^{-1}$. If the exhaustion is suitably chosen the genus and number of contours of the components of $G_{n+1}-\bar{G}_n$ remain fairly small. It is for this reason that (11) may be regarded as an effective criterion.

Condition (11) is not necessary. One can show, in fact, that there are always exhaustions with a convergent product $\prod_1^\infty \mu_n$ (L. Sario [14], H. Wittich [7], p. 88).

15D. In applications a seemingly weaker form of Theorem 15C is often preferred. Given an exhaustion $\{G_n\}$ we suppose that each boundary β_n is surrounded by an open set B_n which is contained in G_{n+1} and in the complement of G_{n-1}. We denote by a_n and b_n the "inner" and "outer" boundaries of B_n; more precisely, b_n is the boundary of $G_n\cup B_n$, a_n is the boundary of $(W - G_n) \cup B_n$. If μ_n' is the harmonic module between a_n and b_n we have:

Corollary. *W is parabolic if*

$$\prod_1^\infty \mu_n' = \infty. \tag{12}$$

For the proof we need only observe that the harmonic module between β_{2m-1} and β_{2m+1} is trivially $\geqq \mu_{2m}'$, while the module between β_{2m} and β_{2m+2} is $\geqq \mu_{2m+1}'$. The hypothesis (12) implies that either $\Pi \mu_{2m}' = \infty$ or $\Pi \mu_{2m+1}' = \infty$, and the conclusion follows on applying Theorem 15C either to the exhaustion $\{G_{2m}\}$ or to the exhaustion $\{G_{2m+1}\}$.

16. Analytic modules

16A. We have just found a sufficient condition for a surface to be of class O_G. We shall now derive an analogous condition for the class O_{AD}.

For this purpose we consider again a compact bordered surface $\overline{W}$ or a finite union of such surfaces whose contours are divided into α and β. We introduce a new module $k_{\alpha\beta}(\overline{W})$, called the *analytic module*.

If f is analytic on $\overline{W}$ the integrals

$$\frac{i}{2}\int_\alpha f\,d\bar{f},\quad \frac{i}{2}\int_\beta f\,d\bar{f}$$

are evidently real, and their difference is equal to the Dirichlet integral

$$D(f) = \frac{i}{2}\int_{\beta-\alpha} f\,d\bar{f}. \tag{13}$$

We restrict f to the class of functions for which

$$\frac{i}{2}\int_\alpha f\,d\bar{f} > 0, \tag{14}$$

and we set

$$k_{\alpha\beta}(\overline{W}) = \inf \frac{\int_\beta f\,d\bar{f}}{\int_\alpha f\,d\bar{f}}$$

when f ranges over this class. It is not difficult to show that there are indeed functions which satisfy (14). At present this is not important, for if the class of competing functions were empty we could set $k_{\alpha\beta} = \infty$.

Because of (14) it is clear that $k_{\alpha\beta} \geqq 1$. Moreover, $k_{\alpha\beta}$ is evidently invariant under conformal mappings. The invariant $k_{\alpha\beta}$ was first considered by A. Pfluger [2].

Geometrically, the problem is to minimize the area enclosed by $f(\beta)$ when the area enclosed by $f(\alpha)$ is given. However, this simplified statement is meaningful only when f is univalent, and there is no guarantee that the extremal function will be univalent.

16B. The nature of the invariant $k_{\alpha\beta}$ has never been successfully investigated. In fact, there is not even a reasonable conjecture as to the characteristic properties of the extremal function. As a consequence, tests based on $k_{\alpha\beta}$ have no practical value. Nevertheless, the analogues of Theorems 15B and 15C that we are going to prove show that $k_{\alpha\beta}$ is a theoretically important invariant connected with the class AD.

In the following statement we refer again to an exhaustion $\{G_n\}$. We denote by $k_{\beta_1\beta_n}$ and k_n the analytic modules associated with $\bar{G}_n - G_1$ and $\bar{G}_{n+1} - G_n$.

Theorem. *W is of class O_{AD} if* $\lim_{n\to\infty} k_{\beta_1\beta_n} = \infty$, *and this condition is fulfilled if*

$$\prod_1^\infty k_n = \infty.$$

Note that the analogy with Theorem 15B is not complete, for we do not assert that the first condition is also necessary.

To prove the theorem, suppose that f is analytic and nonconstant on W. Then

$$D_{G_n}(f) = \frac{i}{2}\int_{\beta_n} f\,d\bar{f},$$

and the definition of analytic module implies

$$D_{G_n}(f) \geqq k_{\beta_1\beta_n} D_{G_1}(f).$$

Here $D_{G_1}(f) > 0$, and if $k_{\beta_1\beta_n} \to \infty$ we conclude that $D(f)$ cannot be finite. The second assertion follows by the obvious inequality

$$k_{\beta_1\beta_n} \geqq k_1 k_2 \cdots k_{n-1}.$$

16C. As already remarked, Theorem 16B is purely theoretical. We can, however, make practical applications by virtue of an inequality that connects k_n and μ_n.

Lemma. *The analytic and harmonic modules satisfy the inequality* $k_{\alpha\beta} \geqq \mu_{\alpha\beta}^{4\pi}$.

For the purpose of orientation we remark that the equality $k_{\alpha\beta} = \mu_{\alpha\beta}^{4\pi}$ holds for a circular annulus, and hence for any doubly connected region.

We denote again by u the harmonic function which is 0 on α, 1 on β, and by f an analytic function. Let $\gamma(t)$ be the level curve $u = t$, $0 \leqq t \leqq 1$.

Although $\gamma(t)$ may consist of several closed curves, we can introduce a parameter $y=u^*$ which runs from 0 to $d=D(u)$. We write:

$$m(t) = \frac{i}{2}\int_{u=t} f\,d\bar{f} = \frac{1}{2}\int_0^d f\bar{f}'\,dy\,,$$

where the differentiation is with respect to the local variable $u+iu^*$.

The Fourier developments of f and f' have the form

$$f \sim \sum_{-\infty}^{\infty} c_n e^{n\frac{2\pi i}{d}y}$$

$$f' \sim \frac{2\pi}{d}\sum_{-\infty}^{\infty} nc_n e^{n\frac{2\pi i}{d}y},$$

and Parseval's formula yields

$$m(t) = \pi\sum_{-\infty}^{\infty} n|c_n|^2$$

$$\int_0^d |f'|^2\,dy = \frac{4\pi^2}{d}\sum_{-\infty}^{\infty} n^2|c_n|^2.$$

On comparing these equations we find that

$$m(t) \leqq \frac{d}{4\pi}\int_0^d |f'|^2\,dy.$$

On the other hand

$$D(t) = m(t)-m(0)$$

represents the Dirichlet integral of f over the area $u<t$. It is clear that

$$D'(t) = \int_0^d |f'|^2\,dy,$$

and we have thus proved the inequality

$$m(t) \leqq \frac{d}{4\pi}\,m'(t).$$

Here $m(t) \geqq m(0) \geqq 0$, by hypothesis. When this differential inequality is integrated between $t=0$ and $t=1$ we obtain

$$\log\frac{m(1)}{m(0)} \geqq \frac{4\pi}{d} = 4\pi\lambda_{\alpha\beta},$$

and thus

$$k_{\alpha\beta} = \inf\frac{m(1)}{m(0)} \geqq e^{4\pi\lambda_{\alpha\beta}} = \mu_{\alpha\beta}^{4\pi}.$$

16D. Direct application of the lemma to the modules k_n leads to nothing new, for if $\Pi\mu_n$ diverges we have the stronger conclusion that W is parabolic. The way to obtain a more significant result is to apply the lemma to the components G_{nj} of each $G_{n+1}-G_n$.

Denote the analytic module of G_{nj} by k_{nj}. We show that k_n, in contrast to μ_n, is simply equal to $\min_j k_{nj}$.

If β_{nj}, $\beta_{n+1,\,j}$ are the inner and outer contours of G_{nj} we have

$$\frac{i}{2}\int_{\beta_n} f\,d\bar{f} = \sum_j \frac{i}{2}\int_{\beta_{nj}} f\,d\bar{f}$$

$$\frac{i}{2}\int_{\beta_{n+1}} f\,d\bar{f} = \sum_j \frac{i}{2}\int_{\beta_{n+1,\,j}} f\,d\bar{f}.$$

By the definition of k_{nj},

$$\frac{i}{2}\int_{\beta_{n+1,j}} f\,d\bar{f} \geqq k_{nj}\left(\frac{i}{2}\int_{\beta_{nj}} f\,d\bar{f}\right)$$

whenever the integral on the right is positive. If the integral is negative or zero we use

$$\frac{i}{2}\int_{\beta_{n+1,j}} f\,d\bar{f} \geqq \frac{i}{2}\int_{\beta_{nj}} f\,d\bar{f}.$$

In the first case we can replace k_{nj} by $\min_j k_{nj}$. The same inequality will hold in the second case, for $\min_j k_{nj}$ is known to be >1, and we are multiplying by a negative factor. On adding the inequalities it follows that

$$\frac{i}{2}\int_{\beta_{n+1}} f\,d\bar{f} \geqq (\min_j k_{nj})\,\frac{i}{2}\int_{\beta_n} f\,d\bar{f}.$$

This is evidently the best inequality that can be proved, for we can choose f to be identically zero in all G_{nj} except the one for which k_{nj} is a minimum. We conclude that $k_n = \min_j k_{nj}$, as asserted.

On using the lemma we have $\min_j k_{nj} = (\min_j \mu_{nj})^{4\pi}$ where μ_{nj} is the harmonic module of G_{nj}. We obtain:

Theorem. *W is of class O_{AD} as soon as there exists an exhaustion with*

$$\prod_{n=1}^{\infty} (\min_j \mu_{nj}) = \infty.$$

As in the case of Theorem 15C we are free to replace $\min_j \mu_{nj}$ by $\min_j \mu'_{nj}$ where μ'_{nj} refers to a region B_n, determined as in 15D.

Remark. The criterion in Theorem 16D was first derived by L. Sario [1, 14]. On the basis of Sario's paper A. Pfluger [2] introduced the module

k and proved Lemma 16C. Earlier, a weaker inequality had been established by R. Nevanlinna [11]. The method of proof in Lemma 16C goes back to T. Carleman [1].

§5. EXPLICIT TESTS

The value of the modular tests depends on our ability to evaluate or at least estimate the modules. This practical question has a meaning only if the surface W is given explicity in one way or another. Our next aim is to derive tests which depend on the given data in an enumerative rather than quantitative manner.

17. Deep coverings

17A. As soon as a Riemann surface W is given in a concrete manner it is possible to construct an explicit covering by parametric disks with the property that any compact set meets only a finite number of the disks. We shall require that this covering is in a certain sense sufficiently deep.

Let the parametric disks be denoted by Δ_i. In order to introduce a hypothesis which guarantees that the overlappings are sufficiently deep we associate with each Δ_i a fixed local variable z_i, that is to say, a fixed conformal mapping of Δ_i on $|z_i| < 1$. We choose once and for all a positive number $q < 1$ and let Δ_i' denote the inverse image of $|z_i| < q$ under the parameter mapping. We shall say that the Δ_i form a *deep covering* if the Δ_i' form a covering, that is, if every point of W belongs to at least one Δ_i'.

17B. For any given deep covering choose a disk Δ_0 as the disk of generation 0. A disk $\Delta_i \neq \Delta_0$ will be said to be of generation 1 if and only if $\Delta_i \cap \Delta_0 \neq 0$. Inductively, Δ_i will be of generation n if it is not of lower generation and if $\Delta_i \cap \Delta_j \neq 0$ for some Δ_j of generation $n-1$.

It is readily seen that every Δ_i belongs to some generation. In fact, the union of all Δ_i that belong to a generation is an open set, and so is the union of the Δ_i that do not belong to any generation. These sets are disjoint and comprise all of W. Because of the connectedness the second set must be empty.

17C. We shall denote by W_n the union of all disks Δ_i of generation $\leq n$. Each W_n is relatively compact and its boundary consists of analytic arcs and, exceptionally, isolated points. We are going to apply Theorems 15C and 16D to the exhaustion defined by $G_n = W_{2n}$; in order to conform with our earlier conventions it is agreed that isolated boundary points of W_{2n} will be added to G_n.

The following remark will play an essential part in the proof that follows: a Δ_i that meets the boundary of G_n is exactly of generation $2n+1$.

17D. Let ν_k denote the number of disks of generation k. The following criterion will be proved:

Theorem. *A Riemann surface W is parabolic if there exists a deep covering which satisfies*

$$\sum_{k=1}^{\infty} \frac{1}{\nu_{2k-1}+\nu_{2k}+\nu_{2k+1}} = \infty. \tag{15}$$

If $\Delta_i \subset G_{n+1}-\bar{G}_n$ it is of generation $2n+2$ or, occasionally, of generation $2n+3$, namely if it contains an isolated boundary point of W_{2n+2}. If it meets the boundary β_n of G_n it is of generation $2n+1$, and if it meets β_{n+1} it is of generation $2n+3$. Hence all Δ_i that meet $\bar{G}_{n+1}-G_n$ are of generation $2n+1$, $2n+2$ or $2n+3$.

Consider all Δ_i of the generations ν_{2n+1}, ν_{2n+2}, ν_{2n+3}. At any point in Δ_i, let z be a local variable, and set $\rho_i = \left|\frac{dz_i}{dz}\right|$. Note that the z_i are fixed functions defined in Δ_i, independently of the local variable. The ρ_i are linear densities in Δ_i, and we can define a linear density ρ throughout $G_{n+1}-\bar{G}_n$ by setting, at a given point and for a given local variable at that point, $\rho = \max \rho_i$, the maximum being with respect to all ρ_i whose domain Δ_i contains the point under consideration. To see that ρ is lower semicontinuous, suppose that $\rho(p_0) = \rho_i(p_0)$, $p_0 \in \Delta_i$. We use the same local variable throughout a neighborhood $V(p_0) \subset \Delta_i$. Then $\rho(p) \geqq \rho_i(p)$ for $p \in V(p_0)$ and we conclude that $\varliminf_{p\to p_0} \rho(p) \geqq \lim_{p\to p_0} \rho_i(p) = \rho(p_0)$, so that ρ is indeed lower semicontinuous.

Let γ be an arc in $\bar{G}_{n+1}-G_n$ that connects β_n and β_{n+1}. Suppose that the end point on β_n is contained in Δ_i' (see 17A) and the end point on β_{n+1} in Δ_j'. Then Δ_i is of generation $2n+1$ and Δ_j is of generation $2n+3$. Hence Δ_i and Δ_j are disjoint. It follows that γ contains a subarc which joins the boundary of Δ_i' to that of Δ_i. The ρ_i-length of this subarc is at least $1-q$, and the ρ-length is greater than or equal to the ρ_i-length. We conclude that $L(\gamma, \rho) \geqq 1-q$ and hence $L(\Gamma, \rho) \geqq 1-q$ where Γ is the family of curves used to define the module λ_n.

On the other hand the ρ-area of $G_{n+1}-\bar{G}_n$ is at most equal to the sum of the ρ_i-areas of all disks that meet $\bar{G}_{n+1}-G_n$, for $\rho^2 \leqq \sum \rho_i^2$ at each point. Each disk has the ρ_i-area π. Hence $A(G_{n+1}-\bar{G}_n, \rho) \leqq (\nu_{2n+1}+\nu_{2n+2}+\nu_{2n+3})\pi$, and together with the estimate $L(\Gamma, \rho) \geqq 1-q$ we obtain

$$\lambda_n \geqq \frac{(1-q)^2}{(\nu_{2n+1}+\nu_{2n+2}+\nu_{2n+3})\pi}.$$

Condition (15) will therefore imply the divergence of $\sum \lambda_n$, and by Theorem 15C the surface must be parabolic.

Remark. A result of the type of Theorem 17D was first proved by R. Nevanlinna [6].

17E. There is a similar application of Theorem 16D. Consider the union of all disks Δ_i of generations $k-1$, k and $k+1$. Let δ_k denote the maximum number of disks in a component of this union.

Theorem. *W is of class O_{AD} if there exists a deep covering which satisfies*

$$\sum_{k=1}^{\infty} \frac{1}{\delta_{2k}} = \infty. \tag{16}$$

We have to estimate the modules of the components of $G_{n+1}-\bar{G}_n$ which have part of their boundary on β_n and part on β_{n+1}. The estimate $L(\Gamma, \rho) \geqq 1-q$ remains valid. The disks Δ_i which meet a component form a connected set. Therefore their number is at most δ_{2n+2}. We obtain

$$\lambda_{nj} \geqq \frac{(1-q)^2}{\pi\delta_{2n+2}},$$

and by Theorem 16D condition (16) implies that W is of class O_{AD}.

17F. A weaker form of Theorem 17D is sometimes easier to apply. Let $N_k = \nu_0 + \nu_1 + \cdots + \nu_k$ be the total number of disks of generation $\leqq k$. Then

$$\sum_{i=1}^{k} (\nu_{2i-1} + \nu_{2i} + \nu_{2i+1}) \leqq 2N_{2k+1}$$

and hence, by Cauchy's inequality,

$$\sum_{i=1}^{k} \frac{1}{\nu_{2i-1} + \nu_{2i} + \nu_{2i+1}} \geq \frac{k^2}{2N_{2k+1}}.$$

By this estimate we find:

Corollary. *W is parabolic if $N_k = o(k^2)$.*

18. Tests by triangulation

18A. It is desirable to possess tests which depend only on the incidence relations in a triangulation. If the triangles are arbitrary it is clear that strong distortion could occur, and it would be impossible to draw any conclusions of a general nature. We must therefore introduce a special assumption which guarantees a certain degree of rigidity.

A very simple condition of this sort is obtained by considering the quadrilaterals that are formed by pairs of adjacent triangles. It was shown in Theorem 13C that the extremal distances between opposite sides in a quadrilateral are reciprocal numbers m and $1/m$. We shall assume that for all quadrilaterals formed by adjacent triangles

$$\frac{1}{K} \leqq m \leqq K \tag{17}$$

with a fixed $K < \infty$. When this condition is fulfilled we say that the triangulation has *bounded distortion.*

18B. As in 17B we begin by defining generations of triangles. We choose an initial vertex as the vertex of generation 0. Inductively, we say that a vertex is of generation n if it is not of lower generation and if it is joined by a side to a vertex of generation $n-1$. The generation of a side or of a triangle is defined as the minimum generation of the end points or the vertices.

Let P_n be the union of all closed triangles of generation $\leqq n$. The boundary of P_n consists entirely of vertices and sides which are exactly of generation $n+1$. Indeed, each vertex on the boundary belongs to at least one triangle of generation $\leqq n$ and at least one of generation $\geqq n+1$; this is possible only so that it is exactly of generation $n+1$.

We choose G_n as the interior of P_n. Theorems 15C and 16D are applicable to the exhaustion $\{G_n\}$.

18C. Let γ be an arc in $\bar{G}_{n+1} - G_n$ which joins β_n to β_{n+1}. We can easily convince ourselves that γ must pass at least once between opposite sides of a quadrilateral formed by a pair of adjacent triangles (Fig. 10). Indeed,

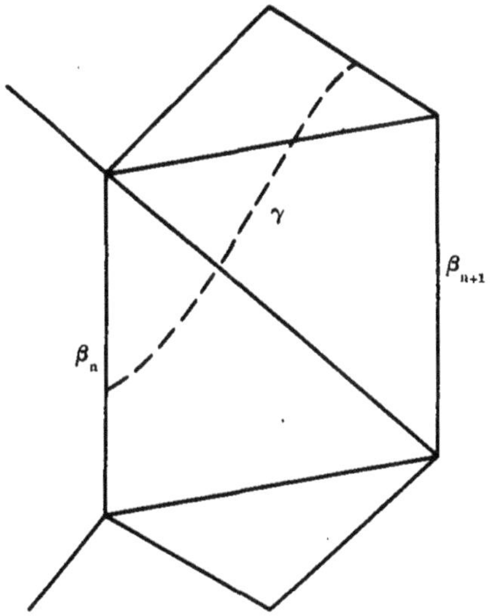

FIG. 10

if this were not so γ would necessarily remain inside the open star of a vertex of generation $n+1$, and hence could not attain β_{n+1}. We remark, in view of a future application (19C), that this conclusion remains valid even if the triangulation is degenerate in the sense that the two end points of a side might coincide, or that two distinct sides could have the same end points.

Each quadrilateral M_i can be mapped conformally on a rectangle $0 \leqq \mathrm{Re}\, z_i \leqq m_i$, $0 \leqq \mathrm{Im}\, z_i \leqq 1$. We may choose the mapping so that $m_i \leqq 1$, and if (17) is fulfilled we have also $m_i \geqq 1/K$. For a given choice of the local variable z we write again $\rho_i = \left|\frac{dz_i}{dz}\right|$ and $\rho = \max \rho_i$, where the maximum is with respect to all ρ_i that are defined at the point in question. By what we have said it is clear that the ρ-length of any γ is $\geqq 1/K$.

On the other hand, the ρ-area of $G_{n+1} - \bar{G}_n$ is at most equal to the sum of the ρ_i-areas of all quadrilaterals M_i which meet the region. Each ρ_i-area is $\leqq 1$. An M_i that meets the region between β_n and β_{n+1} contains at least one triangle of generation $n+1$, and such a triangle belongs to three different M_i. If σ_k denotes the number of triangles of generation k, it follows that the total ρ-area is $\leqq 3\sigma_{n+1}$. The module λ_n of $G_{n+1} - \bar{G}_n$ satisfies

$$\lambda_n \geqq \frac{1}{3K^2\sigma_{n+1}},$$

and we have proved:

Theorem. *W is parabolic if*

$$\sum_{k=1}^{\infty} \frac{1}{\sigma_k} = \infty$$

for a triangulation with bounded distortion.

18D. The proof works equally well for the class O_{AD}. We break the set of triangles of generation k into its strongly connected components, that is, we count two triangles to the same component if they can be connected by a chain of triangles in the same generation in such a way that consecutive triangles have a common side. If τ_k denotes the maximum number of triangles in a component we obtain:

Theorem. *W is of class O_{AD} if*

$$\sum_{k=1}^{\infty} \frac{1}{\tau_k} = \infty$$

for a triangulation with bounded distortion.

§6. APPLICATIONS

We are now going to apply some of the preceding theorems to derive a few results of special interest. In these examples we do not strive for maximum generality. The purpose is primarily to illustrate the methods.

19. Regular covering surfaces

19A. We consider the general case of a Riemann surface W which is given as a regular covering surface (Ch. I, 14D) of a closed surface W_0. If W_0 is a sphere W is identical with W_0, and if W_0 is a torus W is either a cylinder, conformally equivalent to the punctured plane, or the plane. Since these cases are trivial we are going to assume that W_0 is of genus $g>1$.

The fundamental group $\mathscr{F}$ of W_0 can be generated by $2g$ elements $a_1, b_1, \cdots, a_g, b_g$ which satisfy the relation $a_1b_1a_1^{-1}b_1^{-1}\cdots a_gb_ga_g^{-1}b_g^{-1}\approx 1$. The regular covering surfaces are associated with the subgroups $\mathscr{D}$, or rather sets of conjugate subgroups, of $\mathscr{F}$ (Ch. I, 17D).

19B. We represent W_0 as a canonical polygon P with identified sides, as illustrated in Fig. 11. As origin for the fundamental group we choose the point on W_0 which is represented by the vertices of P. The covering surface W is paved by copies cP of P, associated with the elements $c \in \mathscr{F}$, and $cP = c'P$ if and only if $cc'^{-1} \in \mathscr{D}$.

In detail, the points of W are given by paths from the origin, with the understanding that γ_1, γ_2 determine the same point if and only if the homotopy class of $\gamma_1\gamma_2^{-1}$ belongs to $\mathscr{D}$. We agree that cP shall consist of all points represented by paths $c\gamma$ where γ begins at the initial point of the side marked a_1 and stays in the interior of P, except for the terminal point which may lie on the perimeter. The ploygon $1P$ can be identified with P. One of its vertices is represented by $c_0 = 1$, the others by $c_1 = a_1$, $c_2 = a_1b_1$, $c_3 = a_1b_1a_1^{-1}$ etc. up to $c_{4g-1} = a_1b_1a_1^{-1}b_1^{-1}\cdots a_gb_ga_g^{-1}$. As points of W these vertices are not necessarily distinct.

19C. We make the simplifying assumption that W is a normal covering surface (Ch. I, 19B), that is to say, that $\mathscr{D}$ is a normal subgroup. This has the effect that cP remains unaltered if, in the expression of c through the generators, we suppress or insert factors that represent elements of $\mathscr{D}$. Indeed, if $d \in \mathscr{D}$, then $cdc'P = (cdc^{-1})cc'P = cc'P$. In the sequel, all equations are to be interpreted modulo $\mathscr{D}$.

Let us determine the conditions under which two polygons cP and $c'P$ have a common vertex. Obviously, it is necessary and sufficient that $cc_k = c'c_l$ for some choice of c_k and c_l. Hence we must have $c' = cc_kc_l^{-1}$.

In order that cP and $c'P$ have a common side it must be possible to find c_k and c_l so that $cc_k = c'c_l$ and $cc_{k+1} = c'c_{l-1}$. These conditions yield $c_k c_l^{-1} = c_{k+1} c_{l-1}^{-1}$, or $c_{k+1}^{-1} c_k = c_{l-1}^{-1} c_l$. On writing, for a moment, $c_i = c_{i-1} e_i$ we have thus $e_{k+1}^{-1} = e_l$. This will happen if $k+1$, l is either a pair $4n+1$, $4n+3$ or $4n+2$, $4n+4$. One finds that $c^{-1}c'$ must be equal to $c_{4n+2} c_{4n+1}^{-1}$, $c_{4n+3} c_{4n+2}^{-1}$, or their inverses.

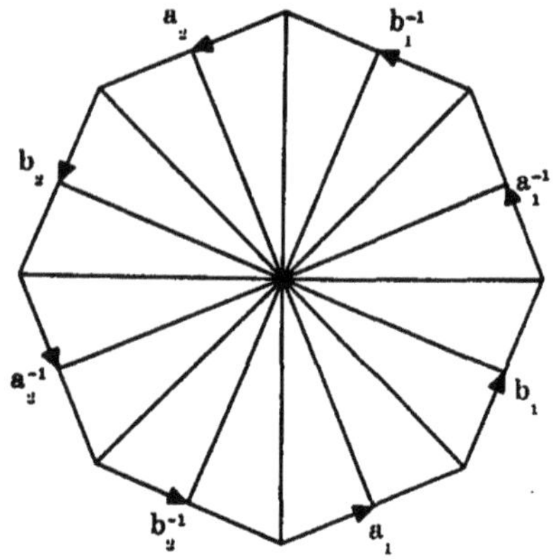

Fig. 11

We triangulate P as suggested in Fig. 11 and all copies cP in the corresponding manner so that each triangle on W projects into a triangle on W_0. In the resulting triangulation on W there are then only a finite number of conformally inequivalent quadrilaterals. Therefore the triangulation is automatically one of bounded distortion (18A). It follows that Theorems 18C and 18D are applicable to this triangulation.

We remark that the triangulation may be degenerate, for many vertices become identified on W. It was pointed out in 18C that this does not impair the validity of the theorems.

19D. For the purpose of counting generations of triangles we take the center of P as the vertex of generation 0. In the present case it is easier, and for most purposes equivalent, to count generations of polygons cP. On choosing P as the polygon of generation 0 we say that $cP \neq P$ is of generation 1 if cP and P have at least a vertex in common. Generally, cP is of generation n if it is not of lower generation and has a common vertex with a polygon of generation $n-1$.

We have seen that two polygons have a common vertex if the corresponding paths differ by $c_k c_l^{-1}$. Therefore, the generation of cP is the least n such that it is possible to write $c = d_1 \cdots d_n$ where each d_i has the form $c_k c_l^{-1}$. We shall also refer to n as the *length* of c.

19E. If cP is of generation n it is quite evident that the triangles in cP are of generation $\leqq 2n$. Conversely, consider a triangle of generation m. One vertex of the triangle can be joined to the center of P by a path consisting of m segments whose end points are either centers of polygons, vertices, or midpoints of sides. Because the path is minimal there are no segments between a vertex and a midpoint. Therefore, every second end point is a center, and the polygons with consecutive centers have at least a vertex in common. If m is even we conclude that a triangle of generation m is contained in a polygon of generation $\leqq \frac{m}{2}$, and if m is odd it is contained in a polygon of generation $\leqq \frac{m+1}{2}$.

We have shown that a polygon of generation n contains only triangles of generation $2n-1$ and $2n$. For the purpose of comparison, let $\bar{\sigma}_n$ be the number of polygons of generation n while σ_n still denotes the number of triangles in the generation n. There are $8g$ triangles in each polygon. Therefore $\sigma_{2n} \leqq 8g\bar{\sigma}_n$, and the divergence of $\sum \frac{1}{\bar{\sigma}_n}$ implies the divergence of $\sum \frac{1}{\sigma_n}$.

Similarly, let $\bar{\tau}_n$ be the maximum number of polygons in a component of generation n (two polygons are in the same component if they can be joined by a chain of polygons, all of the same generation, such that consecutive polygons are adjacent along a side). Two adjacent triangles are contained in the same or adjacent polygons. Therefore, a component formed by triangles of generation $2n$ is contained in a component of polygons of generation n. This proves that $\tau_{2n} \leqq 8g\bar{\tau}_n$, and in order to conclude that $W \in O_{AD}$ it is thus sufficient to show that $\sum \frac{1}{\bar{\tau}_n} = \infty$.

19F. We have reduced the determination of $\bar{\sigma}_n$ and $\bar{\tau}_n$ to a purely combinatorial problem. As an example, we investigate the following situation:

Let the integers $1, \cdots, g$ be divided into two sets I, J, and denote by $\mathcal{D}(I, J)$ the smallest normal subgroup of $\mathcal{F}$ which contains all a_i, $i \in I$, and all $a_j b_j a_j^{-1} b_j^{-1}$, $j \in J$ (in the sequel it is tacitly understood that any index denoted by i belongs to I, and any index j to J). If J is empty the corresponding covering surface is called a *Schottky covering surface*; for arbitrary I, J we speak of a covering surface of Schottky type. We are going to prove:

Theorem. *Every covering surface of Schottky type is of class O_{AD}.*

The quotient group $\mathcal{F}/\mathcal{D}(I, J)$ is generated by b_i and a_j, b_j; a_j and b_j commute. Nevertheless, if $g > 1$ the group is not Abelian; for instance, b_1 and b_2 do not commute.

The vertices of P, that is, the c_k, are seen to form quadruples $1, b_i, b_i, 1$ and $a_j, a_jb_j, b_j, 1$. The distinct c_k are thus $1, b_i, a_j, b_j$ and a_jb_j. Furthermore, on applying the result in 19C we find that cP and $c'P$ are adjacent along a side if and only if $c' = cb_i^{\pm 1}$, $ca_j^{\pm 1}$ or $cb_j^{\pm 1}$.

We have to investigate the circumstances in which adjacent polygons belong to the same generation. Let f_k denote elements $b_i^{\pm 1}, a_j^{\pm 1}, b_j^{\pm 1}$. We wish to compare cP with its direct and indirect neighbors determined by cf_1, cf_1f_2, etc. For the moment we assume that the last element in c does not commute with f_1. We contend that c and $cf_1f_2f_3f_4$ cannot have the same length (provided that $f_1f_2f_3f_4$ cannot be reduced by cancellations). Indeed, because f_1 does not commute with the preceding factor the lengths can be equal only if $f_1f_2f_3f_4$ can be absorbed in the last factor of $c = d_1 \cdots d_n$. But an equation $d_nf_1f_2f_3f_4 = d_n'$ is impossible, for the expression on the left has at least five factors which cannot be cancelled against each other, and the one on the right has at most four (for instance, $d_n' = (a_ib_i)(a_jb_j)^{-1}$ has four factors).

If c ends with $b_i^{\pm 1}$ this preliminary consideration shows that cP belongs to a component which consists of at most $(4g)^3 + 1$ polygons.

Let us now determine the length of a c-path $a_j^rb_j^s$. Clearly, a minimal representation will involve only factors a_j, b_j, a_jb_j, $a_jb_j^{-1}$ and their inverses. If we represent $a_j^rb_j^s$ by the lattice point (r, s) it becomes geometrically evident that the paths of length $\leqq n$ correspond to lattice points in the square $|r| \leqq n$, $|s| \leqq n$. In other words, the length of $a_j^rb_j^s$ is equal to $\max(|r|, |s|)$.

There are $8n$ lattice points on the rim of the square. Neighboring points correspond to adjacent polygons. Hence all $8n$ polygons corresponding to these lattice points belong to the same component. In addition, each polygon has at most $64g^3$ direct or indirect neighbors (formed with $f_1 \neq a_j^{\pm 1}, b_j^{\pm 1}$) which may belong to the same component. We conclude that the total number of polygons in the component is at most $8n(64g^3 + 1)$.

It is easy to generalize this reasoning to a path $ca_j^rb_j^s$ where $|r| + |s| > 0$ and c does not end with $a_j^{\pm 1}, b_j^{\pm 1}$. At most two elements $a_j^{\pm 1}, b_j^{\pm 1}$ can be absorbed in the last factor of c. Therefore, if c has length m and $ca_j^rb_j^s$ has length n we have $n - m \leqq \max(|r|, |s|) \leqq n - m + 2$. There are $24(n - m + 1) \leqq 24n$ lattice points which satisfy the condition. We conclude that the component contains at most $24n(64g^3 + 1)$ polygons.

We have shown that $\bar{\tau}_n \leqq 24n(64g^3 + 1)$. Hence $\sum \frac{1}{\bar{\tau}_n} = \infty$, and it follows that W is of class O_{AD}.

19G. The Schottky surfaces are particularly interesting because, as an easy reasoning shows, they are planar. A Schottky surface can thus be represented as a plane region. Because the surface is of class O_{AD} the cover transformations, which are conformal self-mappings, reduce to linear transformations of the plane (Theorem 2D). The group of these transformations is a free group with g generators. Indeed, $\mathscr{F}/\mathscr{D}$ is generated by $b_1, \cdots, b_g$, and there are no relations between the generators.

We can recover W_0 from W by identifying points which are equivalent under the linear cover transformations. In other words, W_0 is known as soon as we know the linear transformations associated with $b_1, \cdots, b_g$. One of these can be chosen arbitrarily, and the others depend on three complex constants each. We have thus a way of describing W_0 by $3g-3$ complex constants. The weakness of this attempt to parametrize the closed Riemann surfaces of genus g is that the constants are subject to conditions which are difficult to formulate.

20. Ramified coverings of the sphere

20A. Suppose that W is an arbitrary covering surface of the Riemann sphere W_0. Let E be a finite point set on W_0, and remove from W all points which project into E. The remaining surface W' is a covering surface of $W_0'=W_0-E$. We make the specific assumption that W' is a regular covering surface of W_0'. In less precise language, W has branch points of finite and infinite order which project into a finite number of points.

Our aim is to find conditions which guarantee that W is parabolic or of class O_{AD}.

20B. We pass a simple closed curve C through the points $z_1, \cdots, z_q$ of E. The arc from z_i to z_{i+1} will be denoted by C_i (in cyclic order). We assume that the C_i are piecewise analytic and that C_i, C_{i+1} form a nonzero angle at z_{i+1}.

The curve C divides the sphere into two regions Δ_1 and Δ_2. The regions on W' which project into Δ_1 or Δ_2 may be called "cells". We choose a cell of generation 0 and say that a cell is of generation n if it is not of lower generation and has a common side with a cell of generation $n-1$.

Let P_n be the union of all cells of generation $\leqq n$, together with their sides. Although the vertices are not in W' we agree to include them in P_n so as to obtain a polyhedron whose faces are cells. The border vertices are chosen in such a way that each one belongs to a single sequence of adjacent faces. If a border vertex belongs to exactly m cells in P_n we say that it has the order m.

We denote by h_n the total number of cells in P_n which have at least one vertex on the border. Similarly, h_{nj} will be the number of cells which touch a particular contour, and k_n will be the maximum of h_{nj} for a given n. We intend to prove:

Theorem. *W is parabolic if* $\sum \frac{1}{h_n}=\infty$ *and of class* O_{AD} *if* $\sum \frac{1}{k_n}=\infty$.

20C. It may be assumed that all the z_i are finite. We denote by $S_i(r)$ the open disk of center z_i and radius r. Choose r_0 so small that the disks $S_i(r_0)$ are at positive distance from each other.

In order to construct a suitable exhaustion of W we modify each P_n as follows: Consider a border vertex of order m which projects into z_i. We omit from all cells with this vertex the part that lies over $S_i(r_m)$ with $r_m = 2^{-m}r_0$. The remaining part of P_n is denoted by Q_n, its boundary by β_n, and the individual contours by β_{nj}.

If a border vertex of P_n has order m it is either not on the border of P_{n-1}, or it has at most order $m-2$ with respect to P_{n-1}. More accurately, it may split into several border vertices of P_{n-1}, each of order $\leqq m-2$. This observation shows that $Q_{n-1} \subset Q_n$, and it follows that the Q_n form an exhaustion of W.

20D. In order to find a lower bound for the extremal distance between β_{n-1} and β_n we are going to replace Q_{n-1} by a region Q'_{n-1} between Q_{n-1} and Q_n which is more suitable for our purpose. The boundary of Q'_{n-1} will be denoted by β'_{n-1}, and it will be sufficient to estimate the extremal distance between β'_{n-1} and β_n.

For the construction of Q'_{n-1} we need a triangulation of P_n. To this end we triangulate W_0 by choosing points a_1, a_2 in Δ_1, Δ_2, outside of all $S_i(r_0)$, which we join to the points z_i by nonintersecting piecewise analytic arcs C_{i1}, C_{i2}. These arcs will be chosen so that they form positive angles with the C_i. The triangulation is carried over to P_n in the obvious manner.

The first step is now to remove from P_n all triangles which have a side on the border. Except for $n=0$, a case that we exclude, no cell will be completely removed.

Consider now a border vertex of P_n of order $m \geqq 2$, with projection z_i. From the remaining triangles we omit the part that projects into $S_i(r_{m-2})$. The portion of P_n that remains after this second step will be chosen as Q'_{n-1}. Since we have certainly not omitted more than in the construction of Q_{n-1} we see that $Q_{n-1} \subset Q'_{n-1} \subset Q_n$.

For orientation we refer to the slightly schematic Fig. 12.

20E. We choose on W_0 a linear density ρ, or a metric $\rho|dz|$, such that $\rho(z) = 1/|z-z_i|$ in $S_i(r_0)$. The values outside of the disks are immaterial as long as we take care that the total ρ-area of the complement is finite and that completely disjoint arcs used in the triangulation are at positive ρ-distance from each other. The particular merit of our choice is that even the arcs with common end point z_i are at positive ρ-distance from each other.

We transfer ρ to the covering surface and use it to estimate the extremal distance between β'_{n-1} and β_n. Note that ρ does not become infinite in $Q_n - Q'_{n-1}$.

Consider an arc γ between β_n and β'_n, and suppose first that its initial point lies on a side of a triangle. Then γ must meet one of the other two sides of the same triangle, and it follows by our choice of ρ that the ρ-length of γ exceeds a fixed positive number. Assume next that the initial point lies on one of the circular arcs on β_n. If the center of this arc is a vertex of order 1, then γ must meet the opposite side of at least one of the two triangles with this vertex, and we see that the length of γ cannot be

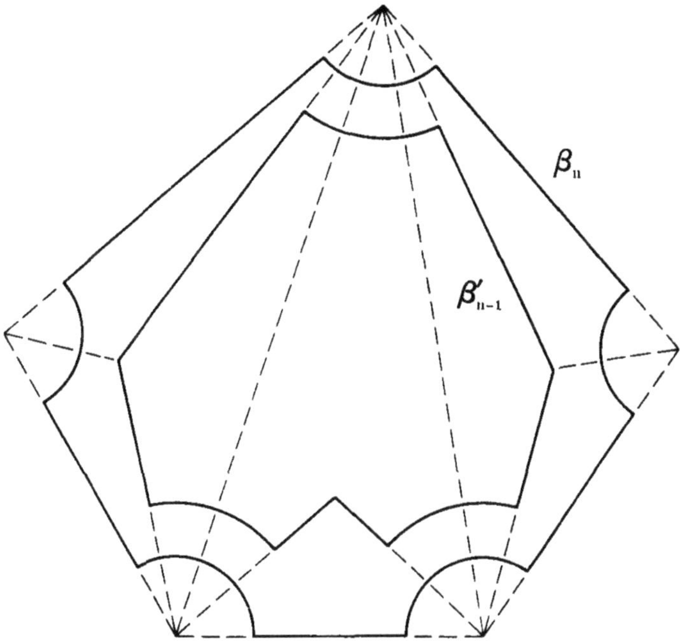

FIG. 12

arbitrarily small. Finally, if the vertex is of order $m \geqq 2$, then the projection of γ, which starts on a circle $|z - z_i| = r_m$ must reach the circle $|z - z_i| = r_{m-2}$, and hence its length is at least $2 \log 2$. In all possible cases we have reached the conclusion that the length of γ is greater than a fixed positive number d, independent of n.

Next, we estimate the ρ-area of $Q_n - Q'_{n-1}$. It is clear that $Q_n - Q'_n$ is covered by the cells which have a vertex on the border of P_n. The number of such cells has been denoted by h_n, and we conclude that the part of $Q_n - Q'_{n-1}$ whose projection lies outside of $\cup S_i(r_0)$ has an area $\leqq h_n A$ where A is a finite constant. Consider now the area that lies over $\cup S_i(r_0)$. For each vertex of order 1 we have to add at most the area of $S_t(r_0)-S_t(r_1)$, that is, $2\pi \log 2$. For a vertex of order $m \geqq 2$ we have to add at most m times the area of $S_t(r_m)-S_t(r_{m-1})$, that is, an amount of $4\pi m \log 2$. The sum of all orders m is $\leqq qh_n$. We find that the total area of $Q_n - Q_n'$ is less than $h_n B$ for a finite B. It follows that

$$\lambda(\beta_{n-1}, \beta_n) \geqq \lambda(\beta'_{n-1}, \beta_n) \geqq \frac{d^2}{h_n B}.$$

The first part of our theorem is now a consequence of Theorem 15C.

The second part follows quite similarly from Theorem 16D, for it is evident that the part of $Q_n - Q'_{n-1}$ which is adjacent to a contour of Q_n is connected.

Remark. The test for parabolicity is due to H. Wittich [2] and R. Nevanlinna [4]. The corresponding test for the class O_{AD} was derived by L. Sario [1] to illustrate Theorem 16D.

§7. PLANE REGIONS

Since every planar Riemann surface can be mapped on a plane region and every plane region is a planar Riemann surface, it would seem that there is no need to distinguish between the two cases. However, when we deal with plane regions we are interested not only in intrinsic properties, but also in properties of the imbedding in the plane. It is therefore important to discuss the relations between these two kinds of properties.

To some extent this was done already in §1, notably through Theorems 2B and 2D.

Remark. A detailed study of plane regions and their complementary sets was made in L. Ahlfors, A. Beurling [2], which is the source of Theorems 23A, 24E, and 24F.

21. Mass distributions

21A. For plane regions the Green's function is closely related to a logarithmic potential. To define potential we need the notion of a mass distribution. It seems advisable to recall the few fundamental facts that will be needed.

Let E be a compact set in the plane. A positive mass distribution μ on E is a functional $\mu(f)$, defined for all real continuous functions f in the plane, with the following properties:

(**A1**) $\mu(c_1 f_1 + c_2 f_2) = c_1 \mu(f_1) + c_2 \mu(f_2)$,

(**A2**) $\mu(f) \geqq 0$ if $f \geqq 0$,

(**A3**) $\mu(f) = 0$ if $f \equiv 0$ on E.

The value $\mu(1)$ is the *total mass* distributed on E.

21B. We shall use the notation $f \cap g = \min(f, g)$; this function is continuous together with f and g.

Suppose that $\{f_n\}$ is a decreasing sequence of continuous functions with the limit $0 : f_n \searrow 0$. The convergence is automatically uniform on any compact set. Hence, given $\epsilon > 0$, we have $f_n \cap \epsilon = f_n$ on E as soon as n is sufficiently large. By (A3) we have thus $\mu(f_n) = \mu(f_n \cap \epsilon) \leqq \epsilon\mu(1)$, and we have proved that $\mu(f_n) \to 0$.

This property makes it possible to extend μ to lower semicontinuous functions, i.e., to limits of increasing sequences of continuous functions. If $f_n \nearrow f$ we set $\mu(f) = \lim_{n\to\infty} \mu(f_n)$. Suppose that we have also $g_m \nearrow f$. Then $g_m \cap f_n \nearrow g_m$ as $n \to \infty$, and the previous remark yields $\mu(g_m \cap f_n) \nearrow \mu(g_m)$. Since $\mu(f) \geqq \mu(f_n) \geqq \mu(g_m \cap f_n)$ we get $\mu(f) \geqq \lim_{m\to\infty} \mu(g_m)$, and it follows that the definition is independent of the choice of the approximating sequence $\{f_n\}$.

The conditions (A1)–(A3) remain in force, provided $c_1, c_2 \geqq 0$ in (A1). It may happen, however, that $\mu(f) = +\infty$. Finally, μ can be extended to differences $f = f_1 - f_2$ of lower semicontinuous functions, provided that $\mu(f_1)$ and $\mu(f_2)$ are both finite. The value $\mu(f) = \mu(f_1) - \mu(f_2)$ is independent of the particular representation, and the functional remains linear.

21C. The characteristic function χ of an open set is lower semicontinuous, that of a closed set is upper semicontinuous. In both cases $\mu(\chi)$ is defined and finite. Explicitly, if the set is closed, $\mu(\chi)$ is defined by the relation $\mu(\chi) = \mu(1) - \mu(1 - \chi)$.

Let Q be a square with sides parallel to the coordinate axes which contains all of E. We agree that Q shall be semiopen in the sense that it does not include the points on the upper and right hand sides. Then Q can be subdivided into smaller squares Q_i, and if f is continuous it is easy to see that we can write

$$\mu(f) = \lim \sum f(z_i)\mu(\chi_i)$$

in the manner of a Riemann integral. This justifies the alternate notation

$$\mu(f) = \int f(z) d\mu(z)$$

which we shall use even when f is only semicontinuous.

21D. Consider now a family M of distributions μ. The following selection lemma is quite easy to prove:

Lemma. *If the total masses $\mu(1)$, $\mu \in M$, are bounded there exists a sequence of mass distributions $\mu_n \in M$ which converges to a limit distribution ν in the sense that $\mu_n(f) \to \nu(f)$ for every continuous f.*

By use of the diagonal process we can choose the μ_n so that $\nu(\chi_i) = \lim \mu_n(\chi_i)$ exists for all squares Q_i. If $\epsilon > 0$ is given, it follows by the uniform continuity of f that it is possible to construct a fixed subdivision of Q into squares Q_i with the property that

$$|\mu_n(f) - \sum f(z_i)\mu_n(\chi_i)| < \epsilon\mu_n(1)$$

for all n. We conclude easily that $\nu(f) = \lim_{n\to\infty} \mu_n(f)$ exists, and it is obvious that ν satisfies (A1)–(A3).

If f is merely lower semicontinuous and $f_m \nearrow f$ we have $\nu(f_m) = \lim_{n\to\infty} \mu_n(f_m) \leq \varliminf_{n\to\infty} \mu_n(f)$, and thus

$$\nu(f) \leq \varliminf_{n\to\infty} \mu_n(f). \tag{18}$$

For upper semicontinuous f the inequality is reversed.

22. The logarithmic potential

22A. For fixed ζ the function $\log \dfrac{1}{|z-\zeta|}$ is lower semicontinuous in z. We can therefore define

$$p_\mu(\zeta) = \int \log \frac{1}{|z-\zeta|}\, d\mu(z),$$

although the value may be $+\infty$. The function p_μ is the *logarithmic potential* of μ.

The potential is a lower semicontinuous function of ζ. Indeed, we have $p_\mu^{(n)} \nearrow p_\mu$ where

$$p_\mu^{(n)}(\zeta) = \mu\left(\log \frac{1}{|z-\zeta|} \cap n\right).$$

It is easy to see that each $p_\mu^{(n)}$ is continuous, and therefore p_μ is lower semicontinuous.

Moreover, p_μ is superharmonic in the whole plane and harmonic outside of E. The proof is very simple and will be omitted.

22B. Let V_μ be the least upper bound of p_μ in the whole plane. We set $V(E) = \inf V_\mu$ where μ ranges over all mass distributions on E with total mass 1. The *logarithmic capacity* of the set E is defined by cap $E = e^{-V(E)}$.

Denote by W the unbounded complementary component of E. If W is hyperbolic, let $g(z)$ be the Green's function with pole at ∞, and set $k = \lim_{z\to\infty} (g(z) - \log|z|)$. In Ch. III, 14, 15, we defined the capacity of the ideal boundary as $c(\beta) = e^{-k}$, where $k = k(\beta)$. We shall now complete the results of that chapter by proving:

Theorem. *W is parabolic if and only if E has zero logarithmic capacity. If the capacity is positive there is a mass distribution μ on E with total mass* 1 *such that $p_\mu(z)=k-g(z)$ in W. Moreover, cap $E=c(\beta)$ and $V_\mu=V(E)$.*

The first part of the theorem is an explicit geometric criterion for parabolicity. Indeed, because cap E depends only on the mutual distances $|z-\zeta|$ between points on E it can be regarded as a purely geometric quantity.

22C. For the proof, suppose first that μ is a normalized mass distribution with $V_\mu<\infty$. Then $V_\mu-p_\mu(z)$ is a positive harmonic function on W with the singularity $\log|z|$ at ∞. This implies the existence of the Green's function, and since $g(z)$ is the least positive harmonic function with the singularity $\log|z|$ we have $g(z)\leqq V_\mu-p_\mu(z)$. On the other hand, $p_\mu(z)+\log|z|\to 0$ for $z\to\infty$, and we obtain $k=\lim\,(g(z)-\log|z|)\leqq V_\mu$. Since this is true for all μ we have proved that $k\leqq V(E)$, or $c_\beta\geqq$ cap E.

To prove the opposite inequality we exhaust W by regular regions Ω. If g_Ω is the corresponding Green's function one obtains, by use of Green's formula (Ch. II, 8C),

$$g_\Omega(\zeta) = k_\Omega+\frac{1}{2\pi}\int\limits_{\beta(\Omega)}\log\frac{1}{|z-\zeta|}\,dg_\Omega^*(z) \tag{19}$$

for $\zeta\in\Omega$ and

$$0 = k_\Omega+\frac{1}{2\pi}\int\limits_{\beta(\Omega)}\log\frac{1}{|z-\zeta|}\,dg_\Omega^*(z) \tag{20}$$

if ζ is in the complement of Ω, including the boundary $\beta(\Omega)$. In these formulas $\beta(\Omega)$ is described in the positive direction with respect to Ω, and hence in the direction of decreasing g_Ω^*.

We define μ_Ω by means of

$$\mu_\Omega(f) = -\frac{1}{2\pi}\int\limits_{\beta(\Omega)} f\,dg_\Omega^*.$$

This is clearly a positive mass distribution on the complement of Ω, and the total mass is 1. If the potential of μ_Ω is denoted by P_Ω the formulas (19) and (20) become

$$\begin{aligned} g_\Omega(\zeta) &= k_\Omega-p_\Omega(\zeta)\\ 0 &= k_\Omega-p_\Omega(\zeta). \end{aligned} \tag{21}$$

Now let $\Omega\to W$ through a sequence $\{\Omega_n\}$. By Lemma 21D there exists a subsequence, again denoted by $\{\Omega_n\}$, with the property that μ_{Ω_n} converges to a limit distribution μ. Clearly, the total mass of μ is 1. Moreoever, μ is a distribution on E. For if $f=0$ on E, then $|f|\leqq\epsilon$ on the complement of Ω_n

for all sufficiently large n. This implies $|\mu_{\Omega_n}(f)| \leqq \epsilon$ and hence, in the limit, $\mu(f)=0$.

We suppose that the Green's function exists. Then $g_{\Omega_n}(\zeta)\to g(\zeta)$ and $k_{\Omega_n}\to k$. Moreover, if $\zeta \in W$ we have $p_{\Omega_n}(\zeta)\to p(\zeta)$ where p is the potential of μ. To reach this conclusion we need only observe that $\log \dfrac{1}{|z-\zeta|}$, although not continuous in the whole plane, is at least continuous on a compact neighborhood E' of E, and that, from a certain n on, all μ_{Ω_n} as well as μ are distributed on E'. On using these results the first equation (21) leads to

$$g(\zeta) = k-p(\zeta),$$

and we have shown that $k-g(\zeta)$ is indeed a potential.

As we try to utilize the second equation (21) we have no longer the advantage of continuity. However, $\log \dfrac{1}{|z-\zeta|}$ is lower semicontinuous, and by (18) we are allowed to conclude that

$$k = \lim p_{\Omega_n}(\zeta) \geqq p(\zeta).$$

Hence $p(\zeta) \leqq k$ throughout the plane, that is, $V_\mu \leqq k$, whence $V(E) \leqq k$ and $\operatorname{cap} E \geqq c_\beta$. Together with our previous result we have proved that $\operatorname{cap} E = c_\beta$, and also that $V_\mu = V(E)$.

A more detailed investigation would show that μ is the only mass distribution which satisfies the condition $V_\mu = V(E)$. It is often referred to as the *equilibrium distribution*.

23. The classes O_{SB} and O_{SD}

23A. For plane regions we denote by S the class of analytic functions F which are univalent ("schlicht"), that is, such that $F(z_1) \neq F(z_2)$ when $z_1 \neq z_2$. The classes O_{SB} and O_{SD} consist of all regions which are not conformally equivalent to a bounded region or to a region of finite area respectively. It turns out that these classes are identical:

Theorem. $O_{SB} = O_{SD}$.

The inclusion $O_{SD} \subset O_{SB}$ is trivial. For the opposite inclusion we utilize the disk mappings introduced in Ch. III, 16. Suppose that F is an SD-function on the plane region W. We exhaust W by regular regions Ω. For each Ω the mapping by F determines one contour $\gamma(\Omega)$ as the outer contour of Ω. The $\gamma(\Omega)$ define a boundary component γ of W. We are going to show that the capacity $c(\gamma)$ with respect to a point $z_0 \in W$ is necessarily positive. By Theorem 16D, Ch. III, there will then exist a mapping of W onto a finite disk with concentric slits, and hence an SB-function. This will imply the assertion $O_{SB} \subset O_{SD}$.

23B. We have to prove that $c_\Omega(\gamma)$, the capacity of $\gamma(\Omega)$, is bounded away from zero. To this end, let Ω be mapped on a disk with concentric slits; z_0 corresponds to the origin and $\gamma(\Omega)$ to the rim. The mapping function is normalized by its derivative at z_0, and when this is so the radius of the disk will be $R = c_\Omega(\gamma)^{-1}$.

For simplicity we shall identify Ω with the slit disk and let F stand for the composite mapping function. Consider a circle $|z| = r$ in Ω which does not meet the slits, and write

$$L(r) = \int_{|z|=r} |F'| r \, d\Theta,$$

$$I(r) = \iint_{|z| \leq r} |F'|^2 \, r \, dr \, d\Theta.$$

The Schwarz inequality implies $L(r)^2 \leqq 2\pi r I'(r)$, and the isoperimetric inequality, applied to the image of $|z| \leqq r$ and its outer contour, leads to $L(r)^2 \geqq 4\pi I(r)$. Together, these inequalities give

$$2I(r) \leqq rI'(r),$$

and on integrating between r_0 and R we find that

$$\log \frac{I(R)}{I(r_0)} \geqq 2 \log \frac{R}{r_0},$$

or

(22) $$\frac{I(R)}{R^2} \geqq \frac{I(r_0)}{r_0^2}.$$

On letting r_0 tend to 0 we have

$$\lim_{r_0 \to 0} \frac{I(r_0)}{r_0^2} = \pi |F'(0)|^2$$

where, because of the normalization, $|F'(0)|$ is independent of Ω. On the other hand, $I(R)$ is bounded, by hypothesis. It follows by (22) that R lies under a finite bound, and hence $c_\Omega(\gamma)$ is bounded away from 0. This is what we had to prove.

§8. COUNTEREXAMPLES

We recall that we have proved the inclusions $O_G \subset O_{HP} \subset O_{HB} \subset O_{HD}$ and also $O_{AB} \subset O_{AD}$. Together with the trivial relations $O_{HB} \subset O_{AB}$, $O_{HD} \subset O_{AD}$ we have thus the following system of inclusions:

$$O_G \subset O_{HP} \subset O_{HB} \begin{matrix} \subset O_{HD} \subset \\ \\ \subset O_{AB} \subset \end{matrix} O_{AD}.$$

For many years it was not known whether these classes are strictly different. To prove that the inclusions are strict one must construct surfaces which are contained in one of the classes but not in the preceding one. Such constructions are fairly difficult, but they serve the useful purpose of demonstrating that certain Riemann surfaces have rather unexpected properties.

24. The case of plane regions

24A. For plane regions we have shown that $O_G = O_{HP} = O_{HB} = O_{HD}$ (Theorem 7E). Thus, by restricting our attention to plane regions we can only hope to prove that $O_G < O_{AB} < O_{AD}$. On the other hand, if we prove that these inclusions are strict already in the case of plane regions the conclusion will be stronger than if we had made use of nonplanar surfaces.

To prove that $O_G < O_{AB}$ in the plane case we shall first derive a simple sufficient condition for a region W to be of class O_{AB}. The complement of W will be denoted by E. We say that E has linear measure zero if it can be enclosed in a finite number of circular disks whose radii have an arbitrarily small sum.

Theorem. *If E has linear measure zero, then $W \in O_{AB}$.*

Let γ_ν be the peripheries, of total length $< \epsilon$, of circles that enclose E. If f is bounded and analytic in W we obtain, at any point ζ outside of the circles,

$$f'(\zeta) = -\frac{1}{2\pi i} \sum \int_{\gamma_\nu} \frac{f\,dz}{(z-\zeta)^2},$$

and it follows immediately that $f'(\zeta) = 0$. Hence f is constant, and $W \in O_{AB}$.

24B. To construct a region W which is of class O_{AB} but not of class O_G we shall make use of *generalized Cantor sets*. Let $\{q_i\}_1^\infty$ be a sequence of real numbers, $0 < q_i < 1$. We denote by $E\{q_i\}$ the point set constructed as follows:

Let E_0 be the closed line segment from 0 to 1. We construct, inductively, sets $E_n(q_1 \cdots q_n)$ consisting of 2^n disjoint closed intervals. To pass from $E_n(q_1 \cdots q_n)$ to $E_{n+1}(q_1 \cdots q_{n+1})$ we remove from each interval, symmetrically about the midpoint, a subinterval whose length has the ratio q_{n+1} to the original interval. The set $E = E\{q_i\}$ is defined as $\bigcap_1^\infty E_n(q_1 \cdots q_n)$.

It is evident that the length of E is $\prod_1^\infty (1 - q_i)$. Thus the length is 0 if and only if $\sum q_i = \infty$.

Let E_1' and E_1'' be the two intervals of $E_1(q_1)$. The sets $E_n' = E_1' \cap E_n(q_1 \cdots q_n)$ and $E_n'' = E_1'' \cap E_n(q_1 \cdots q_n)$, $n \geqq 1$, are congruent and similar to

$E_{n-1}(q_2 \cdots q_n)$ by the factor $\frac{1}{2}(1-q_1)$. We write $c_{n0} = \text{cap}\, E_n(q_1 \cdots q_n)$ and $c_{n1} = \text{cap}\, E_{n-1}(q_2 \cdots q_n)$. The sets E'_n and E''_n have capacity $\frac{1}{2}(1-q_1)$ c_{n1}.

Let unit masses μ', μ'' be distributed on E'_n, E''_n so that each has a potential $\leqq V$ in the whole plane. Then $\mu = \frac{1}{2}(\mu' + \mu'')$ is distributed on $E_n(q_1 \cdots q_n)$. Every point has a distance $\geqq q_1/2$ from at least one of the sets E'_n, E''_n. Therefore, the potential of μ is everywhere $\leqq \frac{1}{2}V + \frac{1}{2}\log \frac{2}{q_1}$. This implies

$$-\log c_{n0} \leqq \tfrac{1}{2}V + \tfrac{1}{2}\log \frac{2}{q_1},$$

and since V can be chosen arbitrarily close to $-\log[\frac{1}{2}(1-q_1)c_{n1}]$ we obtain

$$\log c_{n0} \geqq \tfrac{1}{2}\log c_{n1} + \tfrac{1}{2}\log \frac{q_1(1-q_1)}{4}.$$

If we write $c_{nk} = \text{cap}\, E_{n-k}(q_{k+1} \cdots q_n)$ we have, by the same result,

$$\log c_{nk} \geqq \tfrac{1}{2}\log c_{n,\,k+1} + \tfrac{1}{2}\log \frac{q_{k+1}(1-q_{k+1})}{4}.$$

On eliminating c_{nk}, $k = 1, \cdots, n-1$ and observing that $c_{nn} = \text{cap}\, E_0 = \frac{1}{4}$ we find that

$$\log c_{n0} \geqq \sum_{k=1}^{n} \frac{1}{2^k}\log \frac{q_k(1-q_k)}{4} - \frac{1}{2^n}\log 4.$$

Since $\text{cap}\, E = \lim_{n\to\infty} c_{n0}$ we have proved:

Lemma. *$E\{q_i\}$ has positive capacity if the series*

$$\sum_{1}^{\infty} \frac{1}{2^k}\log \frac{1}{q_k(1-q_k)} \tag{23}$$

converges.

We can choose the q_i so that (23) converges while $\sum q_i$ diverges, for instance by letting all q_i be equal. The corresponding E has positive capacity and linear measure 0. The complement W is thus hyperbolic and of class O_{AB}. We have reached the desired conclusion:

Theorem. *There exist plane regions which are of class O_{AB} but not of class O_G.*

Remark. The proof of Lemma 24B is taken from R. Nevanlinna [24], pp. 152–155.

24C. We shall now investigate the AB- and AD-character of sets E that lie on the unit circle $|z| = 1$. By Theorem 24A the complement W is of class O_{AB} if E has zero length. In analogy with Theorem 3B we show that this condition is also necessary.

Theorem. *If E has positive length, then W is not of class O_{AB}.*

We construct the function

$$F(\zeta) = \frac{1}{2\pi}\int_E \frac{e^{i\Theta}+\zeta}{e^{i\Theta}-\zeta}\,d\Theta.$$

It is analytic in W and it is not a constant for $F(0) = -F(\infty) \neq 0$.

One finds by easy computation

$$\operatorname{Re} F(\zeta) = \frac{1}{2\pi}\int_E \frac{1-|\zeta|^2}{|e^{i\Theta}-\zeta|^2}\,d\Theta,$$

and this formula yields $0 < \operatorname{Re} F(\zeta) < 1$ for $|\zeta| < 1$, $0 > \operatorname{Re} F(\zeta) > -1$ for $|\zeta| > 1$, and $\operatorname{Re} F(\zeta) = 0$ if $|\zeta| = 1$, ζ not on E. Hence all values of $F(\zeta)$ lie in the strip $-1 < \operatorname{Re} F(\zeta) < 1$, and we find, for instance, that the function

$$\tan\frac{\pi F}{4} = \frac{1}{i}\,\frac{e^{\pi i\frac{F}{2}}-1}{e^{\pi i\frac{F}{2}}+1}$$

is bounded. Consequently, W is not of class O_{AB}.

24D. Let us now suppose, temporarily, that E consist of finitely many arcs, and let E' be formed by the complementary arcs together with their end points. Consider the Green's function $g(z)$ with pole at ∞ of W', the complement of E' with respect to the extended plane. As in 22B we set $\lim_{z\to\infty}(g(z)-\log|z|) = k = -\log(\operatorname{cap} E')$.

The function $g(z)-g(1/\bar{z})-\log|z|$ has no singularity and vanishes on E'. Therefore it is identically 0, a fact expressed by the symmetry relation $2g(z)-\log|z| = 2g(1/\bar{z})-\log(1/|z|)$. This relation shows that $2g(z)-\log|z|$ has normal derivative 0 at interior points of E.

Let us now define

$$p_0(z) = \begin{cases} 2g(z)-\log|z| & \text{for} \quad |z|>1 \\ -2g(z)+\log|z| & \text{for} \quad |z|<1. \end{cases}$$

Because it vanishes on E' and has opposite values at symmetric points it remains harmonic at interior points of E', that is, $p_0(z)$ is harmonic in the complement W of E except for the singularities $\log|z|$ at ∞ and $\log|z|$ at 0. In view of the vanishing normal derivatives p_0 is nothing else than one of the principal functions corresponding to these singularities (see Ch. III, 9A). The other principal function on W is obviously $p_1 = \log|z|$.

At $z=\infty$ we have $p_0(z) = \log|z|+2k+\cdots$, and at $z=0$, by virtue of the symmetry, $p_0(z) = \log|z|-2k$. We refer the reader to Theorem 10G,

Ch. III. In this theorem, as applied to the present case, the class of admissible functions is formed by the real parts of single-valued analytic functions in W. Moreover, $q=p_0-p_1$ has the Dirichlet integral $D(q)=2\pi(q(\infty)-q(0))=8\pi k$. The theorem yields

$$|F(\infty)-F(0)| \leqq \sqrt{\frac{2kD(F)}{\pi}} \tag{24}$$

for any analytic F with equality when $\operatorname{Re} F=q$. We recall that $k=-\log(\operatorname{cap} E')$; this is a positive quantity, as seen by the proof, or by the fact that E' is a subset of the whole unit circle whose capacity is 1.

24E. In the case of an arbitrary closed set E on $|z|=1$, we denote by E' a set on the unit circle which is composed of a finite number of closed arcs and contained in the complement of E. The upper bound of cap E', for all such sets E', is called the *inner capacity* of the complement of E. Suppose that F is analytic on W, the complement of E. Then (24) is applicable to the restriction of F to the region formed by $|z|<1$, $|z|>1$, and the interior points of E'.

Theorem. *$W \in O_{AD}$ if and only if the inner capacity of the complement of E with respect to $|z|=1$ equals* 1.

First, if the inner capacity is 1 it follows by (24) that $F(0)=F(\infty)$ for all AD-functions. To see that this implies $W \in O_{AD}$, let f be a univalent function on W. On combining f with a linear function we can achieve that $f(0)=0$, $f(\infty)=\infty$. For an arbitrary z_0 it is easy to see that

$$\frac{f'(z_0)}{f(z)-f(z_0)} - \frac{1}{z-z_0}$$

is an AD-function. It follows that

$$\frac{f'(z_0)}{f(z_0)} = \frac{1}{z_0}$$

and this differential equation implies $f(z)=cz$. Hence f is linear, and by Theorem 2D W must be of class O_{AD}.

Second, if the inner capacity is <1 we form for each E' the corresponding extremal function F with $\operatorname{Re} F=q$. It is clear that a suitable sequence of these functions converges to a nontrivial AD-function.

24F. To construct an example which proves that $O_{AB}<O_{AD}$ for plane regions we show that it is possible to construct a set E of positive length whose complement has inner capacity 1.

Let $E_1'(\alpha)$ be the arc defined by $|\Theta| \leqq \alpha$. Its capacity equals $\sin\frac{\alpha}{2}$, as found by use of an elementary conformal mapping. If $g(z)$ is the Green's function for the complement of E_1' one finds that $\frac{1}{n}g(z^n)$ is the Green's function for the complement of $E_n'(\alpha)$ where $E_n'(\alpha)$ consists of the arcs

$$|\Theta-\frac{m}{n}\cdot 2\pi|\leqq \frac{\alpha}{n}, \quad m = 0,\cdots,n-1.$$

From the developments

$$g(z) = \log|z|-\log\sin\frac{\alpha}{2}+\cdots$$

$$\frac{1}{n}g(z^n) = \log|z|-\frac{1}{n}\log\sin\frac{\alpha}{2}+\cdots$$

we conclude that cap $E_n'(\alpha)=\left(\sin\frac{\alpha}{2}\right)^{1/n}$.

Choose $\alpha_n=\frac{1}{n^2}$, say. Let E' be the union of the interiors of the sets $E_n'(\alpha_n)$, $n\geqq 1$. Then E' has length $\leqq 2\sum_1^\infty n^{-2}=\pi^2/3<2\pi$. Its inner capacity must be at least equal to $\left(\sin\frac{\alpha_n}{2}\right)^{1/n}$ for each n, and since these quantities tend to 1 we conclude that the inner capacity is 1. The complement E of E' is a set with the desired property, for its length is at least $2\pi-\pi^2/3$.

Theorem. *There exist plane regions W which are of class O_{AD} but not of class O_{AB}.*

24G. For plane regions we have also introduced, in 23, the class $O_{SB}=O_{SD}$ which obviously includes all plane regions of class O_{AD}. For a proof that O_{SB} is not included in O_{AD} we refer the reader to the paper of L. Ahlfors and A. Beurling [2].

25. Proof of $O_G<O_{HP}<O_{HB}$

25A. The surface that we are going to construct will be obtained from the unit disk $\Delta:|z|<1$ so that we draw an infinite number of radial slits whose edges will be identified pairwise.

To illustrate the procedure, let s_1 and s_2 be two radial slits of Δ, formed by the points $re^{i\Theta_1}$ and $re^{i\Theta_2}$ respectively with $a\leqq r\leqq b$; we assume that $0<a<b<1$. Each slit s_k, $k=1,2$, has a left edge s_k^+ corresponding to $\Theta=\Theta_k+0$ and a right edge s_k^- corresponding to $\Theta=\Theta_k-0$ (Fig. 13).

We identify s_1^+ with s_2^- and s_2^+ with s_1^-. Without going into details it is clear that this process defines a Riemann surface. For instance, the end point $ae^{i\Theta_1}=ae^{i\Theta_2}$ will have a neighborhood which is represented by two full circular disks in the z-plane. They are identified in the manner of a two-sheeted covering surface with a branch point, and we obtain a local uniformizer by extracting a square root.

A little more generally we can consider a cyclic identification of any finite number of radial slits $s_1, \cdots, s_h$, all extending between $|z|=a$ and $|z|=b$. In this case s_1^+ is identified with s_2^-, s_2^+ with s_3^-, etc., and finally s_h^+ with s_1^-. The end points will have neighborhoods consisting of h disks.

Such identifications may be performed simultaneously for several pairs or cycles, even for infinitely many, provided that they do not intersect or accumulate inside Δ. For formal reasons we will occasionally identify a

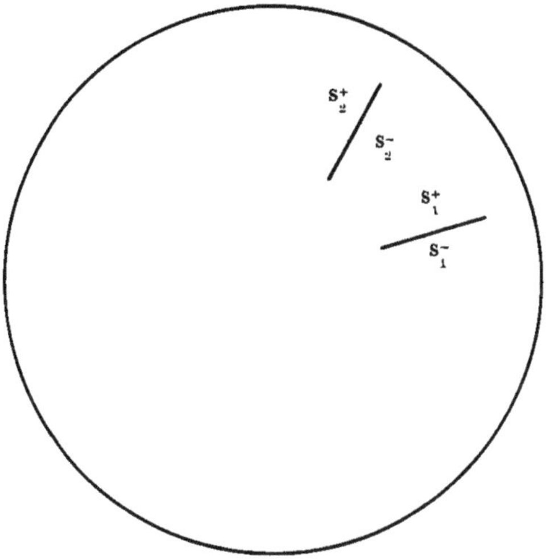

FIG. 13

slit with itself (a cycle with $h=1$). Naturally, this results in no change at all.

Let $\tilde{\Delta}$ be the resulting Riemann surface. The identified slits form a set C on $\tilde{\Delta}$ which is a union of isolated simple arcs with only end points in common. Observe that z will be a well-defined analytic function on $\tilde{\Delta}-C$, but it does not stay continuous on C. It is seen, however, that $\log |z|$ is defined and harmonic on all of $\tilde{\Delta}$, except for the singularity at the origin. Since it tends to 0 as we approach the ideal boundary it is clear that $-\log |z|$ represents the Green's function of $\tilde{\Delta}$ with pole at the origin. In other words, regardless of the particular choice of identifications, $\tilde{\Delta}$ will always be hyperbolic.

25B. We will now introduce a specific rule for constructing $\tilde{\Delta}$. It will depend on infinitely many parameters, given in the form of a strictly increasing sequence $\{r_\nu\}$ of positive numbers with the limit 1, and also on an infinite sequence $\{n_\nu\}$ of positive integers. These sequences will later be chosen so that certain conditions are fulfilled.

Every natural number has a unique representation in the form $\nu = \nu(h, k) = (2h+1)2^k$ where h and k are nonnegative integers. With each ν we are going to associate 2^{k+n_ν} radial slits with end points on $|z| = r_{2\nu}$ and $|z| = r_{2\nu+1}$. These slits will be equally spaced and one will lie on the positive real axis. For convenience we will say that a slit associated with $\nu = \nu(h, k)$ is of *rank* ν and *type* k.

We write $\Theta_k = 2^{-k} \cdot 2\pi$. The sectors $i\Theta_k \leqq \Theta \leqq (i+1)\Theta_k$, $0 \leqq i < 2^k$, will be denoted by S_{ik}. The slits of type k which lie on the rays $\Theta = i\Theta_k$ will be identified cyclically. The remaining slits of the same type will be identified pairwise within each sector S_{ik}, symmetrically about its bisecting ray.

More explicitly, the identification rule can be expressed as follows: Let Θ', Θ'' be the arguments of two slits of rank ν and type k; they may and will be chosen so that $0 \leqq \Theta'' - \Theta' < 2\pi$. The edge $\Theta' + 0$ will be identified with $\Theta'' - 0$, and $\Theta' - 0$ will be identified with $\Theta'' + 0$ if the following two conditions are satisfied: (1) $\Theta'' - \Theta' < \Theta_k$, (2) $\Theta'' + \Theta'$ is an odd multiple of Θ_k. If the second condition is fulfilled and $\Theta'' - \Theta' = \Theta_k$, then only the edges $\Theta' + 0$ and $\Theta'' - 0$ are identified.

This rule completely describes $\tilde{\Delta}$. It has already been remarked that $\tilde{\Delta}$ is a hyperbolic surface.

25C. A harmonic function u on $\tilde{\Delta}$ can be regarded as a function $u(r, \Theta)$ of $z = re^{i\Theta}$ provided that we distinguish, on the slits, between $u(r, \Theta - 0)$ and $u(r, \Theta + 0)$. Conversely, suppose that $u(r, \Theta)$ is harmonic on the complement of the slits, regarded as a bordered surface. In order to conclude that u is harmonic on $\tilde{\Delta}$ it must first of all be known that the values on identified edges are equal. In addition it must be required that the one-sided derivatives $\partial u/\partial\Theta$ at identified points $(r, \Theta' + 0)$ and $(r, \Theta'' - 0)$ coincide.

To prove the sufficiency of this hypothesis we consider the function $v(r, \tau)$ which is equal to $u(r, \Theta' + \tau)$ for $\tau \geqq 0$ and $u(r, \Theta'' + \tau)$ for $\tau \leqq 0$. It follows by use of the reflection principle that $v(r, \tau) - v(r, -\tau)$ and $v(r, \tau) + v(r, -\tau)$ are both harmonic for $\tau = 0$. Hence $v(r, \tau)$ is harmonic, and we conclude that u is harmonic on $\tilde{\Delta}$ at interior points of the slits. The end points are removable singularities.

25D. We say that u, harmonic on $\tilde{\Delta}$, is symmetric of order $m \geqq 1$ if it is symmetric with respect to all directions whose arguments are multiples of Θ_m. In other words, we require that

$$u(r, i\Theta_m + \tau - 0) = u(r, i\Theta_m - \tau + 0)$$

for $i = 1, \cdots, 2^m$ and all r, τ.

Lemma. *A harmonic function with symmetry of order m has normal derivative* 0 *on all slits of type* $\leqq m-1$.

Let $\Theta'+0$ and $\Theta''-0$ correspond to identified edges of type $k \leqq m-1$. Because u is harmonic on $\tilde{\Delta}$ the one-sided derivatives at Θ' and Θ'' are equal. On the other hand, Θ' and Θ'' are symmetrically placed with respect to a multiple of Θ_{k+1}, and the condition $k \leqq m-1$ implies that Θ_{k+1} is a multiple of Θ_m. It follows from the symmetry of order m that the normal derivatives are also opposite, and hence equal to zero.

25E. We construct a new Riemann surface $\tilde{\Delta}_m$ as follows: Retain all slits of type $k \geqq m$ and identify their edges in the same way as before. Of the remaining slits we remove those with $k+n_\nu \leqq m$ and re-identify the others symmetrically to all multiples of Θ_m. This means that we identify edges with arguments $i\Theta_m - \tau + 0$ and $i\Theta_m + \tau - 0$ if $-\Theta_{m+1} < \tau \leqq \Theta_{m+1}$.

It is easy to verify that $\tilde{\Delta}_m$ is symmetric with respect to the direction Θ_{m+1}. Indeed, let $(2i+1)\Theta_{k+1} - \tau + 0$ and $(2i+1)\Theta_{k+1} + \tau - 0$ correspond to identified edges of type $k \geqq m$. The reflected edges have arguments $\Theta_m - (2i+1)\Theta_{k+1} \pm \tau \mp 0$. Because Θ_m is an even multiple of Θ_{k+1} these edges are identified. Similarly, the reflection carries the re-identified edges $i\Theta_m \pm \tau \mp 0$ into identified edges $(1-i)\Theta_m \mp \tau \pm 0$.

Lemma. *A harmonic function on $\tilde{\Delta}$ with symmetry of order m may be regarded as a harmonic function on $\tilde{\Delta}_m$.*

We show first that u is harmonic across the slits with $k \leqq m-1$, $k+n_\nu \leqq m$. The argument of any such slit is a multiple of Θ_m. Therefore the slit lies on a line of symmetry, and hence u has equal values on the edges of the slit. By Lemma 25D the normal derivative vanishes on both sides of the slit. Hence the slit can be removed.

The new identifications are all symmetric to the multiples of Θ_m, and therefore u has equal values on corresponding edges. The normal derivatives vanish for the same reason as above, and we conclude that u is harmonic on $\tilde{\Delta}_m$.

25F. It is consistent with the definition of $\tilde{\Delta}_m$ to set $\tilde{\Delta}_0 = \tilde{\Delta}$. The same reasoning as for $m \geqq 1$ shows that $\tilde{\Delta}_0$ is symmetric with respect to $\Theta_1 = \pi$. The lemma is trivially true without any symmetry condition at all, and for this reason it is natural to say that any function is symmetric of order 0. This convention makes it possible to include the case $m=0$ in the following final lemma:

Lemma. *It is possible to choose the sequence $\{n_\nu\}$ so that, for all $m \geqq 0$, any positive harmonic function on $\tilde{\Delta}$ which is symmetric of order m is also symmetric of order $m+1$.*

Let u be positive, harmonic on $\tilde{\Delta}$, and symmetric of order m. By Lemma 25E u can be regarded as a harmonic function on $\tilde{\Delta}_m$. Because $\tilde{\Delta}_m$ is symmetric with respect to the direction Θ_{m+1} the function $u_m(r, \Theta) = u(r, \Theta_m - \Theta)$ is also harmonic on $\tilde{\Delta}_m$. The difference $U = u_m - u$ vanishes on all slits which are identified symmetrically about Θ_{m+1}, and therefore, by virtue of the periodicity, on all slits of type m. We will show that the existence of sufficiently many slits of this type will cause U to vanish identically. This will prove that u is symmetric of order $m+1$.

The mean value formula

$$u(0) = \frac{1}{2\pi} \int_0^{2\pi} u(r, \Theta)\, d\Theta$$

remains valid on $\tilde{\Delta}_m$, regardless of the identifications, for it depends only on the fact that $\log r$ is harmonic. Since it is also valid for u_m we obtain

$$(25) \qquad \frac{1}{2\pi} \int_0^{2\pi} |U|\, d\Theta \leqq \frac{1}{2\pi} \int_0^{2\pi} (u + u_m)\, d\Theta = 2u(0).$$

Choose $\nu = \nu(h, m) = (2h+1)2^m$, and consider U in the annulus $r_{2\nu-1} < r < r_{2\nu+2}$. In this annulus there are 2^{m+n_ν} slits with end points $r = r_{2\nu}$ and $r = r_{2\nu+1}$. Since U vanishes continuously on the slits, $|U|$ is subharmonic in the whole annulus.

By its subharmonic character $|U|$ is majorized by the harmonic function which is equal to $|U|$ on $|z| = r_{2\nu-1}$ and $|z| = r_{2\nu+2}$. The latter is in turn less than the sum of two ordinary Poisson integrals, one with respect to the interior of the outer circle, the other one with respect to the exterior of the inner circle. Explicitly,

$$|U(re^{i\Theta})| \leqq \frac{1}{2\pi} \int_0^{2\pi} \frac{r_{2\nu+2}^2 - r^2}{|re^{i\Theta} - r_{2\nu+2}e^{i\varphi}|^2}\, U(r_{2\nu+2}e^{i\varphi})\, d\varphi$$

$$+ \frac{1}{2\pi} \int_0^{2\pi} \frac{r^2 - r_{2\nu-1}^2}{|re^{i\Theta} - r_{2\nu-1}e^{i\varphi}|^2}\, U(r_{2\nu-1}e^{i\varphi})\, d\varphi.$$

By aid of (25) we are thus led to the estimate

$$|U(re^{i\Theta})| \leqq \left(\frac{r_{2\nu+2} + r}{r_{2\nu+2} - r} + \frac{r + r_{2\nu-1}}{r - r_{2\nu-1}}\right) 2u(0).$$

There is hence a constant K_ν, independent of u, such that $|U| \leqq K_\nu u(0)$ for $r_{2\nu} \leqq r \leqq r_{2\nu+1}$.

Consider one of the regions bounded by arcs $|z| = r_{2\nu}$, $|z| = r_{2\nu+1}$ and two consecutive slits. It is mapped logarithmically on a rectangle of horizontal dimension $\log r_{2\nu+1} - \log r_{2\nu}$ and vertical dimension $\epsilon_\nu = 2^{-m-n_\nu} \cdot 2\pi$. In this rectangle, let ω be the function which at each point is equal to the sum of the angles subtended by the vertical sides. Since $\omega > 0$ on the horizontal sides and $\omega > \pi$ on the vertical sides it follows by use of the maximum principle that

$$|U| \leq K_\nu u(0) \frac{\omega}{\pi}$$

throughout the rectangle. On the middle line $\log r = \frac{1}{2}(\log r_{2\nu} + \log r_{2\nu+2})$ we have

$$\omega < 4 \arctan \left[\epsilon_\nu / \log \frac{r_{2\nu+1}}{r_{2\nu}}\right],$$

and we conclude that

$$|U(re^{i\Theta})| < \frac{4K_\nu}{\pi} u(0) \arctan \left[\epsilon_\nu / \log \frac{r_{2\nu+1}}{r_{2\nu}}\right]$$

on the full circle $r = (r_{2\nu} r_{2\nu+1})^{\frac{1}{2}}$.

In the bound on the right only ϵ_ν depends on n_ν, and it tends to zero with ϵ_ν, that is to say if $n_\nu \to \infty$. It is therefore possible to choose the n_ν so that the bounds tend to 0 for $h \to \infty$. By application of the maximum principle on $\tilde{\Delta}_m$ it follows that $U = 0$. Since the values of n_ν for different m are completely independent of each other it is possible to choose these numbers so that the assertion of the lemma holds simultaneously for all m.

25G. It remains to draw the conclusion:

Theorem. *There exists a hyperbolic surface on which every positive harmonic function reduces to a constant.*

We construct $\tilde{\Delta}$ according to Lemma 25F. For $m = 0$ the hypothesis of the lemma is trivially fulfilled for all positive harmonic functions on $\tilde{\Delta}$. It follows by induction that such a function is symmetric of all orders, and hence that it must reduce to a constant.

25H. The proof of $O_{HP} < O_{HB}$ is now almost trivial. Delete the origin from our surface $\tilde{\Delta}$. Then $-\log r$ is a positive harmonic function on the resulting surface $\tilde{\Delta}'$. On the other hand, if u were a bounded harmonic function on $\tilde{\Delta}'$, it could be extended to $\tilde{\Delta}$, for the origin would be a removable singularity. It follows that u would have to be constant. Hence $\tilde{\Delta}'$ is an example of a surface which has nonconstant positive harmonic functions, but no nonconstant bounded harmonic functions.

26. Proof of $O_{HB} < O_{HD}$

26A. We have proved earlier (see 5C) that $O_{HD} = O_{HBD}$. Therefore it will be sufficient to construct a surface which admits nonconstant functions of class HB, but none of class HBD. Actually, we shall strive for a little more, namely to construct a surface with functions of class AB, but not of class HBD.

26B. The surface will be constructed as a covering surface of the unit disk $\Delta: |z|<1$. For this reason the function z will be of class AB on the surface.

The branch points will be projected into points $z_{\nu k}$ which we choose as follows:

$$\log |z_{\nu k}| = -2^{-\nu}, \quad \nu = 0, 1, \cdots$$

$$\arg z_{\nu k} = k \cdot 2^{-\nu} \cdot 2\pi, \quad 0 < k \leqq 2^{\nu}.$$

The disk Δ, punctured at the points $z_{\nu k}$, will be denoted by Δ'. Its fundamental group $\mathscr{F}$ is a free group with one generator $a_{\nu k}$ for each $z_{\nu k}$. It does not matter much how we choose the $a_{\nu k}$, but for definiteness we let $a_{\nu k}$ be the homotopy class of a closed curve from the origin which follows a ray of argument just a little less than $\arg z_{\nu k}$ (to avoid other excluded points), circles $z_{\nu k}$ in the positive direction, and returns along the same radius.

A regular covering surface of Δ' is determined by a subgroup $\mathscr{D}$ of $\mathscr{F}$ (Ch. I, 17). This surface can be extended to a ramified covering of Δ by adding points over the $z_{\nu k}$. Explicitly, let s be an element of $\mathscr{F}$. If there exists a smallest $p>0$ such that $s(a_{\nu k})^p s^{-1} \in \mathscr{D}$, then we can associate with s a point of multiplicity p which lies over $z_{\nu k}$.

26C. As in the preceding section we write $\nu=(2h+1)2^m$, $h \geqq 0$, $m \geqq 0$. Since this representation is unique it determines a function $m=m(\nu)$.

In order to define $\mathscr{D}$ we are going to construct a homomorphism α of $\mathscr{F}$ into a group P of permutations of the natural numbers. P will be generated by permutations π_m, one for each integer $m \geqq 0$. To define π_m we represent each positive integer in the form $n=q \cdot 2^m+r$ with $0<r \leqq 2^m$ and set $\pi_m(n)=q \cdot 2^m+2^m-r+1$. In other words, π_m reverses the order within blocks of length 2^m. With the help of this interpretation it is very easy to see that $\pi_m^2=1$ and $\pi_m\pi_n=\pi_n\pi_m$. Thus P is an Abelian group all of whose elements have order 2.

The homomorphism α is completely determined by setting $\alpha(a_{\nu k})=\pi_{m(\nu)}$. We choose $\mathscr{D}$ to be the kernel of this homomorphism. The corresponding covering surface will be denoted by $\Delta(P)$. It is quite clear that all points over the $z_{\nu k}$ are branch points of multiplicity two, except for $m=0$.

26D. We recall (Ch. I, 19) that a regular covering surface has a group of cover transformations which is isomorphic to $\mathscr{F}/\mathscr{D}$. In our case the cover transformations can be extended to the ramified surface $\Delta(P)$, and $\mathscr{F}/\mathscr{D}$ is isomorphic to P. We shall denote the cover transformation that corresponds to π_m by T_m. It is, of course, a directly conformal self-mapping. The fixed points of T_m are the branch points over all $z_{\nu k}$ with $m(\nu)=m$.

To investigate the structure of $\Delta(P)$ let us determine the components that lie over $|z|<\rho<1$. We know that the components are complete covering surfaces, and that each component covers all points the same number of times (Ch. I, 21B). The points over the origin which lie in the component of the initial point are reached by paths which wind only around points $z_{\nu k}$ with $|z_{\nu k}|<\rho$. There are only a finite number of such points, and we find that the closed paths which stay within $|z|<\rho$ are mapped by α onto a subgroup of P which is generated by finitely many π_m. Since this subgroup is finite we may conclude that each component has only a finite number of sheets, and hence that it is relatively compact.

26E. We shall need the following special case of Lemma 3B, Ch. III:

Let W be a Riemann surface, p_0 a point, and A a compact set on W. Let u be a bounded harmonic function on W, $|u|\leqq M$, which vanishes at p_0. Then $|u|\leqq qM$ on A, where $q<1$ is a constant that depends only on W, A and p_0.

This result follows on replacing the set A of the lemma by $A\cup p_0$.

26F. We return to our surface $\Delta(P)$ and assume that u is harmonic and bounded on $\Delta(P)$. Consider the function $U_m(p)=u(p)-u(T_m p)$ which is also harmonic and bounded. We are going to show that U_m is identically zero.

Choose a point $z_{\nu k}$ with $\nu=(2h+1)2^m$. It is contained in the quadrilateral $Q_{\nu k}$ defined by

$$-2^{-\nu+1}<\log|z|<-2^{-\nu-1}$$

$$(k-1)2^{-\nu}\cdot 2\pi<\arg z<(k+1)2^{-\nu}\cdot 2\pi.$$

Each component of $\Delta(P)$ over $Q_{\nu k}$ is a two-sheeted covering surface with a single branch point over $z_{\nu k}$. The transformation T_m interchanges the sheets of each component, and the function U_m vanishes at the branch point.

The logarithmic image of $Q_{\nu k}$ is a rectangle of dimensions $\frac{3}{2}\cdot 2^{-\nu}$ and $4\pi\cdot 2^{-\nu}$; $z_{\nu k}$ corresponds to a point whose relative position in $Q_{\nu k}$ is independent of ν and k. Since all rectangles are similar the two-sheeted regions over the $Q_{\nu k}$ are conformally equivalent.

Let M be the least upper bound of $|U_m|$ on the whole surface $\Delta(P)$. We apply 26E to the components over $Q_{\nu k}$, and are able to conclude that $|U_m|\leqq qM$, with a fixed $q<1$, at all points which lie over the arc $\log|z|=-2^{-\nu}$, $(k-\frac{1}{2})2^{-\nu}\cdot 2\pi\leqq\arg z\leqq(k+\frac{1}{2})2^{-\nu}\cdot 2\pi$. These arcs cover the whole circle, and we find that $|U_m|\leqq qM$ at all points whose projection lies on the circle $\log|z|=-2^{-\nu}$.

Finally, because the components over a disk are relatively compact we may conclude, by the maximum principle, that $|U_m| \leqq qM$ at all points that project inside the circle. On letting h tend to ∞ we find that the same inequality holds throughout $\Delta(P)$. Since M was the least upper bound, we obtain $M \leqq qM$. This is impossible unless $M = 0$, and we have proved that U_m vanishes identically. This means that $u(T_m p) = u(p)$.

26G. We have shown that u has the same values on all sheets, of which there are infinitely many. It is therefore impossible to have $D(u) < \infty$, except in the trivial case of a constant function. Hence $\Delta(P)$ is of class $O_{HBD} = O_{HD}$ while evidently not of class O_{AB}, since it carries the function z, and still less of class $O_{HB} \subset O_{AB}$.

Theorem. *There exists a surface of class O_{HD} which is not of class O_{HB}.*

26H. We have actually proved more, namely that O_{HD} is not contained in O_{AB}. Conversely, O_{AB} cannot be contained in O_{HD}, by Theorem 24B. We list this as a separate result:

Theorem. *There is no inclusion relation between O_{HD} and O_{AB}.*

Remark. Except for slight changes the counter examples in 25 and 26 are identical with the ones introduced by Y. Tôki [1, 2]; see also L. Sario [19]. Earlier, the weaker result $O_G < O_{HD}$ had been proved by L. Ahlfors and H. Royden [1].

CHAPTER V

Differentials on Riemann Surfaces

In Chapters III and IV the emphasis has been on single-valued harmonic and analytic functions on a given Riemann surface. In itself, this is not a serious loss of generality, for multiple-valued functions, when they occur, can be regarded as single-valued functions on a suitable covering surface. However, this is not always a convenient way, and a more direct approach is likely to lead to a much better understanding of the problems that are involved.

Complex integration leads to functions whose branches differ by constants. In particular, if the integrand is algebraic, an Abelian integral is obtained. For multiple-valued functions of this kind the single-valuedness can be restored by focussing the attention on the integrand, that is to say the differential, rather than on the integrated function. The differential possesses "periods" along the cycles on the Riemann surface, and these periods can be studied directly without referring to the covering surface.

From this point of view the single-valued analytic functions correspond to differentials whose periods are zero, and the harmonic functions are associated with differentials whose periods are purely imaginary.

§1. ELEMENTARY PROPERTIES OF DIFFERENTIALS

The first section deals only with differentials that satisfy certain conditions of regularity. A number of elementary properties will be proved, but it is found that many pertinent questions are very difficult to answer within the narrow frame to which we are confining ourselves. For this reason the present section is of a preliminary nature. It is designed to give a general orientation, and to point out the difficulties that lie ahead.

1. Differential calculus

1A. In a preliminary way the notion of differential was introduced in Ch. II, 6E. We wish now to make the definitions sharper and a little more general. Consider a surface $W=(W, \Phi)$ with a structure of class C^2. In order to define a first order differential on W we consider a collection of linear forms $a\,dx+b\,dy$, one for each $h \in \Phi$, whose coefficients a, b are complex-valued functions on $h(V)$, V=domain of h. If h_1, h_2 have overlapping domains, the corresponding forms $a_1\,dx_1+b_1\,dy_1$, $a_2\,dx_2+b_2\,dy_2$ shall be connected by the relations

$$
(1)\qquad
\begin{aligned}
a_1 &= a_2\frac{\partial x_2}{\partial x_1}+b_2\frac{\partial y_2}{\partial x_1}\\
b_1 &= a_2\frac{\partial x_2}{\partial y_1}+b_2\frac{\partial y_2}{\partial y_1}.
\end{aligned}
$$

When these relations hold the collection is said to determine a differential ω, and we write in generic notation

$$(2)\qquad \omega = a\,dx+b\,dy.$$

We note that it is sufficient to define a and b for the mappings h that belong to a structural basis (cf. Ch. II, 1F). We say that $\omega\in C^1$ if all coefficients a, b are of class C^1. Since Φ is of class C^2 it suffices to make this assumption for a basis.

1B. The transformation rules (1) are linear with respect to a, b. For this reason differentials can be added in the obvious manner, and a differential can be multiplied with a function f to give

$$f\omega = \omega f = fa\,dx+fb\,dy.$$

In other words, the differentials on W form a vector space over the ring of complex functions.

The *complex conjugate* of ω is

$$\bar\omega = \bar a\,dx+\bar b\,dy.$$

It is a differential, for the transformation (1) has real coefficients.

1C. The differential of a function $f\in C^1$ is defined to be

$$(3)\qquad \omega = df = \frac{\partial f}{\partial x}dx+\frac{\partial f}{\partial y}dy.$$

By virtue of the rule for forming the derivatives of a composite function it satisfies the invariance requirement (1). This makes the definition legitimate.

Any differential of the form (3) is said to be *exact*. It is clear that the sum of two exact differentials and a constant multiple of an exact differential are exact. Hence the exact differentials form a linear subspace of all differentials when the latter are considered as a vector space over the complex constants.

1D. We pass to the definition of a second order differential. To each local variable there corresponds an expression

$$\Omega = c\,dxdy,$$

and the transformation rule reads

$$c_1 = c_2 \frac{\partial(x_2, y_2)}{\partial(x_1, y_1)}. \tag{4}$$

There are obvious conventions as to the sum of two second order differentials and the product of a function and a second order differential.

The exterior product of two first order differentials

$$\omega_1 = a_1\,dx + b_1\,dy, \quad \omega_2 = a_2\,dx + b_2\,dy$$

is defined as

$$\omega_1\omega_2 = (a_1b_2 - a_2b_1)\,dxdy. \tag{5}$$

One verifies at once that the invariance condition (4) is satisfied.

Multiplication is distributive and anticommutative, $\omega_2\omega_1 = -\omega_1\omega_2$. The formula (5) results on setting $dxdx = dydy = 0$ and $dydx = -dxdy$.

1E. The symbolic differential

$$d = \frac{\partial}{\partial x}\,dx + \frac{\partial}{\partial y}\,dy$$

satisfies the transformation law (1), and this is why df is a first order differential. For the same reason the symbolic product

$$d\omega = \left(\frac{\partial b}{\partial x} - \frac{\partial a}{\partial y}\right) dxdy \tag{6}$$

represents a second order differential, provided that $\omega \in C^1$. We may regard $d\omega$ as the differential of a differential, just as df is the differential of a function.

One verifies the formulas

$$d(f\omega) = (df)\omega + f\,d\omega \tag{7}$$

and

$$d(df) = 0. \tag{8}$$

The latter formula is valid for $f \in C^2$ and expresses the equality of the mixed derivatives.

1F. Until further notice we assume that all functions on W are of class C^2, all first order differentials are of class C^1, and all second order differentials are continuous. Under these conditions all operations that we have introduced are applicable, and all formulas are valid.

We say that ω is *closed* if $d\omega = 0$. It follows from (8) that all exact differentials are closed. In other words, the exact differentials form a subspace of the closed differentials. A closed differential is locally exact, i.e., we can write $\omega = df$ provided that we restrict ω to a sufficiently small neighborhood of a point. In contrast, exactness is a global property.

2. Integration

2A. We recall that a singular 1-simplex is a continuous mapping $t \to f(t)$ of the closed unit interval into W (Ch. I, 33A). It is said to be differentiable if $f(t) = x(t) + iy(t)$ is of class C^1 in terms of the local variables, and the integral of ω along a differentiable 1-simplex σ is defined as

$$\int_\sigma \omega = \int_0^1 (ax'(t) + by'(t))\, dt.$$

More precisely, the definition is possible in a single step only if σ is contained in a parametric region V, and then the value is independent of the choice of the local variable. If σ is not contained in a single parametric region it is necessary to use a subdivision of the interval, and one shows that the result does not depend on the subdivision. Finally, the definition is extended to arbitrary differentiable 1-chains by linearity.

The following theorem is important:

Theorem. *A differential ω is exact if and only if $\int_\gamma \omega = 0$ for every cycle γ.*

We have suppressed as self-evident the condition that γ and ω must be differentiable (see 1F). The necessity follows from

$$\int_\sigma df = f(p_2) - f(p_1)$$

if $\partial\sigma = p_2 - p_1$. Since γ is a cycle the values at the end points will cancel against each other.

The sufficiency follows on setting

$$f(z) = \int_{z_0}^{z} \omega$$

where the integral is taken along an arc from z_0 to z. If the integral along any cycle is zero, $f(z)$ is well-defined, and one verifies that $\omega = df$.

2B. A singular 2-simplex is a mapping of the triangle $\Delta: 0 \leqq u \leqq t \leqq 1$ into W. It is differentiable if the mapping is of class C^1, and we consider only such simplices. The integral of a second order differential $\Omega = c\,dxdy$ is defined by

$$\int_\Delta \Omega = \iint_\Delta c \frac{\partial(x, y)}{\partial(t, u)}\, dtdu$$

provided that the image of Δ is contained in a single V. Due to the composition law for Jacobians the definition does not depend on the choice of the local variable in V. Note that no question of orientation is involved. It is again possible to extend the definition to arbitrary differentiable 2-simplices by subdivision of Δ, and to arbitrary 2-chains by linearity.

The basic duality between chains and differentials is expressed by the relation

$$\int_X d\omega = \int_{\partial X} \omega. \tag{9}$$

With the definitions that we have given the proof is evident, for the equation needs to be verified only for the triangle Δ.

By use of (7) we obtain the more general formulas for *partial integration*

$$\int_X (df)\omega = \int_{\partial X} f\omega - \int_X f\, d\omega. \tag{10}$$

2C. From (9) we obtain the following characterization of closed differentials:

Theorem. *A differential ω is closed if and only if $\int_\gamma \omega = 0$ for every cycle γ that is homologous to* 0.

The necessity is immediate. For the sufficiency we write $d\omega = c\,dxdy$ and recall that c is by assumption continuous. Thus, if $c \neq 0$ at a point it is either positive or negative throughout a whole neighborhood. There exists a differentiable one to one mapping of Δ into this neighborhood, and by hypothesis we should have

$$\iint_\Delta c \frac{\partial(x, y)}{\partial(t, u)}\, dtdu = 0.$$

But this is impossible since neither c nor the Jacobian changes sign. It follows that $d\omega = 0$.

If ω is closed the integral $\int_\gamma \omega$ will be referred to as the *period* of ω along γ. The theorem that we have just proved implies that the period depends only on the homology class of γ. By Theorem 2A a closed differential is exact if and only if all its periods are zero.

2D. We have slurred over a difficulty that needs some attention. Since we are now interested only in differentiable chains it is conceivable that the original definition of homology cannot be used. In order to make sure that the modified definition leads to the same homology group we must show (1) that every cycle is homologous to a differentiable cycle, (2) that every differentiable boundary is the boundary of a differentiable 2-chain. The proof is based on the technique of simplicial approximation and is a repetition of the argument used in Ch. I, 34.

2E. In addition to defining the integral of a second order differential over a 2-chain we shall also need to define its integral over the whole surface W. For this purpose it is essential to assume that the surface is orientable, and we begin by considering the case of a compact bordered or closed surface. The whole surface can then be considered as a 2-chain whose boundary is the positively oriented border. We define

$$\int_W c\, dxdy$$

as the integral over this 2-chain.

In the case of a noncompact surface we can set

$$\int_W c\, dxdy = \lim_{\Omega \to W} \int_\Omega c\, dxdy \tag{11}$$

provided that the limit exists. Due to the orientability, $|c|dxdy$ is also a second order differential, for if the Jacobian is positive (4) remains valid when c_1, c_2 are replaced by their absolute values. We find in the usual manner that (11) exists if and only if the limit

$$\int_W |c|\, dxdy = \lim_{\Omega \to W} \int_\Omega |c|\, dxdy \tag{12}$$

is finite. Consequently, the integral of $c\, dxdy$ is defined if and only if it is absolutely convergent.

3. Conjugate differentials

3A. We suppose now that $W = (W, \Phi)$ is a Riemann surface. The invariance condition (1) for a first order differential can then be written in the alternative form

$$\begin{aligned} -b_1 &= -b_2 \frac{\partial x_2}{\partial x_1} + a_2 \frac{\partial y_2}{\partial x_1} \\ a_1 &= -b_2 \frac{\partial x_2}{\partial y_1} + a_2 \frac{\partial y_2}{\partial y_1}. \end{aligned}$$

We see that the invariance of $\omega = a\,dx + b\,dy$ implies the invariance of

$$\omega^* = -b\,dx + a\,dy. \tag{13}$$

In the presence of a conformal structure it is thus possible to define a first order differential ω^*, called the *conjugate differential* of ω. It should not be confused with the complex conjugate $\bar{\omega}$.

We note the important formulas

$$\omega^{**} = -\omega, \quad \omega_1^* \omega_2^* = \omega_1 \omega_2. \tag{14}$$

3B. Let u be a harmonic function on W. Then

$$du = \frac{\partial u}{\partial x} dx + \frac{\partial u}{\partial y} dy$$

$$du^* = -\frac{\partial u}{\partial y} dx + \frac{\partial u}{\partial x} dy,$$

and Laplace's equation $\Delta u = 0$ shows that du^* is closed.

This provides a motivation for the following terminology:

Definition. *A differential ω is said to be harmonic if ω and ω^* are both closed.*

It is convenient to say that ω is coclosed if ω^* is closed. Thus a harmonic differential is one which is simultaneously closed and coclosed.

In the example, $\omega = du$ was exact. In the general case a harmonic differential ω is only *locally* the differential of a harmonic function u, and ω^* is locally the differential of a conjugate harmonic function u^*. Earlier, we have considered only real harmonic functions. In the present connection we are of course speaking of complex harmonic functions, that is to say combinations $u_1 + iu_2$ of two real harmonic functions. To guard against a misunderstanding that might be caused by the notation we stress that u_2 is *not* the conjugate harmonic function of u_1.

3C. Because of the first equation (14) the differentials ω and ω^* are simultaneously harmonic. The linear combination $\omega + i\omega^*$ has the special property

$$(\omega + i\omega^*)^* = -i(\omega + i\omega^*).$$

Any differential φ which satisfies $\varphi^* = -i\varphi$ is said to be *pure*, and a pure differential which is also harmonic will be called *analytic*. To emphasize this definition we state it in the following equivalent form:

Definition. *A differential φ is said to be analytic if it is closed and satisfies the condition $\varphi^* = -i\varphi$.*

A pure differential is of the form $a(dx + i\,dy) = a\,dz$, and it is analytic if and only if $\frac{\partial a}{\partial x} = -i\frac{\partial a}{\partial y}$. This is the Cauchy-Riemann condition which expresses that a is locally an analytic function of z.

The complex conjugate $\bar{\varphi}$ of an analytic differential is sometimes said to be antianalytic. If φ_1, φ_2 are analytic, then $\varphi_1 + \bar{\varphi}_2$ is harmonic. Conversely, starting from a harmonic differential ω we can set

$$\varphi_1 = \tfrac{1}{2}(\omega + i\omega^*)$$

$$\varphi_2 = \tfrac{1}{2}(\bar{\omega} + i\bar{\omega}^*)$$

and obtain $\omega = \varphi_1 + \bar{\varphi}_2$. In other words, every harmonic differential can be written as the sum of an analytic and an antianalytic differential. The representation is unique, for if $\varphi_1 = -\bar{\omega}_2$ we may pass to the conjugates to obtain $-i\varphi_1 = -i\bar{\varphi}_2$, and consequently $\varphi_1 = \varphi_2 = 0$.

4. The inner product

4A. From $\omega = a\,dx + b\,dy$ we obtain

$$\omega\bar{\omega}^* = (|a|^2 + |b|^2)\,dxdy.$$

This is a second order differential with a nonnegative coefficient, and we can hence form the integral

$$\int_W \omega\bar{\omega}^*$$

which is either finite or $+\infty$.

The positive square root of this integral is denoted by $\|\omega\|$, and we call it the *norm* of ω. It vanishes if and only if $\omega = 0$. One shows in the usual manner that the differentials with finite norm form a vector space over the complex numbers.

4B. If W is a closed surface all differentials have finite norm. In the case of an open surface the differentials with finite norm form only a subspace of all first order differentials.

In the remainder of this chapter we are exclusively concerned with differentials of finite norm. We maintain the convention whereby all first order differentials shall be of class C^1, and we introduce the notation $\Gamma^1(W)$, or Γ^1, for the space of all differentials $\omega \in C^1$ on W with $\|\omega\| < \infty$; the superscript 1 serves as a reminder that the differentials are of class C^1. If W is the interior of a bordered surface $\overline{W}$, it is convenient to distinguish between $\Gamma^1(W)$ and $\Gamma^1(\overline{W})$; the differentials in the latter class are supposed to remain of class C^1 on the border. The differentials of infinite norm are inaccessible to the methods we are going to use, and for this reason we refrain from introducing a special notation for the space of all differentials.

The closed, exact, and harmonic differentials in Γ^1 form linear subspaces which are conveniently denoted by Γ^1_c, Γ^1_e, Γ^1_h. Similarly, Γ^{1*}_c and Γ^{1*}_e will refer to the subspaces formed by all conjugate differentials of differentials in Γ^1_c and Γ^1_e respectively. The notation Γ^{1*}_h is superfluous, for it would coincide with Γ^1_h.

We have shown that $\Gamma^1_e \subset \Gamma^1_c$, and the definition of harmonic differentials implies $\Gamma^1_h = \Gamma^1_c \cap \Gamma^{1*}_c$.

4C. For elements of Γ^1 the integral

$$(\omega_1, \omega_2) = \int_W \omega_1 \bar{\omega}_2^* = \int_W (a_1\bar{a}_2 + b_1\bar{b}_2)\, dxdy$$

converges and is called the *inner product* of ω_1 and ω_2. By virtue of (14), 3A, it satisfies

$$(\omega_2, \omega_1) = \int_W \omega_2\bar{\omega}_1^* = -\int_W \omega_2^*\bar{\omega}_1 = \int_W \bar{\omega}_1\omega_2^* = (\bar{\omega}_1, \bar{\omega}_2) \tag{15}$$

and

$$(\omega_1^*, \omega_2^*) = -\int_W \omega_1^*\bar{\omega}_2 = (\omega_1, \omega_2). \tag{16}$$

The norm is subsumed under the definition of inner product by means of $\|\omega\|^2 = (\omega, \omega)$. For future reference we note Schwarz's inequality

$$|(\omega_1, \omega_2)| \leqq \|\omega_1\| \cdot \|\omega_2\|$$

and the triangle inequality

$$\|\omega_1 + \omega_2\| \leqq \|\omega_1\| + \|\omega_2\|$$

which results from

$$\begin{aligned}\|\omega_1+\omega_2\|^2 &= \|\omega_1\|^2 + \|\omega_2\|^2 + 2\,\mathrm{Re}\,(\omega_1, \omega_2)\\ &\leqq \|\omega_1\|^2 + \|\omega_2\|^2 + 2\|\omega_1\|\ \|\omega_2\|.\end{aligned}$$

4D. Two differentials $\omega_1, \omega_2 \in \Gamma^1$ are said to be *orthogonal* if $(\omega_1, \omega_2) = 0$. We say that ω is orthogonal to a subspace Γ^1_1 if $(\omega, \omega_1) = 0$ for all $\omega_1 \in \Gamma^1_1$, and that Γ^1_1 is orthogonal to Γ^1_2 if $(\omega_1, \omega_2) = 0$ for all $\omega_1 \in \Gamma^1_1$, $\omega_2 \in \Gamma^1_2$. We shall also say that Γ^1_1 is the orthogonal complement of Γ^1_2 if Γ^1_1 consists of exactly those elements which are orthogonal to Γ^1_2.

In the case of a closed surface W the following orthogonality relation holds:

Theorem. *On a closed surface Γ^1_c is the orthogonal complement of Γ^{1*}_e (and Γ^{1*}_c is the orthogonal complement of Γ^1_e).*

From

$$(\omega, df^*) = -\int_W \omega\, \overline{df} = -\int_W \bar{f}\, d\omega \tag{17}$$

we see that $d\omega=0$ implies $(\omega, df^*)=0$ and hence $\omega \perp \Gamma_e^{1*}$. Conversely, suppose that

$$\int_W \bar{f}\, d\omega = 0$$

for all functions $f \in C^2$. Consider a parametric disk V and assume that f is identically zero outside of a compact set in V. We map V on $|z|<1$ and write $d\omega = c\, dxdy$ in terms of this parameter. Then

$$\iint_{|z|<1} \bar{f}c\, dxdy = 0$$

for all $f \in C^2$ which are identically zero outside of a compact set in $|z|<1$.

It is elementary to show that this condition implies $c=0$. For the sake of completeness we indicate a proof. Choose $0<\rho<1$ and set

$$f(z) = (\rho^2 - |z|^2)^3 \quad \text{for} \quad |z| \leqq \rho$$
$$f(z) = 0 \quad \text{for} \quad |z| > \rho.$$

Then $f \in C^2$ and we obtain, in polar coordinates,

$$I(\rho) = \iint_{r<\rho} c(\rho^2 - r^2)^3 r\, dr d\Theta = 0.$$

This identity in ρ, if differentiated four times with respect to ρ^2, yields

$$\int_{r=\rho} c\, d\Theta = 0.$$

It follows by continuity that $c(0)=0$, and since $z=0$ can be identified with any point on W we have proved $d\omega=0$. In other words, ω is a closed differential.

4E. The reader must be careful to notice that by proving Theorem 4D we have by no means shown, conversely, that Γ_e^1 is the full orthogonal complement of Γ_c^{1*}. This converse is true, but we have a long way to go before we can prove it.

5. Differentials on bordered surfaces

5A. We shall say that a differential $\omega = a\, dx + b\, dy$ vanishes *along* a differentiable curve $z=z(t)$ if $ax'(t)+by'(t)=0$ for all t. We use this terminology to make it clear that we do not require ω to vanish *on* the curve, which would obviously mean that a and b vanish at all points of the curve. In particular, consider the border β of a bordered Riemann surface. At each point of the border there is, by definition, a local parameter $z=x+iy$

such that $y=0$ on β. In terms of this parameter $\omega = a\,dx + b\,dy$ vanishes along β if and only if $a=0$ on β.

For a compact bordered Riemann surface $\overline{W}$ we introduce two important subclasses of $\Gamma_c^1(\overline{W})$ and $\Gamma_e^1(\overline{W})$. We shall say that a closed differential ω belongs to the subclass $\Gamma_{c0}^1(W)$ if $\omega=0$ along β. Similarly, $\omega=df$ will be said to be of class $\Gamma_{e0}^1(W)$ if $f=0$ on β. The latter condition obviously implies $\omega=0$ along β, but conversely, if $df=0$ along β we can merely conclude that f is constant on each contour.

With this terminology we obtain two orthogonal relations which may be considered as generalizations of Theorem 4D.

Theorem. *On a compact bordered surface* Γ_{c0}^1 *is the orthogonal complement of* Γ_e^{1*}, *and* Γ_c^1 *is the orthogonal complement of* Γ_{e0}^{1*}.

For the proof we replace equation (17) by

$$(\omega, df^*) = \int_\beta \bar{f}\omega - \iint_W \bar{f}\,d\omega. \tag{18}$$

It follows immediately that $\omega \perp df^*$ if $d\omega=0$ and either $f=0$ on β or $\omega=0$ along β. In other words, $\Gamma_{c0}^1 \perp \Gamma_e^{1*}$ and $\Gamma_c^1 \perp \Gamma_{e0}^{1*}$.

Conversely, if $\omega \perp \Gamma_{e0}^{1*}$ it is evident that the reasoning used in 4D can be repeated to show that $d\omega=0$. The same will be true if $\omega \perp \Gamma_e^{1*}$, and consequently this hypothesis implies $\int_\beta \bar{f}\omega = 0$ for all $f \in C^2$.

We consider again a parametric mapping which carries an arc of β into the real axis. In terms of this parameter we have $\omega = a\,dx$ on β, and the condition on ω yields

$$\int_{-\infty}^{\infty} \bar{f}a\,dx = 0$$

provided that f vanishes identically outside of a certain compact set. From this fact it is elementary to conclude that $a=0$, and hence that $\omega \in \Gamma_{c0}^1$.

5B. There is an intermediate class between Γ_c^1 and Γ_{c0}^1 which will be of considerable importance. We are referring to the class of closed differentials whose periods along the contours of $\overline{W}$ are zero. For lack of a better name we shall call these differentials *semiexact*, and the class of semiexact differentials will be denoted by Γ_{se}^1.

For better orientation we draw the following inclusion diagram:

$$\Gamma_c^1 \supset \Gamma_{se}^1 \begin{array}{c} \supset \;\Gamma_e^1\; \supset \\ \\ \supset \;\Gamma_{c0}^1\; \supset \end{array} (\Gamma_{c0}^1 \cap \Gamma_e^1) \supset \Gamma_{e0}^1.$$

No special name is introduced for the intersection $\Gamma^1_{c0} \cap \Gamma^1_e$; obviously, it consists of the differentials of functions that are constant on the contours.

5C. We complete the orthogonality relations by proving:

Theorem. *Γ^1_{se} is the orthogonal complement of $\Gamma^{1*}_{c0} \cap \Gamma^{1*}_e$.*

First of all, if $\omega \in \Gamma^1_{se}$ and $df \in \Gamma^1_{c0}$ it follows by (18) that $\omega \perp df^*$. Hence $\Gamma^1_{se} \perp (\Gamma^{1*}_{c0} \cap \Gamma^{1*}_e)$. On the other hand, $\omega \perp (\Gamma^{1*}_{c0} \cap \Gamma^{1*}_e)$ implies $\omega \perp \Gamma^{1*}_{e0}$ and hence, by the second part of Theorem 5A, $\omega \in \Gamma^1_c$. Now, let f be a function which is 1 on one of the contours and 0 on the others. Then $df \in \Gamma^1_{c0} \cap \Gamma^1_e$, and from $\omega \perp df^*$ we are able to conclude by (18) that ω has zero period along the contour that was singled out. We find that $\omega \in \Gamma^1_{se}$ as asserted.

6. Differentials on open surfaces

6A. For an open Riemann surface W we wish to define classes $\Gamma^1_{c0}(W)$ and $\Gamma^1_{e0}(W)$ which are as closely analogous as possible to the corresponding classes on a compact bordered surface. It is natural to base the definitions on a zero behavior near the ideal boundary, but there are many different ways to postulate such a behavior. Fortunately, the specific choice of a definition is not very important due to the fact that the space Γ^1 and its subspaces play merely an auxiliary role in the complete theory of square integrable differentials.

It will be convenient to introduce the notion of *support*. By definition, the support of a function f on W is the closure of the set on which f is different from zero. Equivalently, the support of f is the intersection of all closed sets such that f is identically zero outside of the closed sets. We shall be particularly interested in functions with compact support, that is, functions which vanish outside of a compact set.

The notion of support can also be applied to a differential ω. Indeed, the vanishing of a differential at a point is independent of the choice of local variable. For this reason the support of ω can be defined as the smallest closed set outside of which ω is identically zero.

6B. We choose to adopt the following precise definitions:

Definition. *A closed differential $\omega \in \Gamma^1_c$ is said to belong to the subclass Γ^1_{c0} if and only if it has compact support.*

An exact differential $\omega \in \Gamma^1_e$ is said to belong to the subclass Γ^1_{e0} if and only if $\omega = df$ for a function f with compact support.

The inclusion relations $\Gamma^1_{c0} \subset \Gamma^1_c$, $\Gamma^1_{e0} \subset \Gamma^1_e$ and $\Gamma^1_{e0} \subset \Gamma^1_{c0}$ are immediate consequences of the definition. For closed surfaces $\Gamma^1_{c0} = \Gamma^1_c$ and $\Gamma^1_{e0} = \Gamma^1_e$.

6C. With the definitions we have chosen it is not possible to prove a complete analogue of Theorem 5A. However, the following is true:

Theorem. *On any Riemann surface* Γ^1_{c0} *is orthogonal to* Γ^{1*}_e, *and* Γ^1_c *is orthogonal to* Γ^{1*}_{e0}. *Moreover,* $\omega \perp \Gamma^{1*}_{e0}$ *implies* $\omega \in \Gamma^1_c$.

To prove the orthogonality we need only apply (18) to a subregion which contains the support of ω or the support of f, as the case may be. To show, in addition, that Γ^1_c is the full orthogonal complement of Γ^{1*}_{e0} we can repeat the argument in 4D. Indeed, the hypothesis $\omega \perp \Gamma^{1*}_{e0}$ implies that

$$\int_W \bar{f}\, d\omega = 0$$

for any f whose support is contained in a parametric disk V. This is sufficient, according to our earlier reasoning, to conclude that $d\omega = 0$, i.e., that $\omega \in \Gamma^1_c$.

§2. THE METHOD OF ORTHOGONAL PROJECTION

A traditional method to prove existence theorems on Riemann surfaces is by means of the Dirichlet principle. As is well known the earliest applications of this method were deficient in rigor. The first correct proof based on Dirichlet's principle was presented by Hilbert, and soon after a smoother approach was found by Weyl.

Although these proofs were strictly elementary, they were technically rather complicated. It is only in relatively recent years that it has become clear that the existence proofs for harmonic differentials are direct consequences of elementary results in Hilbert space theory, combined with simple facts in integration theory. As we develop this method in detail the reader should bear in mind that it differs only in technical respects from the original method of Riemann.

7. The completion of Γ^1

7A. The norm $\|\omega\|$ defined Γ^1 as a metric space in which the distance between ω_1 and ω_2 is $\|\omega_1 - \omega_2\|$. Accordingly, the sequence $\{\omega_n\}$ is said to converge to ω if and only if $\|\omega - \omega_n\| \to 0$.

A *Cauchy sequence* is one for which $\|\omega_m - \omega_n\| \to 0$ as $m, n \to \infty$. A normed space is *complete* if every Cauchy sequence is convergent.

The space Γ^1 is *not* complete. However, like any metric space it can easily be completed. In a concrete manner the completion can be accomplished by extending Γ^1 to the linear space Γ of all differentials $\omega = a\,dx + b\,dy$ whose coefficients in terms of local coordinates are no longer supposed to be of class C^1, but merely measurable in the sense of Lebesgue. The norm $\|\omega\|$ of a measurable differential can be defined in the same manner as before, and we agree that Γ shall comprise only the differentials with finite norm.

As usual, two differentials are identified if their coefficients differ only on a set of measure zero (in each coordinate system). With this convention Γ becomes a complete metric space, and Γ^1 is identified with a linear subspace of Γ. The completeness of Γ is equivalent to the Riesz-Fischer theorem with which the reader is assumed to be familiar (see F. Riesz-B. Sz. Nagy [1], p. 59).

7B. The definition of inner product can be extended to Γ in the same way as the norm. With the introduction of an inner product Γ becomes a *Hilbert space*. We shall need a few simple facts from the theory of Hilbert space, and in the interest of completeness brief proofs will be included.

Let A and B be linear subspaces of Γ. We shall say that Γ is the *direct sum* of A and B, and we write $\Gamma = A \dotplus B$, if and only if each element $\omega \in \Gamma$ has a unique representation $\omega = \alpha + \beta$ with $\alpha \in A$, $\beta \in B$. The notion of direct sum can be extended to any finite number of subspaces.

Theorem. *If $\Gamma = A \dotplus B$ and $A \perp B$, then A and B are mutually orthogonal complements of each other.*

It must be proved that $\omega \perp B$ implies $\omega \in A$. But this is immediate, for from $\omega = \alpha + \beta$ we obtain $\|\beta\|^2 = (\omega, \beta) - (\alpha, \beta) = 0$ and hence $\beta = 0$, $\omega = \alpha \in A$.

7C. If E is any subset of Γ we denote by $E^\perp$ the orthogonal complement of E, i.e., the set of all elements orthogonal to E. It is clear that $E^\perp$ is a linear subspace. Moreover, $E^\perp$ is closed in Γ regarded as a metric space. In fact, for fixed α the inner product (ω, α) is a continuous function of ω. Therefore, the set of all ω with $(\omega, \alpha) = 0$ is closed, and $E^\perp$ is the intersection of all such sets as α runs through E. It follows that $E^\perp$ is closed.

Theorem 7B shows that any representation of Γ as the direct sum of orthogonal subspaces is a decomposition into closed subspaces. Conversely, we shall show that there is a decomposition $\Gamma = A \dotplus A^\perp$ corresponding to any closed linear subspace A.

7D. An equivalent formulation of the afore-mentioned property is the following:

Theorem. *Let A be any closed linear subspace of Γ. Then there exists, for any $\omega \in \Gamma$, a unique $\alpha \in A$ such that $\omega - \alpha \perp A$.*

The uniqueness is clear, for if α_1, α_2 both have this property, then $\alpha_1 - \alpha_2$ is orthogonal to itself, and hence $\alpha_1 = \alpha_2$.

To prove the existence, let d be the greatest lower bound of $\|\omega - \Theta\|$ for $\Theta \in A$. There exists a sequence $\{\Theta_n\}$, $\Theta_n \in A$, such that $\|\omega - \Theta_n\| \to d$. We make use of the identity

$$\|2\omega - \Theta_m - \Theta_n\|^2 + \|\Theta_m - \Theta_n\|^2 = 2\|\omega - \Theta_m\|^2 + 2\|\omega - \Theta_n\|^2.$$

Here

$$\|2\omega-\Theta_m-\Theta_n\|^2 = 4\left\|\omega-\frac{\Theta_m+\Theta_n}{2}\right\|^2 \geqq 4d^2,$$

and we obtain

$$\|\Theta_m-\Theta_n\|^2 \leqq 2\|\omega-\Theta_m\|^2+2\|\omega-\Theta_n\|^2-4d^2.$$

It follows that $\|\Theta_m-\Theta_n\|\to 0$ for $m, n\to\infty$.

By the completeness of Γ, Θ_n converges to a limit α, and since A is closed, $\alpha \in A$. By continuity, $\|\omega-\alpha\| = d$.

Let Θ be any element of A. If $\Theta \neq 0$ we obtain by computation

$$\left\|\omega-\alpha-\frac{(\omega-\alpha,\Theta)}{\|\Theta\|^2}\Theta\right\|^2 = d^2-\frac{|(\omega-\alpha,\Theta)|^2}{\|\Theta\|^2}.$$

By the minimum property of d the left hand member is $\geqq d^2$. Hence $(\omega-\alpha,\Theta)=0$, and we have proved that $\omega-\alpha \perp A$.

7E. If A is a closed linear subspace we have just shown that $\Gamma = A \dotplus A^\perp$, whence it follows by Theorem 7B that $A^{\perp\perp} = A$. More generally, if A is an arbitrary linear subspace it is trivial that $A \subset A^{\perp\perp}$. If the closure is denoted by Cl A we have, on the other hand, $A \subset \mathrm{Cl}\, A$ and therefore $A^\perp \supset (\mathrm{Cl}\, A)^\perp$, $A^{\perp\perp} \subset (\mathrm{Cl}\, A)^{\perp\perp} = \mathrm{Cl}\, A$. But $A^{\perp\perp}$ is closed. Hence the inclusion $A \subset A^{\perp\perp} \subset \mathrm{Cl}\, A$ yields $A^{\perp\perp} = \mathrm{Cl}\, A$. We conclude further that $(\mathrm{Cl}\, A)^\perp = A^{\perp\perp\perp} = A^\perp$ so that A and its closure have the same orthogonal complement.

Corollary. *Any linear subspace $A \subset \Gamma$ satisfies $A^{\perp\perp} = \mathrm{Cl}\, A$ and $A^\perp = (\mathrm{Cl}\, A)^\perp$.*

7F. If A and B are any two linear subspaces, closed or not, we denote by $A+B$ the set of all elements that have a representation, not necessarily unique, of the form $\alpha+\beta$, $\alpha \in A$, $\beta \in B$. If the representation is unique, that is, if $A \cap B = 0$, we indicate this by the notation $A \dotplus B$. We call $A+B$ the *vector sum*.

Lemma. $(A+B)^\perp = A^\perp \cap B^\perp$.

$A \subset A+B$ implies $(A+B)^\perp \subset A^\perp$. Similarly, $(A+B)^\perp \subset B^\perp$, and hence $(A+B)^\perp \subset A^\perp \cap B^\perp$. In the opposite direction, $A^\perp \cap B^\perp \subset A^\perp$ yields $(A^\perp \cap B^\perp)^\perp \supset A^{\perp\perp} \supset A$, and on interchanging A and B, $(A^\perp \cap B^\perp)^\perp \supset B$. Hence $A+B \subset (A^\perp \cap B^\perp)^\perp$, $(A+B)^\perp \supset (A^\perp \cap B^\perp)^{\perp\perp} = A^\perp \cap B^\perp$. The lemma is proved.

7G. If A and B are closed it does not necessarily follow that $A+B$ is closed (see P. Halmos [1], p. 28). However, if A and B are orthogonal to each other the conclusion is correct.

Lemma. *If A, B are closed and $A \perp B$, then $A+B$ is closed.*

According to Corollary 7E and Lemma 7F it suffices to show that $(A^{\perp} \cap B^{\perp})^{\perp} \subset A+B$. Suppose that $\omega \perp A^{\perp} \cap B^{\perp}$. We can write $\omega = \alpha + \alpha^{\perp} = \beta + \beta^{\perp}$ with $\alpha \in A$, $\alpha^{\perp} \in A^{\perp}$, $\beta \in B$, $\beta^{\perp} \in B^{\perp}$. Then $\omega - \alpha - \beta = \alpha^{\perp} - \beta = \beta^{\perp} - \alpha \in A^{\perp} \cap B^{\perp}$. On the other hand, $\omega \perp A^{\perp} \cap B^{\perp}$ by assumption, and $\alpha + \beta \perp A^{\perp} \cap B^{\perp}$ by Lemma 7F. It follows that $\omega - \alpha - \beta$ is orthogonal to itself, $\omega = \alpha + \beta$. This is what we were required to prove.

8. The subclasses of Γ

8A. It is our intention to extend to Γ the notions of exact and closed differentials. At the same time we will define subspaces corresponding to Γ^1_{e0} and Γ^1_{c0}. To some extent the choice of definitions is a matter of convenience. We find it expedient to base the definitions partly on completion and partly on orthogonality.

Definition. *The subspaces Γ_e and Γ_{e0} are the closures in Γ of Γ^1_e and Γ^1_{e0} respectively. In contrast, we define Γ_c and Γ_{c0} as the orthogonal complements of Γ^*_{e0} and Γ^*_e.*

We will continue to refer to differentials $\omega \in \Gamma_e$ or $\omega \in \Gamma_c$ as exact or closed, but we are no longer able to express these properties through conditions of the form $\omega = df$ or $d\omega = 0$.

By the definitions of Γ_c and Γ_{c0} we have *postulated* the four orthogonal decompositions

$$(19) \qquad \begin{aligned} \Gamma &= \Gamma_c \dotplus \Gamma^*_{0e} = \Gamma^*_c \dotplus \Gamma_{e0} \\ \Gamma &= \Gamma_e \dotplus \Gamma^*_{c0} = \Gamma^*_e \dotplus \Gamma_{c0}. \end{aligned}$$

It is clear that the differentials of class Γ_{c0} have, implicitly, a zero behavior near the ideal boundary, but we have refrained from characterizing this behavior in explicit terms.

8B. The inclusion $\Gamma^1_{e0} \subset \Gamma^1_e$ implies $\Gamma_{e0} \subset \Gamma_e$, and on passing to the orthogonal complements we obtain $\Gamma_{c0} \subset \Gamma_c$. By Theorem 6C we have, in particular, $\Gamma^1_e \perp \Gamma^{1*}_{e0}$. Hence $\Gamma_e \perp \Gamma^*_{e0}$, and this orthogonality relation implies $\Gamma_e \subset \Gamma_c$ as well as $\Gamma_{e0} \subset \Gamma_{c0}$. In other words, all inclusion relations remain valid.

As further consequences of Theorem 6C we have evidently $\Gamma^1_c \subset \Gamma_c$, $\Gamma^1_{c0} \subset \Gamma_{c0}$. The last part of the same theorem implies

$$(20) \qquad \Gamma_c \cap \Gamma^1 = \Gamma^1_c.$$

The corresponding relation

$$(21) \qquad \Gamma_e \cap \Gamma_1 = \Gamma^1_e$$

is also true, but will require a nontrivial proof (see 11F).

9. Weyl's lemma

9A. The key theorem which governs the passage from Γ^1 to Γ is the following:

Theorem. *A differential which is simultaneously in Γ_c and in Γ_c^* is almost everywhere equal to a differential of class C^1, and hence equivalent to a harmonic differential.*

The last conclusion, namely that the differential will be harmonic is an immediate consequence of (20), for if ω is almost everywhere equal to $\omega^1 \in C^1$, then ω^1 belongs to $\Gamma_c \cap \Gamma_c^*$ at the same time as ω, and hence $\omega^1 \in \Gamma_c^1 \cap \Gamma_c^{1*} = \Gamma_h^1$. Henceforth we write Γ_h in place of Γ_h^1 and express the statement of the theorem through $\Gamma_c \cap \Gamma_c^* = \Gamma_h$.

9B. The theorem is of purely local character. Indeed, if ω is in $\Gamma_c \cap \Gamma_c^*$, that is, if it is orthogonal to Γ_{e0}^1 and Γ_{e0}^{1*}, then its restriction to a parametric disk V is orthogonal to $\Gamma_{e0}^1(V)$ and $\Gamma_{e0}^{1*}(V)$, so that the restriction belongs to $\Gamma_c(V) \cap \Gamma_c^*(V)$. If the theorem is proved for V it follows that $\omega \in C^1$ in V, and since V is arbitrary we can conclude that $\omega \in C^1$ on the whole surface.

The theorem will thus be established if we prove the following lemma that has become known as *Weyl's lemma*:

Lemma. *Suppose that ω is square integrable in $\Delta: |z| < 1$, and that $(\omega, df) = (\omega, df^*) = 0$ for all functions $f \in C^2$ with compact support in Δ. Then ω is equivalent to a differential of class C^1.*

9C. We shall denote by M_δ the operator which replaces an integrable function f by its mean value

$$M_\delta f(z) = \frac{1}{\pi\delta^2} \int_0^\delta \int_0^{2\pi} f(z + re^{i\Theta})\, r\, dr d\Theta$$

over a disk of radius δ. All functions f are supposed to be defined over the whole plane; if this is not originally so we set $f = 0$ outside the domain of definition.

Our proof will make essential use of the fact that $M_\delta f$ is more regular than f. To be explicit, $M_\delta f$ is continuous as soon as f is integrable, and it is of class C^1 if f is continuous.

Any two operators M_δ commute: $M_{\delta_1} M_{\delta_2} f = M_{\delta_2} M_{\delta_1} f$. Also, if either f or g has compact support

$$\iint (M_\delta f)\, g\, dxdy = \iint f (M_\delta g)\, dxdy \tag{22}$$

where the integrals are over the whole plane.

For a differential $\omega = a\,dx + b\,dy$ we write $M_\delta\omega = (M_\delta a)\,dx + (M_\delta b)\,dy$. With this notation, if $f \in C^1$ it is found that $M_\delta(df) = d(M_\delta f)$, and with the help of (22) one verifies that

$$(M_\delta\omega, df) = (\omega, M_\delta\, df) = (\omega, dM_\delta f) \tag{23}$$

where the inner products are again over the whole plane.

A somewhat deeper property is the fact that

$$f(z) = \lim_{\delta\to 0} M_\delta f(z) \tag{24}$$

almost everywhere for any integrable f. For the proof of this property we refer to standard text books on integration, for instance E. J. McShane [1], p. 376, or L. M. Graves [1], p. 258.

9D. Suppose now that ω satisfies the hypotheses of the lemma. Consider a function $f \in C^2$ whose support lies in $|z| < 1 - \delta_1 - \delta_2$. Then the support of $M_{\delta_1}M_{\delta_2}f$ is contained in Δ, and it follows by (23) and the hypothesis that

$$(M_{\delta_1}M_{\delta_2}\omega, df) = (\omega, dM_{\delta_1}M_{\delta_2}f) = 0.$$

Here $M_{\delta_1}M_{\delta_2}\omega$ is of class C^1, and we conclude that the restriction of $M_{\delta_1}M_{\delta_2}\omega$ to $|z| < 1 - \delta_1 - \delta_2$ belongs to Γ_c^{1*}. Precisely the same reasoning, applied to ω^*, shows that $M_{\delta_1}M_{\delta_2}\omega$ is also of class Γ_c^1, and we conclude that $M_{\delta_1}M_{\delta_2}\omega$ is harmonic for $|z| < 1 - \delta_1 - \delta_2$.

A harmonic differential has harmonic coefficients and is therefore equal to its mean. Because of this property we obtain

$$M_{\delta_1}M_{\delta_2}\omega = M_\delta M_{\delta_1}M_{\delta_2}\omega = M_{\delta_1}M_{\delta_2}M_\delta\omega$$

in $|z| < 1 - \delta - \delta_1 - \delta_2$. On letting δ_1 tend to 0 we find, by continuity,

$$M_{\delta_2}\omega = M_{\delta_2}M_\delta\omega$$

and for $\delta_2 \to 0$, by (24),

$$\omega = M_\delta\omega$$

almost everywhere in $|z| < 1 - \delta$. Iteration yields

$$\omega = M_\delta M_\delta\omega$$

almost everywhere in $|z| < 1 - 2\delta$. The iterated mean is of class C^1, and we have proved Weyl's lemma.

10. Orthogonal decompositions

10A. Because $\Gamma_{e0} \subset \Gamma_c$ the spaces Γ_{e0} and Γ_{e0}^* are orthogonal to each other. According to Theorem 7F and by (19) we obtain

$$\Gamma = \Gamma_{e0} + \Gamma_{e0}^* + (\Gamma_c \cap \Gamma_c^*)$$

where, by Theorem 9A, the last component is identical with Γ_h. On comparing with (19) it follows that we have also $\Gamma_c = \Gamma_h \dotplus \Gamma_{e0}$.

Theorem. *The whole space* Γ *has the orthogonal decomposition* $\Gamma = \Gamma_h \dotplus \Gamma_{e0} \dotplus \Gamma_{e0}^*$, *and the subspace* Γ_c *has the decomposition* $\Gamma_c = \Gamma_h \dotplus \Gamma_{e0}$.

It is convenient to represent the decomposition of a single element in the form $\omega = \omega_h + \omega_{e0} + \omega_{e0}^*$ where the subscripts serve to indicate the subspaces to which the components belong. It must be remembered, however, that in this notation ω_{e0}^* need not be the conjugate of ω_{e0}.

Another helpful device is to visualize the decomposition through a diagram (Fig. 14).

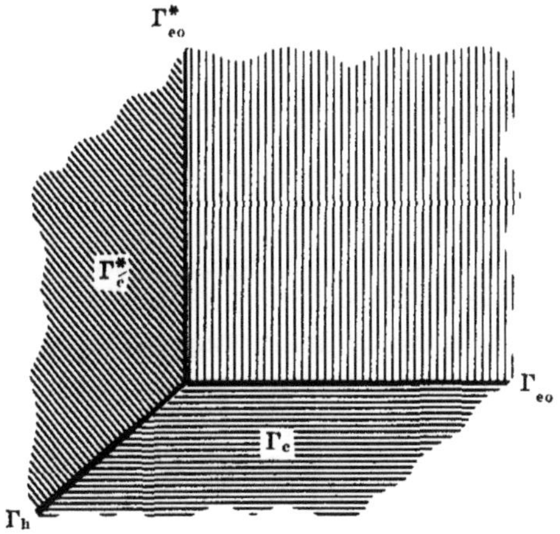

FIG. 14

Remark. The three-way decomposition is analogous to the one introduced by G. de Rham [1] in the theory of more-dimensional differential forms on compact manifolds.

10B. Suppose that ω_e is exact. From the decomposition $\omega_e = \omega_h + \omega_{e0}$ we read off that ω_h is exact, $\omega_h \in \Gamma_h \cap \Gamma_e$. Similarly, if $\omega_{c0} \in \Gamma_{c0}$ we can write $\omega_{c0} = \omega_h + \omega_{e0}$ and conclude this time that $\omega_h \in \Gamma_h \cap \Gamma_{c0}$. With the convenient notations $\Gamma_{he} = \Gamma_h \cap \Gamma_e$, $\Gamma_{h0} = \Gamma_h \cap \Gamma_{c0}$ we have thus

$$(25) \qquad \begin{aligned} \Gamma_e &= \Gamma_{he} \dotplus \Gamma_{e0} \\ \Gamma_{c0} &= \Gamma_{h0} \dotplus \Gamma_{e0}. \end{aligned}$$

It must be recalled that a differential in Γ_{he}, although exact in the sense of belonging to Γ_e, is not yet known to be exact in the original sense of having zero periods. In fact, formula (21) which would permit this conclusion is still unproved.

10C. With the help of (25) we obtain

$$\Gamma = \Gamma_e + \Gamma_{e0}^* = \Gamma_{he} + \Gamma_{h0}^* + \Gamma_{e0} + \Gamma_{e0}^*,$$

and on comparing with de Rham's decomposition (Theorem 10A)

$$\Gamma_h = \Gamma_{he} + \Gamma_{h0}^* = \Gamma_{h0} + \Gamma_{he}^*. \tag{26}$$

10D. The decomposition $\Gamma_c = \Gamma_h + \Gamma_{e0}$ permits us to prove:

Theorem. *Γ_c is the closure of Γ_c^1.*

We have already remarked that $\Gamma_c^1 \subset \Gamma_c$ (8B), and hence $\mathrm{Cl}\,\Gamma_c^1 \subset \Gamma_c$. On the other hand $\Gamma_{e0} = \mathrm{Cl}\,\Gamma_{e0}^1 \subset \mathrm{Cl}\,\Gamma_c^1$ and $\Gamma_h = \Gamma_h^1 \subset \mathrm{Cl}\,\Gamma_c^1$, so that $\Gamma_c = \Gamma_h + \Gamma_{e0} \subset \mathrm{Cl}\,\Gamma_c^1$. Consequently Γ_c^1 is dense in Γ_c.

11. Periods

11A. If ω is a closed differential of class C^1 the value of the integral

$$\int_\gamma \omega, \tag{27}$$

extended over a differentiable cycle, depends only on the homology class of γ. We shall need to extend the notion of period to arbitrary differentials in Γ_c. This cannot be done by interpreting (27) as a Lebesgue integral, for the coefficients of ω need not even be measurable functions of the curve parameter.

To circumvent this difficulty we regard the period (27) as a function, defined on Γ_c^1, and try to extend it to Γ_c by continuity. Since Γ_c^1 is dense in Γ_c (Theorem 10D) the extension is necessarily unique.

11B. Let Γ_0 be a normed linear space, complete or not. A *linear functional* on Γ_0 is a complex-valued function L, defined on Γ_0, with the property

$$L(c_1\omega_1 + c_2\omega_2) = c_1 L(\omega_1) + c_2 L(\omega_2).$$

The functional is *bounded* if there exists a constant M such that $|L(\omega)| \leqq M\|\omega\|$ for all $\omega \in \Gamma_0$.

It is obvious that a bounded linear functional is continuous, for $\|\omega - \omega_0\| \leqq \epsilon$ implies $|L(\omega) - L(\omega_0)| \leqq M\epsilon$. Conversely, suppose that L is continuous. There exists a $\delta > 0$ such that $\|\omega\| \leqq \delta$ implies $|L(\omega)| \leqq 1$. For an arbitrary $\omega \neq 0$ we have then

$$|L(\delta\|\omega\|^{-1}\omega)| \leqq 1$$

and hence

$$|L(\omega)| \leqq \frac{1}{\delta}\|\omega\|.$$

We conclude that a linear functional is bounded if and only if it is continuous.

Suppose now that Γ_0 is a linear subspace of a larger space Γ. A bounded linear functional L on Γ_0 can immediately be extended to the closure of Γ_0 in Γ. Indeed, any $\omega \in \mathrm{Cl}\,\Gamma_0$ is the limit of elements $\omega_n \in \Gamma_0$. From the fact that $|L(\omega_m)-L(\omega_n)| \leqq M\|\omega_m-\omega_n\|$ it follows that $\lim L(\omega_n)$ exists. It is independent of the sequence $\{\omega_n\}$, and if we set $L(\omega)= \lim L(\omega_n)$ it is an easy matter to verify that L is linear and bounded.

11C. In the case under consideration the period is a linear functional on Γ_c^1. If we can show that it is bounded it can be extended to $\Gamma_c = \mathrm{Cl}\,\Gamma_c^1$, and our aim will be achieved.

The proof of the boundedness is particularly simple if γ is an analytic Jordan curve. If that is so, γ can be covered by a finite number of parametric disks V_i with the property that a parametric mapping h_i of V_i takes $V_i \cap \gamma$ into the real axis. We shall construct a partition of unity (see Ch. II, 7B) which is adapted to this covering.

The principle of construction is the same as in Ch. II, 7C, but the details are slightly different. We wish to find functions $e_i \in C^2$ with compact support such that $\sum e_i = 1$ on γ. For this purpose, assume that h_i maps V_i onto $|z|<1$ with $V_i \cap \gamma$ corresponding to the real diameter. We set $g(z)=(1-|z|^2)^3$ and $g_i = g \circ h_i$ in V_i, $g_i = 0$ outside of V_i. Then each g_i is of class C^2.

This time we cannot divide by $\sum g_i$, for the sum is zero outside of $\cup V_i$. To meet this difficulty we cover the compact boundary of $\cup V_i$ by a finite number of parametric disks V_j' which do not meet γ. If h_j' maps V_j' on $|z|<1$, set $g_j' = g \circ h_j'$. Then $g_j' = 0$ on γ and $\sum g_i + \sum g_j' > 0$ throughout $\cup V_i$. We choose

$$e_i = \frac{g_i}{\sum g_i + \sum g_j'}$$

in V_i, $e_i = 0$ outside of V_i. Then $\sum e_i = 1$ on γ, and each e_i is of class C^2.

11D. We find now that

$$\int_\gamma \omega = \sum_i \int_\gamma e_i\omega.$$

Each integral in the right hand sum can be considered as an integral over the whole real axis in the parametric plane. Since $d\omega = 0$ we obtain

$$\int_\gamma e_i\omega = \int (de_i)\omega \tag{28}$$

where the integral on the right is extended over the upper half-plane. This leads to an inequality

$$\left|\int_\gamma e_i\omega\right| \leqq \|de_i\| \cdot \|\omega\| \tag{29}$$

and hence, by addition, to an estimate

$$\left| \int_{\gamma} \omega \right| \leq M(\gamma) \|\omega\|,$$

where $M(\gamma)$ depends only on γ.

With minor modifications the same proof applies if γ is a closed analytic curve with self-intersections.

11E. It is permissible to replace γ by any cycle in the same homology class. Therefore, the boundedness would be proved if we knew that each homology class contains a cycle which is the sum of a finite number of closed analytic curves. This fact, however, would require a fairly long proof.

What we do know is that each cycle is homologous to a sum of piecewise analytic closed curves. Therefore, we need only modify our procedure so as to take care of this case. At a vertex we choose a parametric mapping which transforms the meeting arcs into two analytic arcs from the origin. If the parametric disk is sufficiently small these arcs can be represented by equations of the form $z = re^{i\Theta_1(r)}$ and $z = re^{i\Theta_2(r)}$. They can be transformed into segments of the positive and negative real axis by a nonconformal mapping $z \to z'$ which is such that $|z'| = |z|$ while $\arg z'$, for constant $|z| = r$, varies linearly with $\arg z$ between $\Theta_1(r)$ and $\Theta_2(r)$. Although the mapping fails to be differentiable at the origin it is quite easy to see that (28) remains valid with the double integral extended over the upper half of the z'-plane. The estimate (29) will be the same as before, and the desired conclusion can be drawn.

11F. We have shown that the period along a fixed cycle γ is a bounded linear functional on Γ_c^1 which can be extended to Γ_c. This permits us to prove that $\Gamma_e \cap \Gamma^1 = \Gamma_e^1$, as predicted in 8B (21).

Lemma. $\Gamma_e \cap \Gamma^1 = \Gamma_e^1$.

Suppose that $\omega \in \Gamma_e \cap \Gamma^1$. Then ω is also in $\Gamma_c \cap \Gamma^1 = \Gamma_c^1$, and therefore ω has a period along any cycle γ, defined by the integral $\int_\gamma \omega$. By definition, ω is the limit, with respect to norm, of differentials $\omega_n \in \Gamma_e^1$. Each ω_n has the period 0 along γ, and because the period is a continuous functional it follows that ω has periods 0 along all γ. We conclude by Theorem 2A that $\omega \in \Gamma_e^1$, as asserted.

Let us note that we can also complement Theorem 4D (see 4E). If, on a closed surface, $\omega \in C^1$ is orthogonal to Γ_c^1, then ω is also orthogonal to the closure Γ_c. Hence $\omega \in \Gamma_{e0}^* = \Gamma_e^*$, and by the lemma $\omega \in \Gamma_e^{1*}$. Together with this result Theorem 4D shows that Γ_c^1 and Γ_e^{1*} are mutual orthogonal complements of Γ^1, provided that the surface is compact.

Due to the strong restrictions imposed on differentials of class Γ^1_{c0} and Γ^1_{e0} it cannot be expected that corresponding results would hold for open surfaces.

11G. It follows from $\Gamma_e \cap \Gamma^1 = \Gamma^1_e$ that $\Gamma_{he} = \Gamma_h \cap \Gamma_e = \Gamma_h \cap \Gamma^1_e$. In other words, the differentials of class Γ_{he} are exact in the original sense.

According to the terminology introduced in Ch. IV, 5A a surface is of class O_{HD} if there are no harmonic functions, other than constants, with a finite Dirichlet integral. This will be the case if and only if $\Gamma_{he}=0$. The decomposition $\Gamma_h = \Gamma_{h0} + \Gamma^*_{he}$ shows that $\Gamma_{he}=0$ is equivalent to $\Gamma_h = \Gamma_{h0}$. When $\Gamma_{he}=0$ it follows further, by (25), that $\Gamma_e = \Gamma_{e0}$ and $\Gamma_{c0} = \Gamma_{h0} + \Gamma_{e0} = \Gamma_h + \Gamma_{e0} = \Gamma_c$. Conversely, each of the conditions $\Gamma_e = \Gamma_{e0}$ or $\Gamma_c = \Gamma_{c0}$ implies $\Gamma_{he}=0$.

Theorem. *The three conditions $\Gamma_h = \Gamma_{h0}$, $\Gamma_c = \Gamma_{c0}$, $\Gamma_e = \Gamma_{e0}$ are equivalent to each other, and they are fulfilled if and only if the surface is of class O_{HD}.*

Consequently, the surfaces of class O_{HD} are precisely the surfaces on which there is no need to distinguish between Γ_c and Γ_{c0} or between Γ_e and Γ_{e0}.

12. Reproducing differentials

12A. The following is an elementary theorem in Hilbert space theory:

Theorem. *On a Hilbert space Γ every bounded linear functional is of the form $L(\omega) = (\omega, \sigma)$ for some fixed $\sigma \in \Gamma$.*

We note first that (ω, σ) is a linear functional. It is bounded, for $|(\sigma, \omega)| \leqq \|\sigma\| \, \|\omega\|$.

Suppose that L is a bounded linear functional on Γ. Since L is continuous, $L(\omega)$ vanishes on a closed linear subspace Γ_0. If $L(\omega)$ is not identically zero, let Θ be an element with $L(\Theta) \neq 0$. According to Theorem 7D we can write $\Theta = \Theta_0 + \Theta_1$ with $\Theta_0 \in \Gamma_0$, $\Theta_1 \perp \Gamma_0$. Then $L(\Theta) = L(\Theta_1) \neq 0$, so that $\Theta_1 \neq 0$. We can therefore set

$$\sigma = \overline{L(\Theta_1)} \|\Theta_1\|^{-2} \, \Theta_1.$$

It is evident that $\sigma \perp \Gamma_0$ and $L(\sigma) = \|\sigma\|^2$.

The element $L(\omega)\sigma - L(\sigma)\omega$ belongs to Γ_0 for every ω. It is hence orthogonal to σ, and we find

$$L(\omega)\|\sigma\|^2 - L(\sigma)(\omega, \sigma) = 0,$$

whence $L(\omega) = (\omega, \sigma)$. This proves the theorem except in the trivial case that $L(\omega)$ is identically zero. In this case the assertion is true with $\sigma = 0$.

Needless to say, σ is unique, for if $(\omega, \sigma_1) = (\omega, \sigma_2)$ for all ω, then $\sigma_1 = \sigma_2$.

12B. We apply the preceding theorem to the space Γ_c and the bounded functional $\int_\gamma \omega$, extended to Γ_c. For reasons of symmetry we prefer to denote the differential whose existence is asserted by σ^* rather than σ. Thus, there exists a $\sigma \in \Gamma_c^*$ such that

$$\int_\gamma \omega = (\omega, \sigma^*)$$

for all $\omega \in \Gamma_c$. But the period vanishes for all exact ω. Hence σ^* is orthogonal to Γ_e, and by definition this means that $\sigma^* \in \Gamma_{c0}^*$, $\sigma \in \Gamma_{c0}$. We have consequently $\sigma \in \Gamma_{c0} \cap \Gamma_c^* = \Gamma_{h0}$.

Theorem. *To every cycle γ there corresponds a unique real harmonic differential σ of class Γ_{h0}, such that*

$$\int_\gamma \omega = (\omega, \sigma^*) \tag{30}$$

for all closed differentials ω.

All that remains to be proved is that σ is real. For this purpose we need only note that the period of $\bar{\omega}$ is the complex conjugate of the period of ω. Hence

$$(\omega, \sigma^*) = (\bar{\omega}, \sigma^*)^- = (\omega, \bar{\sigma}^*),$$

and we conclude by the uniqueness that $\sigma = \bar{\sigma}$.

When (30) is fulfilled we say that σ is the *reproducing differential* associated with γ, and we write frequently $\sigma = \sigma(\gamma)$. It is evident that $\sigma(\gamma_1 + \gamma_2) = \sigma(\gamma_1) + \sigma(\gamma_2)$.

13. Compact bordered surfaces

13A. We will now take up a closer study of the differentials on a compact bordered Riemann surface $\overline{W}$. It will be recalled that a differential in $\Gamma^1(\overline{W})$ is required to remain of class C^1 on the border, and that the subclasses $\Gamma_{c0}^1(\overline{W})$ and $\Gamma_{e0}^1(\overline{W})$ were defined in a way that refers to the behavior on the border itself rather than near the border (see 5A).

Differentials on $\overline{W}$ can be restricted to W, and in this sense $\Gamma^1(\overline{W})$ can be regarded as a subspace of $\Gamma^1(W)$. In terms of this identification it is evident that $\Gamma_{c0}^1(W) \subset \Gamma_{c0}^1(\overline{W})$ and $\Gamma_{e0}^1(W) \subset \Gamma_{e0}^1(\overline{W})$. We shall derive various other relationships between the subclasses of $\Gamma(W)$ and $\Gamma^1(\overline{W})$.

13B. The first of these results can be regarded as a justification for the definitions of $\Gamma_{e0}(W)$ and $\Gamma_{c0}(W)$.

Lemma. $\Gamma_{e0}^1(\overline{W}) \subset \Gamma_{e0}(W)$ and $\Gamma_{c0}^1(\overline{W}) \subset \Gamma_{c0}(W)$.

Because of the orthogonal decompositions $\Gamma(W) = \Gamma_{e0}(W) \dotplus \Gamma_c^*(W) = \Gamma_{c0}(W) \dotplus \Gamma_e^*(W)$ it is sufficient to prove that $\Gamma_{e0}^1(\overline{W}) \perp \Gamma_c^*(W)$ and $\Gamma_{c0}^1(\overline{W}) \perp \Gamma_e^*(W)$. For the purpose of proving these orthogonality relations we can even replace $\Gamma_c^*(W)$ and $\Gamma_e^*(W)$ by the dense subspaces $\Gamma_c^{1*}(W)$ and $\Gamma_e^{1*}(W)$.

To prove the first relation, suppose that $df \in \Gamma^1_{e0}(\overline{W})$ and $\omega = a\,dx + b\,dy \in \Gamma^1_c(W)$. By means of a partition of unity it is possible to write f as a finite sum $\sum f_i$ where the support of each f_i is contained in a parametric disk or half-disk. The relation $(df_i, \omega^*) = 0$ that we have to prove is trivial in the case of a full disk. For this reason the whole proof can be localized to a half-disk $|z| < 1$, $y \geqq 0$ which contains the support of f_i; in the following we write f for f_i.

We find at once that

$$(31) \qquad (df, \omega^*) = -\lim_{\eta \to 0} \int_{y=\eta} f\bar{a}\,dx.$$

Therefore, we have also

$$(32) \qquad (df, \omega^*) = -\lim_{\eta \to 0} \frac{1}{\eta} \iint_{0<y<\eta} f\bar{a}\,dxdy.$$

But $f = 0$ for $y = 0$, and since f has continuous derivatives $|f| \leqq My$ with constant M. On using this estimate we obtain

$$\left| \frac{1}{\eta} \iint_{0<y<\eta} f\bar{a}\,dxdy \right| \leqq M \iint_{0<y<\eta} |a|\,dxdy.$$

The integral on the right tends to 0 because a is square integrable, and we conclude that $(df, \omega^*) = 0$.

13C. To prove the second part of the lemma we assume that $df \in \Gamma^1_e(W)$ and $\omega \in \Gamma^1_{c0}(\overline{W})$. As before, we localize by setting $f = \sum f_i$, $f_i = e_i f$. It must be shown that $\|df_i\| < \infty$, and that the limit in (31) vanishes for $f = f_i$. Since $df_i = e_i\,df + f\,de_i$ the first assertion follows if f is square integrable over the support of e_i.

For $|x| < 1$, $0 < y < 1$ we have

$$|f(x, y)| \leqq \int_y^1 \left|\frac{\partial f}{\partial y}\right| dy + A,$$

where A is an upper bound of $|f(x, 1)|$. It follows by use of the Schwarz inequality that

$$\int_{-1}^{1} |f(x, y)|^2\,dx \leqq 2(\|df\|^2 + 2A^2).$$

When integrated with respect to y this inequality shows that f is square integrable over the rectangle $|x| < 1$, $0 < y < 1$. At the same time we conclude that $\int_{-1}^{1} |f_t|\, dx \leqq \int_{-1}^{1} |f|\, dx$ is bounded. Because a tends uniformly to 0 for $y \to 0$ it follows that $\int_{-1}^{1} f_t(x, y)\, \bar{a}\, dx \to 0$, and the lemma is proved.

13D. The purpose of the lemma is to lead to the following characterization of the class $\Gamma_{h0}(W)$, introduced in 10B:

Theorem. $\Gamma_{h0}(W) = \Gamma^1_{c0}(\overline{W}) \cap \Gamma^{1*}_c(\overline{W})$.

It is appropriate to denote the class on the right by $\Gamma_{h0}(\overline{W})$, so that the theorem reads: $\Gamma_{h0}(W) = \Gamma_{h0}(\overline{W})$. The inclusion $\Gamma_{h0}(\overline{W}) \subset \Gamma_{h0}(W)$ is a direct consequence of Lemma 13B. In fact, it follows from $\Gamma^1_{c0}(\overline{W}) \subset \Gamma_{c0}(W)$ and $\Gamma^1_c(\overline{W}) \subset \Gamma^1_c(W) \subset \Gamma_c(W)$ (see 8B) that $\Gamma_{h0}(\overline{W}) \subset \Gamma_{c0}(W) \cap \Gamma^*_c(W) = \Gamma_{h0}(W)$.

13E. The proof of the opposite inclusion will depend on symmetrization. To explain this idea, let us consider the double $\hat{W}$ of W (Ch. II, 3E). The involutory mapping j of $\hat{W}$ on itself has the following property: If $p \to h(p)$ is a parametric mapping with domain V, then $p \to \bar{h}(j(p))$ is a parametric mapping with domain $j(V)$ ($\bar{h}$ is the complex conjugate of h).

With every differential ω on $\hat{W}$ we associate another differential $\omega^{\sim}$ as follows: If $\omega = a(z)dx + b(z)dy$ in terms of the variable $z = h(p)$ in V, then $\omega^{\sim} = a(\bar{z})dx - b(\bar{z})dy$ in terms of $z = \bar{h}(j(p))$ in $j(V)$. It is clear that $\omega^{\sim}$ satisfies the consistency relations, for it is obtained from ω by the transformation $z \to \bar{z}$. For the same reason differentiation and exterior multiplication are invariant, but because the mapping is not directly conformal conjugation is not invariant. We find indeed that $\omega^{*\sim} = -b(\bar{z})\, dx - a(\bar{z})\, dy = -\omega^{\sim *}$.

13F. We say that ω is *even* if $\omega^{\sim} = \omega$, and *odd* if $\omega^{\sim} = -\omega$. Thus, if ω is even, ω^* is odd.

Let ω be odd, and let σ be an arbitrary differential in $\Gamma(\hat{W})$. On taking into account that the involutory mapping j is sense-reversing we obtain

$$
(33) \qquad
\begin{aligned}
(\omega, \sigma)_{\hat{W}} &= \int_{W \cup j(W)} \omega\bar{\sigma}^* = \int_W \omega\bar{\sigma}^* - \omega^{\sim}\bar{\sigma}^{*\sim} \\
&= \int_W \omega(\bar{\sigma}^* - \bar{\sigma}^{\sim *}) = (\omega, \sigma - \sigma^{\sim})_W
\end{aligned}
$$

where the subscripts indicate that the first inner product is over $\hat{W}$ and the last over W.

13G. Suppose now that $\omega \in \Gamma_{h0}(W)$ is given. We extend ω to $\hat{W}$ so that it becomes odd; the extension, which we continue to denote by ω, is undefined on the border, but it determines a unique element of $\Gamma(\hat{W})$. We claim that ω will be harmonic on $\hat{W}$. It can then be defined by continuity on the border. Because it is odd it will vanish along the border, and this will prove that ω belongs to $\Gamma_{h0}(\overline{W})$.

Consider a differential $\sigma \in \Gamma_e^1(\hat{W})$. Clearly, $\sigma - \sigma^{\sim}$ is in $\Gamma_{e0}^1(\overline{W})$ and hence, by Lemma 13B, in $\Gamma_{e0}(W)$. Because $\omega \in \Gamma_c^*(W)$ it follows that $\omega \perp \sigma - \sigma^{\sim}$ on W and thus, by (33), that $(\omega, \sigma)_{\hat{W}} = 0$. On the other hand $(\omega, \sigma^*)_{\hat{W}} = (\omega, \sigma^* + \sigma^{\sim *})_W = 0$ because $\omega \in \Gamma_{c0}(W)$ and $\sigma^* + \sigma^{\sim *} \in \Gamma_e^*(W)$. We have proved that the extension ω is orthogonal to $\Gamma_e^1(\hat{W})$ as well as $\Gamma_e^{1*}(\hat{W})$, and hence harmonic. By our previous remark this implies $\Gamma_{h0}(W) \subset \Gamma_{h0}(\overline{W})$, and the proof of Theorem 13D is complete.

14. Schottky differentials

14A. We continue to investigate the relations between differentials on a compact bordered surface $\overline{W}$ and on its double.

Any differential ω on $\hat{W}$ can be decomposed into an even and an odd part. In fact, if we set

$$\omega_1 = \tfrac{1}{2}(\omega + \omega^{\sim}),\ \omega_2 = -\tfrac{1}{2}(\omega^* + \omega^{*\sim}),$$

then ω_1 and ω_2 are even, and $\omega = \omega_1 + \omega_2^*$. It is obvious that there is only one such decomposition with even ω_1 and ω_2.

14B. Suppose that ω is harmonic on $\overline{W}$. Then ω_1 and ω_2 are harmonic. But we know more, for since ω_1^* and ω_2^* are odd they vanish along the border of $\overline{W}$, and this means that their restrictions to $\overline{W}$ or W are of class $\Gamma_{h0}(\overline{W}) = \Gamma_{h0}(W)$. In other words, in the decomposition $\omega = \omega_1 + \omega_2^*$ one of the components is of class $\Gamma_{h0}(W)$ and the other of class $\Gamma_{h0}^*(W)$.

Conversely, any differential of class $\Gamma_{h0}(\overline{W})$ or $\Gamma_{h0}^*(\overline{W})$ can be continued to $\hat{W}$ by reflection so that it remains harmonic. Hence the whole class of differentials on W which have a harmonic extension to $\hat{W}$ is given by the direct sum $\Gamma_{h0}(W) \dotplus \Gamma_{h0}^*(W)$. Such differentials will be called *Schottky differentials* in honor of Schottky who was the first to study Riemann surfaces with a symmetry.

Theorem. *The class of Schottky differentials is identical with the direct sum* $\Gamma_{h0}(W) \dotplus \Gamma_{h0}^*(W)$.

Note that $\Gamma_{h0}(W)$ and $\Gamma_{h0}^*(W)$ are not orthogonal as subspaces of $\Gamma(W)$ but their elements have extensions which are odd and even respectively, and which are therefore orthogonal on $\hat{W}$. It is for this reason that $\Gamma_{h0}(W) \cap \Gamma_{h0}^*(W)$ reduces to 0 and $\Gamma_{h0}(W) \dotplus \Gamma_{h0}^*(W)$ is closed (see Lemma 7G).

14C. Let us return to the case of an arbitrary open Riemann surface W which we exhaust by regular subregions Ω. We show that any differential $\omega_{h0} \in \Gamma_{h0}(W)$ can be approximated by differentials $\omega_{h0\Omega}$ of class $\Gamma_{h0}(\Omega)$; the approximation is in terms of norm convergence.

The restriction of ω_{h0} to Ω has a decomposition of the form

$$\omega_{h0} = \omega_{h0\Omega} + \omega^*_{he\Omega} \tag{34}$$

where the subscripts are self-explanatory. Suppose that $\Omega \subset \Omega'$. Then $\omega_{h0\Omega} - \omega_{h0\Omega'} = \omega^*_{he\Omega'} - \omega^*_{he\Omega}$ belongs to $\Gamma^*_{he}(\Omega)$ and is therefore orthogonal to $\omega_{h0\Omega}$ on Ω. It follows that

$$\|\omega_{h0\Omega} - \omega_{h0\Omega'}\|^2_\Omega = \|\omega_{h0\Omega'}\|^2_\Omega - \|\omega_{h0\Omega}\|^2_\Omega, \tag{35}$$

and we conclude that $\|\omega_{h0\Omega}\|_\Omega \leqq \|\omega_{h0\Omega'}\|_\Omega \leqq \|\omega_{h0\Omega'}\|_{\Omega'}$. In other words, the quantity $\|\omega_{h0\Omega}\|_\Omega$ increases with Ω. But it is also bounded, for the orthogonal decomposition (34) shows that $\|\omega_{h0\Omega}\|_\Omega \leqq \|\omega_{h0}\|_\Omega \leqq \|\omega_{h0}\|$.

We find that $\|\omega_{h0\Omega}\|_\Omega$ has a finite limit, and this implies, by (35), that

$$\|\omega_{h0\Omega} - \omega_{h0\Omega'}\|_\Omega \to 0 \tag{36}$$

as Ω and $\Omega' \supset \Omega$ tend to W. For a fixed Ω_0 it follows by use of the triangle inequality that

$$\|\omega_{h0\Omega'} - \omega_{h0\Omega''}\|_{\Omega_0} \to 0$$

as $\Omega', \Omega'' \to W$, independently of each other. We may conclude, for instance by Ch. II, Theorem 13C, that $\omega_{h0\Omega}$ tends to a harmonic limit differential which we denote, temporarily, by ω_{h0W}. It follows further, by (36), that

$$\|\omega_{h0\Omega} - \omega_{h0W}\|_\Omega \to 0. \tag{37}$$

If σ is exact on W, $\sigma \in \Gamma_e$, we have $(\omega_{h0\Omega}, \sigma^*)_\Omega = 0$ and hence

$$(\omega_{h0W}, \sigma^*)_\Omega = (\omega_{h0W} - \omega_{h0\Omega}, \sigma^*)_\Omega.$$

By (37) the inner product on the right tends to 0, and we conclude that $(\omega_{h0W}, \sigma^*) = 0$. It follows that $\omega_{h0W} \in \Gamma_{h0}(W)$. With the help of (34) we see that the difference $\omega_{h0} - \omega_{h0W}$ is both in $\Gamma^*_{he}(W)$ and in $\Gamma_{h0}(W)$. These subspaces are orthogonal complements, and we must have $\omega_{h0} = \omega_{h0W}$.

Suppose conversely that $\omega = \lim \omega_{h0\Omega}$ in the sense that $\|\omega - \omega_{h0\Omega}\|_\Omega \to 0$. For any $\sigma \in \Gamma_e$ we get

$$(\omega, \sigma^*) = \lim_{\Omega \to W} (\omega, \sigma^*)_\Omega = \lim_{\Omega \to W} (\omega - \omega_{h0\Omega}, \sigma^*)_\Omega = 0,$$

and we conclude that $\omega \in \Gamma_{h0}(W)$. Thus $\Gamma_{h0}(W)$ is formed precisely by those differentials which can be approximated by differentials of class $\Gamma_{h0}(\Omega)$.

We record this result as a theorem:

Theorem. *$\Gamma_{h0}(W)$ is the limit of $\Gamma_{h0}(\Omega)$ for $\Omega \to W$ in the sense that $\omega \in \Gamma_{h0}(W)$ if and only if there exist differentials $\omega_{h0\Omega} \in \Gamma_{h0}(\Omega)$ such that $\|\omega - \omega_{h0\Omega}\|_{\Omega} \to 0$.*

14D. We are interested in determining all differentials on W which can be approximated, in the sense of norm, by Schottky differentials, i.e., by differentials in $\Gamma_{h0}(\Omega) + \Gamma^*_{h0}(\Omega)$. Such differentials can appropriately be called Schottky differentials on W, and we denote the linear subspace that they form by $\Gamma_S(W)$.

Theorem 14C asserts that $\Gamma_{h0}(W) \subset \Gamma_S(W)$ and hence also $\Gamma^*_{h0}(W) \subset \Gamma_S(W)$. It follows that $\Gamma_S(W)$ contains the linear subspace spanned by $\Gamma_{h0}(W)$ and $\Gamma^*_{h0}(W)$. Because $\Gamma_S(W)$ is closed, by the nature of its definition, it will even contain the smallest *closed* linear subspace that contains $\Gamma_{h0}(W)$ and $\Gamma^*_{h0}(W)$.

With the terminology introduced in 7E we have thus $\Gamma_S \supset \mathrm{Cl}(\Gamma_{h0} + \Gamma^*_{h0})$. We claim that the inclusion is an equality:

Theorem. $\Gamma_S = \mathrm{Cl}(\Gamma_{h0} + \Gamma^*_{h0})$.

In the following all orthogonal complements will be with respect to Γ_h. Then $\Gamma_{h0}^{\perp} = \Gamma^*_{he}$, $\Gamma^{*\perp}_{h0} = \Gamma_{he}$, and by Corollary 7E, together with Lemma 7F,

$$\mathrm{Cl}(\Gamma_{h0} + \Gamma^*_{h0}) = (\Gamma_{h0} + \Gamma^*_{h0})^{\perp\perp} = (\Gamma_{he} \cap \Gamma^*_{he})^{\perp}.$$

We have to show that $\Gamma_S \perp \Gamma_{he} \cap \Gamma^*_{he}$. Suppose, therefore, that $\omega \in \Gamma_S$ and $\sigma \in \Gamma_{he} \cap \Gamma^*_{he}$. We approximate ω by differentials $\omega_{S\Omega} = \omega_{h0\Omega} + \omega^*_{h0\Omega}$ in such a way that $\|\omega - \omega_{S\Omega}\|_{\Omega} \to 0$. Because σ and σ^* are exact, $(\omega_{S\Omega}, \sigma)_{\Omega} = 0$. Hence

$$(\omega, \sigma) = \lim_{\Omega \to W} (\omega, \sigma)_{\Omega} = \lim (\omega - \omega_{S\Omega}, \sigma)_{\Omega} = 0,$$

which is the desired conclusion.

14E. The space $\Gamma_{he} \cap \Gamma^*_{he}$ has another representation. If ω and ω^* are both exact and harmonic, then $\omega - i\omega^*$ and $\bar{\omega} + i\bar{\omega}^*$ are exact analytic differentials. In other words, we can write $\omega = df + d\bar{g}$ where f and g are analytic functions. Conversely, any such differential belongs to $\Gamma_{he} \cap \Gamma^*_{he}$. Thus $\Gamma_{he} \cap \Gamma^*_{he} = \Gamma_{ae} + \bar{\Gamma}_{ae}$, and this is an orthogonal decomposition, for an analytic and an antianalytic differential are always orthogonal to each other.

We obtain the following decomposition:

Theorem. *On any Riemann surface $\Gamma_h = \Gamma_S + \Gamma_{ae} + \bar{\Gamma}_{ae}$.*

As a particularly interesting consequence we note that $\Gamma_h = \Gamma_S$ if and only if $\Gamma_{ae} = 0$, that is to say if the surface does not allow any nonconstant analytic functions with finite Dirichlet integral. In the terminology of Ch. IV we have thus:

Corollary. *A surface is of class O_{AD} if and only if all square integrable harmonic differentials are Schottky differentials.*

14F. We have found that $\Gamma_{h0} \cap \Gamma^*_{h0} = 0$ if W is the interior of a compact bordered surface $\overline{W}$, while $\Gamma_{h0} = \Gamma^*_{h0} = \Gamma_h$ if W is of class O_{HD} (Theorem 11G). The condition $\Gamma_{h0} \cap \Gamma^*_{h0} = 0$ seems thus to be indicative of a strong boundary. Other results in this direction are not known.

15. Harmonic measures

15A. For the case of a compact bordered surface $\overline{W}$ we proved in 5C that $\Gamma^1_{se}(\overline{W})$, the space of semiexact differentials, is the orthogonal complement of $\Gamma^{1*}_{c0}(\overline{W}) \cap \Gamma^{1*}_e(\overline{W})$. If we limit ourselves to harmonic differentials, then the corresponding result is that $\Gamma_{hse}(\overline{W})$, the space of harmonic semiexact differentials, is the orthogonal complement of $\Gamma^*_{h0}(\overline{W}) \cap \Gamma^*_{he}(\overline{W})$.

The space $\Gamma_{h0} \cap \Gamma_{he}$ has a very simple meaning. Indeed, if ω is in this space, then ω is exact, $\omega = du$, and u is a harmonic function which is constant on each contour. As such u can be written as a linear combination of harmonic functions which are 1 on one contour, 0 on all others. Harmonic functions with boundary values 0 and 1 are commonly known as harmonic measures. We shall borrow this terminology and refer to $\Gamma_{h0} \cap \Gamma_{he}$ as the space of *harmonic measures* with the abbreviation Γ_{hm}. Theorem 5C asserts that $\Gamma_{hse}(\overline{W})$ and $\Gamma^*_{hm}(\overline{W})$ are orthogonal complements.

15B. This result will now be generalized to open surfaces. First of all, the definition of Γ_{hse} is quite easy to extend. In Ch. I, 27C, 30D we introduced the notion of a dividing cycle. In singular homology theory a cycle γ is a dividing cycle if and only if, for any compact set E, γ is homologous to a cycle that lies completely outside of E. Equivalently, γ divides W if it is homologous to a cycle on the boundary of any regular region Ω that contains γ.

We define Γ_{hse} to be the class of all harmonic differentials whose periods along the dividing cycles of W are zero, and a differential ω with this property will be called semiexact. Consider the restriction of a semiexact differential ω to a subregion Ω. Evidently, the restriction will be semiexact on Ω provided that the contours of Ω are dividing cycles, that is, if Ω is a canonical region in the terminology of Ch. I, 35A.

15C. As for the definition of $\Gamma_{hm}(W)$ we proceed as in the case of Schottky differentials, except that we consider only canonical regions Ω. Consequently, a differential of class $\Gamma_{hm}(W)$ will be defined as one which can be approximated in norm by differentials of class $\Gamma_{hm}(\Omega)$ *with respect* *to canonical regions* Ω. Precisely, $\omega \in \Gamma_{hm}(W)$ if and only if for every $\epsilon > 0$ and every compact set E there exists a canonical region $\Omega \supset E$ and a harmonic measure $\omega_{hm\Omega} \in \Gamma_{hm}(\Omega)$ such that $\|\omega - \omega_{hm\Omega}\|_\Omega < \epsilon$.

15D. Using this terminology we prove:

Theorem. *The subspaces* $\Gamma_{hse}(W)$ *and* $\Gamma^*_{hm}(W)$ *are orthogonal complements in* $\Gamma_h(W)$.

It is immediate that the spaces are orthogonal. We suppose that $\sigma \in \Gamma_{hse}(W)$ and $\omega \in \Gamma_{hm}(W)$. Then, because Ω is canonical, σ is semiexact on Ω, so that $(\sigma, \omega^*_{hm\Omega})_\Omega = 0$. We have thus

$$(\sigma, \omega^*)_\Omega = (\sigma, \omega^* - \omega^*_{hm\Omega})_\Omega,$$

and this inner product can be made arbitrarily small, while Ω is arbitrarily large. This implies $(\sigma, \omega^*) = 0$, and the orthogonality is proved.

Suppose now that ω is harmonic and orthogonal to $\Gamma^*_{hse}(W)$. For a canonical region Ω, let $\omega_{hm\Omega}$ be the projection of ω, restricted to Ω, on $\Gamma_{hm}(\Omega)$. Then $\omega - \omega_{hm\Omega} \in \Gamma^*_{hse}(\Omega)$. If $\Omega \subset \Omega'$ we conclude that $\omega_{hm\Omega} - \omega_{hm\Omega'} \in \Gamma^*_{hse}(\Omega)$, and hence that $\omega_{hm\Omega} - \omega_{hm\Omega'} \perp \omega_{hm\Omega}$. Therefore,

$$\begin{aligned}\|\omega_{hm\Omega} - \omega_{hm\Omega'}\|^2_\Omega &= \|\omega_{hm\Omega'}\|^2_\Omega - \|\omega_{hm\Omega}\|^2_\Omega \\ &\leqq \|\omega_{hm\Omega'}\|^2_{\Omega'} - \|\omega_{hm\Omega}\|^2_\Omega.\end{aligned}$$

We see that $\|\omega_{hm\Omega}\|_\Omega$ increases with Ω. Since $\|\omega_{hm\Omega}\|_\Omega \leqq \|\omega\|$ the limit for $\Omega \to W$ is finite, and we conclude that $\omega_{hm} = \lim \omega_{hm\Omega}$ exists and lies in $\Gamma_{hm}(W)$. Furthermore, because $\omega - \omega_{hm\Omega} \in \Gamma^*_{hse}(\Omega)$ and every dividing cycle lies in an Ω it follows that $\omega - \omega_{hm} \in \Gamma^*_{hse}$. On the other hand $\omega \perp \Gamma^*_{hse}$ by assumption and $\omega_{hm} \perp \Gamma^*_{hse}$ by the part of the theorem that we have already proved. We conclude that $\omega = \omega_{hm}$, and hence that $\Gamma_{hm} = \Gamma^{*\perp}_{hse}$. Since Γ_{hse} is obviously closed it is also true that $\Gamma^*_{hse} = \Gamma^\perp_{hm}$.

The proof requires a comment. We have actually never shown that every compact set is contained in a canonical regular region. It is not difficult to see, however, that the proof is valid for canonical regions with piecewise analytic boundaries. The existence of an exhaustion by such regions follows on applying Ch. I, 29 to an analytic triangulation (compare with the construction in Ch. II, 12D).

15E. The content of the last theorem can also be expressed through the orthogonal decomposition

$$\Gamma_h = \Gamma_{hse} \dotplus \Gamma^*_{hm} = \Gamma^*_{hse} \dotplus \Gamma_{hm}. \tag{38}$$

The spaces Γ_{hm} and Γ^*_{hm} are orthogonal to each other, for $\Gamma_{hm} \subset \Gamma_{he} \subset \Gamma_{hse}$. Therefore we have also the three-way decomposition

$$\Gamma_h = \Gamma_{hse} \cap \Gamma^*_{hse} \dotplus \Gamma_{hm} \dotplus \Gamma^*_{hm}. \tag{39}$$

On passing to the orthogonal complements, the string of inclusions $\Gamma_{hm} \subset \Gamma_{he} \subset \Gamma_{hse}$ implies $\Gamma_{hse} \supset \Gamma_{h0} \supset \Gamma_{hm}$. We have thus, in particular, $\Gamma_{hm} \subset \Gamma_{h0} \cap \Gamma_{he}$. It can be expected that $\Gamma_{hm} = \Gamma_{h0} \cap \Gamma_{he}$ for a large class of surfaces, but this property is probably not universal.

15F. We can use Theorem 15D to strengthen Theorem 12B. The following additional information is obtained:

Theorem. *The reproducing differential $\sigma(\gamma)$ associated with a dividing cycle γ is of class Γ_{hm}.*

In fact, $(\omega, \sigma(\gamma)^*)$ is the period of ω along γ. If ω is semiexact, this period is 0. Hence $\sigma(\gamma)^* \perp \Gamma_{hse}$, and we have seen that this implies $\sigma(\gamma) \in \Gamma_{hm}$.

15G. Various generalizations suggest themselves, but because these questions have not been subject to systematic research we shall give only a brief indication. As in Ch. III, 7D we can fix our attention on the cycles which are dividing with respect to a given regular partition P of the ideal boundary (Ch. I, 38). We are led to introduce a class $(P)\Gamma_{hse}$ which consists of all harmonic differentials with zero periods along cycles that are dividing with respect to P.

For a regular subregion Ω we consider the partition of its contours which is induced by P, and we let the class $(P)\Gamma_{hm}(\Omega)$ be formed by all harmonic differentials du in Ω such that u has the same constant value on all contours that belong to the same part in the induced partition. For the open surface we define $(P)\Gamma_{hm}$ as the limit of $(P)\Gamma_{hm}(\Omega)$ in the sense of norm convergence. With these definitions one shows, exactly as above, that $(P)\Gamma_{hm}$ and $(P)\Gamma_{hse}^*$ are orthogonal complements.

16. Analytic differentials

16A. The orthogonal decompositions that we have considered appear under a slightly different aspect when attention is restricted to analytic differentials.

Let us first examine the decomposition

$$\Gamma_h = \Gamma_{ae} \dotplus \bar{\Gamma}_{ae} \dotplus \Gamma_S$$

derived in 14E. It is clear that an analytic differential has no component in $\bar{\Gamma}_{ae}$, and its component in Γ_S is analytic. In other words, if we write $\Gamma_a \cap \Gamma_S = \Gamma_{aS}$ we obtain

$$\Gamma_a = \Gamma_{ae} \dotplus \Gamma_{aS}.$$

It is thus seen that the orthogonal complement of Γ_{ae} with respect to the analytic differentials is formed by the class of all analytic Schottky differentials. In a certain sense Γ_{aS} takes over the role played by Γ_{h0}.

16B. In view of the importance which must now be attached to the class of analytic Schottky differentials it is worth while to study them in greater detail. Let us first examine the case of a compact bordered Riemann surface $\overline{W}$.

If φ is an analytic Schottky differential we can write

$$\varphi = \omega_1 + \omega_2^*$$

where $\omega_1, \omega_2 \in \Gamma_{h0}$. Since $\varphi^* = -i\varphi$ we have necessarily

$$\omega_1^* - \omega_2 = -i(\omega_1 + \omega_2^*)$$

whence

$$\omega_1 + i\omega_2 = i\omega_1^* - \omega_2^*.$$

But for a compact bordered surface we know that $\Gamma_{h0} \cap \Gamma_{h0}^* = 0$. Hence $\omega_1 + i\omega_2 = 0$ and we obtain

$$\varphi = \omega_1 + i\omega_1^*.$$

Let us now divide ω_1 in its real and imaginary parts: $\omega_1 = \alpha + i\beta$. We obtain

$$\varphi = (\alpha + i\alpha^*) + i(\beta + i\beta^*).$$

Here α and β are known to vanish along the border of $\overline{W}$. With a change of notation we can therefore write

$$\varphi = \varphi_1 + i\varphi_2$$

where φ_1, φ_2 are analytic, and real along the border.

We can state the result as follows:

An analytic Schottky differential on $\overline{W}$ can be written as the sum of an analytic differential which is real along the border and one which is purely imaginary along the border.

The representation is obviously unique. Moreover, φ_1 and φ_2 are Schottky differentials, for

$$\varphi_1 = \frac{\varphi_1 - \bar{\varphi}_1}{2} + i\left(\frac{\varphi_1 - \bar{\varphi}_1}{2}\right)^*$$

and φ_2 has a similar representation by differentials in Γ_{h0} and Γ_{h0}^*.

16C. An analytic Schottky differential φ on an open surface W is by definition a limit of Schottky differentials $\omega_{S\Omega}$. In particular, the Schottky differential of best approximation is obtained from the orthogonal decomposition in Ω,

$$\varphi = \omega_{ae\Omega} + \omega_{aS\Omega}.$$

Hence $\varphi = \lim \omega_{aS\Omega}$, and we see that φ is necessarily a limit of analytic Schottky differentials. By use of the preceding result it follows that we can write

$$\varphi = \lim_{\Omega \to W} (\varphi_{1\Omega} + i\varphi_{2\Omega})$$

where $\varphi_{1\Omega}$ and $\varphi_{2\Omega}$ are real along the border of $\bar{\Omega}$. We are not able to assert, however, that $\varphi_{1\Omega}$ and $\varphi_{2\Omega}$ tend separately to limits.

16D. We proceed to a study of the class Γ_{ase} of analytic semiexact differentials. We start from the decomposition (39) in 15E,

$$\Gamma_h = \Gamma_{hse} \cap \Gamma^*_{hse} + \Gamma_{hm} + \Gamma^*_{hm},$$

and remark that

$$\Gamma_{hse} \cap \Gamma^*_{hse} = \Gamma_{ase} + \bar{\Gamma}^*_{ase}.$$

For an analytic differential the component in $\bar{\Gamma}_{ase}$ is 0, and we have thus

$$\Gamma_a = \Gamma_{ase} + \Gamma_a \cap (\Gamma_{hm} + \Gamma^*_{hm}).$$

In order to simplify the second component, let us suppose that $\omega_1, \omega_2 \in \Gamma_{hm}$ while $\omega_1 + \omega_2^*$ is analytic. We find that $\omega_1^* - \omega_2 = -i\omega_1 - i\omega_2^*$ or $\omega_2 - i\omega_1 = \omega_1^* + i\omega_2^*$, and since Γ_{hm} is orthogonal to Γ^*_{hm} it follows that $\omega_2 = i\omega_1$. We conclude:

Any analytic differential φ_a has a unique representation in the form

$$\varphi_a = \varphi_{ase} + \omega_{hm} + i\omega^*_{hm} \tag{40}$$

where φ_{ase} is semiexact analytic and ω_{hm} is a harmonic measure.

§3. PERIODS AND SINGULARITIES

We shall now show how to include harmonic and analytic differentials with singularities in our treatment. This is important not only because of the greater generality, but above all because there is an interesting duality between periods and singularities. In last analysis this duality depends on topological properties. It is therefore not surprising that part of our efforts have to be spent on the construction of closed differentials with given periods and singularities, a problem that is much closer to topology than to analysis.

In the classical case of closed Riemann surfaces the main problem is to construct harmonic and analytic differentials from their periods and singularities. When we try to generalize the classical results to open surfaces our emphasis will be on restrictive conditions which guarantee that at least part of the classical theory can be extended in a nontrivial manner. These restrictions will not bear on the surfaces, but merely on the differentials that are brought under consideration.

17. Singular differentials

17A. Consider a point p_0 on a Riemann surface W. We say that a differential Θ_0 defines a singularity at p_0 if it is defined in a punctured neighborhood of p_0. Another differential Θ, also defined in a punctured neighborhood of p_0, is said to have the singularity Θ_0, or to define the same singularity as Θ_0, if and only if $\Theta - \Theta_0$ is square integrable over a punctured neighborhood. If Θ_0 is itself square integrable over a punctured neighborhood the singularity which it defines may be called removable.

17B. A *harmonic singularity* is one that can be described by a Θ_0 which is harmonic in a punctured neighborhood of p_0. Similarly, we speak of an *analytic singularity* if Θ_0 is an analytic differential. Every harmonic differential Θ_0, even if defined only in a punctured neighborhood, has a standard representation $\Theta_0 = \varphi_0 + \bar{\psi}_0$ where φ_0 and ψ_0 are analytic (see 3C). Since $\|\Theta_0\|^2 = \|\varphi_0\|^2 + \|\psi_0\|^2$ we see that Θ_0 is a removable singularity if and only if φ_0 and ψ_0 define removable singularities.

Our main problem will be to construct globally defined harmonic differentials with given harmonic singularities. For this purpose it is evidently sufficient to construct a harmonic differential with the singularity φ_0, and one with the singularity $\bar{\psi}_0$. In other words, we lose no generality by assuming that Θ_0 is analytic. We shall find this a significant simplification.

17C. The possible isolated analytic singularities are easily determined. Suppose that the singular point p_0 corresponds to $z=0$ in terms of a local variable which maps a parametric disk onto $|z|<1$. We can write $\Theta_0 = f dz$ where f is analytic for $0<|z|<1$.

The Laurent development of f has the form

$$f(z) = \sum_{m=0}^{\infty} a_m z^m + \sum_{n=1}^{\infty} b_n z^{-n}.$$

The square norm $\|f dz\|^2$, extended over an annulus $r<|z|<1$, is found to be

$$4\pi\left[\sum_{m=0}^{\infty} \frac{1}{2m+2}(1-r^{2m+2})|a_m|^2 + |b_1|^2 \log\frac{1}{r} + \sum_{n=2}^{\infty} \frac{1}{2n-2}(r^{2-2n}-1)|b_n|^2\right].$$

Hence the singularity is removable if and only if all b_n are 0. The most general analytic singularity is thus represented by the singular part

$$\sum_{n=1}^{\infty} b_n z^{-n}\, dz \tag{41}$$

of a Laurent development. If only a finite number of the b_n are different from zero we say that the singularity is a *pole*. We speak also of a *harmonic pole* in the corresponding sense.

17D. It is desirable to interpret the problem of singularities in a very general manner so that it includes the problem of singularities and boundary behavior. The following key theorem reduces the problem to one that is concerned only with closed differentials:

Theorem. *Let Θ be a closed differential with a finite number of analytic singularities which is analytic outside of a compact set. Then there exists a unique harmonic differential τ such that $\tau - \Theta$ is of class Γ_{e0}.*

The hypothesis shall mean that Θ is of class C^1 and satisfies $d\Theta = 0$ except at the singular points. The differential τ will be harmonic except for having the same singularities as Θ. In addition, τ will have the periods of Θ, and in a certain sense τ has the same boundary behavior as Θ.

The uniqueness is clear, for if τ_1 and τ_2 are two solutions, then $\tau_1 - \tau_2$ will be harmonic and of class Γ_{e0}. Such a differential is necessarily 0 (Theorem 10A).

To prove the existence we note that $\Theta - i\Theta^*$ is identically zero near the singularities and near the ideal boundary. Hence it is square integrable and we can use the decomposition theorem (Theorem 10A) to obtain

$$\Theta - i\Theta^* = \omega_h + \omega_{e0} + \omega_{e0}^*$$

where the terms on the right belong respectively to Γ_h, Γ_{e0} and Γ_{e0}^*. On rewriting the equation in the form

$$\Theta - \omega_{e0} = i\Theta^* + \omega_h + \omega_{e0}^*$$

we find that the differential on the left is closed and the differential on the right coclosed on any region that does not contain any singular points. Hence $\tau = \Theta - \omega_{e0}$ is harmonic, and the theorem is proved.

17E. In all applications Θ will be identically 0 outside of a compact set. This has the consequence that Θ, and hence τ, will have only a finite number of nonzero periods. More precisely, there is a homology basis such that τ has zero period for all but a finite number of cycles in the basis. It is by limitations of this nature that we single out a special class of singular differentials which can be made the object of more detailed study.

18. Differentials of the second kind

18A. Traditionally, an analytic differential is said to be of the first kind if it has no singularities, of the second kind if the singularities are poles with vanishing residues, that is, with $b_1 = 0$, and of the third kind if the singularities are arbitrary poles. For the study of differentials of the second kind we have thus to consider poles of the form

$$\frac{dz}{(z-\zeta)^{m+2}}, \tag{42}$$

$m \geqq 0$. In this notation z is a local variable with range $|z| < 1$, and ζ is the value of z at the singular point. If we were concerned only with the existence of a harmonic differential with the singularity (42) we could just as well take $\zeta = 0$. Actually, it is one of our main objectives to study the solution in its dependence on ζ.

18B. Let r_1, r_2 be chosen so that $|\zeta| < r_1 < r_2 < 1$. We construct an auxiliary function $e(z)$ of class C^2 on $|z| < 1$ which is identically 1 for $|z| < r_1$ and identically 0 for $|z| > r_2$. It can be extended to the whole surface W by setting $e = 0$ outside of the disk Δ represented by $|z| < 1$.

Define

$$v_m = -\frac{1}{m+1} \cdot \frac{1}{(z-\zeta)^{m+1}}$$

on Δ, and set

$$\Theta_m = d(ev_m) = v_m de + \frac{e\,dz}{(z-\zeta)^{m+2}}$$

on Δ, $\Theta_m = 0$ on $W - \Delta$. Then Θ_m is closed, of class C^1, with the analytic singularity (42). It is also exact and vanishes outside of a compact set.

By Theorem 17D there exists a unique harmonic differential τ_m such that $\tau_m - \Theta_m \in \Gamma_{e0}$. We shall use the standard representation

$$\tau_m = \varphi_m + \bar{\psi}_m$$

where φ_m and ψ_m are analytic. It is found that φ_m has the singularity (42) while ψ_m is regular and square integrable. Observe that $\varphi_m + \psi_m$ has purely imaginary periods and $\varphi_m - \psi_m$ has real periods.

In order to indicate the dependence on ζ we write

$$\varphi_m = h_m(z, \zeta)dz$$

$$\psi_m = k_m(z, \zeta)dz.$$

It must be carefully noted, however, that h_m and k_m depend not only on the value of ζ, but also on the choice of local variable.

Observe that τ_m, φ_m and ψ_m do not depend on the choice of the auxiliary function $e(z)$. Indeed, if we change e to e', Θ_m will be replaced by $\Theta'_m = d(e'v_m)$. The difference $\Theta'_m - \Theta_m = d[(e'-e)v_m]$ is of class Γ^1_{e0}. Therefore there is no change in τ_m.

18C. Let $\alpha = a dz$ be a square integrable analytic differential. We shall prove:

Lemma. $(\alpha, \psi_m) = \dfrac{2\pi}{(m+1)!} a^{(m)}(\zeta).$

We have $(\alpha, \psi_m) = (\alpha, \bar{\tau}_m - \bar{\varphi}_m)$. If it were not for the singularity we would have $(\alpha, \bar{\varphi}_m) = 0$, for the integrand $\alpha\varphi_m^* = -i\alpha\varphi_m$ is identically zero. To make legitimate use of this observation we must avoid the singularity by excluding a disk δ about ζ of radius r. With this precaution we obtain

$$(\alpha, \psi_m) = \lim_{r\to 0} (\alpha, \bar{\tau}_m)_{W-\delta}.$$

For convenience the limit will be written as an improper inner product $(\alpha, \bar{\tau}_m)$, and this simplified notation will be used in similar cases whenever the limit is known to exist.

Because $\tau_m - \Theta_m \in \Gamma_{e0}$ we have $(\alpha, \bar{\tau}_m - \bar{\Theta}_m) = 0$ (see 10A), and thus $(\alpha, \bar{\tau}_m) = (\alpha, \bar{\Theta}_m) = (\Theta_m, \bar{\alpha})$. But $\Theta_m \alpha$ vanishes for $|z| < r_1$ because Θ_m is analytic, and for $|z| > r_2$ because $\Theta_m = 0$. Hence

$$(\Theta_m, \bar{\alpha}) = -i \iint\limits_{r_1 \leq |z| \leq r_2} d(ev_m)\alpha = i \int\limits_{|z| = r_1} v_m \alpha$$

$$= -\frac{i}{m+1} \int\limits_{|z|=r_1} \frac{a\, dz}{(z-\zeta)^{m+1}},$$

and by use of Cauchy's integral formula we obtain

$$(\alpha, \psi_m) = (\Theta_m, \bar{\alpha}) = \frac{2\pi}{(m+1)!} a^{(m)}(\zeta).$$

18D. In the special case $m = 0$ the lemma yields

$$(\alpha, \psi_0) = 2\pi\, a(\zeta) \tag{43}$$

or, more explicitly,

$$\frac{i}{2\pi} \int\limits_W a(z)\, \overline{k_0(z, \zeta)}\, dz d\bar{z} = a(\zeta).$$

With the more familiar notation $dz d\bar{z} = -2i\, dx dy$ this relation takes the form

$$\frac{1}{\pi} \iint\limits_W a(z)\, \overline{k_0(z, \zeta)}\, dx dy = a(\zeta). \tag{44}$$

We say that $\overline{k_0(z, \zeta)}$ is the *reproducing kernel* for analytic differentials (Bergman's kernel function, S. Bergman [1]).

In particular, if ζ' is another point on W we can apply (44) with $a(z) = k_0(z, \zeta')$ and find

$$\frac{1}{\pi} \iint\limits_W k_0(z, \zeta')\, \overline{k_0(z, \zeta)}\, dx dy = k_0(\zeta, \zeta').$$

On interchanging ζ and ζ' we obtain $k_0(\zeta', \zeta) = \overline{k_0(\zeta, \zeta')}$ or, with a change of notation,

$$k_0(z, \zeta) = \overline{k_0(\zeta, z)}. \tag{45}$$

This law of symmetry shows above all that $k_0(z, \zeta)$ is antianalytic in the variable ζ. Moreover, since $k_0(\zeta, z)\, d\zeta$ is independent of the choice of the local variable at ζ, the same is true of $k_0(z, \zeta)\, d\zeta$. We can express this result by saying that the double differential

$$k_0(z, \zeta)\, dz d\zeta \tag{46}$$

is invariant with respect to changes of both variables.

The notation is now free from ambiguity. It is well to emphasize, however, that (46) is *not* a second order differential and should not be regarded as an exterior product.

18E. Let us now apply (43) to $\alpha = \psi'_m = k_m(z, \zeta')\, dz$. We find

$$(\psi'_m, \psi_0) = 2\pi\, k_m(\zeta, \zeta').$$

On the other hand, the general form of Lemma 18C applied to $\alpha = \psi_0$ yields

$$(\psi_0, \psi'_m) = \frac{2\pi}{(m+1)!} \frac{d^m}{d\zeta'^m} k_0(\zeta', \zeta).$$

On comparing the two expressions and changing the notation we obtain

$$k_m(z, \zeta) = \frac{1}{(m+1)!} \frac{d^m}{dz^m} k_0(z, \zeta). \tag{47}$$

The relations (47) show that k_m is obtained from k_0 by repeated differentiations. In particular, k_m is antianalytic in ζ.

18F. Similar results hold for $\varphi_m = h_m(z, \zeta)\, dz$. We show first that (α, φ_m) exists and equals 0. We have $(\alpha, \varphi_m) = (\alpha, \varphi_m + \bar{\psi}_m) = (\alpha, \tau_m) = (\alpha, \Theta_m)$, provided that the integrals have a meaning. The product (α, Θ_m) is extended over $|z| \leq r_2$, but we have to exclude a disk about ζ of radius r. If the circumference of the disk is denoted by γ we obtain

$$(\alpha, \Theta_m) = (\bar{\Theta}_m, \bar{\alpha}) = -i \iint \bar{\Theta}_m \alpha$$
$$= -i \iint d(\bar{v}_m \bar{e}) = \lim_{r \to 0} \left[-\frac{i}{m+1} \int_\gamma \frac{a\, dz}{(\bar{z} - \bar{\zeta})^{m+1}} \right].$$

But $\bar{z} - \bar{\zeta} = r^2/(z - \zeta)$ on γ. Hence the integral vanishes, and we have proved that

$$(\alpha, \varphi_m) = 0. \tag{48}$$

For this formula it is essential that the inner product be defined as a Cauchy limit, i.e., by excluding circular disks centered at ζ.

18G. When ζ is replaced by ζ' we add a prime to all notations. In order to express φ_m in terms of φ_0 we note that $(\psi'_m, \varphi_0) = (\psi'_0, \varphi_m) = 0$ implies

$(\tau_m, \bar{\tau}_0'^*) = (\varphi_m + \bar{\psi}_m, i\bar{\varphi}_0' - i\psi_0') = 0$. On the other hand $(\tau_m - \Theta_m, \bar{\tau}_0'^* - \bar{\Theta}_0'^*) = 0$ because $\tau_m - \Theta_m \in \Gamma_{e0}$, $\bar{\tau}_0'^* - \bar{\Theta}_0'^* \in \Gamma_c^*$, and $(\Theta_m, \bar{\Theta}_0'^*) = 0$, provided that we use nonoverlapping neighborhoods of ζ, ζ'. We have thus

$$(\tau_m, \bar{\Theta}_0'^*) + (\Theta_m, \bar{\tau}_0'^*) = 0,$$

or

$$(\Theta_m, \bar{\tau}_0'^*) = (\Theta_0, \bar{\tau}_m^*).$$

For the first of these inner products one obtains by partial integration

$$\begin{aligned}(\Theta_m, \bar{\tau}_0'^*) &= \lim_{r\to 0} \int_\gamma v_m \bar{\tau}_0' \\ &= \lim_{r\to 0} \left[-\frac{1}{m+1} \int_\gamma \frac{h_0(z, \zeta')dz + \overline{k_0(z, \zeta')d\bar{z}}}{(z-\zeta)^{m+1}} \right] \\ &= \frac{-2\pi i}{(m+1)!} \frac{d^m}{d\zeta^m} h_0(\zeta, \zeta').\end{aligned}$$

Similarly,

$$\begin{aligned}(\Theta_0', \bar{\tau}_m^*) &= \lim_{r\to 0} \left[-\int_{\gamma'} \frac{h_m(z, \zeta)dz + \overline{k_m(z, \zeta)d\bar{z}}}{z-\zeta'} \right] \\ &= -2\pi i h_m(\zeta', \zeta).\end{aligned}$$

We have shown that

$$h_m(z, \zeta) = \frac{1}{(m+1)!} \frac{d^m}{d\zeta^m} h_0(\zeta, z).$$

For $m = 0$ the formula shows that h_0 is symmetric,

$$h_0(z, \zeta) = h_0(\zeta, z)$$

and for arbitrary m we obtain, in analogy with (47),

$$h_m(z, \zeta) = \frac{1}{(m+1)!} \frac{d^m}{dz^m} h_0(z, \zeta). \tag{49}$$

In particular, $h_m(z, \zeta)$ depends analytically on ζ, and $h_0(z, \zeta)\, dz d\zeta$ has invariant meaning.

18H. We summarize the main results in a theorem:

Theorem. *With every singularity of the form $(z-\zeta)^{-m-2}dz$, $m \geq 0$, there is associated, through an explicit construction, an exact differential Θ_m with compact support which exhibits the given singularity. This Θ_m is determined up to a differential of class Γ_{e0}^1, and there exists a unique harmonic differential τ_m such that $\tau_m - \Theta_m \in \Gamma_{e0}$.*

On writing $\tau_m = \varphi_m + \bar{\psi}_m$, $\varphi_m = h_m(z, \zeta)\, dz$, $\psi_m = k_m(z, \zeta)\, dz$ *the analytic differentials* φ_m, ψ_m *are connected by*

$$h_m(z, \zeta) = \frac{1}{(m+1)!} \frac{d^m}{dz^m} h_0(z, \zeta)$$

$$k_m(z, \zeta) = \frac{1}{(m+1)!} \frac{d^m}{dz^m} k_0(z, \zeta),$$

and h_0, k_0 *satisfy the symmetry relations*

$$h_0(z, \zeta) = h_0(\zeta, z), \quad k_0(z, \zeta) = \overline{k_0(\zeta, z)}.$$

The differentials ψ_m *have the reproducing property*

$$(\alpha, \psi_m) = \frac{2\pi}{(m+1)!} a^{(m)}(\zeta)$$

for all $\alpha = a dz \in \Gamma_a$, *while the* φ_m *satisfy* $(\alpha, \varphi_m) = 0$, *provided that the inner product is interpreted as a Cauchy limit.*

19. Differentials and chains

19A. In the Laurent development

$$\sum_1^\infty b_n(z-\zeta)^{-n}\, dz$$

of a singular differential Θ the coefficient b_1 is called the *residue* at ζ. It is independent of the choice of the local variable, for

$$b_1 = \frac{1}{2\pi i} \int_\gamma \Theta$$

where γ is a closed curve that encircles ζ. Suppose that Θ is a closed differential with finitely many analytic singularities ζ_j, residues b_{1j}. If Θ has compact support, we obtain, for sufficiently large Ω,

$$\sum b_{1j} = \sum \frac{1}{2\pi i} \int_{\gamma_j} \Theta = \frac{1}{2\pi i} \int_{\beta(\Omega)} \Theta = 0.$$

Hence, in order that there exist a closed differential Θ with compact support and given singularities, *it is necessary that the sum of the residues be* 0.

19B. The simplest case is represented by a singularity

$$\left(\frac{1}{z-\zeta_2} - \frac{1}{z-\zeta_1}\right) dz$$

where ζ_1, ζ_2 are points in the same parametric disk Δ, mapped on $|z|<1$. We assume that $|\zeta_1|, |\zeta_2|<r_1<r_2<1$ and choose the auxiliary function $e(z)$ as in 18B. We define a single-valued branch of

$$v = \log \frac{z-\zeta_2}{z-\zeta_1}$$

in $r_1<|z|<1$. Using this branch we set

$$\Theta = \begin{cases} \left(\frac{1}{z-\zeta_2} - \frac{1}{z-\zeta_1}\right)dz & \text{for} \quad |z| \leqq r_1 \\ d(ev) \quad \text{for} \quad r_1 < |z| < 1 \\ 0 \quad \text{outside of } \Delta. \end{cases} \tag{50}$$

It is verified that $\Theta \in C^1$ and $d\Theta=0$, except at ζ_1, ζ_2.

19C. By Theorem 17D there exists a unique harmonic differential such that $\tau-\Theta \in \Gamma_{e0}$.

This time Θ and τ are exact on $W-\Delta$, but not on W. In order to determine the periods we observe that any cycle γ can be written in the form $\gamma_0+\gamma_1$ where γ_0 lies on $\bar{\Delta}$ and γ_1 on $W-\Delta$. The period along γ is equal to the period along γ_0. But γ_0 can be regarded as a cycle on $|z|\leqq 1$, and we find (Ch. I, 10C) that the period of Θ, and hence of τ, equals $2\pi i[n(\gamma_0, \zeta_2)-n(\gamma_0, \zeta_1)]$.

To express this result in a different way we join ζ_1 to ζ_2 by an arc c in Δ. It can be assumed that c belongs to a triangulation K, and γ_0 to its dual K^*. Then the intersection number $c\times\gamma_0$ is defined. On recalling that $a\times b$ is positive if b crosses a from left to right (Ch. I, 31A) it is seen that $n(\gamma_0, \zeta_2)-n(\gamma_0, \zeta_1)=c\times\gamma_0=c\times\gamma$. *The period of Θ (or τ) along any cycle γ on $W-\{\zeta_1, \zeta_2\}$ is thus equal to $2\pi i(c\times\gamma)$.*

More generally, γ can be replaced by any finite chain whose boundary lies outside of Δ. There is still a representation $\gamma=\gamma_0+\gamma_1$ where γ_0 is a cycle on $\bar{\Delta}$. We find again that the integral of Θ along γ is given by $2\pi i(c\times\gamma_0)$ or, if we prefer, $2\pi i(c\times\gamma)$. Naturally, there is no corresponding result for τ.

19D. The construction of Θ depends on Δ, but the periods depend only on c, and in fact only on the homology class of c (for fixed ζ_1, ζ_2). Suppose that we replace Δ by Δ' and c by $c'\sim c$, contained in Δ'. We construct Θ' in the same way as Θ. Then $\Theta'-\Theta$ is exact, $\Theta'-\Theta=ds$, and s is constant in each component of $W-(\Delta\cup\Delta')$.

Let Ω be a canonical region (see Ch. I, 35A) which contains $\Delta\cup\Delta'$. Any two components of $W-\Omega$ can be joined by a path γ which avoids Δ. Then $c\times\gamma=0$ and, by a slight modification of Theorem 31B, Ch. I,

$c' \times \gamma = 0$. It follows by the reasoning at the end of 19C that s has the same value in all components of $W - \Omega$, and hence that $\Theta' - \Theta \in \Gamma^1_{e0}$. We conclude that τ does not change. In particular, the choice of local variable has no influence on τ.

In order to indicate the dependence on c we shall write $\tau = \tau(c)$, and also, as in 18B, $\tau = \varphi + \bar{\psi}$ with $\varphi = \varphi(c) = h(z, c)dz$, $\psi = \psi(c) = k(z, c)dz$. Because the singularities are analytic ψ is regular.

We extend the definition of $\tau(c)$ to arbitrary finite chains c, whether simplicial or singular, by making the mapping $c \to \tau(c)$ linear. In other words, we postulate that $\tau(c_1 + c_2) = \tau(c_1) + \tau(c_2)$. The notations $\varphi(c)$ and $\psi(c)$ are extended in the same manner. We can also choose, by linearity, a $\Theta(c)$ for each c, but it must be remembered that $\Theta(c)$ is determined only up to a differential of class Γ^1_{e0}.

The singularities depend only on the boundary ∂c. It is found, indeed, that Θ, τ and φ have simple poles with residues equal to the coefficients in ∂c. The differentials are regular if and only if c is a cycle.

19E. As in 18 we wish to compute (α, φ) and (α, ψ) when α is an analytic differential. It is sufficient to consider the original case where c is an arc in Δ.

We have again $(\alpha, \psi) = (\alpha, \bar{\tau}) = (\alpha, \bar{\Theta})$, and since $\alpha\Theta = 0$ for $|z| < r_1$ the last inner product needs to be extended only over $r_1 < |z| < r_2$. We obtain

$$(\alpha, \bar{\Theta}) = (\Theta, \bar{\alpha}) = -i \iint\limits_{r_1 < |z| < r_2} d(ev)\alpha = i \int\limits_{|z| = r_1} v\alpha$$

$$= i \int\limits_{|z|=r_1} v\, da = -i \int\limits_{|z|=r_1} a\, dv = -i \int\limits_{|z|=r_1} a\left(\frac{dz}{z - \zeta_2} - \frac{dz}{z - \zeta_1}\right)$$

$$= 2\pi[a(\zeta_2) - a(\zeta_1)] = 2\pi \int_c \alpha.$$

We have shown that

$$(\alpha, \psi) = 2\pi \int_c \alpha. \tag{51}$$

This is the reproducing property of ψ.

19F. Let us choose $\alpha = \psi_0 = k_0(z, \zeta)\, dz$ in (51). We obtain

$$(\psi_0, \psi) = 2\pi \int_c k_0(z, \zeta)\, dz$$

while, by (43),

$$(\psi, \psi_0) = 2\pi\, k(\zeta, c).$$

Thus

$$\overline{k(\zeta, c)} = \int_c k_0(z, \zeta)\, dz$$

or, with a change of notation,

$$k(z, c) = \int_c k_0(z, \zeta) d\bar{\zeta}. \tag{52}$$

We see that $k(z, c)$ can be obtained from $k_0(z, \zeta)$ by integration just as k_m was obtained by differentiation.

We remark that formulas (51) and (52) are of course valid for arbitrary chains.

19G. The computation of (α, φ) is quite similar. We have $(\alpha, \varphi) = (\alpha, \tau) = (\alpha, \Theta)$, and if γ_1, γ_2 denote small circles about ζ_1, ζ_2 we obtain

$$(\alpha, \Theta) = i \iint d\alpha \bar{\Theta} = \lim \left[-i \int_{\gamma_1+\gamma_2} a\bar{\Theta} \right]$$

$$= \lim \left[-i \int_{\gamma_1+\gamma_2} a\left(\frac{d\bar{z}}{\bar{z}-\bar{\zeta}_2} - \frac{d\bar{z}}{\bar{z}-\bar{\zeta}_1} \right) \right] = \lim \left[i \int_{\gamma_1+\gamma_2} a\left(\frac{dz}{z-\zeta_2} - \frac{dz}{z-\zeta_1} \right) \right]$$

$$= -2\pi \left[a(\zeta_2) - a(\zeta_1) \right] = -2\pi \int_c \alpha.$$

The final formula reads

$$(\alpha, \varphi) = -2\pi \int_c \alpha. \tag{53}$$

By combining (51) and (53) we can compute (ω, τ^*) for an arbitrary harmonic differential ω. If $\omega = \alpha + \bar{\beta}$ we find $(\omega, \tau^*) = (\alpha + \bar{\beta}, -i\varphi + i\bar{\psi}) = i(\alpha, \varphi) - i(\bar{\beta}, \bar{\psi})$, and hence

$$(\omega, \tau^*) = -2\pi i \int_c \omega. \tag{54}$$

19H. Finally, we derive an integral representation for $h(z, c)$ similar to (52). By (48) and (53) we have first

$$(\tau_0, \bar{\tau}^*) = (\varphi_0 + \bar{\psi}_0, i\bar{\varphi} - i\psi) = i(\varphi_0, \psi) - i(\bar{\psi}_0, \bar{\varphi})$$

$$= 2\pi i \int_c \bar{\psi}_0.$$

On the other hand, $(\tau_0 - \Theta_0, \bar{\tau}^* - \bar{\Theta}^*) = 0$ by virtue of the relation

$\Gamma_c \perp \Gamma_{e0}^*$. It is legitimate to assume that Θ_0 and Θ have disjoint supports, in which case $(\Theta_0, \bar{\Theta}^*)=0$. The orthogonality relation can therefore be written in the form

$$\begin{aligned}(\tau_0, \bar{\tau}^*) &= (\tau_0, \bar{\Theta}^*)+(\Theta_0, \bar{\tau}^*)\\ &= -i(\varphi_0, \bar{\Theta})-i(\bar{\psi}_0, \bar{\Theta})-i(\varphi, \bar{\Theta}_0)+i(\bar{\psi}, \bar{\Theta}_0).\end{aligned}$$

These inner products can be evaluated by the methods in 19E, 19G, 18C, 18F respectively. Indeed, it is found that singularities outside the support of Θ or Θ_0 have no influence on the result. We find

$$(\varphi_0, \bar{\Theta}) = 2\pi \int_c \varphi_0$$

$$(\bar{\psi}_0, \bar{\Theta}) = -2\pi \int_c \bar{\psi}_0$$

$$(\varphi, \bar{\Theta}_0) = 2\pi\, h(\zeta, c)$$

$$(\bar{\psi}, \bar{\Theta}_0) = 0.$$

When these results are combined we obtain

$$h(\zeta, c) = -\int_c \varphi_0$$

or, in different notation,

$$h(z, c) = -\int_c h_0(z, \zeta)d\zeta. \tag{55}$$

19I. We give again a summary of the most important results:

Theorem. *To each finite chain c there is assigned a closed differential $\Theta=\Theta(c)$ with compact support. It has simple poles with residues equal to the coefficients in ∂c, and it is unique up to differentials of class Γ_{e0}^1. Corresponding to $\Theta(c)$ there exists a unique harmonic differential $\tau=\tau(c)$ such that $\tau-\Theta \in \Gamma_{e0}$.*

On writing $\tau(c)=\varphi(c)+\overline{\psi(c)}$, $\varphi(c)=h(z, c)dz$, $\psi(c)=k(z, c)dz$ we have

$$h(z, c) = -\int_c h_0(z, \zeta)d\zeta$$

$$k(z, c) = \int_c k_0(z, \zeta)d\bar{\zeta}$$

and

$$(\alpha, \psi) = -(\alpha, \varphi) = 2\pi \int_c \alpha$$

for analytic α.

The formula

$$(\omega, \tau^*) = -2\pi i \int_c \omega$$

holds for all harmonic ω.

20. Differentials and periods

20A. We examine in greater detail the case where c is a cycle. It has already been pointed out that $\tau(c)$, $\varphi(c)$ and $\psi(c)$ are all three regular.

Formula (54) gives the representation of a period of any harmonic differential ω as an inner product. If we compare it with the standard formula (30), 12B,

$$\int_c \omega = (\omega, \sigma(c)^*)$$

it is seen that

(56) $$\tau(c) = 2\pi i \sigma(c).$$

But $\sigma(c)$ is real. Hence $\tau(c) = \varphi(c) + \overline{\psi(c)}$ is purely imaginary, that is, $\mathrm{Re}\{\varphi(c) + \psi(c)\} = 0$. Because $\varphi(c) + \psi(c)$ is analytic, we have $\psi(c) = -\varphi(c)$ and

(57) $$\sigma(c) = \frac{1}{\pi} \operatorname{Im} \varphi(c) = -\frac{1}{\pi} \operatorname{Im} \psi(c).$$

20B. It was shown in 19C that the period of $\tau(c)$ along a cycle γ is

$$\int_\gamma \tau(c) = 2\pi i (c \times \gamma).$$

If we set $c = c_1$, $\gamma = c_2$ we have

$$(\sigma(c_1), \sigma(c_2)^*) = c_1 \times c_2.$$

Theorem. *If c is a finite cycle, then $\varphi(c) = -\psi(c)$ and $\sigma(c) = \frac{1}{2\pi i} \tau(c) = \frac{1}{\pi} \operatorname{Im} \varphi(c)$. For any pair of finite cycles $(\sigma(c_1), \sigma(c_2)^*) = c_1 \times c_2$.*

Above all, the periods of $\sigma(c)$ are integers.

20C. If d is a dividing cycle we recall that $\sigma(d)$ is of class Γ_{hm} (Theorem 15F).

Theorem. *Let c run through all cycles and let d run through all dividing cycles. Then the $\sigma(c)$ span Γ_{h0} and the $\sigma(d)$ span Γ_{hm}.*

If the first assertion were not true there would exist (Theorem 7D) a nonzero $\omega \in \Gamma_{h0}$ which is orthogonal to all $\sigma(c)$. But $(\omega, \sigma(c)) = \omega^*, \sigma(c)^*) = 0$ for all c would imply that ω^* is exact. This is a contradiction, for $\Gamma_{h0} \perp \Gamma_{he}^*$.

Similarly, $(\omega, \sigma(d)) = 0$ for all dividing cycles d would mean that ω^* is semiexact, $\omega \in \Gamma^*_{hse}$. By Theorem 15D Γ^*_{hse} is orthogonal to Γ_{hm}, and we conclude as above that the $\sigma(d)$ must span all of Γ_{hm}.

20D. In Ch. I, 31G we introduced a canonical homology basis modulo dividing cycles consisting of cycles A_n, B_n whose only nonzero intersection numbers are $A_n \times B_n = 1$. It follows from Theorem 20B that $\sigma(A_n)$ has period 1 over B_n, $\sigma(B_n)$ has period -1 over A_n, and otherwise all periods of $\sigma(A_n)$, $\sigma(B_n)$ over cycles that belong to the homology basis are 0.

The following result on the existence of harmonic differentials with given periods is obtained:

Theorem. *There exists a semiexact harmonic differential with arbitrarily prescribed periods over finitely many cycles A_n, B_n and periods 0 over all other cycles in a canonical basis. The differential is uniquely determined if it is required to lie in Γ^*_{h0}.*

It is clear that $\tau = \sum(y_n\sigma(A_n) - x_n\sigma(B_n))$ has periods x_n, y_n over A_n, B_n. We can write $\tau = \tau_{he} + \tau^*_{h0}$, and it is seen that τ^*_{h0} is the unique solution in Γ^*_{h0}.

20E. We have reached the point where it is trivial to prove the converse of Theorem 15F. The full theorem is stated as follows:

Theorem. *A cycle d is dividing if and only if $\sigma(d)$ is exact. If $\sigma(d)$ is exact it is even of class Γ_{hm}.*

We have already proved that $\sigma(d) \in \Gamma_{hm} \subset \Gamma_{he}$ whenever d is dividing. Any cycle d is homologous to a finite linear combination $\sum (x_nA_n + y_nB_n)$ modulo dividing cycles. If $\sigma(d)$ is exact it follows that $\sum (x_n\sigma(A_n) + y_n\sigma(B_n))$ is exact. But this finite sum has periods x_n along B_n and $-y_n$ along A_n. Hence all x_n, y_n are 0, and we conclude that d is a dividing cycle.

Corollary. *There exists a nonzero square integrable harmonic differential on any nonplanar Riemann surface.*

By assumption there exists a nondividing cycle c. The theorem shows that $\sigma(c)$ is not exact and thus not identically 0. In other words, $\Gamma_h \neq 0$.

21. The main existence theorem

21A. The preceding considerations make it clear that it is possible to construct a closed differential Θ with compact support, with a finite number of prescribed polar singularities and a finite number of given periods (in the sense of Theorem 20D), provided merely that the sum of the residues is zero. Indeed, we can start by constructing a linear combination of differentials Θ_m whose singularities correspond to the terms of order greater than one in the given Laurent developments. Next, because the sum of the residues is zero, there exists a finite linear combination of differentials $\Theta(c)$ whose singularities are simple poles with the prescribed residues. It remains to find an everywhere regular Θ with given periods x_n, y_n over A_n, B_n. Provided that only a finite number of x_n, y_n are $\neq 0$ an explicit solution is given by

$$\Theta = \frac{1}{2\pi i} \sum [y_n \Theta(A_n) - x_n \Theta(B_n)].$$

The desired differential is the sum of the three differentials obtained in these steps.

21B. If Θ is a solution of the aforementioned problem there exists a unique harmonic τ such that $\tau - \Theta \in \Gamma_{e0}$, either directly by Theorem 17D, or by application of Theorems 18H and 19I. This τ is a harmonic differential with the given singularities and periods. However, τ is in general not the only solution, even if we agree to consider only differentials which are square integrable in a neighborhood of the ideal boundary. Indeed, the general solution is of course given by $\tau + \omega_{he}$, $\omega_{he} \in \Gamma_{he}$, and $\Gamma_{he} = 0$ only if W is of class O_{HD}.

In order to select a distinguished unique solution we let ourselves be guided by the uniqueness result in Theorem 20D. Because the solution has singularities we can not require that it belong to Γ_{h0}^*, but we can at least fix our attention on the most important characteristic of differentials in Γ_{h0}, namely the vanishing of the periods over all dividing cycles.

21C. Let us re-examine the proof of Theorem 17D. Given Θ we applied the decomposition theorem to write

$$\Theta - i\Theta^* = \omega_h + \omega_{e0} + \omega_{e0}^*,$$

and we found that $\tau = \Theta - \omega_{e0}$ is harmonic. Let us also decompose ω_h by use of Theorem 15D:

$$\omega_h = \omega_{hm} + \omega_{hse}^*.$$

It follows from these equations that

$$\tau - \omega_{hm} = i\Theta^* + \omega_{hse}^* + \omega_{e0}^*.$$

Since Θ vanishes outside of a compact set we recognize that $(\tau - \omega_{hm})^*$ is semiexact (disregarding periods around poles with nonvanishing residues). Consequently,

(58) $$\bar{\tau} = \tau - \omega_{hm}$$

is a solution whose conjugate $\bar{\tau}^*$ is semiexact.

21D. The preceding condition does not yet make $\tilde{\tau}$ unique, for if $\tilde{\tau}'$ is another solution, we can merely assert that $\tilde{\tau}-\tilde{\tau}' \in \Gamma^*_{hse}$, and Γ^*_{hse} is not orthogonal to Γ_{he}. However, $\tilde{\tau}$ satisfies a much stronger condition which finds its expression in the fact that $\tilde{\tau}=\Theta-\omega_{e0}-\omega_{hm}$ is of the form $\omega_{hm}+\omega_{e0}$ outside of a compact set. Let us therefore introduce the following definition:

Definition. *A harmonic differential $\tilde{\tau}$ whose only singularities are harmonic poles is said to be distinguished if*

(1) *$\tilde{\tau}^*$ has vanishing periods over all dividing cycles which lie outside of a sufficiently large compact set,*

(2) *there exist differentials $\omega_{hm} \in \Gamma_{hm}$, $\omega_{e0} \in \Gamma_{e0} \cap \Gamma^1$ such that $\tilde{\tau}=\omega_{hm}+\omega_{e0}$ outside of a compact set.*

It is a consequence of the definition that $\tilde{\tau}$ has only a finite number of poles and a finite number of nonzero periods with respect to a given homology basis. Note that we have strengthened the condition $\omega_{e0} \in \Gamma_{e0}$ to $\omega_{e0} \in \Gamma_{e0} \cap \Gamma^1$. The stronger condition is fulfilled for $\tilde{\tau}$ defined by (58). Indeed, we have $\tilde{\tau}=\Theta-\omega_{e0}-\omega_{hm}$, and since $\tilde{\tau}-\Theta$ and ω_{hm} are of class C^1 the same is true of ω_{e0}.

21E. The above definition leads to the following existence and uniqueness theorem:

Theorem. *There exists a unique distinguished harmonic differential with a finite number of harmonic poles with given Laurent developments and a finite number of given periods.*

The phrase "a finite number of given periods" means the following: There is given a homology basis $\{A_n, B_n\}$ modulo dividing cycles on the surface punctured at the poles. The differential is to have prescribed periods over these cycles, but only a finite number of the prescribed values are different from zero.

The existence has already been established. To prove the uniqueness, suppose that $\tilde{\tau}$ and $\tilde{\tau}'$ are two solutions. The difference $\omega=\tilde{\tau}-\tilde{\tau}'$ belongs to $\Gamma_{he} \cap \Gamma^*_{hse}$, and $\omega-\omega_{hm}-\omega_{e0}$ has compact carrier for a certain choice of $\omega_{hm} \in \Gamma_{hm}$ and $\omega_{e0} \in \Gamma_{e0} \cap \Gamma^1$. Because of these conditions we can set $\omega-\omega_{hm}-\omega_{e0}=df$ where f is constant in each complementary component of a sufficiently large regular region Ω. Choose any $\sigma \in \Gamma_{hse}$. Then

$$(df, \sigma^*) = (df, \sigma^*)_\Omega = -\int_{\beta(\Omega)} f\bar{\sigma} = 0$$

because f is constant on each part of $\beta(\Omega)$ that bounds a component of $W-\Omega$, and the corresponding periods of $\bar{\sigma}$ are 0. Since it is also true that $(\omega_{hm}, \sigma^*)=0$ and $(\omega_{e0}, \sigma^*)=0$ it follows that $\omega \perp \Gamma^*_{hse}$, and hence that $\omega=0$. The uniqueness is proved.

We take notice of the following obvious corollary:

Corollary. *A regular and exact distinguished harmonic differential is identically zero.*

21F. We use notations $\tilde{\tau}_m$, $\tilde{\tau}(c)$ for the distinguished differentials with the same singularities and periods as τ_m, $\tau(c)$. We shall also use the standard decomposition $\tilde{\tau}=\tilde{\varphi}+(\tilde{\psi})^-$. Here $\tilde{\varphi}=\frac{1}{2}(\tilde{\tau}+i\tilde{\tau}^*)$ and $\tilde{\psi}=\frac{1}{2}(\tilde{\tau}-i\tilde{\tau}^*)^-$ are evidently semiexact. To guard against misunderstandings that may be caused by the notation we emphasize that $\tilde{\varphi}$ and $\tilde{\psi}$ are in general not distinguished.

A regular distinguished differential is always of class Γ_{h0}. Indeed, we have $\tilde{\tau}=\omega_{hm}+\omega_{e0}$ outside of a compact set. This implies $\tilde{\tau}-\omega_{hm}-\omega_{e0}=\omega_{c0}\in\Gamma_{c0}$, and since $\Gamma_{hm}\subset\Gamma_{h0}$ it follows that $\tilde{\tau}\in\Gamma_{h0}$.

Consider now the difference $\tilde{\tau}-\tau$ for any choice of singularities and periods. Because $\tau=\Theta+\omega_{e0}$ we have $\tilde{\tau}-\tau=\omega_{hm}+\omega_{e0}$ outside of a compact set. Since $\tilde{\tau}-\tau$ is exact we can repeat the reasoning used for proving the uniqueness in Theorem 21E to conclude that $\tilde{\tau}-\tau\perp\Gamma^*_{hse}$. This shows that $\tilde{\tau}-\tau\in\Gamma_{hm}$, a stronger result than the trivial statement $\tilde{\tau}-\tau\in\Gamma_{he}$.

Suppose that α is analytic and semiexact. Then $\bar{\alpha}$ and $\alpha^*=-i\alpha$ are also semiexact, and it follows that $(\alpha,\tilde{\tau}-\tau)=(\bar{\alpha},\tilde{\tau}-\tau)=0$ as well as $(\alpha,\tilde{\tau}^*-\tau^*)=(\bar{\alpha},\tilde{\tau}^*-\tau^*)=0$. These results imply $(\alpha,\tilde{\varphi})=(\alpha,\varphi)$ and $(\alpha,\tilde{\psi})=(\alpha,\psi)$. Thus $\tilde{\varphi}$ and $\tilde{\psi}$ have the same reproducing properties as φ and ψ, *but only with respect to semiexact analytic differentials.*

Since $\tilde{\varphi}_m$ and $\tilde{\psi}_m$, $\tilde{\varphi}(c)$ and $\tilde{\psi}(c)$ are themselves semiexact the symmetry properties can be proved in the same way as before. Briefly:

Theorems 18H *and* 19I *remain valid, with modified notations, except that the differentials* α *must be assumed semiexact.*

21G. By passing from τ to $\tilde{\tau}$ we have lost some of the reproducing power, but we have gained a more important advantage, namely unique solutions to the problem of periods and singularities.

The advantage shows up very clearly when we replace $\sigma(c)$, the reproducing differential for a cycle c, by the corresponding $\tilde{\sigma}(c)$. Suppose that $\omega=\alpha+\bar{\beta}\in\Gamma_{hse}\cap\Gamma^*_{hse}$. Then α and β are both semiexact and we obtain, as a counterpart of (54),

$$(\omega,\tilde{\tau}(c)^*) = -2\pi i\int_c\omega.$$

Hence, on setting

$$\tilde{\sigma}(c) = \frac{1}{2\pi i}\tilde{\tau}(c),$$

we have

$$(59) \qquad (\omega, \tilde{\sigma}(c)^*) = \int_c \omega$$

for all $\omega \in \Gamma_{hse} \cap \Gamma^*_{hse}$.

It follows from (59) that $\tilde{\sigma}(c)$ is real, and hence $\tilde{\varphi}(c) = -\tilde{\psi}(c)$,

$$\tilde{\sigma}(c) = -\frac{1}{\pi} \operatorname{Im} \tilde{\psi}(c).$$

We have already pointed out that $\tilde{\tau}(c) - \tau(c) \in \Gamma_{hm}$. Hence $\tilde{\sigma}(c)$ and $\sigma(c)$ have the same periods, and since $\tilde{\sigma}(c) \in \Gamma_{hse} \cap \Gamma^*_{hse}$ we find by (59) and 20B

$$(60) \qquad (\tilde{\sigma}(c_1), \tilde{\sigma}(c_2)^*) = c_1 \times c_2.$$

Actually, since $\tilde{\sigma}(c) - \sigma(c) \in \Gamma_{hm}$ and $\sigma(c) \in \Gamma_{h0}$ (see 20C) we have $\tilde{\sigma}(c) \in \Gamma_{h0} \cap \Gamma^*_{hse}$. The following theorem shows that there is a one to one correspondence between the $\tilde{\sigma}(c)$ and the relative homology classes modulo dividing cycles:

Theorem. *A cycle c is dividing if and only if $\tilde{\sigma}(c) = 0$. If c runs through all cycles the $\tilde{\sigma}(c)$ span $\Gamma_{h0} \cap \Gamma^*_{hse}$.*

First, if $\tilde{\sigma}(c) = 0$, then (60) shows that c has zero intersection with all cycles, and hence c is dividing (Ch. I, 31E). Second, if $\tilde{\sigma}(c) = 0$ then $\sigma(c) \in \Gamma_{hm}$ so that c is dividing, by Theorem 20E. Third, if $\omega \in \Gamma_{h0} \cap \Gamma^*_{hse}$ is orthogonal to all $\tilde{\sigma}(c)$, then $\omega^* \in \Gamma_{he}$, and $\Gamma_{h0} \perp \Gamma^*_{he}$ implies that $\omega = 0$.

21H. It seems worth while to point out that $\tilde{\tau} = \tau$, regardless of the singularities and periods, if and only if $\Gamma_{hm} = 0$ or, equivalently, if $\Gamma_h = \Gamma_{hse}$. The condition is certainly sufficient, for we have shown that $\tilde{\tau} - \tau \in \Gamma_{hm}$. It is also necessary, for if $\tilde{\sigma}(c) = \sigma(c)$ for all cycles we conclude by Theorems 20C and 21G that $\Gamma_{h0} = \Gamma_{h0} \cap \Gamma^*_{hse}$. Hence $\Gamma_{h0} \subset \Gamma^*_{hse}$, and on passing to the orthogonal complements we see that $\Gamma_{hm} \subset \Gamma^*_{he}$. Thus $\Gamma_{hm} \subset \Gamma_{h0} \cap \Gamma^*_{he} = 0$.

In particular, $\tilde{\tau} = \tau$ if $W \in O_{HD}$, for then $\Gamma_{hm} \subset \Gamma_{he} = 0$.

22. Abel's theorem

22A. The classical theory deals only with closed surfaces. A function f which is analytic except for poles is called a rational function, and one of the important problems is to determine conditions on the zeros and poles of rational functions.

For arbitrary meromorphic functions on an open surface there are no such conditions (H. Behnke, K. Stein [2]). However, if the functions are restricted in a suitable manner we will show that it is possible to derive a close analogue of Abel's classical theorem (see, e.g., H. Weyl [4]).

22B. Suppose that f is meromorphic and not identically 0 on the Riemann surface W, open or closed. Then $d \log f$ is an analytic differential with simple poles at the poles and zeros of f; the residues are equal to the corresponding multiplicities, counted positive at the zeros, negative at the poles. Moreover, the periods of $d \log f$ are integral multiples of $2\pi i$.

As a restriction on f we shall require that $d \log f$ is a distinguished harmonic differential in the sense of Definition 21D. Because of the close analogy with rational functions we shall say that a function f which satisfies this condition is *quasi-rational.*

It is an immediate consequence of the definition that f has only a finite number of zeros and poles. Secondly, because $d \log f$ fulfills condition (1) of Definition 21D, the sum of the residues is zero. This means that f has equally many zeros and poles, counted with multiplicities.

We denote the zeros of f by $a_1, \cdots, a_n$ and the poles by $b_1, \cdots, b_n$, multiple zeros and poles being repeated. The 0-chain $\sum (a_j - b_j)$ is called the *divisor* of f.

22C. We are going to prove the following generalization of Abel's theorem:

Theorem. *In order that a given 0-chain be the divisor of a quasi-rational function it is necessary and sufficient that the 0-chain can be written in the form ∂c where c has the property that*

$$\int_c \alpha = 0 \tag{61}$$

for all semiexact square integrable analytic differentials on W.

We suppose first that f is a given quasi-rational function. Because there are equally many zeros and poles the divisor of f can certainly be written as the boundary ∂c of a 1-dimensional chain c, which is determined up to a cycle.

We form $\tilde{\tau}(c) = \tilde{\varphi}(c) + [\tilde{\psi}(c)]^-$. Then $\chi = d \log f - \tilde{\varphi}(c)$ is a square integrable analytic differential. Indeed, it has no singularities, and by assumption $d \log f = \omega_{hm} + \omega_{e0}$ outside of a compact set, so that $d \log f$ is square integrable over a neighborhood of the ideal boundary.

We observe that $\mathrm{Re}(\chi - \tilde{\psi}(c)) = \mathrm{Re}(\chi - [\tilde{\psi}(c)]^-) = d \log |f| - \mathrm{Re}\, \tilde{\tau}(c)$ is regular and distinguished. Moreover, it is exact, for the periods of $\tilde{\tau}(c)$, equal to the periods of $\tau(c)$, are multiples of $2\pi i$ (see 19C). But a regular and exact distinguished differential is identically 0 (Corollary 21E). Hence $\mathrm{Re}(\chi - \tilde{\psi}(c)) = 0$, and since $\chi - \tilde{\psi}(c)$ is analytic this implies $\chi = \tilde{\psi}(c)$.

The equation $\chi = \tilde{\psi}(c)$ can be written in the form

$$d \log f = \tilde{\varphi}(c) + \tilde{\psi}(c) = \tilde{\tau}(c) + 2i \mathrm{Im}\, \tilde{\psi}(c).$$

We conclude that $\mathrm{Im}\, \tilde{\psi}(c)$ is distinguished. For this reason $\mathrm{Im}\, \tilde{\psi}(c)$ has only a finite number of nonzero periods with respect to a given canonical basis. The periods are integral multiples of π. By Theorem 21E, or at least by going back to the proof of Theorem 20D, we can find a cycle c_0 such that $\tilde{\sigma}(c_0) = \frac{1}{2\pi i}\tilde{\tau}(c_0)$ has the periods of $\frac{1}{\pi}\,\mathrm{Im}\,\tilde{\psi}(c)$. Then $i\tilde{\tau}(c_0) - 2\,\mathrm{Im}\,\tilde{\psi}(c)$ is exact and distinguished, and consequently equal to 0. We have thus $2\,\mathrm{Im}\,\tilde{\psi}(c) = i\tilde{\tau}(c_0) = 2\,\mathrm{Im}\,\tilde{\psi}(c_0)$, from which it follows that $\tilde{\psi}(c) = \tilde{\psi}(c_0)$, or $\tilde{\psi}(c - c_0) = 0$.

We conclude by the reproducing property of $\tilde{\psi}$ (see (51), 19E, and 21F) that

$$\int_{c-c_0} \alpha = \frac{1}{2\pi}\,(\alpha, \tilde{\psi}(c-c_0)) = 0$$

for all semiexact analytic α. We have shown that $c - c_0$ has the properties claimed in the theorem.

22D. The converse is very easy. If c satisfies (61) for all semiexact analytic α, then $(\alpha, \tilde{\psi}(c)) = 0$, and since $\tilde{\psi}(c)$ is itself semiexact (21F) we have necessarily $\tilde{\psi}(c) = 0$. Then $\tilde{\tau}(c) = \tilde{\varphi}(c)$, and we see that $\tilde{\varphi}(c)$ is an analytic differential all of whose periods are multiples of $2\pi i$. This means, automatically, that we can write $\tilde{\varphi}(c) = d \log f$ with a single-valued f. The relation $d \log f = \tilde{\tau}(c)$ shows that $d \log f$ is distinguished, i.e., f is quasi-rational, and that its divisor is precisely ∂c. The proof of Theorem 22C is complete.

22E. It can of course happen that there are no quasi-rational functions on W other than constants. In that case Theorem 22C, far from being void of content, informs us that no chain c which is not a cycle can have the property expressed by (61).

For closed surfaces we have the other extreme: all meromorphic functions are quasi-rational and all differentials are semiexact. The theorem reduces to one of the classical formulations of Abel's theorem.

23. The bilinear relation

23A. Under certain circumstances it is possible to express an inner product of the form (ω, σ^*) in terms of the periods of ω and σ. Naturally, ω and σ must be closed, but they must also be subject to additional restrictions, particularly in regard to their behavior near the ideal boundary.

It is no restriction at all to assume that ω and σ are harmonic. Indeed, we can write $\omega = \omega_h + \omega_{e0}$, $\sigma = \sigma_h + \sigma_{e0}$ where the subscripts indicate, as usual, that $\omega_h, \sigma_h \in \Gamma_h$ and $\omega_{e0}, \sigma_{e0} \in \Gamma_{e0}$. With this decomposition we have $(\omega, \sigma^*) = (\omega_h, \sigma_h^*)$. On the other hand, ω and σ have the same periods as ω_h and σ_h. Therefore, if (ω_h, σ_h^*) can be expressed through the periods the same is true of (ω, σ^*).

23B. We make the drastic assumption that σ is semiexact and has only a finite number of nonzero periods (in the usual sense). As for ω it will be assumed that $\omega \in \Gamma_{h0}$.

Let the A_n-periods of σ be denoted by x_n, the B_n-periods by y_n. Then

$$\sigma_0 = \sigma - \sum [y_n\sigma(A_n) - x_n\sigma(B_n)]$$

has no periods whatsoever, that is, σ_0 is exact. Hence $(\omega, \sigma_0^*) = 0$, and we obtain

$$(\omega, \sigma^*) = \sum [\bar{y}_n(\omega, \sigma(A_n)^*) - \bar{x}_n(\omega, \sigma(B_n)^*)].$$

But $(\omega, \sigma(A_n)^*)$ and $(\omega, \sigma(B_n)^*)$ are the periods of ω along A_n and B_n. The formula can be rewritten as

$$(\omega, \sigma^*) = \sum_n \left[\int_{A_n} \omega \int_{B_n} \bar{\sigma} - \int_{B_n} \omega \int_{A_n} \bar{\sigma} \right]. \tag{62}$$

This is Riemann's bilinear relation. We state the conditions under which we have proved it:

Theorem. *The bilinear relation* (62) *holds whenever* ω *is of class* Γ_{c0} *while* σ *is semiexact and possesses only a finite number of nonzero periods.*

We make the observation that the bilinear relation is symmetric in ω and σ. Therefore, ω and σ can be interchanged in the hypothesis.

The conditions of the theorem are sufficiently broad to include the classical case of any two differentials on a closed surface.

23C. It is natural to ask whether the bilinear relation remains valid in the presence of infinitely many periods. Complete results are not known, but interesting partial answers are given in L. Ahlfors [13], A. Pfluger [7, 10], and R. Accola [to be published].

§4. THE CLASSICAL THEORY

It cannot be denied that the classical theory of Abelian integrals has a scope which goes far beyond what one can hope to recapture in the more general setting of open surfaces. The reason is that the algebraic methods do not carry over to the noncompact case.

The theory of closed Riemann surfaces is equivalent to the theory of algebraic curves, or to the theory of algebraic functions of one variable. It is therefore mainly algebraic in character, although it uses function-theoretic methods as an essential tool. In accordance with the plan of this book we shall discuss these transcendental methods, especially in connection with the general existence theorems, but for full appreciation of their implications the reader must be referred to treatises on algebraic geometry.

24. Analytic differentials on closed surfaces

24A. A closed Riemann surface has a canonical homology basis of the form $A_1, B_1, \cdots, A_g, B_g$ where g is the genus of the surface. It is evident that the corresponding harmonic differentials $\sigma(A_1), \sigma(B_1), \cdots, \sigma(A_g), \sigma(B_g)$ with one nonzero period each form a basis for the space Γ_h of all harmonic differentials. Hence Γ_h is a vector space of dimension $2g$ over the complex numbers.

In the classical theory the emphasis is usually on the analytic differentials. Since every harmonic differential has a unique representation $\omega=\varphi+\bar{\psi}$ as a sum of an analytic and an antianalytic differential it is immediate that the space of harmonic differentials has twice the dimension of the space of analytic differentials. We conclude:

Theorem. *The analytic differentials on a closed Riemann surface of genus g form a vector space of dimension g over the complex numbers.*

24B. The space of analytic differentials is denoted by Γ_a. Let the A-periods of a differential $\varphi \in \Gamma_a$ be $x_1, \cdots, x_g$. The mapping which carries φ into the vector $(x_1, \cdots, x_g)$ is linear. Suppose that φ_0 is mapped on $(0, \cdots, 0)$, that is to say that φ_0 has no A-periods. If we apply the bilinear relation (62) to $\omega=\varphi_0$, $\sigma=i\varphi_0$ it follows at once that $\|\varphi_0\|^2=(\omega, \sigma^*)=0$, and hence that $\varphi_0=0$. This shows that the mapping of φ on its A-periods is one to one, and since Γ_a is g-dimensional it is also onto. We have proved:

Theorem. *There exists a unique analytic differential with given A-periods.*

24C. In particular, for every $k=1, \cdots, g$ there exists a unique analytic differential α_k whose A_i-periods are 0 for $i\neq k$ and 1 for $i=k$. We denote the B_l-period of α_k by z_{kl}.

Theorem. *The differentials α_k, $k=1, \cdots, g$, form a basis for Γ_a, and the matrix $Z=\|z_{kl}\|$ has the following properties: Z is nonsingular, symmetric, and* Im *Z is negative definite.*

The first assertion requires no proof. To derive the properties of Z we make use of the bilinear relation. We know that $(\alpha_k, \bar{\alpha}_l)=0$. On the other hand, by (62),

$$(\alpha_k, \bar{\alpha}_l) = i(\alpha_k, \bar{\alpha}_l^*) = \sum_{j=1}^{g} i\left[\int_{A_j} \alpha_k \int_{B_j} \alpha_l - \int_{B_j} \alpha_k \int_{A_j} \alpha_l\right] = i(z_{lk}-z_{kl}).$$

Hence $z_{kl}=z_{lk}$.

By similar calculation

$$(\alpha_k, \alpha_l) = -i(\alpha_k, \alpha_l^*) = -i(\bar{z}_{lk} - z_{kl}) = -2 \operatorname{Im} z_{kl}.$$

But the matrix $\|(\alpha_k, \alpha_l)\|$ is positive definite, for

$$\|t_1\alpha_1 + \cdots + t_g\alpha_g\|^2 = \sum_{k,l} (\alpha_k, \alpha_l) t_k \bar{t}_l > 0$$

except when all t_j are zero. We conclude that Im Z is negative definite.

Suppose that Z were singular. Then there would exist $x_1, \cdots, x_g$, not all 0, such that $\sum_j z_{ij} x_j = 0$ for all i. The resulting equation $\sum_{i,j} z_{ij} \bar{x}_i x_j = 0$ implies $\sum_{i,j} \bar{z}_{ij} \bar{x}_i x_j = 0$ because of the symmetry of Z. These two equations are irreconcilable with the negative definiteness of Im Z. Hence Z is nonsingular.

24D. The matrix Z serves to determine the B-periods of an analytic differential whose A-periods are given. In fact, if the A-periods $x_1, \cdots, x_g$ are considered as components of a vertical vector x, then the B-periods are evidently the components of Zx.

Let us also regard $\alpha = (\alpha_1, \cdots, \alpha_g)$ and $\sigma(A) = (\sigma(A_1), \cdots, \sigma(A_g))$, $\sigma(B) = (\sigma(B_1), \cdots, \sigma(B_g))$ as vertical vectors. Then we have

$$\alpha = -\sigma(B) + Z\sigma(A)$$

for the components of both sides are harmonic differentials with the same periods.

Since $\alpha^* = -i\alpha$ we obtain

$$-\sigma(B)^* + Z\sigma(A)^* = i\sigma(B) - iZ\sigma(A).$$

We write $Z = X + iY$ and separate the real and imaginary parts, taking into account that $\sigma(A)$ and $\sigma(B)$ are real. In this way we find

$$-\sigma(B)^* + X\sigma(A)^* = Y\sigma(A)$$
$$Y\sigma(A)^* = \sigma(B) - X\sigma(A),$$

and hence

$$\sigma(A)^* = Y^{-1}\sigma(B) - Y^{-1}X\sigma(A)$$
$$\sigma(B)^* = XY^{-1}\sigma(B) - (XY^{-1}X + Y)\sigma(A).$$

These relations make it possible to determine the conjugate of any harmonic differential. Indeed, if ω has A-periods $\xi = (\xi_1, \cdots, \xi_g)$ and B-periods $\eta = (\eta_1, \cdots, \eta_g)$, considered as horizontal vectors, then $\omega = -\xi\sigma(B) + \eta\sigma(A)$, and we obtain

$$\omega^* = -(\xi X - \eta)Y^{-1}\sigma(B) + [\xi(XY^{-1}X + Y) - \eta Y^{-1}X]\sigma(A)$$

so that the periods of ω^* are given by $(\xi X-\eta)Y^{-1}$ and $\xi(XY^{-1}X+Y)-\eta Y^{-1}X$ respectively.

24E. It is clear from these considerations that the matrix Z reflects very fundamental properties of the conformal structure of W. Since Z is symmetric it contains $\frac{1}{2}g(g+1)$ complex constants. By other considerations one can show that a Riemann surface of genus $g>1$ depends on $3g-3$ complex parameters, in a sense that can be made precise. For $g>3$ the elements of Z are therefore subject to additional conditions whose precise nature is not known. The discussion of these fascinating questions falls outside the scope of this book.

25. Rational functions

25A. Let $\alpha_1=a_1\,dz$ and $\alpha_2=a_2\,dz$ be analytic differentials on a Riemann surface, the latter not identically zero. Then the quotient $f=a_1/a_2$ has, at each point, a value which is independent of the local variable, and f is thus a single-valued analytic function on the surface. An exception must be made for the points where a_2 is zero. At these points f has at most a pole, and it will be more accurate to describe f as a meromorphic function. If the surface is closed there can be only a finite number of poles, and we say that f is a *rational* function on the surface W. For economy of expression we will agree that a *function on* W means a rational function.

The total number of poles of f, counted with their multiplicities, is called the *order* of f. We know that the order is zero if and only if f reduces to a constant. Unless f is identically zero, $d\log f$ is an analytic differential with simple poles at the zeros and poles of f. The residue at a zero is equal to the multiplicity of the zero, and at a pole the residue is the negative of the multiplicity. Since the sum of the residues must vanish we conclude that f has as many zeros as poles. If c is any constant $f-c$ has the same order as f, and it follows that f takes each complex value c as many times as indicated by the order. This fact can be expressed as follows:

Theorem. *Any nonconstant function f on a closed surface W defines W as a complete covering surface of the Riemann sphere. The number of sheets is equal to the order of f.*

Actually, the analytic proof is a luxury, for we know that W is a covering surface of its image, and since W is compact the covering surface must be complete (Ch. I, 21) and its projection must be the whole sphere.

25B. It is important to show that there are nonconstant functions on every closed W. Clearly; $f=\alpha_1/\alpha_2=a_1/a_2$ is nonconstant if α_1 and α_2 are linearly independent. Here we may allow α_1 and α_2 to have poles. We have shown that there exists an analytic differential with a double pole at a prescribed point. If the poles of α_1, α_2 are chosen at different points it is evident that the differentials are linearly independent, and hence that f is not constant.

From this result we conclude that any closed Riemann surface may be considered as a covering surface of the Riemann sphere. There are of course many such representations, with different numbers of sheets.

25C. We prove the following fundamental theorem:

Theorem. *Any two functions f_1, f_2 on a closed surface W satisfy an algebraic equation $P(f_1, f_2)=0$ where P is an irreducible polynomial in two variables.*

If f_1 is constant there is a trivial relation of the form $f_1-c=0$. Suppose therefore that f_1 is of order $n>0$. Let w_0 be a complex number which is such that the equation $f_1(z)=w_0$ has distinct roots $z_1(w_0), \cdots, z_n(w_0)$. Then there exist neighborhoods V of w_0 and V_i of $z_i(w_0)$ with the property that the equation $f_1(z)=w$, $w \in V$, has one and only one root $z_i(w)$ in each V_i. The functions $f_2(z_i(w))$ are meromorphic in V, and the same is true of the elementary symmetric functions $S(f_2(z_i(w)))$. But the latter are independent of the order of the $z_i(w)$ and are thus well-defined, except at the projections of branch points. If w approaches such a projection w_0 we can easily see that $f_2(z_i(w))$ and consequently $S(f_2(z_i(w)))$ does not grow more rapidly than a negative power of $|w-w_0|$. Hence all singularities are nonessential, and we conclude that the symmetric functions are rational functions of w. We obtain an identity

$$f_2^n+R_1(f_1)f_2^{n-1}+\cdots+R_{n-1}(f_1)f_2+R_n(f_1) = 0,$$

where the R_i are rational functions. If we multiply by the common denominator we obtain a polynomial relation $P(f_1, f_2)=0$.

In order to show that P can be replaced by an irreducible polynomial we decompose P into prime factors $P_1\cdots P_k$. Since $P(f_1(z), f_2(z))=0$ for all $z \in W$ it follows that at least one $P_i(f_1(z), f_2(z))$ must vanish at infinitely many points. Since it is a rational function on W it is identically zero, and the theorem is proved.

We add that the irreducible polynomial P is uniquely determined up to a constant factor. For if P and Q are irreducible polynomials which do not differ merely by a numerical factor it is possible to determine polynomials A and B so that $AP+BQ=1$. It follows that $P(f_1, f_2)$ and $Q(f_1, f_2)$ cannot both be identically zero.

25D. If f_1 is of order n our proof shows that the irreducible polynomial $P(f_1, f_2)$ is always of degree $\leqq n$ in f_2. We show that there exists an f_2 for which the degree is exactly n. For this purpose we choose w_0 so that the equation $f_1(z)=w_0$ has distinct roots $z_1, \cdots, z_n$. If we can determine f_2 so that the values $f_2(z_i)$ are all different and finite, the equation $P(w_0, f_2)=0$ will have n roots and must therefore be of degree $\geqq n$.

To complete the proof we show that there exists an f_2 with arbitrarily prescribed values at $z_1, \cdots, z_n$. Set $\alpha_0 = d\left(\frac{1}{f_1 - w_0}\right)$ and let α_i be an analytic differential with a double pole at z_i, and no other poles. Then α_i/α_0 is 0 at $z_j \neq z_i$ and different from 0 and ∞ at z_i. By choosing f_2 as a linear combination of the α_i/α_0 we can consequently prescribe the values $f_2(z_i)$ at will.

25E. We shall now reverse the procedure and consider an equation $P(z, w) = 0$ between two indeterminates, P being irreducible and of degree $n > 0$ in w. It determines w as an *algebraic function* of z, and we proceed to construct the Riemann surface of this function. The method is very familiar, and an outline will be sufficient.

For every region V on the complex sphere we consider the family Φ_V of all meromorphic functions $w(z)$ in V which satisfy $P(z, w(z)) = 0$. We construct a space W_0 consisting of all couples (z, w) with $z \in V$ and $w \in \Phi_V$. With the help of proper identifications and suitable topology it is possible to consider each component of W_0 as a smooth covering surface of the sphere; for the details of the construction we refer to Ch. I, 16. It becomes a regular covering surface if we exclude all points z for which the equations $P(z, w) = 0$ and $P_w(z, w) = 0$ have a common solution. However, since W_0 has only a finite number of sheets the deleted points can be reintroduced and W_0 can be extended in such a way that its every component becomes a complete covering surface of the sphere, with or without branch points.

W_0 inherits the conformal structure of the sphere, and in this manner each component of W_0 becomes a closed Riemann surface. The functions z and w are rational on each component, and they satisfy an irreducible equation which can only be $P(z, w) = 0$. It follows that each component has n sheets. For the same reason W_0 has n sheets, and we conclude that W_0 is connected.

25F. We return to the situation that was considered in 25D. In other words, f_1 and f_2 are two functions on W which satisfy an irreducible equation $P(f_1, f_2) = 0$ whose degree in f_2 is equal to the order n of the function f_1. By the method in 25E we construct the Riemann surface W_0 which corresponds to the relation $P(z, w) = 0$.

We contend that W_0 and W are conformally equivalent. To define a mapping of W on W_0 we consider a point $p \in W$ which is such that neither $f_1(p)$ nor $f_2(p)$ is a multiple value. Then there exists a unique point $(z, w) \in W_0$ such that $z = f_1(p)$ and $w(z) = f_2(p)$. This point will be the image of p, and at the excluded points the mapping can be uniquely defined by continuity.

The mapping $p \to (z, w)$ is clearly analytic, and it defines W as a covering surface of W_0. On the other hand, the projection $(z, w) \to z$ defines W_0 as a covering surface of the sphere. The combined mapping $p \to z$ is the one defined by $f_1(z)$, and under this mapping W is an n-sheeted covering surface. But W_0 is likewise n-sheeted over the sphere, and hence W is one-sheeted over W_0. This proves that W and W_0 are conformally equivalent.

26. Divisors

26A. A divisor on a closed Riemann surface W is a 0-dimensional chain $D = n_1a_1 + \cdots + n_ka_k$ where the a_i are points on W and the n_i are integers. We say that D is the divisor of a function f if the a_i are the zeros and poles of f and the n_i are their orders, counted positive for zeros and negative for poles.

All divisors form an Abelian group, and the divisors of functions, or *principal divisors*, form a subgroup D_0. The elements of the quotient group D/D_0 are called *divisor classes*. Thus two divisors belong to the same divisor class if and only if their difference is a principal divisor. The class of principal divisors is the *principal class*.

The *degree* of a divisor is the sum of its coefficients. Clearly, the degree of a principal divisor is 0, and any two divisors in the same divisor class have the same degree. The degree of D is denoted as deg D. It can also be considered as the degree of the divisor class which contains D.

We can also speak of the divisors of an analytic differential $\alpha = a\,dz$ which is not identically zero. It is formed by the zeros and poles of a taken with their respective positive and negative multiplicities. Since the quotient of two differentials is a function we conclude that all divisors of differentials are in the same class. It is called the *canonical class*. In particular, all divisors in the canonical class have the same degree, and we will show that the degree is $2g-2$.

26B. A divisor is called an *integral divisor* if all its coefficients are non-negative, and we say that D_1 is a *multiple* of D_2 if $D_1 - D_2$ is an integral divisor. As a simplification of the language we shall also say that a function f is a multiple of D if the divisor of f is a multiple of D.

The functions which are multiples of a given divisor D form a vector space over the complex numbers. The dimension, dim D, of this space is always finite, for it is clear that sufficiently many multiples of D permit a linear combination which is free from poles and must therefore reduce to a constant. In addition, dim D depends only on the divisor class of D. To see this, suppose that $D_1 = D_2 + D_0$ where D_0 is the divisor of a function f_0. If f is a multiple of D_2, then ff_0 is a multiple of D_1, and vice versa. For every linear relation between the multiples f there is a corresponding relation between the multiples ff_0. Hence dim $D_1 =$ dim D_2, and we speak of the dimension of the whole divisor class.

We can also consider the differentials which are, in the same sense, multiples of a given divisor. However, this case can be reduced to the preceding situation. For to say that α is a multiple of D means the same thing as to say that α/α_0 is a multiple of $D-Z$, where Z denotes the divisor of α_0. Hence the number of linearly independent divisors which are multiples of D is given by $\dim(D-Z)$. Here Z stands for any representative of the canonical class.

26C. It is clear that dim $D=0$ whenever deg $D>0$. Indeed, a function which is a multiple of D would have a divisor of positive degree, contrary to the fact that the divisor of any function has degree zero.

A multiple of the divisor 0 is an everywhere regular function, and hence a constant. This shows that dim $0=1$.

A differential is a multiple of the divisor 0 if and only if it is regular. We know that the regular differentials form a vector space of dimension g. Hence $\dim(-Z)=g$.

27. Riemann-Roch's theorem

27A. There is no general method to determine the dimension of a divisor. However, there is an important duality relation known as the *Riemann-Roch theorem*:

Theorem. dim $D=\dim(-D-Z)-\deg D-g+1$.

If this theorem is applied to $D=-Z$ we get $\dim(-Z)=\dim 0+\deg Z-g+1$, and since dim $0=1$, $\dim(-Z)=g$ we find that deg $Z=2g-2$ as already announced. With the aid of this result it is possible to formulate the theorem more symmetrically:

If $-(D_1+D_2)$ *belongs to the canonical class, then* $2\dim D_1+\deg D_1=2\dim D_2+\deg D_2$.

27B. The theorem is a consequence of the symmetry relation $h_0(z,\zeta)=h_0(\zeta,z)$, proved in 18H. The following algebraic lemma will also be needed:

Lemma. *Let U and V be two finite dimensional vector spaces over the complex numbers, and suppose that there is given a complex-valued bilinear product uv defined for all $u\in U$, $v\in V$. Denote by U_0 the subspace formed by all $u\in U$ such that $uv=0$ for all $v\in V$. Similarly, let V_0 be formed by all $v\in V$ such that $uv=0$ for all $u\in U$. Then* $\dim(U/U_0)=\dim(V/V_0)$.

There exists a basis $\{u_1,\cdots,u_m\}$ of U such that a subset $\{u_1,\cdots,u_r\}$ is a basis of U_0. Let $\{v_1,\cdots,v_n\}$ and $\{v_1,\cdots,v_s\}$ be similar bases for V and V_0. Consider the linear equations

$$x_{r+1}(u_{r+1}v_j)+\cdots+x_m(u_mv_j) = 0,\quad j = s+1,\cdots,n.$$

If $m-r$ were $>n-s$ this system would have a nontrivial solution $(x_{r+1}, \cdots, x_m)$. It would follow that

$$(x_{r+1}u_{r+1}+\cdots+x_m u_m)v_j = 0$$

for $j=1, \cdots, n$, and hence that $x_{r+1}u_{r+1}+\cdots+x_m u_m \in U_0$. This would imply that $x_{r+1}u_{r+1}+\cdots+x_m u_m$ could be written as a linear combination of $u_1, \cdots, u_r$. This contradicts the linear independence of $u_1, \cdots, u_m$. Hence $m-r \leqq n-s$, and the reverse inequality can be proved symmetrically. We conclude that $m-r=n-s$, which is the assertion of the lemma.

27C. Consider an integral divisor $A=\sum_1^k m_i a_i$, $m_i \geqq 1$, of degree $d(A)=\sum_1^k m_i$. We shall denote by $U(A)$ the vector space spanned by the differentials $h_m(z, \zeta)\,dz$ with $\zeta=a_i$, $0 \leqq m \leqq m_i-1$. These generating differentials are linearly independent, for they have poles at different points or of different order at the same point. Hence the dimension of $U(A)$ is equal to $d(A)$, the number of generators.

We define also a second vector space $V(A)$ generated by the following generators: (1) differentials $h_m(z, \zeta)\,dz$ with $\zeta=a_i$, $0 \leqq m \leqq m_i-2$, provided that $m_i \geqq 2$, (2) differentials $h(z, c)\,dz$ where c runs through a system of chains $c_2, \cdots, c_k$ with $\partial c_i = a_i - a_1$, (3) differentials $h(z, c)\,dz$ where c runs through the A-cycles $A_1, \cdots, A_g$ in a canonical basis.

To see that these generators are linearly independent we remark first that all the differentials listed under (1) and (2) have different singularities. The only possible linear dependence would thus be between the differentials $h(z, A_i)\,dz=\varphi(A_i)$. But the $\varphi(A_i)$ are known to be linearly independent (as a consequence of the last relation in Theorem 19 I). The dimension of $V(A)$ is thus equal to the number of generators. The latter is $d(A)+g-1$ if $k \geqq 1$ and g if $k=0$.

27D. An arbitrary divisor D can be written in the form $D=B-A$ where $A=\sum_1^k m_i a_i$ and $B=\sum_1^l n_j b_j$ are integral divisors and no a_i coincides with a b_j. We shall consider $U(A)$ and $V(B)$.

In order to define a bilinear product $u \cdot v$ for $u \in U(A)$, $v \in V(B)$ it is sufficient to prescribe its values for the basis elements of $U(A)$ and $V(B)$. The generators of $U(A)$ are all of the form $u=h_m(z, a)\,dz$. If $v=h_n(z, b)\,dz$ we set

$$(63) \qquad u \cdot v = \frac{1}{(m+1)!(n+1)!}\left[\frac{\partial^{m+n}}{\partial z^m \partial \zeta^n} h_0(z, \zeta)\right]_{z=a,\ \zeta=b},$$

and if $v = h(z, c)dz$ we postulate

$$u \cdot v = -\int_c h_m(\zeta, a)\, d\zeta. \tag{64}$$

Through this definition the subspaces $U_0(A)$ and $V_0(B)$ are determined, and by Lemma 27B

$$\dim U(A) - \dim U_0(A) = \dim V(B) - \dim V_0(B).$$

If the known values of $\dim U(A)$ and $\dim V(B)$ are substituted we obtain

$$\dim V_0(B) - \dim U_0(A) = d(B) - d(A) + g - 1 = \deg D + g - 1 \tag{65}$$

if $B \neq 0$ and

$$\dim V_0(B) - \dim U_0(A) = \deg D + g \tag{66}$$

if $B = 0$.

27E. Let us first investigate the meaning of $V_0(B)$. We will show that a differential belongs to $V_0(B)$ if and only if it is a multiple of $A - B$.

Suppose first that $v = h(z)\, dz$ belongs to $V_0(B)$. Then $h(z)$ has poles only at the points b_j, and at most of order n_j. This means that v is a multiple of $-B$. To prove that v is even a multiple of $A - B$ we must show that $h(z)$ has zeros of at least order m_i at the points a_i, that is to say that $h^{(m)}(a_i) = 0$ for $m \leqq m_i - 1$.

Let $\{v_\nu\}$ be the generators of $V(B)$, and assume that $v = \sum y_\nu v_\nu$. Consider a v_ν of the form $h_n(z, b)\, dz$. We have

$$h_n(z, \zeta) = \frac{1}{(n+1)!} \frac{\partial^n}{\partial \zeta^n} h_0(z, \zeta)$$

(see 18H) and hence

$$\left[\frac{\partial^m}{\partial z^m} h_n(z, b)\right]_{z=a} = \frac{1}{(n+1)!} \left[\frac{\partial^{m+n}}{\partial z^m \partial \zeta^n} h_0(z, \zeta)\right]_{z=a,\ \zeta=b}$$

Similarly, if $v_\nu = h(z, c)\, dz$ we have (see 19 I)

$$h(z, c) = -\int_c h_0(\zeta, z)\, d\zeta$$

and

$$\left[\frac{\partial^m}{\partial z^m} h(z, c)\right]_{z=a} = -(m+1)! \int_c h_m(\zeta, a)\, d\zeta.$$

On comparing with (63) and (64) we find that

$$h^{(m)}(a) = (m+1)!\, u \cdot v \tag{67}$$

for $u=h_m(z,a)\,dz$. If $a=a_i$, $m\leqq m_i-1$ this u belongs to $U(A)$, and we conclude that $h^{(m)}(a)=0$. This is what we wanted to show.

Conversely, suppose that the differential φ is a multiple of $A-B$. There exists a differential in $V(B)$ with the same singularities as φ, and because $V(B)$ contains a basis for the regular differentials we conclude that φ is itself in $V(B)$. If we set $\varphi=v=h\,dz$ it follows by (67) and the conditions $h^{(m)}(a)=0$ for $a=a_i$, $m\leqq m_i-1$ that $u\cdot v=0$ for all generators u of $U(A)$. This proves that $\varphi\in V_0(B)$.

We conclude that $\dim(-D-Z)=\dim V_0(B)$.

27F. We show next that $U_0(A)$ consists of the differentials df of functions f which are multiples of $B-A$. Suppose first that $u=h\,d\zeta\in U_0(A)$; the change of notation for the local variable is dictated by convenience. Then u has zero residues and poles of order $\leqq m_i+1$ at the points a_i. If we can show that u is exact, $u=df$, it follows that f is a multiple of $-A$. In order that it be possible to choose f as a multiple of $B-A$ it is sufficient that u have zero integrals over paths c_j between b_1 and b_j, and that h have a zero of at least the order n_j-1 at b_j whenever $n_j>1$.

Suppose that $u=\sum x_\nu u_\nu$ where $\{u_\nu\}$ is a basis of $U(A)$. Each u_ν is of the form $h_m(\zeta,a)\,d\zeta$. We have

$$h_m(\zeta,z)=\frac{1}{(m+1)!}\frac{\partial^m}{\partial z^m}h_0(\zeta,z)=\frac{1}{(m+1)!}\frac{\partial^m}{\partial z^m}h_0(z,\zeta)$$

$$\left[\frac{\partial}{\partial\zeta^n}h_m(\zeta,a)\right]_{\zeta=b}=\frac{1}{(m+1)!}\left[\frac{\partial^{m+n}}{\partial z^m\partial\zeta^n}h_0(z,\zeta)\right]_{z=a,\ \zeta=b}$$

and consequently, by comparison with (63),

$$h^{(n)}(b)=(n+1)!\,u\cdot v \tag{68}$$

for $v=h_n(z,b)\,dz$. The corresponding equation

$$\int_c h(\zeta)\,d\zeta=-\,u\cdot v \tag{69}$$

for $v=h(z,c)\,dz$ follows from (64) without computation. Because $u\in U_0(A)$ we conclude that h vanishes at b_j to the required order, that u has zero A-periods, and that the integral of u over each of the paths c_j used in the definition of $V(B)$ is zero.

It remains to show that u has zero B-periods. To see this, let us write $u_\nu=\varphi^{(\nu)}$, $u=\varphi=\sum x_\nu\varphi^{(\nu)}$. If $\varphi^{(\nu)}=h_m(\zeta,a)\,d\zeta$, set $\psi^{(\nu)}=k_m(\zeta,a)\,d\zeta$ (see 18B) and $\psi=\sum\bar{x}_\nu\psi^{(\nu)}$. Then $\tau=\varphi+\bar{\psi}$ is exact, and ψ is regular. It follows that ψ has zero A-periods, and hence that $\psi=0$. We conclude that $\varphi=u$ is exact, as asserted, and we have completed the proof that all elements of $U_0(A)$ are differentials df of multiples of $B-A$.

Suppose, conversely, that f is a multiple of $B-A$. There exists a $u=\varphi \in U(A)$ with the same singularities as df. As before we can associate with φ a regular analytic differential ψ which makes $\varphi+\bar{\psi}$ exact. Then $\varphi+\bar{\psi}-df$ is a regular exact harmonic differential, and consequently identically 0. From $\bar{\psi}=df-\varphi$ we conclude that $\psi=0$, $\varphi=df$, for there is no differential other than 0 which is at once analytic and antianalytic. From the fact that f is a multiple of $B-A$ it follows by way of (68) and (69) that $u\cdot v=0$ for all $v \in V(B)$, and hence that $u \in U_0(A)$.

If $B\neq 0$ the functions f which are multiples of $B-A$ are in one to one correspondence with their differentials df, for then the integration constant is uniquely determined. In this case we have consequently $\dim D = \dim U_0(A)$. If, however, $B=0$ the constants are also multiples of D, and we obtain $\dim D=\dim U_0(A)+1$.

27G. To collect the results, we have shown that $\dim(-Z-D)=\dim V_0(B)$ in all cases, and $\dim D=\dim U_0(A)$ if $B\neq 0$, $\dim D=\dim U_0(A)+1$ if $B=0$. If these results are substituted in (65) or (66), as the case may be, we obtain in all cases the identity

$$\dim D = \dim(-D-Z)-\deg D-g+1$$

which constitutes the assertion of Riemann-Roch's theorem.

28. Consequences of Riemann-Roch's theorem

28A. If we use the Riemann-Roch theorem merely as an inequality we obtain

$$\dim D \geqq -\deg D-g+1.$$

We find that $\dim D>1$ if $\deg D<-g$.

Theorem. *There exists a nonconstant multiple of D whenever* $\deg D<-g$.

For instance, there will always exist a rational function which has a single pole of order $\leqq g+1$ at a prescribed point.

28B. If $\deg D<-(2g-2)=-\deg Z$ we get $\deg(-D-Z)>0$ and hence $\dim(-D-Z)=0$ (see 26C). For this case the Riemann-Roch theorem yields

(70) $$\dim D = -\deg D-g+1.$$

If a_0 is any prescribed point, the same formula can be applied to $D-a_0$, and we find that

(71) $$\dim(D-a_0) = \dim D+1.$$

Hence there exists a multiple of $D-a_0$ which is not a multiple of D. If we choose D so that a_0 does not occur in D we can make the following conclusion:

Theorem. *There exists a rational function with a simple pole at any prescribed point.*

The reciprocal of the function will have a simple zero at the point in question.

28C. For a fixed point a_0, consider the divisors $-na_0$ with integral n. We compare $\dim[-(n+1)a_0]$ with $\dim(-na_0)$. The former is always at least equal to the latter, and it will be greater than the latter if and only if there exists a rational function whose only singularity is a pole of order $n+1$ at a_0. When this is so $\dim[-(n+1)a_0]$ exceeds $\dim(-na_0)$ by exactly 1.

Starting with $n=0$ we have $\dim 0=1$. For $n=2g-1$ we can apply (70) and find that $\dim[-(2g-1)]a_0=g$. In between there are thus exactly $g-1$ places where the dimension increases by 1 and g places where it remains constant. From $n=2g-1$ on the dimension will always increase by 1 as seen by (71). We conclude:

Theorem. *There are exactly g integers $0<n_1<n_2<\cdots<n_g<2g$ such that no rational function has a pole at a_0 of exactly order n_i and is otherwise regular.*

This theorem is often referred to as Weierstrass's gap theorem. The integers n_i will in general depend on a_0. The smallest possible value of n_g is g. If the minimum is attained, $n_g=g$, we have $\dim(-ga_0)=1$ and there are no rational functions with a single pole at a_0 of order $\leqq g$.

28D. It is the rule rather than the exception that $n_1, \cdots, n_g$ coincides with the sequence $1, 2, \cdots, g$. The points a_0 for which this is not the case are called *Weierstrass points*. We will prove:

Theorem. *There are only a finite number of Weierstrass points.*

If a_0 is a Weierstrass point we have $\dim(-ga_0)\geqq 2$. The Riemann-Roch theorem yields

$$\dim(-ga_0) = \dim(ga_0-Z)+1,$$

and hence $\dim(ga_0-Z)\geqq 1$ for a Weierstrass point. This means that there exists a regular differential φ which has a zero at a_0 of order $\geqq g$.

If $\{\varphi_1, \cdots, \varphi_g\}$ is a basis for the regular differentials we can write $\varphi=x_1\varphi_1+\cdots+x_g\varphi_g$. On setting $\varphi_i=h_i\,dz$, we have

$$\begin{aligned} x_1h_1+\cdots+x_gh_g &= 0\\ x_1h_1'+\cdots+x_gh_g' &= 0\\ x_1h_1^{(g-1)}+\cdots+x_gh_g^{(g-1)} &= 0 \end{aligned}$$

at a_0. This implies

$$W(z) = \begin{vmatrix} h_1 & \cdot & \cdot & \cdot & h_g \\ \cdot & \cdot & \cdot & \cdot & \cdot \\ h_1^{(g-1)} & \cdot & \cdot & \cdot & h_g^{(g-1)} \end{vmatrix} = 0$$

for $z=a_0$. Either a_0 is an isolated zero of $W(z)$, or else $W(z)\equiv 0$ in a neighborhood of a_0. As soon as we can exclude the second possibility we shall have proved that the Weierstrass points are isolated, and hence that they are finite in number.

Consider the vectors $h^{(n)}(a_0)=(h_1^{(n)}(a_0), \cdots, h_g^{(n)}(a_0))$ for all $n\geqq 0$. $W(a_0)=0$ implies that $h(a_0), h'(a_0), \cdots, h^{(g-1)}(a_0)$ are linearly dependent. If $W(z)\equiv 0$ all derivatives $W^{(m)}(a_0)$ are 0. We claim that all $h^{(n)}(a_0)$ lie in a $(g-1)$-dimensional subspace. If not, there would exist a smallest n such that $h(a_0), \cdots, h^{(n)}(a_0)$ span the whole g-dimensional space. The derivative $W^{(n-g+1)}(a_0)$ can be expressed as a sum of determinants whose rows are formed by $h(a_0)$ and its derivatives of order $\leqq n$. The determinants that do not contain $h^{(n)}(a_0)$ are zero by assumption. Therefore the one determinant which contains $h^{(n)}(a_0)$ must also be zero, an obvious contradiction.

Let $(c_1, \cdots, c_g)\neq(0, \cdots, 0)$ be a vector which is orthogonal to the space spanned by all $h^{(n)}(a_0)$. Then $c_1h_1+\cdots+c_gh_g$ vanishes at a_0 together with all its derivatives. Because of the analyticity this implies that $c_1h_1+\cdots+c_gh_g$ is identically zero. Hence $c_1\varphi_1+\cdots+c_g\varphi_g=0$ contrary to the fact that $\{\varphi_1, \cdots, \varphi_g\}$ is a basis. We have shown that $W(z)\not\equiv 0$, and the theorem is proved.

*Bibliography**

Ahlfors, L. V.

[1] *Zur Bestimmung des Typus einer Riemannschen Fläche.* Comment. Math. Helv. 3 (1931), 173–177. FM 57, 370.

[2] *Quelques propriétés des surfaces de Riemann correspondant aux fonctions méromorphes.* Bull. Soc. Math. France 60 (1932), 197–207. FM 58, 368.

[3] *Über die Kreise die von einer Riemannschen Fläche schlicht überdeckt werden.* Comment. Math. Helv. 5 (1933), 28–38. FM 59, 354.

[4] *Über die konforme Abbildung von Überlagerungsflächen.* C.R. Huitième Congr. Math. Scand. Stockholm, 1934, pp. 299–305. FM 61, 365.

[5] *Sur le type d'une surface de Riemann.* C.R. Acad. Sci. Paris 201 (1935), 30–32. FM 61, 365.

[6] *Zur Theorie der Überlagerungsflächen.* Acta Math. 65 (1935), 157–194. FM 61, 365.

[7] *Über eine Klasse von Riemannschen Flächen.* Soc. Sci. Fenn., Comment. Phys.-Math. 9, no. 6 (1936), 5 pp. FM 62, 388.

[8] *Geometrie der Riemannschen Flächen.* C.R. Congr. Intern. Math., Oslo, 1936. I, pp. 239–248. FM 63, 300.

[9] *Über die Anwendung differentialgeometrischer Methoden zur Untersuchung von Überlagerungsflächen.* Acta Soc. Sci. Fenn. Nova Ser. A. II. no. 6 (1937), 17 pp. FM 63, 301.

[10] *Zur Uniformisierungstheorie.* C.R. Neuvième Congr. Math. Scand. Helsinki, 1938, pp. 235–248. FM 65, 345.

[11] *Die Begründung des Dirichletschen Prinzips.* Soc. Sci. Fenn., Comment. Phys.-Math. 11, no. 15 (1943), 15 pp. MR 7, 203.

[12] *Das Dirichletsche Prinzip.* Math. Ann. 120 (1947), 36–42. MR 9, 238.

[13] *Normalintegrale auf offenen Riemannschen Flächen.* Ann. Acad. Sci. Fenn. Ser. A. I. no. 35 (1947), 24 pp. MR 10, 28.

[14] *Bounded analytic functions.* Duke Math. J. 14 (1947), 1–11. MR 9, 24.

[15] *Open Riemann surfaces and extremal problems on compact subregions.* Comment. Math. Helv. 24 (1950), 100–134. MR 12, 90.

[16] *Remarks on the classification of open Riemann surfaces.* Ann. Acad. Sci. Fenn. Ser. A. I. no. 87 (1951), 8 pp. MR 13, 338.

[17] *Remarks on the Neumann-Poincaré integral equation.* Pacific J. Math., 2 (1952), 271–280. MR 14, 182.

[18] *On the characterization of hyperbolic Riemann surfaces.* Ann. Acad. Sci. Fenn. Ser. A.I. no. 125 (1952), 5 pp. MR 14, 970.

[19] *Complex analysis. An introduction to the theory of analytic functions of one complex variable.* McGraw-Hill, New York-Toronto-London, 1953, 247 pp. MR 14, 857.

[20] *Development of the theory of conformal mapping and Riemann surfaces through a century.* Contributions to the theory of Riemann surfaces, pp. 3–13. Princeton Univ. Press, Princeton, 1953. MR 14, 1050.

[21] *On quasiconformal mappings.* J. Analyse Math. 3 (1954), 1–58; correction 207–208. MR 16, 348.

* FM and MR refer to Jahrbuch über die Fortschritte der Mathematik and Mathematical Reviews.

AHLFORS, L. V.
[22] *Remarks on Riemann surfaces.* Lectures on functions of a complex variable, pp. 45–48. Univ. Michigan Press, Ann Arbor, 1955.
[23] *Conformality with respect to Riemannian metrics.* Ann. Acad. Sci. Fenn. Ser. A. I. no. 206 (1955), 22 pp. MR 17, 657.
[24] *Square-integrable differentials on open Riemann surfaces.* Proc. Nat. Acad. Sci. U.S.A. 42 (1956), 758–760. MR 18, 727.
[25] *Extremalprobleme in der Funktionentheorie.* Ann. Acad. Sci. Fenn. Ser. A. I. no. 249/1 (1958), 9 pp. MR 19, 845.
[26] *Abel's theorem for open Riemann surfaces.* Seminars on analytic functions. II, pp. 7–19. Institute for Advanced Study, Princeton, 1958.
[27] *The method of orthogonal decomposition for differentials on open Riemann surfaces.* Ann. Acad. Sci. Fenn. Ser. A. I. no. 249/7 (1958), 15 pp. MR 20, 19.

AHLFORS, L. V., et BEURLING, A.
[1] *Invariants conformes et problèmes extrémaux.* C.R. Dixième Congr. Math. Scand., Copenhague, 1946, pp. 341–351. MR 9, 23.
[2] *Conformal invariants and function-theoretic null-sets.* Acta Math. 83 (1950), 101–129. MR 12, 171.
[3] *Conformal invariants. Construction and applications of conformal maps.* Proceedings of a symposium, pp. 243–245. National Bureau of Standards, Appl. Math. Ser. no. 18, 1952. MR 14, 861.

AHLFORS, L. V., und GRUNSKY, H.
[1] *Über die Blochsche Konstante.* Math. Z. 42 (1937), 671–673. FM 63, 300.

AHLFORS, L. V., and ROYDEN, H. L.
[1] *A counterexample in the classification of open Riemann surfaces.* Ann. Acad. Sci. Fenn. Ser. A. I. no. 120 (1952), 5 pp. MR 14, 864.

AHLFORS, L. V. ET AL. (edited by)
[1] *Contributions to the theory of Riemann surfaces.* Princeton Univ. Press, Princeton, 1953, 264 pp.

ALEXANDROFF, P., und HOPF, H.
[1] *Topologie.* I. Springer, Berlin, 1935, 636 pp. FM 61, 602.

ANDREIAN-CASACU, C.
[1] *Le théorème des disques pour les surfaces de Riemann normalement exhaustibles.* Acad. Rep. Pop. Romaîne. Bull. Sti. Secţ. Şci. Mat. Fiz. 4 (1952), 263–272. (Romanian. Russian and French summaries.) MR 15, 615.
[2] *Relations de structure dans la famille des transformations intérieures.* Ibid. 5 (1953), 431–441. MR 16, 812.
[3] *Rapports entre les surfaces riemanniennes normalement exhaustibles et les surfaces riemanniennes algébriques qui se rapprochent de* ∞ : A_∞. Ibid. 7 (1955), 529–542. MR 17, 473.
[4] *Le théorème d'Iversen pour des surfaces riemanniennes normalement exhaustibles.* Ibid. 5 (1955), 1145–1150. MR 17, 956.
[5] *Über die normal ausschöpfbaren Riemannschen Flächen.* Math. Nachr. 15 (1956), 77–86. MR 18, 647.
[6] *Sur la classe des surfaces de Riemann normalement exhaustibles et ses relations avec d'autres classes.* C.R. Acad. Sci. Paris 242 (1956), 2281–2283. MR 19, 23.

ANDREOTTI, A.
[1] *Une applicazione di un teorema di Cecioni ad un problema di rappresentazione conforme.* Ann. Scuola Norm. Super. Pisa (3) 2 (1948), (1950), 99–103. MR 11, 507.

APPELL, P., et GOURSAT, E.
[1] *Théorie des fonctions algébriques et de leurs intégrales.* I. *Etude des fonctions analytiques sur une surface de Riemann.* Gauthier-Villars, Paris, 1929, 526 pp. FM 55, 224.
[2] *Théorie des fonctions algébriques et des transcendantes qui s'y rattachent.* II. *Fonctions automorphes, par P. Fatou.* Gauthier-Villars, Paris, 1930, 521 pp. FM 56, 330.

D'ARCAIS, F.
[1] *Intro al teorema di Riemann-Roch.* Nota. Ven. Ist. Atti. 63 (8) 6 (1904), 99–103. FM 35, 420.

ARIMA, K.
[1] *On harmonic measure functions in some regions.* Kōdai Math. Sem. Rep. 1950, 75–80. MR 12, 692.
[2] *On uniformizing functions.* Ibid. 1950, 81–83. MR 12, 692.

ARONSZAJN, N.
[1] *Sur les singularités des surfaces de Riemann des fonctions inverses de fonctions entières.* C.R. Acad. Sci. Paris 200 (1935), 1569–1571. FM 61, 366.

BADER, R.
[1] *La théorie du potentiel sur une surface de Riemann.* C.R. Acad. Sci. Paris 228 (1949), 2001–2002. MR 11, 108.
[2] *Différentielles sur une surface de Riemann ouverte.* Ibid. 233 (1951), 1564–1565. MR 13, 643.
[3] *Fonctions à singularités polaires sur des domaines compacts et des surfaces de Riemann ouvertes.* Ann. Sci. Ecole Norm. Sup. (3) 71 (1954), 243–300. MR 16, 1012.
[4] Cf. SÖRENSEN, W., and BADER, R. [1].

BADER, R., et PARREAU, M.
[1] *Domaines non compact et classification des surfaces de Riemann.* C.R. Acad. Sci. Paris 232 (1951), 138–139. MR 12, 603.

BADER, R., et SÖRENSEN, W.
[1] *Sur le problème de Cousin pour une surface de Riemann non compacte.* Ann. Acad. Sci. Fenn. Ser. A. I. no. 250/1 (1958), 9 pp.

BARBILIAN, D.
[1] *Über die Metrik der geschlossenen Riemannschen Flächen.* C.R. Acad. Sci. Roumanie 2 (1938), 201–207. FM 64, 316.

BAUM, W.
[1] *A topological problem originating in the theory of Riemann surfaces.* Lectures on functions of a complex variable, pp. 405–407. Univ. Michigan Press, Ann Arbor, 1955. MR 16, 1140.

BAUR, L.
[1] *Über die Verzweigung der dreiblättrigen Riemannschen Flächen.* J. Math. 119 (1898), 171–174. FM 16, 1140.

BEHNKE, H.
[1] *Klassische Funktionentheorie.* I. Aschendorff'sche Verlag, Münster, 1947, 299 pp. MR 10, 439.
[2] *Klassische Funktionentheorie.* II. Aschendorff'sche Verlag, Münster, 1948, 217 pp. MR 10, 439.

BEHNKE, H., und SOMMER, F.
[1] *Theorie der analytischen Funktionen einer komplexen Veränderlichen.* Springer, Berlin-Göttingen-Heidelberg, 1955, 582 pp. MR 17, 470.

BEHNKE, H., und STEIN, K.
[1] *Die Sätze von Weierstrass und Mittag-Leffler auf Riemannschen Flächen.* Vjschr. Naturf. Ges. Zürich 85 (1940), 178–190. MR 2, 85.
[2] *Entwicklung analytischer Funktionen auf Riemannschen Flächen.* Math. Ann. 120 (1949), 430–461. MR 10, 696.
[3] *Elementarfunktionen auf Riemannschen Flächen als Hilfsmittel für die Funktionentheorie mehrerer Veränderlichen.* Canadian J. Math. 2 (1950), 152–165. MR 11, 652.

BERGMAN, S.
[1] *The kernel function and conformal mapping.* Math. Surveys 5, Amer. Math. Soc., New York, 1950, 161 pp. MR 12, 402.

BERGMAN, S., and SCHIFFER, M.

[1] *Kernel functions and conformal mappings.* Compositio Math. 8 (1951), 205–249. MR 12, 602.

BERS, L.

[1] *On rings of analytic functions.* Bull. Amer. Math. Soc. 54 (1948), 311–315. MR 9, 575.

[2] *Partial differential equations and pseudo-analytic functions on Riemann surfaces.* Contributions to the theory of Riemann surfaces, pp. 157–165. Princeton Univ. Press, Princeton, 1953, MR 15, 304.

[3] *Spaces of Riemann surfaces.* Proc. Intern. Congr. Math., Edinburgh, 1958.

BERTINI, E.

[1] *Osservazioni sulle " Vorlesungen über Riemann's Theorie der Abel'schen Integrale von Dr. C. Neumann."* Palermo Rend. 6 (1892), 165–172. FM 24, 384.

BEURLING, A.

[1] *Études sur un problème de majoration.* Thèse, Upsala, 1935, 109 pp. FM 59, 1042.

[2–4] Cf. AHLFORS, L. V., et BEURLING, A. [1–3].

BIEBERBACH, L.

[1] *Zur Theorie der automorphen Funktionen.* Diss., Göttingen, 1910, 42 pp. FM 41, 483.

[2] *Über die Einordnung des Hauptsatzes der Uniformisierung in die Weierstrassische Funktionentheorie.* Math. Ann. 78 (1918), 312–331. FM 46, 548.

[3] *Neue Untersuchungen über Funktionen von komplexen Variablen.* Enzykl. d. Math. Wiss. II C 4 (1921), 379–532. FM 48, 313.

[4] *Über einen Riemannschen Satz aus der Lehre von der konformen Abbildung.* Sitzungsber. Preuss. Akad. Wiss. Berlin 24 (1925), 6–9. FM 51, 273.

[5] *Lehrbuch der Funktionentheorie.* I. *Elemente der Funktionentheorie.* Teubner, Berlin, 1930, 322 pp. FM 56, 258.

[6] *Lehrbuch der Funktionentheorie.* II. *Moderne Funktionentheorie.* Teubner, Berlin, 1931, 370 pp. FM 57, 340.

[7] *Einführung in die konforme Abbildung.* Gruyter, Berlin-Leipzig, 1927, 131 pp. FM 53, 319.

BLANC, C.

[1] *Le type des surfaces de Riemann simplement connexes.* C.R. Acad. Sci. Paris 202 (1936), 623–625. FM 62, 387.

[2] *Les surfaces de Riemann des fonctions méromorphes.* I. Comment. Math. Helv. 9 (1937), 193–216. FM 63, 302.

[3] *Les surfaces de Riemann des fonctions méromorphes.* II. Ibid. 9 (1937), 335–368. FM 63, 302.

[4] *Une décomposition du problème du type des surfaces de Riemann.* C.R. Acad. Sci. Paris 206 (1938), 1078–1080. FM 64, 317.

[5] *Les demi-surfaces de Riemann. Application au problème du type.* Comment. Math. Helv. 11 (1938), 130–150. FM 64, 318.

BLANC, C., et FIALA, F.

[1] *Le type d'une surface et sa courbure totale.* Ibid. 14 (1942), 230–233. FM 68, 169.

BLOCH, A.

[1] *Les théorèmes de M. Valiron sur les fonctions entières et la théorie de l'uniformisation.* C.R. Acad. Sci. Paris 178 (1924), 2051–2053. FM 50, 217.

BOCHNER, S.

[1] *Fortsetzung Riemannscher Flächen.* Math. Ann. 98 (1927), 406–421. FM 53, 322.

BOUTON, CH. L.

[1] *Examples of the construction of Riemann surfaces for the inverse of rational functions by the method of conformal representation.* Ann. of Math. 12 (1898), 1–26, FM 29, 344.

BRĂZMA, N.

[1] *Über eine Riemannsche Fläche.* Univ. Riga. Wiss. Abh. Kl. Math. Abt. 1 (1943), 1–21. (German. Latvian summary.) MR 9, 506.

BRELOT, M.

[1] *Familles de Perron et problem de Dirichlet.* Acta Szeged 9 (1939), 133–153. FM 65, 418.

[2] *Sur le principe des singularités positives et la topologie de R. S. Martin.* Ann. Univ. Grenoble. Sect. Sci. Math. Phys. (N.S.) 23 (1948), 113–138. MR 10, 192.

[3] *Quelques applications de la topologie de R. S. Martin dans la théorie des fonctions harmoniques.* C.R. Acad. Sci. Paris 226 (1948), 49–51. MR 9, 284.

[4] *Principe et problème de Dirichlet dans les espaces de Green.* Ibid. 235 (1952), 598–600. MR 16, 35.

[5] *Topologies on the boundary and harmonic measure.* Lectures on functions of a complex variable, pp. 85–104. Univ. Michigan Press, Ann Arbor, 1955. MR 16, 1108.

[6] *Topology of R. S. Martin and Green lines.* Ibid., pp. 105–112. MR 16, 1108.

[7] *Lignes de Green et problème de Dirichlet.* C.R. Acad. Sci. Paris 235 (1952), 1595–1597. MR 16, 35.

[8] *Sur l'allure à la frontière des fonctions sous-harmoniques ou holomorphes.* Ann. Acad. Sci. Fenn. Ser. A. I. no. 250/4 (1958), 9 pp.

BRELOT, M., et CHOQUET, G.

[1] *Espaces et lignes de Green.* Ann. Inst. Fourier Grenoble 3 (1951), (1952), 199–263. MR 16, 34.

BRÖDEL, W.

[1] *Fortgesetzte Untersuchungen über Deformationsklassen bei mehrdeutigen topologischen Abbildungen.* Ber. Verh. Sächs. Akad. Wiss. Leipzig 91 (1939), 229–260. MR 1, 211.

BROUWER, L. E. J.

[1] *Über die topologischen Schwierigkeiten des Kontinuitätsbeweises der Existenztheoreme eindeutig umkehrbarer polymorpher Funktionen auf Riemannschen Flächen.* Nachr. Akad. Wiss. Göttingen (1912), 603–606. FM 43, 527.

[2] *Über die Singularitätenfreiheit der Modulmannigfaltigkeit.* Ibid., 803–806. FM 43, 527.

BURKHARDT, H.

[1] *Einführung in die Theorie der analytischen Funktionen einer complexen Veränderlichen.* (*Funktionentheoretische Vorlesungen.* 1. *Teil.*) Weit et Comp., Leipzig, 1897, 213 pp. FM 28, 325.

CACCIOPPOLI, R.

[1] *Funzioni pseudo-analitiche e rappresentazioni pseudo-conformi delle superficie riemanniane.* Ricerche Mat. 2 (1953), 104–127. MR 16, 27.

CALABI, E.

[1] *Metric Riemann surfaces.* Contributions to the theory of Riemann surfaces, pp. 77–85. Princeton Univ. Press, Princeton, 1953. MR 15, 863.

CARATHÉODORY, C.

[1] *Bemerkungen zu den Existenztheoremen der konformen Abbildung.* Bull. Calcutta Math. Soc. 20 (1930), 125–134. FM 56, 295.

[2] *Conformal representation.* Cambridge Univ. Press, London, 1932, 102 pp. FM 58, 354.

[3] *A proof of the first principal theorem on conformal representation.* Studies and essays presented to R. Courant on his 60th birthday, Jan. 8, 1948, pp. 75–83. Interscience Publ., New York, 1948. MR 9, 232.

[4] *Bemerkung über die Definition der Riemannschen Flächen.* Math. Z. 52 (1950), 703–708. MR 12, 251.

[5] *Conformal representation.* 2nd ed., Univ. Press, Cambridge, 1952, 115 pp. MR 13, 734.

[6] *Funktionentheorie.* I. Birkhäuser, Basel, 1950, 288 pp. MR 12, 248.

[7] *Funktionentheorie.* II. Birkhäuser, Basel, 1950, 194 pp. MR 12, 248.

CARLEMAN, T.
[1] *Sur une inégalité différentielle dans la théorie des fonctions analytiques.* C.R. Acad. Sci. Paris 196 (1933), 995–997. FM 59, 327.

CASORATI, F.
[1] *Teoria delle funzioni di variabili complesse.* I. Pavia, 1868. FM 1, 128.
[2] *Sopra le coupures del Sig. Hermite, i Querschnitte e le superficie di Riemann, ed i concetti d'integrazione sì reale che complessa.* Annali di Mat. (2) 15 (1887), 223–234. FM 19, 392.
[3] *Sopra le coupures del Sig. Hermite, i Querschnitte e le superficie di Riemann, ed i concetti d'integrazione sì reale che complessa.* Ibid. 16 (1888), 1–20. FM 20, 402.

CAZZANIGA, T.
[1] *Funzioni olomorfe nel campo ellittico.* Torino Atti 33 (1898), 808–823. FM 29, 354.

ČEBOTARËV, N. G.
[1] *Theory of Algebraic Functions.* OGIZ, Moskow-Leningrad, 1948, 396 pp. (Russian.) MR 10, 697.

CECIONI, F.
[1] *Sulla rappresentazione conforme delle aree pluri-connesse appartenenti a superficie di Riemann.* Rendiconti Accad. d. L. Roma (6) 9 (1929), 149–153. FM 55, 794.

CHERN, S. S.
[1] *An elementary proof of the existence of isothermal parameters on a surface.* Proc. Amer. Math. Soc. 6 (1955), 771–782. MR 17, 657.

CHEVALLEY, C.
[1] *Introduction to the theory of algebraic functions of one variable.* Math. Surveys 6. Amer. Math. Soc., New York, 1951, 188 pp. MR 13, 64.

CHISINI, O.
[1] *Sulle superficie di Riemann multiple, prive di punti di diramazione.* Rom. Acc. L. Rend. (5) 24 (1915), 153–158. FM 45, 658.

CHOQUET, G.
[1] *Fonctions analytiques et surfaces de Riemann.* Enseignement Math. (2) 2 (1956), 1–11. MR 18, 120.
[2] Cf. BRELOT, M., et CHOQUET, G. [1].

CHRISTOFFEL, M.
[1] *Über eine Klasse offener Überlagerungsflächen.* Zürig, Leeman, 1950, 42 pp.

CLEBSCH, A.
[1] *Zur Theorie der Riemann'schen Flächen.* Math. Ann. 6 (1873), 216–230. FM 5, 285.

CLÉMENT, L.
[1] *Étude de la surface de Riemann de* $f(z) = e^{hz}\dfrac{e^z - 1}{z}$, $h > 0$. C.R. Acad. Sci. Paris 227 (1948), 256–257. MR 10, 110.

COMBES, J.
[1] *Familles normales sur une surface de Riemann.* C.R. Acad. Sci. Paris 226 (1948), 379–381. MR 9, 341.
[2] *Sur l'uniformisation des fonctions algébroides.* Ibid. 227 (1948), 1325–1326. MR 11, 24.
[3] *Fonctions uniformes sur une surface de Riemann algébroide.* Ibid. 229 (1949), 14–16. MR 11, 96.
[4] *Sur quelques propriétés des fonctions algébroides.* Ann. Fac. Sci. Univ. Toulouse (4) 12 (1950), 5–76. MR 13, 125.

CONFORTO, F.
[1] *Abelsche Funktionen und algebraische Geometrie.* Springer, Berlin-Göttingen-Heidelberg, 1956, 276 pp. MR 18, 68.

COPSON, E. T.

[1] *An introduction to the theory of functions of a complex variable.* Oxford, Clarendon Press, 1935, 448 pp. FM 61, 301.

COURANT, R.

[1] *Zur Begründung des Dirichletschen Prinzipes.* Nachr. Akad. Wiss. Göttingen 1910, 154–160. FM 41, 855.

[2] *Über die Methode des Dirichletschen Prinzipes.* Math. Ann. 72 (1912), 517–550. FM 43, 490.

[3] *Über konforme Abbildung von Bereichen, welche nicht durch alle Rückkehrschnitte zerstückelt werden, auf schlichte Normalbereiche.* Math. Z. 3 (1919), 114–122. FM 47, 323.

[4] *Conformal mapping of multiply connected domains.* Duke Math. J. 5 (1939), 814–823. MR 1, 111.

[5] *The conformal mapping of Riemann surfaces not of genus zero.* Univ. Nac. Tucumán. Revista A. 2 (1941), 141–149. MR 4, 9.

[6] *Dirichlet's principle, conformal mapping, and minimal surfaces.* Appendix by M. Schiffer. Interscience Publ., New York, 1950, 330 pp. MR 12, 90.

[7] Cf. HURWITZ, A., und COURANT, R. [1].

DEDECKER, P.

[1] *Pseudo-surfaces de Riemann et pseudo-involutions.* I. Acad. Roy. Belgique. Bull. Cl. Sci. (5) 30 (1944), (1945), 120–133. MR 8, 145.

[2] *Pseudo-surfaces de Riemann et pseudo-involutions.* II. Ibid. (5) 30 (1944), (1945), 179–188. MR 8, 145.

DENJOY, A.

[1] *Sur les fonctions analytiques uniformes à singularités discontinues.* C.R. Acad. Sci. Paris 149 (1909), 258–260. FM 40, 442.

DENNEBERG, H.

[1] *Konforme Abbildung einer Klasse unendlich-vielfach zusammenhängender schlichter Bereiche auf Kreisbereiche.* Ber. Verh. Säch. Akad. Wiss. Leipzig 84 (1932), 331–352. FM 58, 1092.

DINGHAS, A.

[1] *Eine Bemerkung zur Ahlforsschen Theorie der Überlagerungsflächen.* Math. Z. 44 (1938), 568–572. FM 64, 316.

DIXON, A. C.

[1] *Prime functions on a Riemann surface.* London Math. Soc. Proc. 33 (1901), 10–26. FM 32, 406.

[2] *On a class of matrices of infinite order, and on the existence of "matricial" functions on a Riemann surface.* Cambr. Trans. 19 (1902), 190–233. FM 33, 448.

DOUGLAS, J.

[1] *Green's function and the problem of Plateau.* Amer. J. Math. 61 (1939), 545–589. MR 1, 19.

[2] *The most general form of the problem of Plateau.* Amer. J. Math. 61 (1939), 590–608. MR 1, 19.

DRAPE, E.

[1] *Über die Darstellung Riemannscher Flächen durch Streckenkomplexe.* Deutsche Math. 1 (1936), 805–824. FM 62, 1218.

DUFRESNOY, J.

[1] *Sur la théorie d'Ahlfors des surfaces de Riemann.* C.R. Acad. Sci. Paris 212 (1941), 595–598. MR 3, 81.

[2] *Sur une nouvelle démonstration d'un théorème d'Ahlfors.* Ibid., 662–665. MR 3, 81.

[3] *Sur certain propriétés nouvelles des fonctions algébroides.* Ibid., 746–749. MR 3, 81.

[4] *Une propriété des surfaces de recouvrement.* Ibid., 215 (1942), 252–253. MR 5, 94.

[5] *Remarques sur les fonctions méromorphes dans le voisinage d'un point singulier essentiel isolé.* Bull. Soc. Math. France 70 (1942), 40–45. MR 6, 206.

Durège, H.

[1] *Zur Analysis situs Riemannscher Flächen.* Sitzungsber. Kaiserl. Akad. Wiss. Wien 69 (1874), 115–120. FM 6, 234.

Dyck, W.

[1] *Über Aufstellung und Untersuchung von Gruppe und Irrationalität regulärer Riemannscher Flächen.* Math. Ann. 17 (1880), 473–510. FM 12, 315.

[2] *Notiz über eine reguläre Riemannsche Fläche vom Geschlechte drei und die zugehörige "Normalcurve" vierter Ordnung.* Ibid. 17 (1880), 510–517. FM 12, 319.

Elfving, G.

[1] *Über eine Klasse von Riemannschen Flächen und ihre Uniformisierung.* Acta Soc. Sci. Fenn. Nova Ser. 3 (1934), 60 pp. FM 60, 1034.

[2] *Über Riemannsche Flächen und Annäherung von meromorphen Funktionen.* Huitième Congr. Math. Scand., Stockholm, 1934, pp. 96–105. FM 61, 1160.

[3] *Zur Flächenstruktur und Wertverteilung. Ein Beispiel.* Acta Acad. Abo. 8, no. 10 (1935), 13 pp. FM 61, 1161.

Endl, K.

[1] *Zum Typenproblem Riemannscher Flächen.* Mitt. Mathem. Sem. Univ. Giessen no. 49 (1954), 35 pp. MR 16, 1012.

Enriques, F.

[1] *Sul gruppo di monodromia delle funzioni algebriche, appartenenti ad una data superficie di Riemann.* Rom. Acc. L. Rend. 13 (1904), 382–384. FM 35, 420.

Ermakow, W. P.

[1] *Die Theorie der Abel'schen Funktionen und Riemann'schen Oberflächen.* Kiew, 1897, 121 pp. FM 28, 371.

Fabian, W.

[1] *The Riemann surfaces of a function and its fractional integral.* Edinburgh Math. Notes no. 39 (1954), 14–16. MR 16, 461.

Fatou, P.

[1] Cf. Appell, P., et Goursat, E. [2].

Fiala, F.

[1] Cf. Blanc, C., et Fiala, F. [1].

Figueiredo, H. M.

[1] *Superficies de Riemann.* Coimbra, 1887. FM 18, 328, FM 19, 390.

Fischer, W.

[1] *Über die Riemann'sche Fläche der Gauss'schen ψ-Funktion und der Mittag-Leffler'schen E_α-Funktionen.* Mitt. Math. Sem. Univ. Giessen no. 37 (1949), 35 pp. MR 12, 90.

Flathe, H.

[1] *Approximation analytischer Funktionen auf nicht geschlossenen Riemannschen Flächen.* Math. Ann. 125 (1952), (1953), 287–306. MR 14, 1076.

Florack, H.

[1] *Reguläre und meromorphe Funktionen auf nicht geschlossenen Riemannschen Flächen.* Schr. Math. Inst. Univ. Münster no. 1 (1948), 34 pp. MR 12, 251.

Ford, L. R.

[1] *Automorphic functions*, 2nd ed., Chelsea, New York, 1951, 342 pp. FM 55, 810.

Forsyth, A. R.

[1] *Theory of functions of a complex variable.* 2nd ed., Univ. Press, Cambridge, 1900, 782 pp. FM 31, 392.

Fourès, L.

[1] *Décomposition en feuillets des surfaces de Riemann de type parabolique.* C.R. Acad. Sci. Paris 228 (1949), 644–646. MR 10, 523.

[2] *Sur les surfaces de Riemann à arbre topologique régulierement ramifié.* Ibid. 230 (1950), 353–355. MR 11, 590.

Fourès, L.

[3] *Sur les surfaces de recouvrement régulièrement ramifiées.* Ibid. 232 (1951), 467–469. MR 13, 25.

[4] *Sur la théorie des surfaces de Riemann.* Ann. Sci. Ecole Norm. Sup. (3) 68 (1951), 1–64. MR 12, 691.

[5] *Recouvrements de surfaces de Riemann.* Ibid. (3) 69 (1952), 183–201. MR 14, 550.

[6] *Le problème des translations isothermes ou construction d'une fonction analytique admettant dans un domain donné une fonction d'automorphie donnée.* Ann. Inst. Fourier Grenoble 3 (1951), (1952), 265–275. MR 14, 462.

[7] *Sur les recouvrements régulièrement ramifiés.* Bull. Sci. Math. (2) 76 (1952), 17–32. MR 13, 833.

[8] *Coverings of Riemann surfaces.* Contributions to the theory of Riemann surfaces, pp. 141–155. Princeton Univ. Press, Princeton, 1953. MR 15, 25.

[9] *Groupes fuchsiens et revêtements.* Ann. Inst. Fourier Grenoble 4 (1952), (1954), 49–71. MR 15, 614.

[10] *Determination of an automorphic function for a given analytic equivalence relation.* Seminars on analytic functions. II, pp. 20–31. Institute for Advanced Study, Princeton, 1958.

Freudenthal, H.

[1] *Über die Enden topologischer Räume und Gruppen.* Math. Z. 33 (1931), 692–713. FM 57, 731.

Freundlich, E.

[1] *Analytische Funktionen mit beliebig vorgeschriebenem unendlichblättrigen Existenzbereiche.* Diss., Göttingen, 1910. FM 41, 483.

Fricke, R.

[1] *Über die Module der algebraischen Gebilde.* Ber. Deutsche Math.-Verein. 3 (1894), 93–96. FM 25, 692.

[2] *Über die Theorie der automorphen Modulgruppen.* Nachr. Akad. Wiss. Göttingen (1896), 91–101. FM 27, 326.

[3] *Kurz-gefasste Vorlesungen über verschiedene Gebiete der höheren Mathematik mit Berücksichtigung der Anwendungen. Analytisch-funktionen-theoretischer Teil.* Teubner, Leipzig, 1900, 520 pp. FM 31, 393.

[4] *Die Ritter'sche Primform auf einer beliebigen Riemannschen Fläche.* Nachr. Akad. Wiss. Göttingen (1900), 314–321. FM 31, 429.

[5] *Beiträge zum Kontinuitätsbeweise der Existenz polymorpher Funktionen auf Riemannschen Flächen.* Math. Ann. 59 (1904), 449–513. FM 35, 427.

[6] *Zur Transformation der automorphen Funktionen.* Nachr. Akad. Wiss. Göttingen (1911), 518–526. FM 42, 455.

[7] *Automorphe Funktionen mit Einschluss der elliptischen Modulfunktionen.* Enzykl. d. Math. Wiss. II 2 (1913), 349–470. FM 44, 504.

Fricke, R., und Klein, F.

[1] *Vorlesungen über die Theorie der automorphen Funktionen.* I. *Die gruppentheoretischen Grundlagen.* Teubner, Leipzig, 1897, 634 pp. FM 28, 334.

[2] *Vorlesungen über die Theorie der automorphen Funktionen.* II. *Die Funktionentheoretischen Ausführungen und die Anwendungen. Erste Lieferung: Engere Theorie der automorphen Funktionen.* Teubner, Leipzig, 1901, 282 pp. FM 32, 430.

[3] *Vorlesungen über die Theorie der automorphen Funktionen.* II. *Die Funktionentheoretischen Ausführungen und die Anwendungen. Zweite Lieferung: Kontinuitätsbetrachtungen im Gebiete der Hauptkreisgruppen.* Teubner, Leipzig, 1911, pp. 283–438. FM 42, 452.

Frostman, O.

[1] *Potentiel d'équilibre et capacité des ensembles avec quelques applications à la théorie des fonctions.* Thèse, Lund, 1935, 115 pp. FM 61, 1262.

Galbură, G.

[1] *Sur le genre d'une coubre algébrique.* Com. Acad. R.P. Romîne 3 (1953), 105–107. (Romanian. Russian and French summaries.) MR 17, 27.

GARABEDIAN, P.
[1] *Schwarz's lemma and the Szegö kernel function.* Trans. Amer. Math. Soc., 67 (1949), 1–35. MR 11, 340.
[2] *Distortion of length in conformal mapping.* Duke Math. J. 16 (1949), 439–459. MR 11, 21.
[3] *The sharp form of the principle of hyperbolic measure.* Ann. of Math. (2) 51 (1950), 360–379. MR 11, 590.
[4] *A remark on the moduli of Riemann surfaces of genus 2.* Proc. Amer. Math. Soc. 1 (1950), 668–673. MR 12, 492.
[5] *Asymptotic identities among periods of integrals of the first kind.* Amer. J. Math. 73 (1951), 107–121. MR 12, 691.

GARABEDIAN, P., and SCHIFFER, M.
[1] *Identities in the theory of conformal mapping.* Trans. Amer. Math. Soc. 65 (1949), 187–238.

GARWICK, J. V.
[1] *Über das Typenproblem.* Arch. Math. Naturvid. 43 (1940), 33–46. MR 2, 84.

GEORGI, K.
[1] *Über die konforme Abbildung gewisser nichtsymmetrischer unendlich-vielfach zusammenhängender schlichter Bereiche auf Kreisbereiche.* Diss., Jena, 1915, 44 pp. FM 45, 673.

GERSTENHABER, M.
[1] *On a theorem of Haupt and Wirtinger concerning the periods of a differential of the first kind, and a related topological theorem.* Proc. Amer. Math. Soc. 4 (1953), 476–481. MR 14, 970.

GERSTENHABER, M., and RAUCH, H. E.
[1] *On extremal quasi-conformal mappings.* I. Proc. Nat. Acad. Sci. U.S.A. 40 (1954), 808–812. MR 16, 349.
[2] *On extremal quasi-conformal mappings.* II. Ibid. 40 (1954), 991–994. MR 16, 349.

GHIKA, A.
[1] *Sur le prolongement d'une fonction analytique sur un domaine riemannien donnéc.* Acad. Roum. Bull. Sect. Sci. 26 (1946), 155–161. MR 10, 110.

GLODEN, A.
[1] *Sur les surfaces de Riemann.* Linden et Hansen, Luxembourg, 1935, 96 pp. FM 61, 1157.

GOLUBEV, V. V.
[1] *Lectures on the analytic theory of differential equations.* 2nd ed. Gosudarstv. Izdat. Tehn.-Teor. Lit., Moscow-Leningrad, 1950, 436 pp. (Russian.) MR 13, 131.

GOLUZIN, G. M.
[1] *Geometrical theory of functions of a complex variable.* Gosudarstv. Izdat. Tehn.-Teor. Lit., Moscow-Leningrad, 1952, 540 pp. (Russian.) MR 15, 112.
[2] *Geometrische Funktionentheorie.* VEB Deutscher Verlag der Wissenschaften, Berlin, 1957, 438 pp. MR 19, 735.

GOTTSCHALK, W. H.
[1] *Conformal mapping of abstract Riemann surfaces.* Published by the author, Univ. of Pennsylvania, Philadelphia, 1949, 77 pp. MR 11, 342.

GOURSAT, E.
[1-2] Cf. APPELL, P., et GOURSAT, E. [1, 2].

GRAEUB, W.
[1] *Über die schwächste Uniformisierende.* Math. Z. 60 (1954), 66–78. MR 15, 787.

GRAVES, L. M.
[1] *The theory of functions of real variables.* 2nd ed. McGraw-Hill, New York, 1956, 375 pp. MR 17, 717.

GRAWE, D.
[1] *Die Riemannschen Oberflächen und die Elektrizitätstheorie.* J. Inst. Math. Kiew, (1935–36/), 3–15. (Ukrainian.) FM 61, 1157.

GRÖTZSCH, H.

[1] *Über einige Extremalprobleme der konformen Abbildung.* I. Ber. Verh. Säch. Acad. Wiss. Leipzig 80 (1928), 367–376. FM 54, 378.

[2] *Über einige Extremalprobleme der konformen Abbildung.* II. Ibid. 80 (1928), 497–502. FM 54, 378.

[3] *Über konforme Abbildung unendlich vielfach zusammenhängender schlichter Bereiche mit endlich vielen Häufungskomponenten.* Ibid. 81 (1929), 51–86. FM 55, 792.

[4] *Über die Verzerrung bei schlichter konformer Abbildung mehrfach zusammenhängender schlichter Bereiche.* I. Ibid. 81 (1929), 38–47. FM 55, 793.

[5] *Über die Verzerrung bei schlichter konformer Abbildung mehrfach zusammenhängender schlichter Bereiche.* II. Ibid. 81 (1929), 217–221. FM 55, 794.

[6] *Über die Verzerrung bei nichtkonformen schlichten Abbildungen mehrfach zusammenhängender schlichter Bereiche.* Ibid. 82 (1930), 69–80. FM 56, 298.

[7] *Zur konformen Abbildung mehrfach zusammenhängender schlichter Bereiche.* (*Iterationsverfahren.*) Ibid. 83 (1931), 67–76. FM 57, 400.

[8] *Zum Parallelschlitztheorem der konformen Abbildung schlichter unendlich-vielfach zusammenhängender Bereiche.* Ibid. 83 (1931), 185–200. FM 57, 401.

[9] *Das Kreisbogenschlitztheorem der konformen Abbildung schlichter Bereiche.* Ibid. 83 (1931), 238–253. FM 57, 401.

[10] *Über die Verschiebung bei schlichter konformer Abbildung schlichter Bereiche.* I. Ibid. 83 (1931), 254–279. FM 57, 402.

[11] *Über die Verzerrung bei schlichter konformer Abbildung mehrfach zusammenhängender schlichter Bereiche.* III. Ibid. 83 (1931), 283–297. FM 57, 402.

[12] *Über Extremalprobleme bei schlichter konformer Abbildung schlichter Bereiche.* Ibid. 84 (1932), 3–14. FM 58, 364.

[13] *Über das Parallelschlitztheorem der konformen Abbildung schlichter Bereiche.* Ibid. 84 (1932), 15–36. FM 58, 364.

[14] *Über die Verschiebung bei schlichter konformer Abbildung schlichter Bereiche.* II. Ibid. 84 (1932), 269–278. FM 58, 365.

[15] *Über zwei Verschiebungsprobleme der konformen Abbildung.* I. Sitzungsber. Preuss. Akad. Wiss. Berlin (1933), 87–100. FM 59, 349.

[16] *Die Werte des Doppelverhältnisses bei schlichter konformer Abbildung.* Ibid. (1933), 501–515. FM 59, 350.

[17] *Über die Geometrie der schlichten konformen Abbildung.* I. Ibid. (1933), 654–671. FM 59, 350.

[18] *Über die Geometrie der schlichten konformen Abbildung.* II. Ibid. (1933), 893–908. FM 60, 287.

[19] *Über die Geometrie der schlichten konformen Abbildung.* III. Ibid. (1934), 434–444. FM 60, 288.

[20] *Verallgemeinerung eines Bieberbachschen Satzes.* Jbr. Deutsche Math.-Verein. 43 (1933), 143–145. FM 59, 351.

[21] *Über Flächensätze der konformen Abbildung.* Ibid. 44 (1934), 266–269. FM 60, 288.

[22] *Einige Bemerkungen zur schlichten konformen Abbildung.* Ibid. 44 (1934), 270–275. FM 60, 289.

[23] *Zur Theorie der konformen Abbildung schlichter Bereiche.* I. Ber. Verh. Säch. Akad. Wiss. Leipzig 87 (1935), 145–158. FM 61, 363.

[24] *Zur Theorie der konformen Abbildung schlichter Bereiche.* II. Ibid. 87 (1935), 159–167. FM 61, 363.

[25] *Eine Bemerkung zum Koebeschen Kreisnormierungsprinzip.* Ibid. 87 (1935), 319–324. FM 61, 1156.

[26] *Zur Theorie der Verschiebung bei schlichter konformer Abbildung.* Comment. Math. Helv. 8 (1936), 382–390. FM 62, 385.

[27] *Konvergenz und Randkonvergenz bei Iterationsverfahren der konformen Abbildung.* Wiss. Z. Martin-Luther-Univ. Halle-Wittenberg. Math.-Nat. Reihe 5 (1955–56), 575–581. MR 19, 845.

GRUNSKY, H.
[1] *Neue Abschätzungen zur konformen Abbildung ein- und mehrfach zusammenhängender Bereiche.* Schriften Sem. Univ. Berlin 1 (1932), 95–140. FM 58, 361.
[2] *Über die konforme Abbildung mehrfach zusammenhängender Bereiche auf mehrblättrige Kreise.* Sitzungsber. Preuss. Akàd. Wiss. Berlin (1937), 40–46. FM 63, 300.
[3] *Über die konforme Abbildung mehrfach zusammenhängender Bereiche auf mehrblättrige Kreise.* II. Ibid. (1941), no. 11, 8 pp. MR 8, 324.
[4] Cf. AHLFORS, L. V., und GRUNSKY, H. [1].

HABSCH, H.
[1] *Die Theorie der Grundkurven und das Äquivalenzproblem bei der Darstellung Riemannscher Flächen.* Mitt. Math. Sem. Univ. Giessen, no. 42 (1952), 51 pp. MR 14, 969.

AF HÄLLSTRÖM, G.
[1] *On the study of algebraic functions of automorphism by help of graphs.* C.R. Dixième Congr. Math. Scand., Copenhague, 1946, pp. 97–107. MR 9, 233.
[2] *Zur Beziehung zwischen den Automorphiefunktionen und dem Flächentypus.* Acta Acad. Abo 20 (1956), no. 10, 12 pp. MR 18, 883.

HALMOS, P. R.
[1] *Introduction to Hilbert Space and the theory of spectral multiplicity.* Chelsea, New York, 1951, 114 pp. MR 13, 563.

HARDY, G. H., LITTLEWOOD, J. E., and PÓLYA, G.
[1] *Inequalities.* Cambridge Univ. Press, Cambridge, 1934, 314 pp. FM 60, 169.

HARTMAN, P.
[1] *On integrating factors and on conformal mappings.* Trans. Amer. Math. Soc. 87 (1958), 387–406.

HARZER, W.
[1] *Bestimmung der Anzahl der dreiblättrigen Riemannschen Flächen mit beliebig gegebenen Windungspunkten und der vierblättrigen mit Windungspunkten gleicher Ordnung.* Diss., Leipzig, 1913, 43 pp. FM 44, 510.

HAVINSON, S. Y.
[1] *On extremal properties of functions mapping a region on a multisheeted circle.* Doklady Akad. Nauk SSSR (N.S.) 88 (1953), 957–959. (Russian.) MR 14, 967.
[2] Cf. TUMARKIN, G., and HAVINSON,, S. [1].

HAYASHI, T.
[1] *On algebraic functions, equally ramified on the same Riemann surface.* Tôhoku Math. J. 4 (1913), 71–74. FM 44, 494.

HEINS, M.
[1] *A generalization of the Aumann-Carathéodory "Starrheitssatz".* Duke Math. J. 8 (1941), 312–316. MR 3, 81.
[2] *On the continuation of a Riemann surface.* Ann. of Math. (2) 43 (1942), 280–297. MR 4, 77.
[3] *The conformal mappings of simply-connected Riemann surfaces.* Ibid. (2) 50 (1949), 686–690. MR 11, 93.
[4] *A lemma on positive harmonic functions.* Ibid. (2) 52 (1950), 568–573. MR 12, 259.
[5] *Interior mapping of an orientable surface into S^2.* Proc. Amer. Math. Soc. 2 (1951), 951–952. MR 13, 547.
[6] *Riemann surfaces of infinite genus.* Ann. of Math. (2) 55 (1952), 296–317. MR 13, 643.
[7] *A problem concerning the continuation of Riemann surfaces.* Contributions to the theory of Riemann surfaces, pp. 55–62. Princeton Univ. Press, Princeton, 1953. MR 15, 25.
[8] *Studies in the conformal mapping of Riemann surfaces.* I. Proc. Nat. Acad. Sci. U.S.A. 39 (1953), 322–324. MR 15, 787.

HEINS, M.

[9] *Studies in the conformal mapping of Riemann surfaces.* II. Ibid. 40 (1954), 302–305. MR 15, 787.

[10] *Remarks on the elliptic case of the mapping theorem for simply connected Riemann surfaces.* Nagoya Math. J. 9 (1955), 17–20. MR 17, 473.

[11] *On the Lindelöf principle.* Ann. of Math. (2) 61 (1955), 440–473. MR 16, 1011.

[12] *Lindelöfian maps.* Ibid. (2) 62 (1955), 418–446. MR 17, 726.

[13] *A theorem concerning the existence of deformable conformal maps.* Seminars on analytic functions. II, pp. 32–38. Institute for Advanced Study, Princeton, 1958.

[14] *A theorem concerning the existence of deformable conformal maps.* Ann. of Math. (2) 67 (1958), 42–44. MR 19, 949.

[15] *Some remarks concerning parabolic Riemann surfaces.* J. Math. Pures Appl. (9) 36 (1957), 305–312. MR 20, 19.

[16] *On certain meromorphic functions of bounded valence.* Rev. Math. Pures Appl. 2 (1957), 263–267. MR 20, 290.

HENSEL, K.

[1] *Bemerkung zu der Abhandlung "On the theory of Riemann's integrals" by H. F. Baker.* Math. Ann. 45 (1894), 598–599. FM 25, 690.

[2] *Über die Verzweigung der drei- und vierblätterigen Riemannschen Flächen.* Monatsber. Preuss. Akad. Wiss. Berlin, (1895), 1103–1114. FM 26, 448.

[3] *Über die Ordnungen der Verzweigungspunkte einer Riemann'schen Fläche.* Ibid. (1895), 933–943. FM 26, 448.

[4] *Zur Theorie der algebraischen Funktionen einer Veränderlichen und der Abelschen Integrale.* Math. Ann. 54 (1901), 437–497. FM 32, 407.

HENSEL, K., und LANDSBERG, G.

[1] *Theorie der algebraischen Funktionen einer Variabeln und ihre Anwendung auf algebraische Kurven und Abelsche Integrale.* Teubner, Leipzig, 1902, 707 pp. FM 33, 427.

HERSCH, J.

[1] *Longueurs extrémales et théorie des fonctions.* Comment. Math. Helv. 29 (1955), 301–337. MR 17, 835.

[2] *Contribution à la théorie des fonctions pseudo-analytiques.* Ibid. 30 (1956), 1–19 (1955).

HERZ, J.

[1] *Über meromorphe transzendente Funktionen auf Riemannschen Flächen.* Diss., München, 1933, 41 pp. FM 59, 1040.

HILBERT, D.

[1] *Sur le principe de Dirichlet.* Nouv. Ann. Math. (3) 19 (1900), 337–344. FM 31, 418.

[2] *Über das Dirichletsche Prinzip.* Jbr. Deutsche Math.-Verein. 8_1 (1900), 184–188. FM 31, 418.

[3] *Über das Dirichletsche Prinzip.* Sonderabdruck Festschr. zur Feier des 150-jähr. Best. d. K. G. d. W. zu Göttingen, 1901, 27 pp. FM 32, 423.

[4] *Über das Dirichletsche Prinzip.* Math. Ann. 59 (1904), 161–186. FM 35, 436.

[5] *Über das Dirichletsche Prinzip.* J. Math. 129 (1905), 63–67. FM 36, 442.

HILL, J. M.

[1] *On the proofs of the properties of Riemann surfaces discovered by Lüroth and Clebsch.* Proc. London Math. Soc. (2) 10 (1911), 191–206. FM 42, 427.

HÖLDER, E.

[1] *Fortsetzung Abelscher Differentiale 1. Gattung ins Nichtlineare.* Ann. Acad. Sci. Fenn. Ser. A. I. no. 250/15 (1958), 5 pp. MR 20, 159.

HOPF, H.

[1] Cf. ALEXANDROFF, P., und HOPF, H. [1].

HORNICH, H.
[1] *Integrale erster Gattung auf speziellen transzendenten Riemannschen Flächen.* Verhandlungen Kongress Zürich, 1932. II, pp. 40–41. FM 58, 353.
[2] *Beschränkte Integrale auf der Riemannschen Fläche von* $\sqrt{\cos \frac{\pi z}{2}}$. Monatsh. Math. 42 (1935), 377–388. FM 61, 412.
[3] *Über transzendente Integrale erster Gattung.* Monatsh. Math. Phys. 47 (1939), 380–387. MR 1, 49.
[4] *Beschränkte Integrale auf speziellen transzendenten Riemannschen Flächen.* Ibid. 53 (1949), 187–201. MR 11, 510.
[5] *Beschränkte Integrale auf speziellen transzendenten Riemannschen Flächen.* II. Ibid. 54 (1950), 37–44. MR 12, 493.
[6] *Lehrbuch der Funktionentheorie.* Springer, Wien, 1950, 216 pp. MR 11, 589.
HOYER, P.
[1] *Über den Zusammenhang in Reihen mit einer Anwendung auf die Theorie der Substitutionen.* Math. Ann. 42 (1893), 58–88. FM 25, 695.
[2] *Über Riemannsche Flächen mit beschränkt veränderlichen Verzweigungspunkten.* Ibid. 47 (1896), 47–71. FM 27, 324.
HUBER, H.
[1] *Über analytische Abbildungen Riemannscher Flächen in sich.* Comment. Math. Helv. 27 (1953), 1–73. MR 14, 862.
[2] *On subharmonic functions and differential geometry in the large.* Comment. Math. Helv. 32 (1957), 13–72. MR 20, 159.
HUCKEMANN, F.
[1] *Bestimmung der Wertverteilung der Gammafunktion aus ihrer Riemannschen Fläche.* Math. Z. 59 (1934), 375–382. MR 15, 695.
[2] *Verschmelzung von Randstellen Riemannscher Flächen.* Mitt. Math. Sem. Univ. Giessen, no. 41 (1952), 36 pp. MR 15, 24.
[3] *Typusänderung bei Riemannschen Flächen durch Verschiebung von Windungspunkten.* Math. Z. 59 (1954), 385–387. MR 15, 695.
[4] *Zur Darstellung von Riemannschen Flächen durch Streckenkomplexe.* Ibid. 65 (1956), 215–239. MR 18, 120.
[5] *Über den Einfluss von Randstellen Riemannscher Flächen auf die Wertverteilung.* Ibid. 65 (1956), 240–282. MR 18, 120.
[6] *Über den defekt von mittelbaren Randstellen auf beschränktartigen Riemannschen Flächen.* Ann. Acad. Sci. Fenn. Ser. A. I. no. 250/16 (1958), 12 pp. MR 20, 291.
HUREWICZ, W. and WALLMAN, H.
[1] *Dimension Theory.* Princeton Univ. Press, Princeton, 1941, 165 pp. MR 3, 312.
HURWITZ, A.
[1] *Über Riemannsche Flächen mit gegebenen Verzweigungspunkten.* Math. Ann. 39 (1891), 1–61. FM 23, 429.
[2] *Über algebraische Gebilde mit eindeutigen Transformationen in sich.* Ibid. 41 (1892), 403–442. FM 24, 380.
[3] *Zur Theorie der Abel'schen Funktionen.* Nachr. Akad. Wiss. Berlin (1892), 247–254. FM 24, 385.
[4] *Über die Anzahl der Riemannschen Flächen mit gegebenen Verzweigungspunkten.* Math. Ann. 55 (1901), 53–66. FM 32, 404.
[5] *Sulle superficie di Riemann con dati punti di diramazione.* Batt. G. 41 (1903), 337–376. FM 34, 458.
HURWITZ, A., und COURANT, R.
[1] *Vorlesungen über allgemeine Funktionentheorie und elliptische Funktionen. Geometrische Funktionentheorie.* Springer, Berlin, 1929, 534 pp. FM 55, 171
IVERSEN, F.
[1] *Recherches sur les fonctions inverses des fonctions méromorphes.* Thèse, Helsinki, 1914. FM 45, 656.

IWASAWA, K.

[1] *Theory of algebraic functions.* Iwanami, Tokyo, 1952, 356 pp. (Japanese.) MR 15, 414.

JENKINS, J. A.

[1] *Some problems in conformal mapping.* Trans. Amer. Math. Soc. 67 (1949), 327–350. MR 11, 341.

[2] *Remarks on "Some problems in conformal mapping."* Proc. Amer. Math. Soc. 3 (1952), 147–151. MR 13, 642.

[3] *Some results related to extremal length.* Contributions to the theory of Riemann surfaces, pp. 87–94. Princeton Univ. Press, Princeton, 1953. MR 15, 115.

[4] *Another remark on "Some problems in conformal mapping."* Proc. Math. Soc. 4 (1953), 978–981. MR 15, 414.

[5] *A general coefficient problem.* Trans. Amer. Math. Soc. 77 (1954), 262–280. MR 16, 232.

[6] *Sur quelques aspects globaux du théorème de Picard.* Ann. Sci. École Norm. Sup. (3) 72 (1955), 151–161. MR 17, 725.

[7] *Some new canonical mappings for multiply-connected domains.* Ann. Math. (2) 65 (1957), 179–196. MR 18, 568.

[8] *On the existence of certain general extremal metrics.* Ibid., 440–453. MR 19, 845.

[9] Cf. JENKINS, J. A., and MORSE, M. [1].

JENKINS, J. A., and MORSE, M.

[1] *Topological methods on Riemann surfaces. Pseudoharmonic functions.* Contributions to the theory of Riemann surfaces, pp. 111–139. Princeton Univ. Press, Princeton, 1953. MR 15, 210.

[2] *Conjugate nets, conformal structure, and interior transformations on open Riemann surfaces.* Proc. Nat. Acad. Sci. U.S.A. 39 (1953), 1261–1268. MR 15, 415.

[3] *Conjugate nets on an open Riemann surface.* Lectures on functions of a complex variable, pp. 123–185. Univ. Michigan Press, Ann Arbor, 1955. MR 16, 1097.

JOHANSSON, S.

[1] *Ein Satz über die konforme Abbildung einfach zusammenhängender Riemannscher Flächen auf den Einheitskreis.* Math. Ann. 62 (1906), 177–183. FM 37, 438.

[2] *Beweis der Existenz linear-polymorpher Funktionen vom Grenzkreistypus auf Riemannschen Flächen.* Ibid. 62 (1906), 184–193. FM 37, 438.

JULIA, G.

[1] *Sur le domaine d'existence d'une fonction implicite définie par une relation entière* $G(x, y)=0$. Bull. Soc. Math. 54 (1926), 26–37. FM 52, 327.

[2] *Principes géométriques d'analyse.* I. Gauthier-Villars, Paris, 1930, 116 pp. FM 56, 294.

[3] *Leçons sur la représentation conforme des aires simplement connexes.* Gauthier-Villars, Paris, 1931, 114 pp. FM 57, 397.

[4] *Principes géométriques d'analyse.* II. Gauthier-Villars, Paris, 1932, 121 pp. FM 58, 355.

[5] *Prolongement d'une surface de Riemann σ correspondant à une aire multiplement connexe.* C.R. Acad. Sci. Paris 194 (1932), 580–583. FM 58, 358.

[6] *Reconstruction d'une surface de Riemann σ correspondant à une aire multiplement connexe.* Ibid. 194 (1932), 423–425. FM 58, 358.

[7] *Essai sur le développement de la théorie des fonctions de variables complexes.* Gauthier-Villars, Paris, 1933, 53 pp. FM 59, 311.

[8] *Leçons sur la représentation conforme des aires multiplement connexes.* Gauthier-Villars, Paris, 1934, 94 pp. FM 60, 285.

[9] *Quelques applications fonctionnelles de la topologie.* Reale Accademia d'Italia, Fondazione A. Volta, Atti dei Convegni, v. 9 (1939), Rome, 1943, 291–306. MR 12, 171.

JURCHESCU, M.

[1] *Au sujet des fonctions analytiques définies par des équations différentielles non algébriques.* Acad. Rep. Pop. Romaîne. Bull. Şti. Secţ. Şci. Mat. Fiz. 7 (1955), 347–354. (Romanian, Russian, and French summaries.) MR 17, 956.

Jurchescu, M.
[2] *Recouvrements riemanniens définis par des équations différentielles du second ordre.* C.R. Acad. Sci. Paris 245 (1957), 627–630. MR 19, 1044.
[3] *Modulus of a boundary component.* Pacific J. Math. 8 (1958), 791–804.
Kagno, I. N.
[1] *On a certain non-separating graph on an orientable surface.* J. Math. Phys. Mass. Inst. Tech. 20 (1941), 370–387. MR 3, 141.
Kakutani, S.
[1] *Einführung einer Metrik auf die Riemannsche Fläche und der Typus der Riemannschen Fläche.* Proc. Imp. Acad. Tokyo 13 (1937), 89–92. FM 63, 302.
[2] *Applications of the theory of pseudo-regular functions to the type-problem of Riemann surfaces.* Japanese J. Math. 13 (1937), 375–392. FM 63, 303.
[3] *Two-dimensional Brownian motion and the type problem of Riemann surfaces.* Proc. Japan Acad. 21 (1949), 138–140. MR 11, 257.
[4] *Random walk and the type problem of Riemann surfaces.* Contributions to the theory of Riemann surfaces, pp. 95–101. Princeton Univ. Press, Princeton, 1953. MR 15, 25.
[5] *Rings of analytic functions.* Lectures on functions of a complex variable, pp. 71–78. Univ. Michigan Press, Ann Arbor, 1955. MR 16, 1125.
Kaplan, W.
[1] *On Gross's star theorem, schlicht functions. logarithmic potentials and Fourier series.* Ann. Acad. Sci. Fenn. Ser. A.I. no. 86 (1951), 23 pp. MR 13, 337.
[2] *Construction of parabolic Riemann surfaces by the general reflection principle.* Contributions to the theory of Riemann surfaces, pp. 103–106. Princeton Univ. Press, Princeton, 1953. MR 15, 24.
[3] *Curve families and Riemann surfaces.* Lectures on functions of a complex variable, pp. 425–432. Univ. Michigan Press, Ann Arbor, 1955. MR 16, 1097.
Kaplan, W. (edited by)
[1] *Lectures on functions of a complex variable.* Univ. Michigan Press, Ann Arbor, 1955, 435 pp.
Kasten, H.
[1] *Zur Theorie der dreiblättrigen Riemannschen Fläche.* Diss., Göttingen, 1876. FM 9, 286.
Keinänen, V.
[1] *Über die vermischte Randwertaufgabe der Halbebene bei unendlich vielen Randintervallen.* Akademische Abhandlung, Helsinki, 1947, 80 pp. MR 9, 343.
Kerékjártó, B.
[1] *Vorlesungen über Topologie.* I. Springer, Berlin, 1923, 270 pp. FM 49, 396.
Klein, F.
[1] *Über Riemann's Theorie der algebraischen Funktionen und ihrer Integrale.* Teubner, Leipzig, 1882, 82 pp. FM 14, 359.
[2] *Riemann'sche Flächen.* Autographierte Vorlesungen, Göttingen, 1891–1892. I. 301 pp. FM 25, 684.
[3] *Riemann'sche Flächen.* Autographierte Vorlesungen, Göttingen, 1891–1892. II. 287 pp. FM 25, 684.
[4] *Mathematische Abhandlungen.* II. Springer, Berlin, 1922, 713 pp.
[5] *Mathematische Abhandlungen.* III. Springer, Berlin, 1923, 744 pp.
[6] Cf. Fricke, R., und Klein, F. [1-3].
Kneser, H.
[1] *Analytische struktur und Abzählbarkeit.* Ann. Acad. Sci. Fenn. Ser. A. I. no. 251/5 (1958), 8 pp.
Knopp, K.
[1] *Funktionentheorie.* I. *Grundlagen der allgemeinen Theorie der analytischen Funktionen.* Sammlung Göschen, Gruyter, Berlin, 1944, 135 pp.
[2] *Funktionentheorie.* II. *Anwendungen und Weiterführung der allgemeinen Theorie.* Sammlung Göschen, Gruyter, Berlin, 1949, 130 pp.

Kobayashi, Z.

[1] *Theorems on the conformal representation of Riemann surfaces*. Sci. Rep. Tokyo Bunrika Daigaku 2 (1935), 125–166. FM 61, 1158.

[2] *On the type of Riemann surfaces*. Ibid. 2 (1935), 217–233. FM 61, 1159.

[3] *Ein Satz über ein Problem von Herrn Speiser*. Ibid. 3 (1936), 29–32. FM 61, 1160.

[4] *The meromorphic function with every complex number as its asymptotic value*. Ibid. 3 (1936), 89–110. FM 62, 387.

[5] *A remark on the type of Riemann surfaces*. Ibid. 3 (1937), 185–193. FM 63, 983.

[6] *On Kakutani's theory of the type of Riemann surfaces*. Ibid. 4 (1940), 9–44. MR 14, 157.

Kodaira, K.

[1] *On the existence of analytic functions on closed analytic surfaces*. Kōdai Math. Sem. Rep. 2 (1949), 21–26. MR 11, 96.

Koebe, P.

[1] *Untersuchung der birationalen Transformationen, durch welche ein algebraisches Gebilde von Range eins in sich selbst übergeht, in bezug auf ihr Verhalten bei der Iteration*. Sitzungsber. Berl. Math. Ges. 5 (1906), 57–64. FM 37, 438.

[2] *Über konforme Abbildung mehrfach zusammenhängender Bereiche, insbesondere solcher Bereiche, deren Begrenzung von Kreisen gebildet wird*. Jbr. Deutsche Math.-Verein. 15 (1906), 142–153.

[3] *Über konforme Abbildung mehrfach zusammenhängender ebener Bereiche*. Ibid. 16 (1907), 116–130. FM 38, 688.

[4] *Über die Uniformisierung reeller algebraischer Kurven*. Nachr. Akad. Wiss. Göttingen (1907), 177–190. FM 38, 453.

[5] *Über die Uniformisierung beliebiger analytischer Kurven*. I. Ibid. (1907), 191–210. FM 38, 454.

[6] *Zur Uniformisierung der algebraischen Kurven*. Ibid. (1907), 410–414. FM 38, 455.

[7] *Über die Uniformisierung beliebiger analytischer Kurven*. II. Ibid. (1907), 633–669. FM 38, 455.

[8] *Über die Uniformisierung der algebraischen Kurven (imaginäre Substitutionsgruppen)*. (*Voranzeige*.) *Mitteilung eines Grenzübergangs durch iterierendes Verfahren*. Ibid. (1908), 112–116. FM 39, 489.

[9] *Über die Uniformisierung beliebiger analytischer Kurven*. III. Ibid. (1908), 337–358. FM 40, 467.

[10] *Konforme Abbildung der Oberfläche einer von endlich vielen regulären analytischen Flächenstücken gebildeten körperlichen Ecken auf die schlichte ebene Fläche eines Kreises*. Nachr. Ges. Wiss. Göttingen (1908), 359–360. FM 39, 491.

[11] *Sur un principe général d'uniformisation*. C.R. Acad. Sci. Paris 148 (1909), 824–828. FM 40, 471.

[12] *Fonction potentielle et fonction analytique ayant un domaine d'existence donné à un nombre quelconque (fini ou infini) de feuillets*. Ibid. 148 (1909), 1446–1448. FM 40, 471.

[13] *Über die Uniformisierung der algebraischen Kurven*. I. Math. Ann. 67 (1909), 145–224. FM 40, 470.

[14] *Über die Uniformisierung der algebraischen Kurven durch automorphe Funktionen mit imaginärer Substitutionsgruppe*. Nachr. Akad. Wiss. Göttingen (1909), 68–76. FM 40, 468.

[15] *Über die Uniformisierung beliebieger analytischer Kurven*. IV. Ibid. (1909), 324–361. FM 40, 468.

[16] *Über ein allgemeines Uniformisierungsprinzip*. Rom. 4. Math. Kongr. 2 (1909), 25–30. FM 40, 466.

[17] *Über die konforme Abbildung mehrfach zusammenhängender Bereiche*. Jbr. Deutsche Math.-Verein. 19 (1910), 339–348. FM 41, 747.

[18] *Über die Uniformisierung der algebraischen Kurven*. II. Math. Ann. 69 (1910), 1–81. FM 41, 480.

KOEBE, P.

[19] *Über die Hilbertsche Uniformisierungsmethode.* Nachr. Akad. Wiss. Göttingen (1910), 59–74. FM 41, 479.

[20] *Über die Uniformisierung der algebraischen Kurven durch automorphe Funktionen mit imaginärer Substitutionsgruppe.* (*Fortsetzung und Schluss.*) Ibid. (1910), 180–189. FM 41, 480.

[21] *Über die Uniformisierung beliebiger analytischer Kurven.* I. *Das allgemeine Uniformisierungsprinzip.* J. Math. 138 (1910), 192–253. FM 41, 482.

[22] *Über die Uniformisierung beliebiger analytischer Kurven.* II. *Die zentralen Uniformisierungsprobleme.* Ibid. 139 (1911), 251–292. FM 42, 449.

[23] *Über die Uniformisierung der algebraischen Kurven.* III. Math. Ann. 72 (1912), 437–516. FM 43, 522.

[24] *Über eine neue Methode der konformen Abbildung und Uniformisierung.* Nachr. Akad. Wiss. Göttingen (1912), 844–848. FM 43, 520.

[25] *Begründung der Kontinuitätsmethode im Gebiete der konformen Abbildung und Uniformisierung.* (*Voranzeige.*) Ibid. (1912), 879–886. FM 43, 521.

[26] *Das Uniformisierungstheorem und seine Bedeutung für Funktionentheorie und nichteuklidische Geometrie.* Annali di Mat. (3) 21 (1913), 57–64. FM 44, 495.

[27] *Zur Theorie der konformen Abbildung und Uniformisierung.* Ber. Verh. Säch. Akad. Wiss. Leipzig 66 (1914), 67–75. FM 45, 670.

[28] *Über die Uniformisierung der algebraischen Kurven.* IV. *Zweiter Existenzbeweis der allgemeinen kanonischen uniformisierenden Variablen: Kontinuitätsmethode.* Math. Ann. 75 (1914), 42–129. FM 45, 669.

[29] *Abhandlungen zur Theorie der konformen Abbildung.* I. *Die Kreisabbildung des allgemeinsten einfach und zweifach zusammenhängenden schlichten Bereichs und die Ränderzuordnung bei konformer Abbildung.* J. Math. 145 (1915), 177–225. FM 45, 667.

[30] *Begründung der Kontinuitätsmethode im Gebiete der konformen Abbildung und Uniformisierung.* (*Voranzeige, zweite Mitteilung.*) Nachr. Akad. Wiss. Göttingen (1916), 266–269. FM 46, 544.

[31] *Abhandlungen zur Theorie der konformen Abbildung.* II. *Die Fundamentalabbildung beliebiger mehrfach zusammenhängender schlichter Bereiche nebst einer Anwendung auf die Bestimmung algebraischer Funktionen zu gegebener Riemannscher Fläche.* Acta Math. 40 (1916), 251–290. FM 46, 544.

[32] *Abhandlungen zur Theorie der konformen Abbildung.* III. *Der allgemeine Fundamentalsatz der konformen Abbildung nebst einer Anwendung auf die konforme Abbildung der Oberfläche einer körperlichen Ecke.* J. Math. 147 (1917), 67–104. FM 46, 545.

[33] *Abhandlungen zur Theorie der konformen Abbildung.* IV. *Abbildung mehrfach zusammenhängender schlichter Bereiche auf Schlitzbereiche.* Acta Math. 41 (1918), 305–344. FM 46, 545.

[34] *Abhandlungen zur Theorie der konformen Abbildung.* V. *Abbildung mehrfach zusammenhängender schlichter Bereiche auf Schlitzbereiche.* (*Erste Fortsetzung.*) Math. Z. 2 (1918), 198–236. FM 46, 546.

[35] *Begründung der Kontinuitätsmethode im Gebiete der konformen Abbildung und Uniformisierung.* (*Voranzeige, dritte Mitteilung.*) Nachr. Akad. Wiss. Göttingen (1918), 57–59. FM 46, 544.

[36] *Zur konformen Abbildung unendlich-vielfach zusammenhängender schlichter Bereiche auf Schlitzbereiche.* Ibid. (1918), 60–71. FM 46, 546.

[37] *Über die Strömungspotentiale und die zugehörenden konformen Abbildungen Riemannscher Flächen.* Ibid. (1919), 1–46. FM 53, 320.

[38] *Abhandlungen zur Theorie der konformen Abbildung.* VI. *Abbildung mehrfach zusammenhängender schlichter Bereiche auf Kreisbereiche. Uniformisierung hyperelliptischer Kurven.* (*Iterationsmethode.*) Math. Z. 7 (1920), 235–301. FM 47, 925.

[39] *Über die konforme Abbildung endlich- und unendlich-vielfach zusammenhängender symmetrischer Bereiche.* Acta Math. 43 (1922), 263–287. FM 48, 1233.

KOEBE, P.

[40] *Allgemeine Theorie der Riemannschen Mannigfaltigkeiten. (Konforme Abbildung und Uniformisierung.)* Acta Math. 50 (1927), 27–157. FM 43, 322.

[41] *Methoden der konformen Abbildung und Uniformisierung.* Atti Congr. Bologna 3 (1930), 195–203. FM 56, 295.

[42] *Iterationstheorie der hyperbolischen Uniformisierungsgrössen vom Geschlecht Null.* Ber. Säch. Akad. Wiss. Leipzig 91 (1939), 135–192. MR 1, 114.

[43] *Zur allgemeinen Iterationstheorie der Uniformisierung algebraischer Funktionen.* Ibid. 93 (1941), 43–66. MR 11, 21.

KOMATU, Y.

[1] *Theory of conformal mappings.* I. Kyoritsu, Tokyo, 1944, 579 pp. (Japanese.)

[2] *Theory of conformal mappings.* II. Kyoritsu, Tokyo, 1949, 409 pp. (Japanese.)

KOMATU, Y., and MORI, A.

[1] *Conformal rigidity of Riemann surfaces.* J. Math. Soc. Japan 4 (1952), 302–309. MR 14, 969.

KONSTANTINESKU, K.

[1] *On the behaviour of analytic functions at boundary elements of Riemann surfaces.* Rev. Math. Pures Appl. 2 (1957), 269–276. (Russian.) MR 20, 290.

KORN, A.

[1] *Application de la méthode de la moyenne arithmétique aux surfaces de Riemann.* C.R. Acad. Sci. Paris 135 (1902), 94–95. FM 33, 430.

[2] *Sur le problème de Dirichlet pour des domaines limités par plusieurs contours (ou surfaces).* Ibid. 135 (1902), 231–232. FM 33, 430.

[3] *Zwei Anwendungen der Methode der sukzessiven Anwendungen.* Schwarz Festschrift, pp. 215–229 (1914). FM 45, 568.

KÜNZI, H.

[1] *Représentation et répartition des valeurs des surfaces de Riemann à extrémités bipériodiques.* C.R. Acad. Sci. Paris 234 (1952), 793–795. MR 13, 643.

[2] *Surfaces de Riemann avec un nombre fini d'extrémités simplement et doublement périodiques.* Ibid. 234 (1952), 1660–1662. MR 13, 833.

[3] *Über ein Teichmüllersches Wertverteilungsproblem.* Arch. Math. 4 (1953), 210–215. MR 15, 116.

[4] *Über periodische Enden mit mehrfach zusammenhängendem Existenzgebiet.* Math. Z. 61 (1954), 200–205. MR 16, 1095.

[5] *Neue Beiträge zur geometrischen Wertverteilungslehre.* Comment. Math. Helv. 29 (1955), 223–257. MR 16, 1095.

[6] *Konstruktion Riemannscher Flächen mit vorgegebener Ordnung der erzeugenden Funktionen.* Math. Ann. 128 (1955), 471–474. MR 16, 1095.

[7] *Zur Theorie der Viertelsenden Riemannscher Flächen.* Comment. Math. Helv. 30 (1956), 107–115. MR 17, 837.

KÜNZI, H., and WITTICH, H.

[1] *Sur le module maximal de quelques fonctions transcendantes entières.* C.R. Acad. Sci. Paris 245 (1957), 1103–1106, MR 19, 845.

KURAMOCHI, Z.

[1] *Potential theory and its applications.* I. Osaka Math. J. 3 (1951), 123–174. MR 13, 650.

[2] *Potential theory and its applications.* II. Ibid. 4 (1952), 87–99. MR 14, 272.

[3] *A remark on the bounded analytic function.* Ibid. 4 (1952), 185–190. MR 14, 742.

[4] *On covering surfaces.* Ibid. 5 (1953), 155–201. MR 15, 518.

[5] *Relations between harmonic dimensions.* Proc. Japan Acad. 30 (1954), 576–580. MR 16, 588.

[6] *Dirichlet problem on Riemann surfaces.* I. *Correspondence of boundaries.* Ibid. 30 (1954), 731–735. MR 16, 1012.

[7] *Dirichlet problem on Riemann surfaces.* II. *Harmonic measures of the set of accessible boundary points.* Ibid. 30 (1954), 825–830. MR 16, 1013.

KURAMOCHI, Z.

[8] *Dirichlet problem on Riemann surfaces.* III. *Types of covering surfaces.* Ibid. 30 (1954), 831–836. MR 16, 1013.

[9] *Dirichlet problem on Riemann surfaces.* IV. *Covering surfaces of finite number of sheets.* Ibid. 30 (1954), 946–950. MR 16, 1013.

[10] *Harmonic measures and capacity of sets of the ideal boundary.* I. Ibid. 30 (1954), 951–956. MR 16, 1013.

[11] *An example of a null-boundary Riemann surface.* Osaka Math. J. 6 (1954), 83–91. MR 16, 233.

[12] *On covering property of abstract Riemann surfaces.* Ibid. 6 (1954), 93–103. MR 16, 26.

[13] *On the existence of harmonic functions on Riemann surfaces.* Ibid. 7 (1955), 23–28. MR 17, 27.

[14] *On the behaviour of analytic functions on abstract Riemann surfaces.* Ibid., 7 (1955), 109–127. MR 17, 26.

[15] *Dirichlet problem on Riemann surfaces.* V. *On covering surfaces.* Proc. Japan Acad. 31 (1955), 20–24. MR 16, 1013.

[16] *Harmonic measures and capacity of sets of the ideal boundary.* II. Ibid. 31 (1955), 25–30. MR 16, 1013.

[17] *Evans's theorem on abstract Riemann surfaces with null-boundaries.* I. Ibid. 32 (1956), 1–6. MR 17, 1072.

[18] *Evans's theorem on abstract Riemann surfaces with null-boundaries.* II. Ibid. 32 (1956), 7–9. MR 17, 1072.

[19] *On estimation of the measure of linear sets.* Ibid. 32 (1956), 105–110. MR 17, 1191.

[20] *Capacity of subsets of the ideal boundary.* Ibid. 32 (1956), 111–116. MR 18, 27.

[21] *Evans-Selberg's theorem on abstract Riemann surfaces with positive boundaries.* I. Ibid. 32 (1956), 228–233. MR 18, 290.

[22] *Evans-Selberg's theorem on abstract Riemann surfaces with positive boundaries.* II. Ibid. 32 (1956), 234–236. MR 18, 290.

[23] *Mass distributions on the ideal boundaries of abstract Riemann surfaces.* I. Osaka Math. J. 8 (1956), 119–137. MR 18, 120.

[24] *Analytic functions in the neighbourhood of the ideal boundary.* Proc. Japan Acad. 33 (1957), 84–86. MR 19, 641.

KURAMOCHI, Z., and KURODA, T.

[1] *A note on the set of logarithmic capacity zero.* Proc. Japan Acad. 30 (1954), 566–569. MR 17, 27.

KURIBAYASHI, A.

[1] *On functions of bounded Dirichlet integral.* Kōdai Math. Sem. Rep. 7 (1955), 30–32. MR 17, 356.

KURODA, T.

[1] *On the type of an open Riemann surface.* Proc. Japan Acad. 27 (1951), 57–60. MR 13, 735.

[2] *Notes on an open Riemann surface.* I. Kōdai Math. Sem. Rep. (1951), 61–63. MR 13, 735.

[3] *Some remarks on an open Riemann surface with null boundary.* Tôhoku Math. J. (2) 3 (1951), 182–186. MR 13, 833.

[4] *Notes on an open Riemann surface.* II. Kōdai Math. Sem. Rep. (1952), 36–38. MR 14, 470.

[5] *A property of some open Riemann surfaces and its application.* Nagoya Math. J. 6 (1953), 77–84. MR 15, 519.

[6] *On the classification of symmetric Fuchsian groups of genus zero.* Proc. Japan Acad. 29 (1953), 431–434. MR 16, 25.

[7] *Theorems of the Phragmén-Lindelöf type on an open Riemann surface.* Osaka Math. J. 6 (1954), 231–241. MR 16, 581.

[8] *On analytic functions on some Riemann surfaces.* Nagoya Math. J. 10 (1956), 27–50. MR 18, 290.

[9] Cf. KURAMOCHI, Z., and KURODA, T. [1].

KUSUNOKI, Y.
[1] *Über Streckenkomplex und Ordnung gebrochener Funktionen.* Mem. Coll. Sci. Univ. Kyoto Ser. A. Math. 26 (1951), 255–269. MR 14, 156.
[2] *Über die hinreichenden Bedingungen dafür, dass eine Riemannsche Fläche nullberandet ist.* Ibid. 27 (1952), 99–108. MR 14, 550.
[3] *Maximum principle for analytic functions on open Riemann surfaces.* Ibid. 28 (1953), 61–66. MR 15, 519.
[4] *Note on the continuation of harmonic and analytic functions.* Ibid. 29 (1955), 11–16. MR 17, 26.
[5] *Some classes of Riemann surfaces characterized by the extremal length.* Proc. Japan Acad. 32 (1956), 406–408. MR 18, 200.
[6] *On Riemann's period relations on open Riemann surfaces.* Mem. Coll. Sci. Univ. Kyoto Ser. A. Math. 30 (1956), 1–22. MR 19, 846.
[7] *Notes on meromorphic covariants.* Ibid. 30 (1957), 243–249. MR 20, 158.
[8] *Contributions to Riemann-Roch's theorem.* Ibid. 31 (1958), 161–180.

LAASONEN, P.
[1] *Über die einfachsten zweifach zusammenhängenden Riemannschen Flächen.* Ann. Acad. Sci. Fenn. Ser A. I. no. 9 (1941), 16 pp. MR 8, 24.
[2] *Zum Typenproblem der Riemannschen Flächen.* Ibid. 11 (1942), 7 pp. MR 8, 24.
[3] *Beiträge zur Theorie der Fuchsoiden Gruppen und zum Typenproblem der Riemannschen Flächen.* Ibid. 25 (1944), 87 pp. MR 8, 24.

LAMBIN, N. V.
[1] *Lines of symmetry of non-simply-connected Riemann surfaces.* Uspehi Mat. Nauk (N.S.) 11 (1956), no. 5 (71), 84–85. (Russian.) MR 19, 23.

LANDFRIEDT, E.
[1] *Theorie der algebraischen Funktionen und ihrer Integrale.* G. J. Göschen, Leipzig, 1902, 294 pp. FM 33, 429.

LANDSBERG, G.
[1] Cf. HENSEL, K., und LANDSBERG, G. [1].

LEFSCHETZ, S.
[1] *Topology.* Amer. Math. Soc. Colloq. Publ. 12, 1930, 410 pp. FM 56, 491.

LEHTO, O.
[1] *Anwendung orthogonaler Systeme auf gewisse funktionentheoretische Extremal- und Abbildungsprobleme.* Ann. Acad. Sci. Fenn. Ser. A. I. no. 59 (1949), 51 pp. MR 11, 170.
[2] *On the existence of analytic functions with a finite Dirichlet integral.* Ibid. no. 67 (1949), 7 pp. MR 11, 338.
[3] *Value distribution and boundary behaviour of a function of bounded characteristic and the Riemann surface of its inverse function.* Ibid. no. 177 (1954), 46 pp. MR 16, 688.

LEHTO, O., and NEVANLINNA, R. (edited by)
[1] *Proceedings of the International Colloquium on the Theory of Functions, Helsinki, 1957.* Suomalainen Tiedeakatemia, Helsinki, 1958.

LEJA, F.
[1] *Teoria funkcji analitycznych.* Państwowe Wydawnictwo Naukowe, Warszawa, 1957, 560 pp. MR 19, 641.

LEONIDOVA, L. M.
[1] *The Riemann surface for the Green's function of a multiply connected region.* Mat. Sbornik N.S. 28 (70), (1951), 621–634. (Russian.) MR 13, 125.

LE-VAN, T.
[1] *Beitrag zum Typenproblem der Riemannschen Flächen.* Comment. Math. Helv. 20 (1947), 270–287. MR 9, 139.
[2] *Über das Umkehrproblem der Wertverteilungslehre.* Ibid. 23 (1949), 26–49. MR 11, 22.

LE-VAN, T.
[3] *Le degré de ramification d'une surface de Riemann et la croissance de la caractéristique de la fonction uniformisante.* C.R. Acad. Sci. Paris 228 (1949), 1192–1195. MR 10, 523.
[4] *Un problème de type généralisé.* Ibid. 228 (1949), 1270–1272. MR 10, 523.

LI CHIAVI, M. S.
[1] *Sulla rappresentazione conforme delle aree pluriconnesse appartenenti a superficie di Riemann su un'opportuna superficie di Riemann su cui siano eseguiti dei tagli paralleli.* Rendiconti Sem. Padova 3 (1932), 95–127. FM 58, 1092.

LICHTENSTEIN, L.
[1] *Zur Theorie der konformen Abbildung nichtanalytischer, singularitätenfreier Flächenstücke auf ebene Gebiete.* Bull. Internat. Acad. Sci. Cracovie, Cl. Sci. Math. Nat. Ser. A. (1916), 192–217. FM 46, 547.

LINDEBERG, J. W.
[1] *Sur l'existence des fonctions d'une variable complexe et de fonctions harmoniques bornées.* Ann. Acad. Sci. Fenn. (A) 11, no. 6 (1918), 27 pp. FM 46, 525.

LINDEMANN, F.
[1] *Untersuchungen über den Riemann-Roch'schen Satz.* Teubner, Leipzig, 1879. FM 12, 323.

LIPPICH, F.
[1] *Untersuchungen über den Zusammenhang der Flächen im Sinne Riemann's.* Math. Ann. 7 (1874), 212–230. FM 6, 234.

LITTLEWOOD, J. E.
[1] *Lectures on the theory of functions.* Oxford Univ. Press, 1944, 243 pp. MR 6, 261.
[2] Cf. HARDY, G. H., LITTLEWOOD, J. E., and PÓLYA, G. [1].

LOCHER, L.
[1] *Ein Satz über die Riemannsche Fläche der Inversen einer im endlichen meromorphen Funktion.* Comment. Math. Helv. 3 (1931), 179–182. FM 57, 370.

LOKKI, O.
[1] *Über Existenzbeweise einiger mit Extremaleigenschaft versehenen analytischen Funktionen.* Ann. Acad. Sci. Fenn. Ser. A. I. no. 76 (1950), 15 pp. MR 12, 401.
[2] *Beiträge zur Theorie der analytischen und harmonischen Funktionen mit endlichem Dirichletintegral.* Ibid. no. 92 (1951), 11 pp. MR 13, 338.
[3] *Über eindeutige analytische Funktionen mit endlichem Dirichletintegral.* Ibid. no. 105 (1951), 13 pp. MR 15, 642.
[4] *Über harmonische Funktionen mit endlichem Dirichletintegral.* C.R. Onzième Congr. Math. Scand., Trondheim, 1949, 239–242. MR 14, 644.

LOTZ, W.
[1] *Zur Streckenkomplexdarstellung Riemannscher Flächen.* Mitt. Math. Sem. Giessen no. 39 (1951), 25 pp. MR 15, 303.

LÜROTH, J.
[1] *Note über Verzweigungsschnitte und Querschnitte in einer Riemann'schen Fläche.* Math. Ann. 3 (1871), 181–184. FM 3, 192.

MACLANE, G. R.
[1] *Concerning the uniformization of certain Riemann surfaces allied to the inverse-cosine and inverse-gamma surfaces.* Trans. Amer. Math. Soc. 62 (1947), 99–113. MR 9, 85.
[2] *Riemann surfaces and asymptotic values associated with real entire functions.* Rice Inst. Pamphlet. Special Issue. The Rice Institute, Houston, Texas, 1952, 93 pp. MR 14, 739.

MARKUŠEVIČ, A. I.
[1] *Theory of analytic functions.* Gosudarstv. Izdat. Tehn.-Teor. Lit., Moskow-Leningrad, 1950, 703 pp. MR 12, 87.

MARTIN, R. S.
[1] *Minimal positive harmonic functions.* Trans. Amer. Math. Soc. 49 (1941), 137–172. MR 2, 292.

MARTIS IN BIDDAU, S.
[1] *I funzionali lineari algebrici e abeliani.* Rend. Circ. Mat. Palermo 62 (1940), 289–358. FM 66, 1285.

MATILDI, P.
[1] *Sulla rappresentazione conforme di domini appartenenti a superficie di Riemann su domini di un tipo canonico assegnato.* Ann. Scuola Norm. Super. Pisa (2) 14 (1945), (1948), 81–90. MR 10, 241.

MCSHANE, E. J.
[1] *Integration.* Princeton Univ. Press, Princeton, 1944, 392 pp. MR 6, 43.

MESCHKOWSKI, H.
[1] *Über die konforme Abbildung gewisser Bereiche von unendlich hohem Zusammenhang auf Vollkreisbereiche.* I. Math. Ann. 123 (1951), 392–405. MR 13, 454.
[2] *Über die konforme Abbildung gewisser Bereiche von unendlich hohem Zusammenhang auf Vollkreisbereiche.* II. Ibid. 124 (1952), 178–181. MR 13, 642.
[3] *Einige Extremalprobleme aus der theorie der konformen Abbildung.* Ann. Acad. Sci. Fenn. Ser. A. I. no. 117 (1952), 12 pp. MR 15, 367.

MIZUMOTO, H.
[1] *On Riemann surfaces with finite spherical area.* Kōdai Math. Sem. Rep. 7 (1957), 87–96. MR 19, 1044.

MONTEL, P.
[1] *Sur quelques propriétés des couples de fonctions uniformisantes.* Memorial volume dedicated to D. A. Grave, Moskow, 1940, pp. 166–171. MR 2, 188.

MOPPERT, K. F.
[1] *Über eine gewisse Klasse von elliptischen Riemannschen Flächen.* Comment. Math. Helv. 23 (1949), 174–176. MR 11, 93.

MORI, A.
[1] *Conformal representation of multiply connected domain on many-sheeted disc.* J. Math. Soc. Japan 2 (1951), 198–209. MR 13, 338.
[2] *On Riemann surfaces, on which no bounded harmonic function exists.* Ibid. 3 (1951), 285–289. MR 14, 367.
[3] *On the existence of harmonic functions on a Riemann surface.* J. Fac. Sci. Univ. Tokyo, Sect. I. 6 (1951), 247–257. MR 13, 735.
[4] *A remark on the class O_{HD} of Riemann surfaces.* Kōdai Math. Sem. Rep. 1952, 57–58. MR 14, 264.
[5] *A remark on the prolongation of Riemann surfaces of finite genus.* J. Math. Soc. Japan 4 (1952), 27–30. MR 14, 263.
[6] *An imbedding theorem on finite covering surfaces of the Riemann sphere.* Ibid. 5 (1953), 263–268. MR 15, 615.
[7] *A note on unramified abelian covering surfaces of a closed Riemann surface.* Ibid. 6 (1954), 162–176. MR 16, 581.
[8] Cf. KOMATU, Y., and MORI, A. [1].

MORI, S., and OTA, M.
[1] *A remark on the ideal boundary of a Riemann surface.* Proc. Japan Acad. 32 (1956), 409–411. MR 18, 290.

MORIYA, M.
[1] *Algebraische Funktionenkörper und Riemannsche Flächen.* J. Fac. Sci. Hokkaido Imp. Univ. Ser. I. 9 (1941), 209–245. MR 3, 83.

MORSE, M.
[1] *La construction topologique d'un réseau isotherme sur une surface ouverte.* J. Math. Pures Appl. (9) 35 (1956), 67–75. MR 17, 1071.
[2–4] Cf. JENKINS, J., and MORSE, M. [1–3].

MORSE, M. *et al.* (edited by)

[1] *Seminars on analytic functions* (Conference on Analytic Functions, Princeton, 1957). Vols. 1, 2. Institute for Advanced Study, and U.S. Air Force, Office of Scientific Research. 345 pp. 319 pp.

MYRBERG, L.

[1] *Normalintegrale auf zweiblättrigen Riemannschen Flächen mit reellen Verzweigungspunkten.* Ann. Acad. Sci. Fenn. Ser. A. I. no. 71 (1950), 51 pp. MR 12, 90.

[2] *Über das Verhalten der Greenschen Funktionen in der Nähe des idealen Randes einer Riemannschen Fläche.* Ibid. no. 139 (1952), 8 pp. MR 14, 744.

[3] *Über die Existenz von positiven harmonischen Funktionen auf Riemannschen Flächen.* Ibid. no. 146 (1953), 6 pp. MR 14, 979.

[4] *Über die Integration der Poissonschen Gleichung auf offenen Riemannschen Flächen.* Ibid. 161 (1953), 10 pp. MR 15, 519.

[5] *Über die Existenz der Greenschen Funktion der Gleichung* $\Delta u = c(P)\cdot u$ *auf Riemannschen Flächen.* Ibid. no. 170 (1954), 8 pp. MR 16, 34.

[6] *Über die Integration der Differentialgleichung* $\Delta u = c(P)\cdot u$ *auf offenen Riemannschen Flächen.* Math. Scand. 2 (1954), 142–152. MR 16, 33.

[7] *Über die Existenz von positiven harmonischen Funktionen auf offenen Riemannschen Flächen.* Douzième Congr. Math. Scand., Lund, 1953, pp. 214–216. MR 16, 471.

[8] *Über das Dirichletsche Problem auf offenen Riemannschen Flächen.* Ann. Acad. Sci. Fenn. Ser. A. I. no. 197 (1955), 11 pp. MR 17, 726.

[9] *Différentielles méromorphes sur des surfaces de Riemann ouvertes.* C.R. Acad. Sci. Paris 241 (1955), 1194–1195, MR 17, 473.

[10] *Über meromorphe Funktionen und Kovarianten auf Riemannschen Flächen.* Ann. Acad. Sci. Fenn. Ser. A. I. no. 244 (1957), 18 pp. MR 20, 19.

MYRBERG, P. J.

[1] *Über die numerische Ausführung der Uniformisierung.* Acta Soc. Sci. Fenn. 48, no. 7 (1920), 53 pp. FM 47, 328.

[2] *L'existence de la fonction de Green pour un domaine plan donné.* C.R. Acad. Sci. Paris 190 (1930), 1372–1374. FM 56, 295.

[3] *Über die Existenz der Greenschen Funktionen auf einer gegebenen Riemannschen Fläche.* Acta Math. 61 (1933), 39–79. FM 59, 355.

[4] *Über die analytische Darstellung der automorphen Funktionen bei hyperelliptischen Riemannschen Flächen.* Ibid. 65 (1935), 195–261. FM 61, 367.

[5] *Sur la détermination du type d'une surface riemannienne simplement connexe.* C.R. Acad. Sci. Paris 200 (1935), 1818–1820. FM 61, 366.

[6] *Über die Bestimmung des Typus einer Riemannschen Fläche.* Ann. Acad. Sci. Fenn. A. 45, no. 3 (1935), 30 pp. FM 61, 1158.

[7] *Über die analytische Darstellung der automorphen Funktionen bei gewissen fuchsoiden Gruppen.* Ibid. 48, no. 1 (1936), 25 pp. FM 62, 393.

[8] *Über die automorphen Funktionen.* C.R. Neuvième Congr. Math. Scand., Helsinki, 1938, pp. 23–38. FM 65, 351.

[9] *Die Kapazität der singulären Menge der linearen Gruppen.* Ann. Acad. Sci. Fenn. Ser. A. I. no. 10 (1941), 19 pp. MR 7, 516.

[10] *Über Integrale auf transzendenten symmetrischen Riemannschen Flächen.* Ibid. 31 (1945), 21 pp. MR 7, 428.

[11] *Über Integrale auf transzendenten symmetrischen Riemannschen Flächen.* Ibid. 31 (1945), 21 pp. MR 7, 428.

[12] *Über analytische Funktionen auf transzendenten zweiblättrigen Riemannschen Flächen mit reellen Verzweigungspunkten.* Acta Math. 76 (1945), 185–224. MR 7, 57.

[13] *Über einige Probleme der Funktionentheorie.* Sitzungsber. Finnisch. Akad. Wiss. 1945 (1946), 229–233. MR 8, 576.

[14] *Über analytische Funktionen auf transzendenten Riemannschen Flächen.* C.R. Dixième Congr. Math. Scand., Copenhague, 1946, pp. 77–96. MR 8, 509.

[15] *Über die analytische Fortsetzung von beschränkten Funktionen.* Ann. Acad. Sci. Fenn. Ser. A. I. no. 58 (1949), 7 pp. MR 10, 441.

Myrberg, P. J.

[16] *Über automorphe Funktionen und Riemannsche Flächen.* C.R. Onzième Congr. Math. Scand., Trondheim, 1949, pp. 24–34. MR 14, 743.

[17] *Über die Existenz von beschränktartigen automorphen Funktionen.* Ann. Acad. Sci. Fenn. Ser. A. I. no. 104 (1950), 16 pp. MR 15, 403.

[18] *Über Primfunktionen auf einer algebraischen Riemannschen Fläche.* Ibid. no. 104 (1951), 16 pp. MR 13, 539.

[19] *Sur les fonctions automorphes.* Ann. Sci. École Norm. Sup. (3) 68 (1951), 383–424. MR 13, 735.

[20] *Über die Iteration von algebraischen Funktionen.* Ann. Acad. Sci. Fenn. Ser. A. I. no. 164 (1954), 9 pp. MR 15, 614.

Nagai, Y.

[1] *On the behaviour of the boundary of Riemann surfaces.* I. Proc. Japan Acad. 26, nos. 2–5 (1950), 111–115. MR 14, 367.

[2] *On the behaviour of the boundary of Riemann surfaces.* II. Ibid. 26, no. 6 (1950), 10–16. MR 14, 367.

Nagura, S.

[1] *Kernel functions on Riemann surfaces.* Kōdai Math. Sem. Rep. 1951, 73–76. MR 13, 547.

Nagy, B. Sz.

[1] Cf. Riesz, F., and Nagy, B. Sz. [1].

Naim, L.

[1] *Sur le rôle de la frontière de R. S. Martin dans la théorie du potentiel.* Thèse, Paris, 1957, 103 pp.

Nehari, Z.

[1] *The kernel function and canonical conformal maps.* Duke Math. J. 16 (1949), 165–178. MR 10, 440.

[2] *On bounded analytic functions.* Proc. Amer. Math. Soc. 1 (1950), 268–275. MR 11, 590.

[3] *Conformal mapping of open Riemann surfaces.* Trans. Amer. Math. Soc. 68 (1950), 258–277. MR 11, 590.

[4] *Conformal mapping.* McGraw-Hill, New York-Toronto-London, 1952, 396 pp. MR 13, 640.

[5] *Dirichlet's principle and some inequalities in the theory of conformal mapping.* Contributions to the theory of Riemann surfaces, pp. 167–175. Princeton Univ. Press, Princeton, 1953. MR 15, 24.

Neumann, C.

[1] *Untersuchungen über das logarithmische und Newtonsche Potential.* Teubner, Leipzig, 1877. FM 10, 658.

[2] *Vorlesungen über Riemann's Theorie der Abel'schen Integrale.* 2. Aufl. Teubner, Leipzig, 1884, 472 pp. FM 16, 336.

[3] *Über die Stetigkeit mehrdeutiger Funktionen.* Ber. Verh. Sächs. Akad. Wiss. Leipzig, 1888, 121–124. FM 20, 387.

[4] *Über die Methode des arithmetischen Mittels insbesondere über die Vervollkommnungen, welche die betreffenden Poincaré'schen Untersuchungen in letzter Zeit durch die Arbeiten von A. Korn und E. R. Neumann erhalten haben.* Math. Ann. 54 (1900), 1–48. FM 31, 416.

Nevanlinna, R.

[1] *Über die Wertverteilung der eindeutigen analytischen Funktionen.* Abh. Math. Sem. Univ. Hamburg 8 (1931), 351–400. FM 57, 367.

[2] *Über die Riemannsche Fläche einer analytischen Funktion.* Verh. Kongr. Zürich, 1, 1932, 1, pp. 221–239. FM 58, 368.

[3] *Über Riemannsche Flächen mit endlich vielen Windungspunkten.* Acta Math. 58 (1932), 295–373. FM 58, 369.

[4] *Ein Satz über die konforme Abbildung Riemannscher Flächen.* Comment. Math. Helv. 5 (1933), 95–107. FM 59, 354.

NEVANLINNA, R.
[5] *Das harmonische Mass von Punktmengen und seine Anwendung in der Funktionentheorie.* C.R. Huitième Congr. Math. Scand., Stockholm, 1934, pp. 116–133. FM 61, 306.
[6] *Ein Satz über offene Riemannsche Flächen.* Ann. Acad. Sci. Fenn. (A) 54, no. 3 (1940), 16 pp. MR 2, 85 and 276.
[7] *Quadratisch integrierbare Differentiale auf einer Riemannschen Mannigfaltigkeit.* Ann. Acad. Sci. Fenn. Ser. A. I. no. 1 (1941), 34 pp. MR 7, 427.
[8] *Über die Konstruktion von analytischen Funktionen auf einer Riemannschen Fläche.* Reale Accademia d'Italia, Fondazione A. Volta, Atti dei Convegni, 9 (1939), Rome, 1943, pp. 307–324. MR 12, 492.
[9] *Eindeutigkeitsfragen in der Theorie der konformen Abbildung.* C.R. Dixième Congr. Math. Scand., Copenhague, 1946, pp. 225–240. MR 8, 509.
[10] *Über das Anwachsen des Dirichletintegrals einer analytischen Funktion auf einer offenen Riemannschen Fläche.* Ann. Acad. Sci. Fenn. Ser. A. I. no. 45 (1948), 9 pp. MR 10, 28.
[11] *Über Mittelwerte von Potentialfunktionen.* Ibid. no. 57 (1949), 12 pp. MR 11, 516.
[12] *Sur l'existence de certaines classes de différentielles analytiques.* C.R. Acad. Sci. Paris 228 (1949), 2002–2004. MR 11, 341.
[13] *Über die Neumannsche Methode zur Konstruktion von Abelschen Integralen.* Comment. Math. Helv. 22 (1949), 302–316. MR 10, 525.
[14] *Über die Randelemente einer Riemannschen Fläche.* Ann. Mat. Pura Appl. (4) 29 (1949), 71–73. MR 12. 492.
[15] *Über die Anwendung einer Klasse von Integralgleichungen für Existenzbeweise in der Potentialtheorie.* Acta Sci. Math. Szeged 12, Leopoldo Fejér et Frederico Riesz 70 annos natis dedicatus, Pars A (1950), 146–160. MR 12, 259.
[16] *Über die Existenz von beschränkten Potentialfunktionen auf Flächen von unendlichen Geschlecht.* Math. Z. 52 (1950), 599–604. MR 12, 493.
[17] *Bemerkungen zur Lösbarkeit der ersten Randwertaufgabe der Potentialtheorie auf allgemeinen Flächen.* Math. Z. 53 (1950), 106–109. MR 13, 36.
[18] *Surfaces de Riemann ouvertes.* Proc. Intern. Congr. Math., Cambridge, Mass., 1950, pp. 247–252. MR 13, 547.
[19] *Beschränktartige Potentiale.* Math. Nachr. 4 (1951), 489–501. MR 12, 603.
[20] *Über den Gauss-Bonnetschen Satz* Festschrift zur feier des zweihundertjährigen Bestehens der Akademie der Wissenschaften in Göttingen. I. Math.-Phys. Kl., pp. 175–178, Springer, Berlin-Göttingen-Heidelberg, 1951. MR 14, 263.
[21] *Beitrag zur Theorie der Abelschen Integrale.* Ann. Acad. Sci. Fenn. A. I. no. 100 (1951), 11 pp. MR 13, 644.
[22] *Über die Polygondarstellung einer Riemannschen Fläche.* Ibid. no. 122 (1952), 9 pp. MR 14, 743.
[23] *Uniformisierung.* Springer, Berlin-Göttingen-Heidelberg, 1953, 391 pp. MR 15, 208.
[24] *Eindeutige analytische Funktionen.* 2. Aufl. Springer, Berlin-Göttingen-Heidelberg, 1953, 379, pp. MR 15, 208.
[25] *Countability of a Riemann surface.* Lectures on functions of a complex variable, pp. 61–64. Univ. Michigan Press, Ann Arbor, 1955. MR 16, 1097.
[26] *Polygonal representation of Riemann surfaces.* Ibid. pp. 65–70. MR 16, 1097.

NEVANLINNA, R., and LEHTO, O. (edited by)
[1] Cf. LEHTO, O., and NEVANLINNA, R. (edited by) [1].

NIINI, R.
[1] *Über eine nichtkonstruierbare Riemannsche Fläche vom Geschlecht Eins.* Ann. Acad. Sci. Fenn. Ser. A. I. no. 132 (1952), 6 pp. MR 14, 744.

NIKODYM, O.
[1] *Sur les surfaces de Riemann des fonctions analytiques d'une variable.* C.R. Congr. Math. Pays Slaves, 1930, 159–165. FM 56, 299.

NOSHIRO, K.

[1] *Survey of function theory*. II. Iwanami, Tokyo, 1944, 130 pp. (Japanese.)

[2] *Contributions to the theory of the singularities of analytic functions*. Japanese J. Math. 19 (1948), 299–327. MR 11, 428.

[3] *Open Riemann surface with null boundary*. Nagoya Math. J. 3 (1951), 73–79. MR 13, 833.

[4] *The modern theory of functions*. Iwanami, Tokyo, 1954, 428 pp. MR 16, 912.

OHTSUKA, M.

[1] *Dirichlet problems on Riemann surfaces and conformal mappings*. Nagoya Math. J. 3 (1951), 91–137. MR 13, 642.

[2] *On the behavior of an analytic function about an isolated boundary point*. Ibid. 4 (1952), 103–108. MR 14, 36.

[3] *On a covering surface over an abstract Riemann surface*. Ibid. 4 (1952), 109–118. MR 14, 36.

[4] *Note on the harmonic measure of the accessible boundary of a covering Riemann surface*. Ibid. 5 (1953), 35–38. MR 14, 862.

[5] *Boundary components of Riemann surfaces*. Ibid. 7 (1954), 65–83. MR 16, 349.

[6] *Théorème étoilés de Gross et leurs applications*. Ann. Inst. Fourier, Grenoble 5 (1953–54), 1–28 (1955). MR 17, 1191.

[7] *Sur les ensembles d'accumulation relatifs à des transformations plus générales que les transformations quasi conformes*. Ibid. 5 (1953–54), 29–37. (1955). MR 17, 1191.

[8] *Sur un théorème étoile de Gross*. Nagoya Math. J. 9 (1955), 191–207. MR 17, 1191.

[9] *Boundary components of abstract Riemann surfaces*. Lectures on functions of a complex variable, pp. 303–307. Univ. Michigan Press, Ann Arbor, 1955. MR 16, 1012.

[10] *On asymptotic values of functions analytic in a circle*. Trans. Amer. Math. Soc. 78 (1955), 294–304. MR 16, 686.

[11] *Generalizations of Montel-Lindelöf's theorem on asymptotic values*. Nagoya Math. J. 10 (1956), 129–163. MR 18, 291.

[12] *Remarks to the paper "On Montel's theorem" by Kawakami*. Ibid. 10 (1956), 165–169. MR 18, 292.

[13] *Sur les ensembles d'accumulation relatifs à des transformations localment pseudo-analytiques au sense de Pfluger-Ahlfors*. Ibid. 11 (1957), 131–144. MR 19, 737.

[14] *Topics on function theory*. Kyoritsu, Tokyo, 1957, 132 pp. (Japanese.)

OIKAWA, K.

[1] *Notes on conformal mappings of a Riemann surface onto itself*. Kōdai Math. Sem. Rep. 8 (1956), 23–30. MR 18, 290.

[2] *A supplement to "Notes on conformal mappings of a Riemann surface onto itself."* Ibid., 8 (1956), 115–116. MR 18, 797.

[3] *On the prolongation of an open Riemann surface of finite genus*. Ibid. 9 (1957), 34–41. MR 19, 258.

ONO, I.

[1] Cf. OZAKI, S., ONO, I., and OZAWA, M. [1].

OSGOOD, W. F.

[1] *On the existence of the Green's function for the most general simply connected plane region*. Amer. Math. Soc. Trans. 1 (1900), 310–314. FM 31, 420.

[2] *Allgemeine Theorie der analytischen Funktionen a) einer und b) mehrerer komplexen Grössen*. Enzykl. d. Math. Wiss. 2_2 (1902), 1–114. FM 33, 389.

[3] *Existenzbeweis betreffend Funktionen, welche zu einer eigentlichen diskontinuierlichen automorphen Gruppe gehören*. Palermo Rend. 35 (1913), 103–106. FM 44, 505.

[4] *Lehrbuch der Funktionentheorie*. I. Teubner, Leipzig, 1928, 818 pp. FM 54, 326.

OSGOOD, W. F.

[5] *Lehrbuch der Funktionentheorie.* II_1. Teubner, Leipzig, 1929, 307 pp. FM 55, 171.

[6] *Lehrbuch der Funktionentheorie.* II_2. Teubner, Leipzig, 1932, pp. 309–686. FM 58, 390.

OSSERMAN, R.

[1] *A hyperbolic surface in 3-space.* Proc. Amer. Math. Soc. 7 (1956), 54–58. MR 17, 837.

[2] *Riemann surfaces of class A.* Trans. Amer. Math. Soc. 82 (1956), 217–245. MR 20, 158.

OTA, M.

[1] Cf. MORI, S., and OTA, M. [1].

OTT, K.

[1] *Über die Konstruktion monogener analytischer Funktionen mit vorgegebenen Unstetigkeitsstellen auf der Riemann'schen Fläche.* Monatsh. Math. 4 (1893), 367–375. FM 25, 683.

OZAKI, S., ONO, I., and OZAWA, M.

[1] *On the pseudo-meromorphic mappings on Riemann surfaces.* Sci. Rep. Tokyo Bunrika Daigaku, Sec. A 4 (1952), 211–213. MR 14, 461.

OZAWA, M.

[1] *On classification of the function-theoretic null-sets on Riemann surfaces of infinite genus.* Kōdai Math. Sem. Rep. (1951), 43–44. MR 13, 547.

[2] *Classification of Riemann surfaces.* Ibid. (1952), 63–76. MR 14, 462.

[3] *Remarks on Mr. Ullemar's second harmonic measure.* Ibid. (1953), 93–96. MR 15, 309.

[4] *On harmonic dimension.* I. Ibid. (1954), 33–37. MR. 16, 245.

[5] *On harmonic dimension.* II. Ibid. (1954), 55–58 and 70. MR 16, 245.

[6] *On a maximality of a class of positive harmonic functions.* Ibid. (1954), 65–70. MR 16, 471.

[7] *Some classes of positive solutions of $\Delta u = Pu$ on Riemann surfaces.* I. Ibid. (1954), 121–126. MR 16, 819.

[8] *Some classes of positive solutions of $\Delta u = Pu$ on Riemann surfaces.* II. Ibid. 7 (1955), 15–20. MR 16, 1109.

[9] *On Riemann surfaces admitting an infinite cyclic conformal transformation group.* Ibid. 8 (1956), 152–157. MR 19, 259.

[10] Cf. OZAKI, S., ONO, I., and OZAWA, M. [1].

PAATERO, V.

[1] *Über die konforme Abbildung mehrblättriger Gebiete von beschränkter Randdrehung.* Ann. Acad. Sci. Fenn. A. I. no. 128 (1952), 14 pp. MR 14, 861.

[2] *Über die Verzerrung bei der Abbildung mehrblättriger Gebiete von beschränkter Randdrehung.* Ibid. no. 147 (1953), 7 pp. MR 15, 303.

[3] *Über die Randdrehung der mehrblättrigen einfach zusammenhängenden Gebiete.* Ibid. no. 194 (1955), 7 pp. MR 17, 601.

PARREAU, M.

[1] *Comportement à la frontière de la fonction de Green d'une surface de Riemann.* C.R. Acad. Sci. Paris 230 (1950), 709–711. MR 11, 426.

[2] *La théorie du potentiel sur les surfaces de Riemann à frontière positive.* Ibid. 230 (1950), 914–916. MR 11, 516.

[3] *Sur les moyennes des fonctions harmoniques et la classification des surfaces de Riemann.* Ibid. 231 (1950), 679–681. MR 12, 259.

[4] *Sur certaines classes de fonctions analytiques uniformes sur les surfaces de Riemann.* Ibid. 231 (1950), 751–753. MR 12, 402.

[5] *Fonctions harmoniques et classification des surfaces de Riemann.* Ibid. 234 (1952), 286–288. MR 13, 735.

PARREAU, M.

[6] *Sur les moyennes des fonctions harmoniques et analytiques et la classification des surfaces de Riemann.* Ann. Inst. Fourier Grenoble 3 (1951), (1952), 103–197. MR 14, 263.

[7] *Théorème de Fatou et problème de Dirichlet pour les lignes de Green de certaines surfaces de Riemann.* Ann. Acad. Sci. Fenn. Ser. A. I. no. 250/25 (1958), 8 pp.

[8] Cf. BADER, R., et PARREAU, M. [1].

PERRON, O.

[1] *Eine neue Behandlung der ersten Randwertaufgabe für $\Delta u = 0$.* Math. Z. 18 (1923), 42–54. FM 49, 340.

DEL PEZZO, P.

[1] *Sulle superficie di Riemann relative alle curve algebraic.* Palermo Rend. 6 (1892), 115–126. FM 24, 380.

PFLUGER, A.

[1] *Sur une propriété de l'application quasi conforme d'une surface de Riemann ouverte.* C.R. Acad. Sci. Paris 227 (1948), 25–26. MR 10, 28.

[2] *Über das Anwachsen eindeutiger analytischer Funktionen auf offenen Riemannschen Flächen.* Ann. Acad. Sci. Fenn. Ser. A. I. no. 64 (1949), 18 pp. MR 11, 342.

[3] *La croissance des fonctions analytiques et uniformes sur une surface de Riemann ouverte.* C.R. Acad. Sci. Paris 229 (1949), 505–507. MR 11, 93.

[4] *Sur l'existence des fonctions non constantes, analytiques, uniformes et bornées sur une surface de Riemann ouverte.* Ibid. 230 (1950), 166–168. MR 11, 342.

[5] *Über das Typenproblem Riemannscher Flächen.* Comment. Math. Helv. 27 (1953), (1954), 346–356. MR 15, 615.

[6] *Extremallängen und Kapazität.* Ibid. 29 (1955), 120–131. MR 16, 810.

[7] *Über die Riemannsche Periodenrelation auf transzendenten hyperelliptischen Flächen* Ibid. 30 (1956), 98–106. MR 17, 725.

[8] *Ein alternierendes Verfahren auf Riemannschen Flächen.* Ibid. 30 (1956), 265–274. MR 17, 1072.

[9] *Ein Approximationssatz für harmonische Funktionen auf Riemannschen Flächen.* Ann. Acad. Sci. Fenn. Ser. A. I. no. 216 (1956), 8 pp. MR 17, 1072.

[10] *Theorie der Riemannschen Flächen.* Springer, Berlin-Göttingen-Heidelberg, 1957, 248 pp. MR 18, 796.

[11] *A direct construction of Abelian differentials on Riemann surfaces.* Seminars on analytic functions. II, pp. 39–48. Institute for Advanced Study, Princeton, 1958.

[12] *Harmonische und analytische Differentiale auf Riemannschen Flächen.* Ann. Acad. Sci. Fenn. Ser. A. I. no. 249/4 (1958), 18 pp.

PICARD, E.

[1] *De l'equation $\Delta u = ke^u$ sur une surface de Riemann fermée.* J. de Math. (4) 9 (1893), 273–291. FM 25, 683.

[2] *Traité d'analyse.* II: *Fonctions harmoniques et fonctions analytiques. Introduction à la théorie des équations différentielles, integrales abéliennes et surfaces de Riemann.* Gauthier-Villars, Paris, 1926, 624 pp. FM 52, 288.

PLEMELJ, J.

[1] *Die Grenzkreis-Uniformisierung analytischer Gebilde.* Monatsh. f. Math. 23 (1912), 297–304. FM 43, 519.

[2] *Theory of analytic functions.* Slavenska Akademija Znanosti in Umetnosti Ljubljana, 1953, 516 pp. MR 15, 693.

POINCARÉ, H.

[1] *Théorie des groupes fuchsiens.* Acta Math. 1 (1882), 1–62. FM 14, 338.

[2] *Mémoire sur les fonctions fuchsiennes.* Ibid. 1 (1882), 193–294. FM 15, 342.

[3] *Sur la méthode de Neumann et le problème de Dirichlet.* C.R. Acad. Sci. Paris 120 (1895), 347–352. FM 26, 441.

[4] *La méthode de Neumann et le problème de Dirichlet.* Acta Math. 20 (1896), 59–142. FM 27, 316.

POINCARÉ, H.
[5] *Sur l'uniformisation des fonctions analytiques.* Acta Math. 31 (1907), 1–64. FM 38, 452.

PÓLYA, G.
[1] Cf. HARDY, G. H., LITTLEWOOD, J. E., and PÓLYA, G. [1].

POPOVĂT, P.
[1] *Sur les régions de monovalence des fonctions rationnelles.* Disquisit. Math. Phys. 2 (1942), 169–251. MR 9, 24.

PÖSCHL, K.
[1] *Über die Wertverteilung der erzeugenden Funktionen Riemannscher Flächen mit endlich vielen periodischen Enden.* Math. Ann. 123 (1951), 79–95. MR 13, 224.

DE POSSEL, R.
[1] *Sur le prolongement des surfaces de Riemann.* C.R. Acad. Sci. Paris 186 (1928), 1092–1095. FM 54, 379.
[2] *Sur le prolongement des surfaces de Riemann.* Ibid. 187 (1928), 98–100. FM 54, 379.
[3] *Zum Parallelschlitztheorem unendlich-vielfach zusammenhängender Gebiete.* Nachr. Akad. Wiss. Göttingen (1931), 199–202. FM 57, 400.
[4] *Sur quelques propriétés de la représentation conforme des domaines multiplement connexes, en relation avec le théorème des fentes parallèles.* Math. Ann. 107 (1932), 496–504. FM 58, 364.
[5] *Quelques problèmes de représentation conforme.* J. École Polyt. (2) 30 (1932), 1–98. FM 58, 1093.
[6] *Sur les ensembles de type maximum, et le prolongement des surfaces de Riemann.* C.R. Acad. Sci. Paris 194 (1932), 585–587. FM 58, 1093.
[7] *Sur la représentation conforme d'un domaine à connexion infinie sur un domaine à fentes parallèles.* J. Math. Pures Appl. 18 (1939), 285–290. MR 1, 111.

POTYAGAĬLO, D. B.
[1] *Condition of hyperbolicity of a class of Riemannian surfaces.* Ukrain. Mat. Žurnal 5 (1953), 459–463. (Russian.) MR 15, 787.

PRATJE, I.
[1] *Iteration der Joukowski-Abbildung und ihre Streckenkomplexe.* Mitt. Math. Sem. Giessen no. 48 (1954), 54 pp. MR 16, 685.

RADÓ, T.
[1] *Bemerkung zur Arbeit des Herrn Bieberbach: Über die Einordnung des Hauptsatzes der Uniformisierung in die Weierstrassche Funktionentheorie (Math. Ann. 78).* Math. Ann. 90 (1923), 30–37. FM 49, 248.
[2] *Über eine nicht fortsetzbare Riemannsche Mannigfaltigkeit.* Math. Z. 20 (1924), 1–6. FM 50, 255.
[3] *Über den Begriff der Riemannschen Fläche.* Acta Szeged 2 (1925), 101–121. FM 51, 273.
[4] *Subharmonic functions.* Springer, Berlin, 1937, 56 pp. FM 63, 458.

RADOJČIĆ, M.
[1] *Über die Zerteilung Riemannscher Flächen in Blätter.* Glas 134 (63), (1929), 63–81. (Serbian with French summary.) FM 55, 199.
[2] *Sur les domaines fondamentaux des fonctions analytiques au voisinage d'une singularité essentielle.* Publ. Math. Univ. Belgrade 4 (1935), 185–200. FM 61, 1132.
[3] *Sur l'allure des fonctions analytiques au voisinage des singularités essentielles.* Bull. Soc. Math. France 64 (1936), 137–146. FM 62, 387.
[4] *Über einen Satz von Herrn Ahlfors.* Publ. Math. Univ. Belgrade 6–7 (1938), 77–83. FM 64, 316.
[5] *Remarque sur le problème des types des surfaces de Riemann.* Acad. Serbe Sci. Publ. Inst. Math. 1 (1947), 97–100. MR 10, 442.
[6] *Sur un problème topologique de la théorie des surfaces de Riemann.* Ibid. 2 (1948), 11–25. (French. Serbian summary.) MR 10, 523.

RADOJČIĆ, M.
[7] *Certains critères concernant le type des surfaces de Riemann à points de ramification algébriques.* Ibid. 3 (1950), 25–52, 305–306. MR 12, 602.
[8] *Sur le discernement des types des surfaces de Riemann.* Premier Congr. Math. Phys. R.P.F.Y., 1949, 2. Communications et Exposés Scient., pp. 163–167. Naučna Knjiga, Belgrade, 1951. (Serbo-Croatian. French summary.) MR 13, 547.
[9] *Sur le problème des types des surfaces de Riemann.* Srpska Akad. Nauka. Zbornik Radova 35. Nat. Inst. 3 (1953), 15–28. (Serbo-Croatian. French summary.) MR 15, 615.
[10] *Sur les séries de fonctions algébriques et les produits infinis analogues, définissant des fonctions analytiques multiformes dans leurs domaines d'existence quelconques.* Acad. Serbe Sci. Publ. Inst. Math. 7 (1954), 95–118. MR 16, 812.
[11] *Entwicklung analytischer Funktionen auf Riemannschen Flächen nach algebraischen oder gewissen endlich vieldeutigen transzendenten Funktionen.* Ibid. 8 (1955), 93–122. MR 17, 1067.
[12] *Über die Weierstrassche Produktentwicklung analytischer Funktionen auf Riemannschen Flächen.* Ann. Acad. Sci. Fenn. Ser. A. I. no. 250/27 (1958), 11 pp.

RAUCH, H. E.
[1] *On the transcendental moduli of algebraic Riemann surfaces.* Proc. Nat. Acad. Sci. U.S.A. 41 (1955), 42–49. MR 17, 251.
[2] *On the moduli in conformal mapping.* Ibid. 41 (1955), 176–180. MR 17, 251.
[3] *On the moduli of Riemann surfaces.* Ibid. 41 (1955), 236–238; errata, 421. MR 17, 251.
[4] *The first variation of the Douglas functional and the periods of the Abelian integrals of the first kind.* Seminars on analytic functions. II, pp. 49–55. Institute for Advanced Study, Princeton, 1958.
[5–6] Cf. GERSTENHABER, M., and RAUCH, H. E. [1–2].

READ, A. H.
[1] *Conjugate extremal problems of class* $p=1$. Ann. Acad. Sci. Fenn. Ser. A. I. no. 250/28 (1958), 8 pp.

RENGEL, E.
[1] *Über einige Schlitz-Theoreme der konformen Abbildung.* Schrift. Math. Sem. Univ. Berlin 1 (1933), 140–162. FM 59, 351.
[2] *Existenzbeweise für schlichte Abbildungen mehrfach zusammenhängender Bereiche auf gewisse Normalbereiche.* Jbr. Deutsch Math.-Verein. 44 (1934), 51–55. FM 60, 286.
[3] *Ein neues Schlitztheorem der konformen Abbildung.* Ibid. 45 (1935), 83–87.

DE RHAM, G.
[1] *Variétés différentiables. Formes, courants, formes harmoniques.* Actualités Sci. Ind., no. 1222 = Publ. Inst. Math. Univ. Nancago. III. Hermann et Cie, Paris, 1955. 196 pp. MR 16, 957.

RIEMANN, B.
[1] *Gesammelte mathematische Werke und wissenschaftlicher Nachlass. Herausgegeben unter Mitwirkung von R. Dedekind und H. Weber.* Teubner, Leipzig, 1876, 526 pp. FM 8, 231.

RIESZ, F., and NAGY, B. Sz.
[1] *Functional analysis.* (Translated by L. F. Boron.) F. Ungar Pub. Co., New York, 1955, 468 pp. MR 17, 175.

RITT, J. F.
[1] *Abel's theorem and a generalization of one-parameter groups.* Trans. Amer. Math. Soc. 67 (1949), 491–497. MR 11, 343.

RITTER, E.
[1] *Ausdehnung des Riemann-Roch'schen Satzes auf Formenscharen, die sich bei Umlaufen auf einer Riemann'schen Fläche linear substituiren.* Nachr. Akad. Wiss. Göttingen (1894), 328–337. FM 25, 715.

RÖHRL, H.
[1] *Zur Theorie der Faberschen Entwicklungen auf geschlossenen Riemannschen Flächen.* Arch. Math. 3 (1952), 93–102. MR 14, 154.
[2] *Funktionenklassen auf geschlossenen Riemannschen Flächen.* Math. Nachr. 6 (1952), 355–384. MR 13, 736.
[3] *Die Elementartheoreme der Funktionenklassen auf geschlossenen Riemannschen Flächen.* Ibid. 7 (1952), 65–84. MR 14, 462.
[4] *Fabersche Entwicklungen und die Sätze von Weierstrass und Mittag-Leffler auf Riemannschen Flächen endlichen Geschlechts.* Arch. Math. 4 (1953), 298–307. MR 15, 415.
[5] *Über gewisse Verallgemeinerungen der Abelschen Integrale.* Math. Nachr. 9 (1953), 23–44. MR 14, 1076.

ROSENBLOOM, P. C.
[1] *Semigroups of transformations of a Riemann surface into itself.* Contributions to the theory of Riemann surfaces, pp. 31–39. Princeton Univ. Press, Princeton, 1953. MR 15, 415.

ROYDEN, H. L.
[1] *Some remarks on open Riemann surfaces.* Ann. Acad. Sci. Fenn. A. I. no. 85 (1951), 8 pp. MR 13, 339.
[2] *Harmonic functions on open Riemann surfaces.* Trans. Amer. Math. Soc. 73 (1952), 40–94. MR 14, 167.
[3] *A modification of the Neumann-Poincaré method for multiply connected regions.* Pacific J. Math. 2 (1952), 385–394. MR 14, 182.
[4] *On the ideal boundary of a Riemann surface.* Contributions to the theory of Riemann surfaces, pp. 107–109. Princeton Univ. Press, Princeton, 1953. MR 15, 25.
[5] *Some counterexamples in the classification of open Riemann surfaces.* Proc. Amer. Math. Soc. 4 (1953), 363–370. MR 14, 864.
[6] *A property of quasi-conformal mapping.* Ibid. 5 (1954), 266–269. MR 15, 695.
[7] *The conformal rigidity of certain subdomains on a Riemann surface.* Trans. Amer. Math. Soc. 76 (1954), 14–25. MR 15, 519.
[8] *Conformal deformation.* Lectures on functions of a complex variable, pp. 309–313. Univ. Michigan Press, Ann Arbor, 1955.
[9] *Rings of analytic and meromorphic functions.* Trans. Amer. Math. Soc. 83 (1956), 269–276. MR 19, 737.
[10] *Rings of meromorphic functions.* Seminars on analytic functions. II, pp. 273–285. Institute for Advanced Study, Princeton, 1958.
[11] *Open Riemann surfaces.* Ann. Acad. Sci. Fenn. Ser. A. I. no. 249/5 (1958), 13 pp.
[12] Cf. AHLFORS, L. V., and ROYDEN, H. L. [1].

RUDIN, W.
[1] *Analyticity, and the maximum modulus principle.* Duke Math. J. 20 (1953), 449–457. MR 15, 21.
[2] *Analytic functions of class H_p.* Trans. Amer. Math. Soc. 78 (1955), 46–66. MR 16, 810.

SAGAWA, A.
[1] *A note on a Riemann surface with null boundary.* Tôhoku Math. J. (2) 3 (1951), 273–276. MR 13, 931.
[2] *On the existence of Green's function.* Ibid. (2) 7 (1955), 136–139. MR 17, 358.

SAKAKIHARA, K.
[1] *Meromorphic approximations on Riemann surfaces.* J. Inst. Polytech. Osaka City Univ. Ser. A. 5 (1954), 63–70. MR 16, 687.

SAKS, S., and ZYGMUND, A.
[1] *Analytic functions.* Warszawa-Wroclaw, 1952, 451 pp. MR 14, 1073.

SANSONE, G.
[1] *Lezioni sulla teoria delle funzioni di une variabile complessa.* CEDAM, Padova, 1947. I. 359 pp. MR 8, 143.

SANSONE, G.

[2] *Lezioni sulla teoria delle funzioni di une variabile complessa.* CEDAM, Padova, 1947. II. 546 pp. MR 8, 143.

SARIO, L.

[1] *Über Riemannsche Flächen mit hebbarem Rand.* Ann. Acad. Sci. Fenn. Ser. A. I. no. 50 (1948), 79 pp. MR 10, 365.

[2] *Sur la classification des surfaces de Riemann.* C.R. Onzième Congr. Math. Scand. Trondheim, 1949, pp. 229–238. MR 14, 863.

[3] *Sur le problème du type des surfaces de Riemann.* C.R. Acad. Sci. Paris 229 (1949), 1109–1111. MR 11, 342.

[4] *Existence des fonctions d'allure donnée sur une surface de Riemann arbitraire.* Ibid. 229 (1949), 1293–1295. MR 11, 342.

[5] *Quelques propriétés à la frontière sa rattachant à la classification des surfaces de Riemann.* Ibid. 230 (1950), 42–44. MR 11, 342.

[6] *Existence des intégrales abéliennes sur les surfaces de Riemann arbitraires.* Ibid. 230 (1950), 168–170. MR 11, 342.

[7] *Questions d'existence au voisinage de la frontière d'une surface de Riemann.* Ibid. 230 (1950), 269–271. MR 11, 342.

[8] *On open Riemann surfaces.* Proc. Intern. Congr. Math., Cambridge, Mass., 1950, 398–399.

[9] *Linear operators on Riemann surfaces.* Bull. Amer. Math. Soc. 57 (1951), 276.

[10] *Principal functions on Riemann surfaces.* Ibid. 58 (1952), 475–476.

[11] *A linear operator method on arbitrary Riemann surfaces.* Trans. Amer. Math. Soc. 72 (1952), 281–295. MR 13, 735.

[12] *An extremal method on arbitrary Riemann surfaces.* Ibid. 73 (1952), 459–470. MR 14, 863.

[13] *Construction of functions with prescribed properties on Riemann surfaces.* Contributions to the theory of Riemann surfaces, pp. 63–76. Princeton Univ. Press, Princeton, 1953. MR 15, 209.

[14] *Modular criteria on Riemann surfaces.* Duke Math. J. 20 (1953), 279–286. MR 14, 969.

[15] *Minimizing operators on subregions.* Proc. Amer. Math. Soc. 4 (1953), 350–355. MR 14, 969.

[16] *Alternating method on arbitrary Riemann surfaces.* Pacific J. Math. 3 (1953), 631–645. MR 15, 209.

[17] *Capacity of the boundary and of a boundary component.* Ann. of Math. (2) 59 (1954), 135–144. MR 15, 518.

[18] *Functionals on Riemann surfaces.* Lectures on functions of a complex variable, pp. 245–256. Univ. Michigan Press, Ann Arbor, 1955.

[19] *Positive harmonic functions.* Ibid., pp. 257–263.

[20] *Extremal problems and harmonic interpolation on open Riemann surfaces.* Trans. Amer. Math. Soc. 79 (1955), 362–377. MR 19, 846.

[21] *On univalent functions.* C.R. Treizième Congr. Math. Scand. Helsinki, 1957, pp. 202–208.

[22] *Countability axiom in the theory of Riemann surfaces.* Ann. Acad. Sci. Fenn. Ser. A. I. no. 250/32 (1958), 7 pp.

[23] *Stability problems on boundary components.* Seminars on analytic functions. II, pp. 55–72. Institute for Advanced Study, Princeton, 1958.

[24] *Strong and weak boundary components.* J. Analyse Math. 5 (1956/57), 389–398. MR 20, 161.

SAVAGE, N.

[1] *Weak boundary components of an open Riemann surface.* Duke Math. J. 24 (1957), 79–95. MR 18, 647.

SCHAEFFER, A. C.

[1] *An extremal boundary value problem.* Contributions to the theory of Riemann surfaces, pp. 41–47, Princeton Univ. Press, Princeton, 1953. MR 15, 24.

SCHIFFER, M.
[1] *A method of variation within the family of simple functions.* Proc. London Math. Soc. (2), 44 (1938), 432–449. FM 64, 307.
[2] *The span of multiply connected domains.* Duke Math. J. 10 (1943), 209–216. MR 4, 271.
[3] *Variation of the Green function and theory of the p-valued functions.* Amer. J. Math. 65 (1943), 341–360. MR 4, 215.
[4] *Hadamard's formula and variation of domain-functions.* Ibid. 68 (1946), 417–448. MR 8, 325.
[5] *An application of orthonormal functions in the theory of conformal mapping.* Ibid. 70 (1948), 147–156. MR 9, 341.
[6] *Variational methods in the theory of Riemann surfaces.* Contributions to the theory of Riemann surfaces, pp. 15–30, Princeton Univ. Press, Princeton, 1953. MR 15, 25.
[7] Cf. COURANT, R. [6].
[8] Cf. GARABEDIAN, P., and SCHIFFER, M. [1].
[9] Cf. BERGMAN, S., and SCHIFFER, M. [1].

SCHIFFER, M., and SPENCER, D. C.
[1] *A variational calculus for Riemann surfaces.* Ann. Acad. Sci. Fenn. Ser. A. I. no. 93 (1951), 9 pp. MR 13, 338.
[2] *On the conformal mapping of one Riemann surface into another.* Ibid. no. 94 (1951), 10 pp. MR 13, 338.
[3] *Functionals of finite Riemann surfaces.* Princeton Univ. Press, Princeton, N.J., 1954, 451 pp. MR 16, 461.

SCHILLING, O. F. G.
[1] *Ideal theory on open Riemann surfaces.* Bull. Amer. Math. Soc. 52 (1946), 945–963. MR 8, 454.

SCHLESINGER, L.
[1] *Sur la théorie des fonctions algébriques.* C.R. Acad. Sci. Paris 135 (1902), 676–678. FM 33, 430.
[2] *Sur la détermination des fonctions algébriques uniformes sur une surface de Riemann donnée.* Ann. École Norm. Sup. (3) 20 (1903), 331–347. FM 34, 457.

SCHÖNFLIES, A.
[1] *Über gewisse geradlinig begrenzte Stücke Riemann'scher Flächen.* Nachr. Akad. Wiss. Göttingen (1892), 257–267. FM 24, 383.

SCHOTTKY, F.
[1] *Über die konforme Abbildung merhfach zusammenhängender ebener Fläche.* Inaug.-Diss. J. Rein. Angew. Math. 83 (1877), 300–351. FM 9, 584.
[2] *Über eindeutige Funktionen mit linearen Transformationen in sich.* Math. Ann. 20 (1882), 293–300. FM 14, 345.

SCHWARTZ, M. H.
[1] *Sur les surfaces de Riemann possédant des points critiques arbitrairement rappochés.* C.R. Acad. Sci. Paris 228 (1949), 154–155. MR 10, 523.

SCHWARZ, H. A.
[1] *Gesammelte mathematische Abhandlungen.* 2 Bände. Springer, Berlin, 1890, 338 pp., 370 pp. FM 22, 31.

SEIBERT, P.
[1] *Flächenbau und Wertverteilung einiger Funktionen, die aus harmonischen Massen entspringen.* Mitt. Math. Sem. Giessen no. 38 (1951), 51 pp. MR 16, 917.
[2] *Flächenbau und Wertverteilung einiger Funktionen, die aus harmonischen Massen entspringen.* Arch. Math. 3 (1952), 87–92. MR 16, 917.
[3] *Über die bei Deformationen Riemannscher Flächen mit ondlich vielen Windungssorten entstehenden Randstellen.* Arch. Math. 5 (1954), 389–400. MR 16, 349.
[4] *Typus und topologische Randstruktur einfach-zusammenhängender Riemannscher Flächen.* Ann. Acad. Sci. Fenn. Ser. A. I. no. 250/34 (1958), 11 pp. MR 19, 949.

SEIFERT, H., and THRELFALL, W.
[1] *Lehrbuch der Topologie.* Teubner, Leipzig, 1934, 353 pp. FM 60, 496.

SELBERG, H. L.
[1] *Algebroide Funktionen und Umkehrfunktionen Abelscher Integrale.* Diss., Oslo, 1934, 72 pp. FM 60, 1020.
[2] *Eine Ungleichung der Potentialtheorie und ihre Anwendung in der Theorie der meromorphen Funktionen.* Comment. Math. Helv. 18 (1946), 309–326. MR 8, 23.

SFORZA, G.
[1] *Origine geometrica delle superficie di Riemann. Annurio 1899–1900 del R. Instituto tecnico Argelo Senti di Reggio Emilia.* Reggio-Emilia: Stab. tip.-lit. degli artigianelli (1901), 5–36. FM 32, 404.

SHIFFMAN, M.
[1] *On Dirichlet's principle.* Contributions to the theory of Riemann surfaces, pp. 49–53, Princeton Univ. Press, Princeton, 1953. MR 15, 220.

SHIRAI, T.
[1] *A remark on Riemann surfaces defined by M. S. Stoïlow.* Mem. Coll. Sci. Kyoto Imp. Univ. Ser. A. 23 (1942), 369–372. MR 11, 93.

SOMMER, F.
[1] Cf. BEHNKE, H., und SOMMER, F. [1].

SÖRENSEN, W.
[1] Cf. BADER, R., et SÖRENSEN, W. [1].

SÖRENSEN, W., et BADER, R.
[1] *Noyaux de Green-de Rham sur les surfaces de Riemann.* C.R. Acad. Sci. Paris 244 (1957), 1309–1311. MR 19, 259.

SPEISER, A.
[1] *Über Riemannsche Flächen.* Comment. Math. Helv. 2 (1930), 284–293. FM 56, 987.
[2] *Riemannsche Flächen vom hyperbolischen Typus.* Ibid. 10 (1938), 232–242. FM 64, 317.
[3] *Über symmetrische analytische Funktionen.* Ibid. 16 (1944), 105–114. MR 5, 234.
[4] *Sulle superficie Riemanniane.* Rond. Sem. Mat. Fis. Milano 18 (1947), (1948), 91–92. MR 10, 697.

SPENCER, D. C.
[1–3] Cf. SCHIFFER, M., and SPENCER, D. C. [1–3].

SPRINGER, G.
[1] *Introduction to Riemann surfaces.* Addison-Wesley Publ., Reading, Mass. 1957, 307 pp. MR 19, 1169.

STAHL, H.
[1] *Abriss einer Theorie der algebraischen Funktionen einer Veränderlichen in neuer Fassung.* Nachgelassene Schrift. Teubner, Leipzig, 1911, 105 pp. FM 42, 413.

STEIN, K.
[1] *Primfunktionen und multiplikative automorphe Funktionen auf nichtgeschlossenen Riemannschen Flächen und Zylindergebieten.* Acta Math. 83 (1950), 165–196. MR 12, 252.
[2–4] Cf. BEHNKE, H., und STEIN, K. [1–3].

STEINER, A.
[1] *Eine direkte Konstruktion der Abelschen Integrale erster Gattung.* Diss., Zürich, 1950, 29 pp. MR 14, 743.

STOÏLOW, S.
[1] *Les propriétés topologiques des fonctions analytiques d'une variable.* Ann. Inst. H. Poincaré 2 (1932), 233–266. FM 58, 1090.
[2] *Sur les fonctions analytiques dont les surfaces de Riemann ont des frontières totalement discontinues.* Mathematica, Cluj. 12 (1936), 123–138. FM 62, 386.

Stoïlow, S.
[3] *Sur les transformations intérieurs et la caractérisation topologique des surfaces de Riemann.* Compositio Math. 3 (1936), 435–440. FM 62, 667.
[4] *Sur la définition des surfaces de Riemann.* C.R. Congr. Intern. Math., Oslo, 1936, 2 (1937), 143–144. FM 63, 305.
[5] *Sur une classe des surfaces de Riemann régulièrement exhaustibles et sur le théorème des disques de M. Ahlfors.* C.R. Acad. Sci. Paris 207 (1938), 517–519. FM 64, 1075.
[6] *Sur les surfaces de Riemann normalement exhaustibles et sur le théorème des disques pour ces surfaces.* Compositio Math. 7 (1940), 428–435. MR 2, 85.
[7] *Sur les singularités des fonctions analytiques multiformes dont la surface de Riemann a sa frontière de measure harmonique nulle.* Mathematica, Timişoara 19 (1943), 126–138. MR 5, 177.
[8] *Remarques sur la définition des points singuliers des fonctions analytiques multiformes.* Acad. Roum. Bull. Sect. Sci. 26 (1944), 671–672. MR 10, 28.
[9] *Quelques remarques sur les éléments frontières des surfaces de Riemann et sur les fonctions correspondant à ces surfaces.* C.R. Acad. Sci. Paris 227 (1948), 1326–1328. MR 10, 442.
[10] *Les surfaces de Riemann à frontière nulle.* VI Sjazd Mat., Warszawa, 1948, 36–37.
[11] *Note sur les fonctions analytiques multiformes.* Ann. Soc. Polon. Math. 25 (1952), (1953), 69–74. MR 14, 1076.
[12] *Theory of functions of a complex variable. I. Notions and fundamental principles.* Edit. Acad. Rep. Pop. Romaine, Bucharest, 1954, 308 pp. MR 16, 458.
[13] *Sur la classification topologique des recouvrements riemanniens.* Rev. Math. Pures Appl. 1 (1956), no. 2, 37–42. MR 18, 290.
[14] *Leçons sur les principes topologiques de la théorie des fonctions analytiques.* Deuxième edition, augmentée de notes sur les fonctions analytiques et leurs surfaces de Riemann. Gauthier-Villars, Paris, 1956, 194 pp. MR 18, 568.
[15] *Sur quelques points de la théorie moderne des surfaces de Riemann.* Rend. Matem. e sue appl. S. V, 16 (1957), 170–196.
[16] *Sur la théorie topologique des recouvrements Riemanniens.* Ann. Acad. Sci. Fenn. Ser. A. I. no. 250/35 (1958), 7 pp.

Strebel, K.
[1] *Eine Bemerkung zur Hebbarkeit des Randes einer Riemannschen Fläche.* Comment. Math. Helv. 23 (1949), 350–352. MR 11, 342.
[2] *Eine Ungleichung für extremale Längen.* Ann. Acad. Sci. Fenn. Ser. A. I. no. 90 (1951), 8 pp. MR 13, 338.
[3] *Über das Kreisnormierungsproblem der konformen Abbildung.* Ibid. 101 (1951), 22 pp. MR 14, 549.
[4] *A remark on the extremal distance of two boundary components.* Proc. Nat. Acad. Sci. U.S.A. 40 (1954), 842–844. MR 16, 917.
[5] *Die extremale Distanz zweier Enden einer Riemannschen Fläche.* Ann. Acad. Sci. Fenn. Ser. A. I. no. 179 (1955), 21 pp. MR 16, 917.

Tamura, J.
[1] *A note on Riemann surfaces and analytic functions.* Sci. Papers Coll. Gen. Ed. Univ. Tokyo 2 (1952), 125–128. MR 14, 743.
[2] *A prolongable Riemann surface.* Ibid. 6 (1956), 123–127. MR 18, 727.
[3] *On the maximal Riemann surface.* Ibid. 7 (1957), 19–22.

Tauber, A.
[1] *Über die Neumann'sche Methode des arithmetischen Mittels.* Monatsh. Math. 5 (1894), 137–150. FM 25, 683.

Taylor, H. E.
[1] *Determination of the type and properties of the mapping function of a class of doubly-connected Riemann surfaces.* Proc. Amer. Math. Soc. 4 (1953), 52–68. MR 14, 744.

TEICHMÜLLER, O.
[1] *Eine Anwendung quasikonformer Abbildungen auf das Typenproblem.* Deutsche Math. 2 (1937), 321–327. FM 63, 303.
[2] *Untersuchungen über konforme und quasikonforme Abbildung.* Ibid. 3 (1938), 621–678. FM 64, 313.
[3] *Erreichbare Randpunkte.* Ibid. 4 (1939), 455–461. MR 1, 8.
[4] *Extremale quasikonforme Abbildungen und quadratische Differentiale.* Abh. Preuss. Akad. Wiss. Math.-Nat. Kl. (1939), no. 22 (1940), 197 pp. MR 2, 187.
[5] *Vollständige Lösung einer Extremalaufgabe der quasikonformen Abbildung.* Ibid. 1941, no. 5 (1941), 18 pp. MR 8, 202.
[6] *Über Extremalprobleme der konformen Geometrie.* Deutsche Math. 6 (1941), 50–77. MR 3, 202.
[7] *Skizze einer Begründung der algebraischen Funktionentheorie durch Uniformisierung.* Ibid. 6 (1941), 257–265. MR 8, 327.
[8] *Bestimmung der extremalen quasikonformen Abbildungen bei geschlossenen orientierten Riemannschen Flächen.* Abh. Preuss. Akad. Wiss. Math.-Nat. Kl. (1943), no. 4 (1943), 42 pp. MR 8, 202.
[9] *Beweis der analytischen Abhängigkeit des konformen Moduls einer analytischen Ringflächenschar von den Parametern.* Deutsche Math. 7 (1944), 309–336. MR 8, 327.
[10] *Veränderliche Riemannsche Flächen.* Ibid. 7 (1944), 344–359. MR 8, 327.
[11] *Einfache Beispiele zur Wertverteilungslehre.* Ibid. 7 (1944), 360–368. MR 8, 327.

TELEMAN, C.
[1] *Une classe de fonctions analytiques d'une surface de Riemann, généralisant les intégrales abéliennes.* Acad. R. P. Romine. Stud. Cerc. Mat. 8 (1957), 163–182. (Romanian. Russian and French summaries.) MR 20, 290.

TERASAKA, H.
[1] *On the division of Riemann surfaces into sheets.* Japanese J. Math. 8 (1932), 309–326. FM 58, 367.

THRELFALL, W.
[1] Cf. SEIFERT, H., und THRELFALL, W. [1].

TIETZ, H.
[1] *Fabersche Entwicklungen auf geschlossenen Riemannschen Flächen.* J. Reine Angew. Math. 190 (1952), 22–33. MR 13, 833.
[2] *Partialbruchzerlegung und Produktdarstellung von Funktionen auf geschlossenen Riemannschen Flächen.* Arch. Math. 4 (1953), 31–38. MR 14, 859.
[3] *Zur Realisierung Riemannscher Flächen.* Math. Ann. 128 (1955), 453–458. MR 16, 688.
[4] *Eine Normalform berandeter Riemannscher Flächen.* Ibid. 129 (1955), 44–49. MR 16, 1012.
[5] *Faber-Theorie auf nicht-kompakten Riemannschen Flächen.* Math. Ann. 132 (1957), 412–429. MR 18, 883.
[6] *Laurent-Trennung und zweifach unendliche Faber-Systeme.* Ibid. 129 (1955), 431–450. MR 17, 251.
[7] *Berichtigung der Arbeit "Zur Realisierung Riemannscher Flächen".* Ibid. 129 (1955), 453–458. MR 17, 27.
[8] *Funktionen mit Cauchyscher Integraldarstellung auf nicht-kompakten Gebieten Riemannscher Flächen.* Ann. Acad. Sci. Fenn. Ser. A. I. no. 250/36 (1958), 10 pp. MR 20, 291.

TIKHOMANDRITZKY, M.
[1] *Résolution d'un problème concernant les surfaces de Riemann.* Bull. Sci. Math. (2) 37 (1913), 55–62. FM 44, 493.

TITCHMARSH, E. C.
[1] *The theory of functions.* Clarendon Press, Oxford, 1932, 454 pp. FM 58, 297.

TÔKI, Y.

[1] *On the classification of open Riemann surfaces.* Osaka Math. J. 4 (1952), 191–201. MR 14, 864.

[2] *On the examples in the classification of open Riemann surfaces.* I. Ibid. 5 (1953), 267–280. MR 15, 519.

TRICOMI, F.

[1] *Funzioni analitiche.* N. Zanichelli (Monografie di matematica applicata), Bologna, 1936, 110 pp. FM 62, 1198.

TROHIMČUK, YU. YU.

[1] *On the theory of sequences of Riemann surfaces.* Ukrain. Mat. Žurnal 4 (1952), 49–56. (Russian.) MR 15, 415.

[2] *On removable boundary sets.* Ibid. 4 (1952), 312–322. (Russian.) MR 15, 303.

[3] *On sequences of analytic functions and Riemann surfaces.* Ibid. 4 (1952), 431–446. (Russian.) MR 15, 415.

TSUJI, M.

[1] *On the Riemann surface of an inverse function of a meromorphic function in the neighborhood of a closed set of capacity zero.* Proc. Imp. Acad. Tokyo 19 (1943), 257–258. MR 8, 508.

[2] *On non-prolongable Riemann surfaces.* Ibid. 19 (1943), 429–430. MR 8, 373.

[3] *On the domain of existence of an implicit function defined by an integral relation* $G(x, y)=0$. Proc. Imp. Acad. Japan 19 (1943), 235–240. MR 8, 508.

[4] *Theory of meromorphic functions in a neighbourhood of a closed set of capacity zero.* Japanese J. Math. 19 (1944), 139–154. MR 8, 508.

[5] *Some metrical theorems on Fuchsian groups.* Kōdai Math. Sem. Rep. (1950), 89–93. MR 13, 125.

[6] *On the uniformization of an algebraic function of genus* $p \geq 2$. Tôhoku Math. J. (2) 3 (1951), 277–281. MR 14, 157.

[7] *Some theorems on open Riemann surfaces.* Nagoya Math. J. 3 (1951), 141–145. MR 13, 338.

[8] *A theorem of Bloch type concerning the Riemann surface of an algebraic function of genus* $p \geq 2$. Kōdai Math. Sem. Rep. (1951), 77. MR 13, 644.

[9] *Maximal continuation of a Riemann surface.* Ibid. 1952, 55–56. MR 14, 462.

[10] *Existence of a potential function with a prescribed singularity on any Riemann surface.* Tôhoku Math. J. (2) 4 (1952), 54–68. MR 14, 168.

[11] *On covering surfaces of a closed Riemann surface of genus* $p \leqq 2$. Ibid. (2) 5 (1953), 185–188. MR 15, 863.

[12] *On the capacity of general Cantor sets.* J. Math. Soc. Japan 5 (1953), 235–252. MR 15, 309.

[13] *Theory of meromorphic functions on an open Riemann surface with null boundary.* Nagoya Math. J. 6 (1953), 137–150. MR 15, 518.

[14] *On Ahlfors' theorem on covering surfaces.* J. Fac. Sci. Univ. Tokyo. Sect. I. 6 (1953), 319–328. MR 14, 969.

[15] *On Royden's theorem on a covering surface of a closed Riemann surface.* J. Math. Soc. Japan 6 (1954), 32–36. MR 15, 948.

[16] *A metrical theorem on the singular set of a linear group of Schottky type.* Ibid. 6 (1954), 115–121. MR 16, 349.

[17] *On Neumann's problem for a domain on a closed Riemann surface.* Ibid. 6 (1954), 122–128. MR 16, 349.

[18] *On the moduli of closed Riemann surfaces.* Comment. Math. Univ. St. Paul. 5 (1956), 25–28. MR 19, 401.

[19] *On a Riemann surface which is conformally equivalent to a Riemann surface with a finite spherical area.* Ibid. 6 (1957), 1–7. MR 19, 1043.

[20] *On Abelian and Schottkyan covering surfaces of a closed Riemann surface.* Ibid. 6 (1957), 8–28. MR 19, 1044.

TSUJI, R.

[1] *On conformal mapping of a hyperelliptic Riemann surface onto itself.* Kōdai Math. Sem. Rep. 10 (1958), 127–136.

TUMARKIN, G., and HAVINSON, S.

[1] *On the removing of singularities for analytic functions of a certain class (class D).* Uspehi Mat. Nauk (N.S.) 12 (1957), no. 4 (76), 193–199. (Russian.) MR 20, 17.

TUMURA, Y.

[1] *Quelques applications de la théorie de M. Ahlfors.* Japanese J. Math. 18 (1942), 303–322. MR 7, 516.

ULLEMAR, L.

[1] *Über die Existenz der automorphen Funktionen mit beschränktem Dirichlet-integral.* Ark. Mat. 2 (1952), 87–97. MR 14, 470.

ULLRICH, E.

[1] *Über ein Problem von Herrn Speiser.* Comment. Math. Helv. 7 (1934), 63–66. FM 60, 289.

[2] *Zum Umkehrproblem der Wertverteilungslehre.* Nachr. Akad. Wiss. Göttingen, Math.-Phys. Kl., FG I. Neue Folge 1 (1936), 135–150. FM 62, 388.

[3] *Flächenbau und Wachstumsordnung bei gebrochenen Funktionen.* Jbr. Deutsche Math.-Verein. 46 (1936), 232–274. FM 62, 386.

ULRICH, F. E.

[1] *The problem of type for a certain class of Riemann surfaces.* Duke Math. J. 5 (1939), 567–589. MR 1, 8.

USKILA, L.

[1] *Über die Existenz der beschränkten automorphen Funktionen.* Ark. Mat. 1 (1949), 1–11. MR 11, 343.

VALIRON, G.

[1] *Sur le théorème de Bloch.* Rendiconti Palermo 54 (1930), 76–82. FM 56, 269.

[2] *Sur le domaine couvert par les valeurs d'une fonction algébroide finie.* Bull. Sci. Math. (2) 64 (1940), 199–206. MR 3, 83.

[3] *Division en feuillets de la surface de Riemann définie par $w=(e^z-1)/z+h$.* J. Math. Pures Appl. (9) 19 (1940), 339–358. MR 2, 358.

[4] *Théorie des fonctions.* Masson et Cie., Paris, 1942, 522 pp. MR 7, 283.

[5] *Des théorèmes de Bloch aux théories d'Ahlfors.* Bull. Sci. Math. (2) 73 (1949), 152–162. MR 11, 572.

[6] *Surfaces de Riemann simplement connexes dont des points de ramification sont donnée.* Ann. Mat. Pura Appl. (4) 29 (1949), 321–326. MR 12, 17.

[7] *Les notions de fonction analytique et de surface de Riemann.* Phil. Sci., Paris, 1949, Phil. Math. Méc., pp. 27–35. Actualités Sci. Ind. no. 1137. Hermann, Paris, 1951. MR 13, 454.

[8] *Fonctions analytiques.* Presses Univ. France, Paris, 1954, 236 pp. MR 15, 861.

VIRTANEN, K. I.

[1] *Über Abelsche Integrale auf nullberandeten Riemannschen Flächen von unendlichem Geschlecht.* Ann. Acad. Sci. Fenn. Ser. A. I. no. 56 (1949), 44 pp. MR 11, 510.

[2] *Über eine Integraldarstellung von quadratisch integrierbaren analytischen Differentialen.* Ibid. no. 69 (1950), 21 pp. MR 12, 493.

[3] *Über die Existenz von beschränkten harmonischen Funktionen auf offenen Riemannschen Flächen.* Ibid. no. 75 (1950), 8 pp. MR 12, 403.

[4] *Bemerkungen zur Theorie der quadratisch integrierbaren analytischen Differentiale* Ibid. no. 78 (1950), 6 pp. MR 12, 493.

[5] *Über Extremalfunktionen auf offenen Riemannschen Flächen.* Ibid. no. 141 (1952), 7 pp. MR 14, 743.

VITALI, G.

[1] *Sulle funzioni analitiche sopra le superficie di Riemann.* Palermo Rend. 14 (1900), 202–208. FM 31, 422.

VOLKOVYSKIĬ, L. I.

[1] *On the problem of type of simply connected Riemann surfaces.* Rec. Math. (Mat. Sbornik) (N.S.) 18 (60), (1946), 185–212. (Russian.) MR 8, 326.

[2] *Investigations on the problem of type for a simply connected Riemann surface.* Uspehi Matem. Nauk (N.S.) 3, no. 3 (25), (1948), 215–216. (Russian.) MR 10, 365.

[3] *The determination of the type of certain classes of simply-connected Riemann surfaces.* Mat. Sbornik (N.S.) 23 (65), (1948), 229–258. (Russian.) MR 10, 364.

[4] *Convergent sequences of Riemann surfaces.* Ibid. 23 (65), (1948), 361–382. (Russian.) MR 10, 365.

[5] *The influence of the closeness of the branch points on the type of a simply connected Riemann surface.* Ibid. 25 (67), (1949), 415–450. (Russian.) MR 12, 17.

[6] *On the type of problem of a simply connected Riemann surface.* Ukrain. Mat. Žurnal 1, no. 1 (1949), 39–48. (Russian.) MR 14, 744.

[7] *An example of a simply connected Riemann surface of hyperbolic type.* Ibid. 1, no. 3 (1949), 60–67. (Russian.) MR 14, 744.

[8] *Investigation of the type problem for a simply connected Riemann surface.* Trudy Mat. Inst. Steklov. 34 (1950), 171 pp. (Russian.) MR 14, 156.

[9] *Quasi-conformal mappings and problems of conformal pasting.* Ukrain. Mat. Žurnal 3 (1951), 39–51. (Russian.) MR 14, 862.

[10] *Contemporary investigations on the theory of Riemann surfaces.* Uspehi Mat. Nauk 6 (1956), 101–105. (Russian.)

VOLTERRA, V.

[1] *Sul principo di Dirichlet.* Palermo Rend. 11 (1897), 83–86. FM 28, 363.

VAN DER WAERDEN, B. L.

[1] *Topologie und Uniformisierung der Riemannschen Flächen.* Ber. Verh. Sächs. Akad. Wiss. Leipzig. Math.-Phys. Kl. 93 (1941), 147–160. MR 11, 22.

WAGNER, H.

[1] *Über eine Klasse Riemannscher Flächen mit endlich vielen nur logarithmischen Windungspunkten.* J. Reine Angew. Math. 175 (1936), 6–49. FM 62, 388.

WALLMAN, H.

[1] Cf. HUREWICZ, W., and WALLMAN, H. [1].

WARSCHAWSKY, S.

[1] *Über einige Konvergenzsatze aus der Theorie der konformen Abbildungen.* Nachr. Ges. Wiss. Göttingen (1930), 344–369. FM 56, 297.

WEBER, H.

[1] *Note zu Riemann's Beweis des Dirichlet'schen Prinzips.* J. Reine Angew. Math. 71 (1870), 29–39. FM 2, 209.

WERMER, J.

[1] *Function rings and Riemann surfaces.* Ann. of Math. (2) 67 (1958), 45–71. MR 20, 19.

WEYL, H.

[1] *Die Idee der Riemannschen Fläche.* Teubner, Leipzig, 1913, 169 pp. FM 44, 492.

[2] *The method of orthogonal projection in potential theory.* Duke Math. J. 7 (1940), 411–444. MR 2, 202.

[3] *Über die kombinatorische und kontinuumsmässige Definition der Überschneidungszahl zweier geschlossener Kurven auf einer Fläche.* Z. Angew. Math. Physik 4 (1953), 471–492. MR 15, 460.

[4] *Die Idee der Riemannschen Fläche.* Dritte vollständig umgearbeitete Auflage. Teubner, Stuttgart, 1955, 162 pp. MR 16, 1097.

WIRTH, E. M.
[1] *Über die Bestimmung des Typus einer Riemannschen Fläche.* Comment. Math. Helv. 31 (1956), 90–107. MR 18, 568.

WIRTINGER, W.
[1] *Algebraische Funktionen und ihre Integrale.* Enzykl. d. math. Wissenschaft. 2_2 (1902), 115–175. FM 33, 425.
[2] *Über die konforme Abbildung der Riemannschen Flächen durch Abelsche Integrale besonders bei $p=1$, 2.* Denkschr., Wien, 1909, 22 pp. FM 40, 475.
[3] *Zur Theorie der konformen Abbildung mehrfach zusammenhängender ebener Flächen.* Abh. Preuss. Akad. Wiss. Math.-Nat. Kl. no. 4 (1942), 9 pp. MR 8, 324.
[4] *Über gewisse mehrdeutige Umkehrprobleme bei Abel'schen Integralen insbesondere beim Geschlechte Vier.* Reale Accademia d'Italia, Fond. A. Volta, Atti dei Convegni, v. 9 (1939), pp. 159–169, Rome 1943. MR 12, 17.
[5] *Integrale dritter Gattung und linear polymorphe Funktionen.* Monatsh. Math. Phys. 51 (1944), 101–114. MR 6, 263.

WITTICH, H.
[1] *Ein Kriterium zur Typenbestimmung von Riemannschen Flächen.* M. Math. Phys. 44 (1936), 85–96. FM 62, 387.
[2] *Über die konforme Abbildung einer Klasse Riemannscher Flächen.* Math. Z. 45 (1939), 642–668. MR 1, 211.
[3] *Über die Wachstumsordnung einer ganzen transzendenten Funktion.* Ibid. 51 (1947), 1–16. MR 9, 180.
[4] *Über eine Klasse meromorpher Funktionen.* Arch. Math. 1 (1948), 160–166. MR 10, 442.
[5] *Über den Einfluss algebraischer Windungspunkte auf die Wachstumsordnung.* Math. Ann. 122 (1950), 37–46. MR 12, 251.
[6] *Bemerkung zum Typenproblem.* Comment. Math. Helv. 26 (1952), 180–183. MR 14, 367.
[7] *Neuere Untersuchungen über eindeutige analytische Funktionen.* Springer, Berlin-Göttingen-Heidelberg, 1955, 163 pp. MR 17, 1067.
[8] *Über eine Klasse Riemannscher Flächen.* Comment. Math. Helv. 30 (1956), 116–123.
[9] *Konforme Abbildung schlichter Gebiete.* Ann. Acad. Sci. Fenn. Ser. A. I. no. 249/6 (1958), 12 pp.
[10] Cf. KÜNZI, H., and WITTICH, H. [1].

WOLONTIS, V.
[1] *Properties of conformal invariants.* Amer. J. Math. 74 (1952), 587–606. MR 14, 36.

YANG, O. T.
[1] *Surfaces de Riemann régulières de points de ramification données.* C.R. Acad. Sci. Paris 213 (1941), 556–558. MR 5, 116.

YANG, T.
[1] *Analyse zur Definition der Riemannschen Flächen.* Tôhoku Math. J. 49 (1943), 208–212. MR 8, 570.
[2] *Ein elementares Potential auf einer geschlossenen konformen Riemannschen Fläche, und seine Anwendungen.* Acta Math. Sinica 4 (1954), 279–294. (Chinese. German summary.) MR 18, 121.
[3] *Über einen Existenzatz.* Ibid. 4 (1954), 295–299. MR 18, 121.
[4] *Einige Eigenschaften nichtfortsetzbarer konformer Riemannscher Flächen.* Ibid. 4 (1954), 301–304. MR 18, 121.

YOSHIDA, Y.
[1] *Theory of functions.* Iwanami, Tokyo, 1938, pp. 320. (Japanese.)

YOSIDA, T.
[1] *On a sufficient condition for a given Riemann surface to be of hyperbolic type.* Sci. Rep. Tokyo Bunrika Daigaku, Sec. A. 4 (1941), 89–92. MR 14, 157.

Yosida, T.
[2] *On the mapping functions of Riemann surfaces.* J. Math. Soc. Japan 2 (1950), 125–128. MR 12, 602.

Yûjôbô, Z.
[1] *On the Riemann surfaces, no Green function of which exists.* Math. Japonicae 2 (1951), 61–68. MR 13, 735.
[2] *An application of Ahlfors's theory of covering surfaces.* J. Math. Soc. Japan 4 (1952), 59–61. MR 14, 549.

Zygmund, A.
[1] Cf. Saks, S., and Zygmund, A. [1].

*Index**

* Italicized section numbers refer to formal definitions.

GPSR Authorized Representative: Easy Access System Europe - Mustamäe tee 50, 10621 Tallinn, Estonia, gpsr.requests@easproject.com

www.ingramcontent.com/pod-product-compliance
Ingram Content Group UK Ltd.
Pitfield, Milton Keynes, MK11 3LW, UK
UKHW021824060425
457147UK00006B/131

* 9 7 8 0 6 9 1 6 2 6 1 2 3 *